Lecture Notes in Computer Science

Lecture Notes in Bioinformatics

16569

The series Lecture Notes in Bioinformatics (LNBI) was established in 2003 as a topical subseries of LNCS devoted to bioinformatics and computational biology.

The series publishes state-of-the-art research results at a high level. As with the LNCS mother series, the mission of the series is to serve the international R & D community by providing an invaluable service, mainly focused on the publication of conference and workshop proceedings and postproceedings.

Manuel Lafond

Editor

Comparative Genomics

23rd RECOMB International Workshop, RECOMB-CG 2026
Thessaloniki, Greece, May 24–25, 2026
Proceedings

 Springer

Editor
Manuel Lafond
Université de Sherbrooke
Sherbrooke, QC, Canada

ISSN 0302-9743   ISSN 1611-3349 (electronic)
Lecture Notes in Bioinformatics
ISBN 978-3-032-26890-7   ISBN 978-3-032-26891-4 (eBook)
https://doi.org/10.1007/978-3-032-26891-4

LNCS Sublibrary: SL8 – Bioinformatics

Preface

The RECOMB Satellite Conference on Comparative Genomics (RECOMB-CG), established in 2003, brings together researchers from mathematics, computer science, and the life sciences to advance the study of comparative genomics. The conference highlights both theoretical and applied research on genome rearrangements, phylogenetics, genome assemblies, and many others, maintaining a strong tradition of interdisciplinary exchange. The 2026 edition continued this mission.

This volume contains the papers presented at the 23rd RECOMB Satellite Conference on Comparative Genomics (RECOMB-CG 2026), held on May 24–25, 2026, in Thessaloniki, Greece. Organized in conjunction with the annual RECOMB conference, RECOMB-CG 2026 offered a focused forum for the presentation and discussion of recent theoretical and applied advances in the field.

A total of 36 manuscripts were submitted for consideration. Each submission was evaluated in a single-blind process by three to four members of the program committee, with the support of external reviewers. Following discussions among reviewers and difficult decisions, 19 papers were selected for inclusion in the conference program and appear in these proceedings.

We are grateful to our keynote speakers, Simona Georgieva (Bulgarian Academy of Sciences) and Pavlos Pavlidis (University of Crete and FORTH-ICS), for sharing their expertise and perspectives with the community. We also thank all the presenters at the conference.

We sincerely thank the members of the program committee and the external reviewers for their careful assessments and constructive feedback. We also acknowledge the continued guidance of the Steering Committee, with special thanks to Krister M. Swenson for assisting the decision-making process. We also thank the RECOMB chair Rayan Chikhi and Ezgi Ebren from the RECOMB organizing committee for their responsiveness. Last but certainly not least, we are especially appreciative of the local organizers for their dedicated efforts in hosting and supporting the conference, and thank Christos Ouzounis, Ilias Kappas, and Asimina Kournoutou for their help.

Finally, we thank all authors who contributed their work, as well as the participants whose engagement helped make RECOMB-CG 2026 a stimulating and rewarding event.

May 2026   Manuel Lafond

Organization

Program Committee Chair

Manuel Lafond — Université de Sherbrooke, Canada

Steering Committee

Dannie Durand — Carnegie Mellon University, USA
Jens Lagergren — KTH Royal Institute of Technology, Sweden
Luay Nakhleh — Rice University, USA
David Sankoff — University of Ottawa, Canada
Krister M. Swenson — CNRS, LIRMM, Université de Montpellier, France

Program Committee

Lars Arvestad — Stockholm University, Sweden
Mukul S. Bansal — University of Connecticut, USA
Anne Bergeron — Université du Québec à Montréal, Canada
Paola Bonizzoni — University of Milano-Bicocca, Italy
Sèverine Bérard — Université de Montpellier, France
Cedric Chauve — Simon Fraser University, Canada
Yao-Ban Chan — University of Melbourne, Australia
Miklos Csuros — Université de Montréal, Canada
Nadia El-Mabrouk — Université de Montréal, Canada
Oliver Eulenstein — Iowa State University, USA
Guillaume Fertin — CNRS, LS2N, University of Nantes, France
Martin Frith — AIST, University of Tokyo, Japan
Pawel Gorecki — University of Warsaw, Poland
Maribel Hernandez-Rosales — CINVESTAV Irapuato Unit, Mexico
Asif Javed — Chulabhorn Research Institute, Thailand
Jaebum Kim — Konkuk University, South Korea
Lingling Jin — University of Saskatchewan, Canada
Kevin Liu — Michigan State University, USA
Istvan Miklos — Rényi Institute, Hungary
Siavash Mirarab — University of California, San Diego, USA

Aida Ouangraoua	Université de Sherbrooke, Canada
Fabio Pardi	LIRMM - CNRS, France
Teresa Przytycka	National Center of Biotechnology Information, USA
Aakrosh Ratan	University of Virginia, USA
Michael Sammeth	Coburg University of Applied Sciences, Germany
Celine Scornavacca	ISE-M, Université Montpellier, CNRS, France
Sagi Snir	University of Haifa, Israel
Giltae Song	Pusan National University, South Korea
Yanni Sun	University of Hong Kong, China
Wing-Kin Sung	University of Hong Kong, China
Nadia Tahiri	University of Sherbrooke, Canada
Olivier Tremblay-Savard	University of Manitoba, Canada
Tamir Tuller	Tel Aviv University, Israel
Fábio Viduani Martinez	Universidade Federal de Mato Grosso do Sul, Brazil
Tomas Vinar	Comenius University in Bratislava, Slovakia
Yong Wang	Academy of Mathematics and Systems Science, China
Yufeng Wu	University of Connecticut, USA
Louxin Zhang	National University of Singapore, Singapore
Jie Zheng	Shanghai Tech University, China

Contents

Phylogenetics II: Comparing and Evaluating Phylogenies

Machine Learning in Genomics

Combinatorics of Genome Rearrangements

Time Complexity and Relaxation Gap for the Rank Median of Three Genomes

Victor de Moraes⬤ and Joao Meidanis⁽✉⁾⬤

Institute of Computing, University of Campinas, Campinas, Brazil
`meidanis@unicamp.br`

Abstract. In this paper, we study the genomic median of 3 problem with respect to the rank distance, which looks for a genome that minimizes the sum of the rank distances to three given genomes. We advance the knowledge on the mathematical properties of this problem, and settle its computational complexity, showing it is NP-hard. We also prove that the gap between exact and relaxed solutions of the problem can be arbitrarily large.

1 Introduction

The genome median problem is the problem of finding a genome that minimizes the sum of its distances to three other genomes, given some metric on the genomes. Many metrics have been proposed for genome comparison, with the two most relevant to this paper being the double cut and join (DCJ) distance [10] and the rank distance [11], that focus on large genome rearrangements rather than on point mutations.

Both distances belong to a line of research that aims to achieve a balance between biological significance and computational tractability. Thus, these distances compare multichromosomal genomes, contemplating both linear and circular chromosomes, and are compatible with biologically observed rearrangements (also called operations in the computational context) such as inversions, transpositions, translocations, fusions, and fissions, among others. The weights assigned to these operations approximate well, for the most part, biological frequency measurements on real genomes [3]. On the other hand, they are efficiently computable given the two genomes.

From the start, a connection between genome rearrangement theory, permutations, and matchings was clear. In 2000, Meidanis and Dias proposed an algebraic formalism for genome rearrangement problems [7], which led to an algebraic distance for circular genomes [3]. Years later, Yancopoulos et al. [10] extended it to encompass linear chromosomes as well, creating the DCJ distance, that quickly became a favorite among researchers. Tannier et al. showed in 2009 that the DCJ median problem is NP-hard. In 2013, Feijão introduced yet another way of extending the algebraic distance to linear chromosomes [5], pointed out that it is similar to but distinct from DCJ, and conjectured that the median

M. Lafond (Ed.): RECOMB-CG 2026, LNBI 16569, pp. 3–28, 2026.
https://doi.org/10.1007/978-3-032-26891-4_1

problem would be NP-hard for it too [4]. Zanetti et al. later replaced the permutations that represent genomes by permutation matrices, creating the rank distance, which is just Feijao's algebraic distance in another guise and multiplied by 2 [11, Sec.2.3.2].

In summary, unlike the DCJ median problem, which is known to be NP-hard [8], the complexity of the median problem under the rank distance is not known, to the best of our knowledge. In this paper, we prove that finding genomic rank medians is NP-hard as well.

In the rank model, genomes are represented as binary, symmetric, orthogonal matrices. A relaxation of the genome median problem, where arbitrary, real-valued matrices are allowed, can be solved in polynomial time [1,6]. Although these developments may give rise to heuristics for the genomic version through, for example, maximum weight matchings, little is known about the gap between optimum scores for the genomic and relaxed medians. We show here that this gap can be arbitrarily large, prompting the search for approximation algorithms. Zanetti et al. [11] mention a trivial, 4/3-approximation given by one of the input matrices. To the best of our knowledge, no better approximation bound is known.

This article is structured as follows. In Sect. 2, we formally define the rank median problem. In Sect. 3, we introduce some properties of this problem, with the most notable ones being a composition rule for medians of related instances (Lemma 9), and a result on 4-cycles (Theorem 2), which will be used for the two main contributions of this paper: an example of a class of instances where the optimum genomic score is arbitrarily greater than the optimum relaxed score, and a proof that the rank median problem is NP-hard, given in Sects. 4 and 5. Finally, our conclusions and plans for future work are in Sect. 6.

2   Definitions

The rank median model, originally defined by Zanetti et al. [11], works as follows. Genomes are represented by binary, symmetric, orthogonal matrices that encode the adjacencies among gene extemities. For instance, in Fig. 1 we have a genome A consisting of one circular chromosome with genes a and b, another circular chromosome with genes c and d, and a linear chromosome with genes e and f.

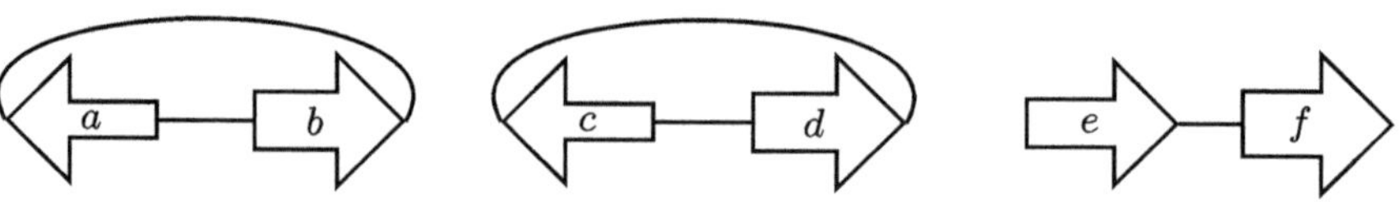

Fig. 1. Genome A.

In Fig. 2, the same genome is depicted as a matching, showing just the adjacencies between gene extremities, that are marked with subscripts h for *head* and t for *tail*.

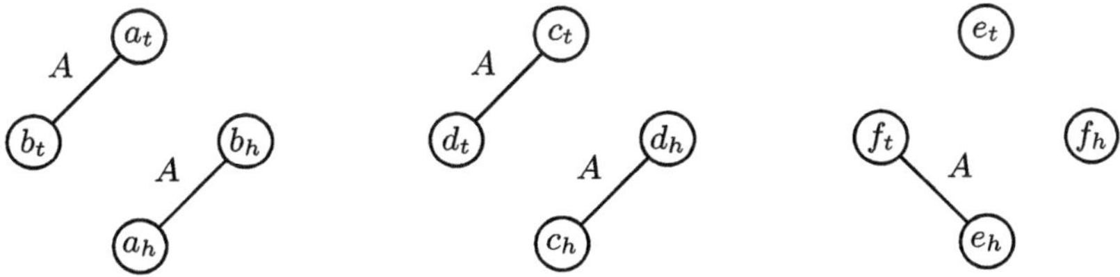

Fig. 2. Genome A as a matching.

The same genome A can be represented as a symmetric, orthogonal, 0-1 matrix as follows, with 1's outside the main diagonal indicating the adjacencies between gene extremities and 1's in the main diagonal indicating free genome extremities in the end of linear chromosomes.

$$
\begin{array}{c}
\begin{array}{ccccccccccccc}
& a_t & a_h & b_t & b_h & c_t & c_h & d_t & d_h & e_t & e_h & f_t & f_h
\end{array} \\
\begin{array}{c}
a_t \\ a_h \\ b_t \\ b_h \\ c_t \\ c_h \\ d_t \\ d_h \\ e_t \\ e_h \\ f_t \\ f_h
\end{array}
\left[
\begin{array}{cccccccccccc}
0 & 0 & 1 & 0 & 0 & 0 & 0 & 0 & 0 & 0 & 0 & 0 \\
0 & 0 & 0 & 1 & 0 & 0 & 0 & 0 & 0 & 0 & 0 & 0 \\
1 & 0 & 0 & 0 & 0 & 0 & 0 & 0 & 0 & 0 & 0 & 0 \\
0 & 1 & 0 & 0 & 0 & 0 & 0 & 0 & 0 & 0 & 0 & 0 \\
0 & 0 & 0 & 0 & 0 & 0 & 1 & 0 & 0 & 0 & 0 & 0 \\
0 & 0 & 0 & 0 & 0 & 0 & 0 & 1 & 0 & 0 & 0 & 0 \\
0 & 0 & 0 & 0 & 1 & 0 & 0 & 0 & 0 & 0 & 0 & 0 \\
0 & 0 & 0 & 0 & 0 & 1 & 0 & 0 & 0 & 0 & 0 & 0 \\
0 & 0 & 0 & 0 & 0 & 0 & 0 & 0 & 1 & 0 & 0 & 0 \\
0 & 0 & 0 & 0 & 0 & 0 & 0 & 0 & 0 & 0 & 1 & 0 \\
0 & 0 & 0 & 0 & 0 & 0 & 0 & 0 & 0 & 1 & 0 & 0 \\
0 & 0 & 0 & 0 & 0 & 0 & 0 & 0 & 0 & 0 & 0 & 1
\end{array}
\right]
\end{array}
$$

In this setting, the **distance** between two genomes A and B is given by the rank of their difference as matrices:

$$d(A, B) = r(B - A).$$

In the rank median problem, the input consists of a triple of genomes $T = (A, B, C)$ and we are interested in finding genomes M that minimize the **score**

$$d(M; T) = d(M; A, B, C) = d(M, A) + d(M, B) + d(M, C). \tag{1}$$

We say that a genome M is a **median** of T, when it minimizes the score given by Eq. 1. A **relaxed median** is any real-valued matrix minimizing the score. Sometimes we use the term **genomic median** to refer to a median, to stress the fact that we are dealing with a genomic matrix.

Because d satisfies the triangle inequality, for any matrix M we have

$$d(M; A, B, C) \geq \frac{d(A, B) + d(B, C) + d(C, A)}{2}, \tag{2}$$

and so the right hand side provides a lower bound for the score of a median. It is known that relaxed medians always meet this lower bound [2]. However, this is far from true for genomic medians, as we show in Sect. 4.

2.1   Relationship with Algebraic and DCJ Distances

We recall the basic details of algebraic and DCJ distances here. Algebraic distance, as mentioned, is exactly rank distance divided by two. As for DCJ distance, its original formulation involved capping the ends of linear chromosomes, thus creating fictitious genes, and then equalizing the number of genes in both genomes by adding null chromosomes (chromosomes with just two caps each— more fictitious genes), and closing the paths to cycles. After that, we find a minimum series of double-cut-and-join operations (see Fig. 3) that transforms one genome into the other. The number of operations in this series is the DCJ distance.

It turns out that the same effect can be achieved without adding caps or null chromosomes, as long as the repertoire of operations is enlarged to include single-cut-and-join (Fig. 4), as well as cuts and joins. These two last operations are just the destruction and creation of adjacencies, respectively (see Fig. 5).

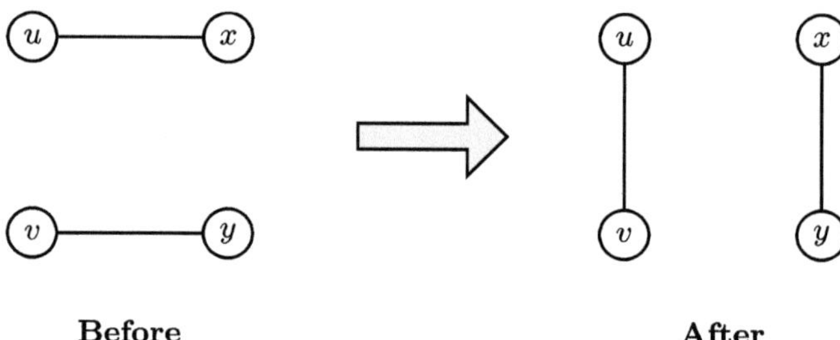

Fig. 3. Double-cut-and-join.

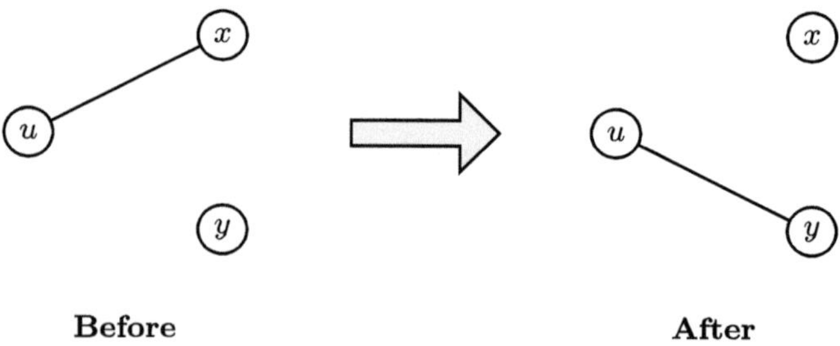

Fig. 4. Single-cut-and-join.

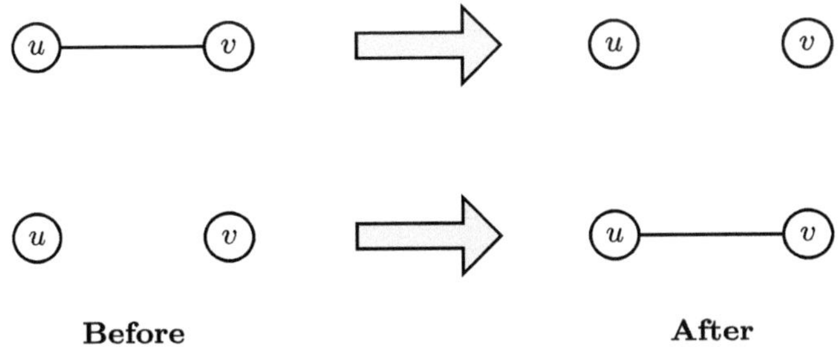

Fig. 5. Above: cut. Below: join.

Biologically, this is appealing, since these basic operations implement real, observable genome large mutations such as inversions, translocations, fusions, fission, and others. The same operations can be used to define DCJ, algebraic, and rank distance, only the weights change, and we always look for a minimum weight series of operations transforming one genome into the other. Table 1 shows the weights given by each one of the distances mentioned to the basic operations.

Table 1. Weights given to basic operations by DCJ, algebraic, and rank distances.

	DCJ	Algebraic	Rank
Cut / Join	1	0.5	1
Single-cut-and-join	1	1	2
Double-cut-and-join	1	1	2

Some aspects of the relationship among these distances are important in the NP-hardness proof, so we summarize them in the following theorem. Let d_{alg} and d_{DCJ} denote the algebraic and DCJ distances, respectively.

Theorem 1. *For any two genomes (matchings) A and B in G, we have*

$$\frac{d(A, B)}{2} = d_{alg}(A, B) \leq d_{DCJ}(A, B), \tag{3}$$

with equality between d_{alg} and d_{DCJ} when both A and B are circular genomes (perfect matchings).

Proof. The equality in Eq. 3 comes from Zanetti et al. [11, Sec. 2.3.2]. The inequality comes from Feijao and Meidanis [4, Eq. 7], which also shows why we have equality for circular genomes. □

2.2   Matchings and Partial Medians

In this paper, we take a graph oriented approach. It is known that genomes, as defined previously, correspond to matchings of gene extremities [2]. Therefore, from now on, we will equate genomes to matchings of a complete graph G whose vertices are the gene extremities, and proceed to study the problem from this viewpoint. This is the same approach taken by Tannier et al. [8].

We follow definitions from West [9]. We say that two edges are **independent** when they do not share endpoints. A **matching** is a set of mutually independent edges. A matching A **saturates** a vertex v, and v is called A-**saturated**, when v is incident to an edge in A. Otherwise, we call v A-**unsaturated**. An edge e is **compatible** with a matching L when e is independent from every edge in L. Two matchings L and N are **compatible** when every edge of L is compatible with N. Notice that this happens if an only if $L \cup N$ is also a matching.

Given two matchings A and B in G, we denote by $A + B$ the multigraph with the vertices of G and all edges from A and B, where edges that belong to both A and B are duplicated in $A + B$. Therefore, the connected components of $A + B$ are paths and cycles, including the 2-cycles of duplicated edges.

We translate the notion of distance to matchings. Chindelevitch et al. [2] point out that, when A and B are matchings corresponding to genomes, we have:

$$d(A, B) = |A| + |B| - 2\mathcal{C}(A, B), \tag{4}$$

where $|X|$ denotes the size of set X (in the case of matchings, their number of edges) and $\mathcal{C}(A, B)$ is the number of cycles in $A + B$, including 2-cycles.

We say that a matching L is a **partial median** with respect to an instance $T = (A, B, C)$ of the rank median problem when there exists a matching $M \supseteq L$ such that M is a median of T.

3   Preliminary Results

3.1   Modifying Matchings I

Throughout this document we will often have to inspect how the score of a matching is affected as we change its edges. For this, we need to know how distances are affected when we change edges.

We will now examine how $d(A, L)$ changes when we add a compatible edge e to an arbitrary matching L, given a fixed genome A. A first observation is that e will belong to at most one cycle in $A + (L \cup \{e\})$. Thus, e can either belong to exactly one cycle in $A + (L \cup \{e\})$ or to no cycle at all. The next result tells us how $d(A, L \cup \{e\})$ is related to $d(A, L)$ in each case.

Lemma 1. *Let A and L be matchings, and e an edge compatible with L. Then:*

$$d(A, L \cup \{e\}) = \begin{cases} d(A, L) - 1 & \text{if } \mathcal{C}(A, L \cup \{e\}) = \mathcal{C}(A, L) + 1, \\ d(A, L) + 1 & \text{otherwise.} \end{cases}$$

Proof. In the conditions of the lemma, if $\mathcal{C}(A, L \cup \{e\}) = \mathcal{C}(A, L) + 1$ we have, according to Eq. 4,

$$d(A, L \cup \{e\}) = |A| + |L \cup \{e\}| - 2\mathcal{C}(A, L \cup \{e\})$$
$$= |A| + |L| + 1 - -2(\mathcal{C}(A, L) + 1)$$
$$= |A| + |L| - 2\mathcal{C}(A, L) - 1$$
$$= d(A, L) - 1.$$

Otherwise, the number of cycles does not change, that is, $\mathcal{C}(A, L \cup \{e\}) = \mathcal{C}(A, L)$, and a similar reasoning leads to $d(A, L \cup \{e\}) = d(A, L) + 1$. $\square$

This impacts the calculation of scores. Since the possible change in distance to a specific genome when adding a compatible edge e to L can be either -1 or $+1$, determined by whether e belongs to a cycle or not, we have the following corollary.

Corollary 1. *Let A, B, C, and L be matchings, and e an edge compatible with L. Then:*
$$d(L \cup \{e\}; A, B, C) = d(L; A, B, C) + s,$$
where $s = -3, -1, +1, +3$ in the cases where e belongs to a total of 3, 2, 1, or 0 cycles, respectively, in $A + L$, $B + L$, and $C + L$.

3.2 Collapsed Instances

Motivated by Lemma 1, we define the **collapse** of a matching A by a matching L, denoted by $K(A, L)$, as the set of edges that, when added to L, get it closer to A:

$$K(A, L) = \{e \, compatible \, with \, L \mid \mathcal{C}(A, L \cup \{e\}) = \mathcal{C}(A, L) + 1\}.$$

The term "collapse" comes from the fact that edges of A adjacent to edges in L will disappear in $K(A, L)$. The other edges will remain, and edges not in A can be present in $K(A, L)$ as well, although every vertex saturated by $K(A, L)$ is also saturated by A. Below we list a few properties of the collapse.

Lemma 2. *If A and L are matchings, the collapse $K(A, L)$ satisfies:*

1. $K(A, L)$ is a matching.
2. $K(A, L)$ is compatible with L.
3. $K(A, S) = A \setminus S$ when $S \subseteq A$.
4. $K(A, L) = A$ when L is compatible with A.
5. $L \subseteq M$ implies $K(A, M) \subseteq K(A, L)$.
6. M compatible with A implies M compatible with $K(A, L)$.

Proof. We will prove only Item 3, which will be needed later. By definition:

$$K(A, S) = \{e \text{ compatible with } S \mid \mathcal{C}(A, S \cup \{e\}) = \mathcal{C}(A, S) + 1\}.$$

Notice that $\mathcal{C}(A, S) = |S|$, since every edge of S belongs to A and will form a 2-cycle. An edge e compatible with S will only form an extra cycle if it also belongs to A. Hence, $K(A, S) = A \setminus S$, because edges in S are not compatible with it. $\square$

We also define the collapse of an instance $T = (A, B, C)$ by a matching L as the instance $K(T, L) = (K(A, L), K(B, L), K(C, L))$.

3.3   Modifying Matchings II

Sometimes we want to force the presence of an edge e into a matching L, even when e is not compatible with L. The straightforward way to do this would be to just remove the edges adjacent to e first. However, there is a smarter way to go about it, using collapses.

Definition 1 (Smart edge addition). *Let L be a matching and e an edge such that $e \notin L$ and incompatible with L. The* smart edge addition *of e to L is defined as*

$$L \oplus e = K(L, \{e\}) \cup \{e\}.$$

The next result parallels Lemma 1 when e is not compatible with L.

Lemma 3. *Let A and L be matchings, and e an edge not in L and incompatible with L. Then:*

$$d(A, L \oplus e) = \begin{cases} d(A, L) - 2 & if\ e \in A, \\ d(A, L) + s & otherwise, \end{cases}$$

where $s \leq 2$.

Proof. See Appendix A. $\square$

We state a first consequence of these observations.

Lemma 4. *Let A, B, and C be three matchings. If e belongs to at least two of these matchings, then $\{e\}$ is a partial median of $T = (A, B, C)$.*

Proof. Let e be as in the statement, and let M be a median of T. Suppose $e \notin M$. Then, e can be either compatible with M or not.

If e is compatible with M, then $M \cup \{e\}$ has a smaller score than M, due to Lemma 1. If e is not compatible with M, then $M \oplus e$ will have a smaller score than M, due to Lemma 3. In both cases, the new score will be decreased by the contributions of the two genomes it belongs to, and the possible increase from the third genome will not be enough to offset this.

As a result, since M is a median, we rule out these possibilities and the only option left is $e \in M$, proving the claim. $\square$

3.4   Medians in Collapsed Instances

The following result relates distances in a genome and its collapse.

Lemma 5. *If A and L are matchings, and e is an edge compatible with L, we have:*

$$d(L \cup \{e\}, A) - d(L, A) = d(\{e\}, K(A, L)) - d(\emptyset, K(A, L)).$$

Proof. According to Lemma 1 and the definition of collapse, the left hand side is equal to -1 if $e \in K(A, L)$ or to $+1$ otherwise. Likewise, the right hand side is equal to -1 if $e \in K(K(A, L), \emptyset)$ and $+1$ otherwise. However, $K(K(A, L), \emptyset) = K(A, L)$, since collapsing by $\emptyset$ does not modify a matching, as per Lemma 2, Item 3. $\qquad\square$

What this means is that for any matching L and any compatible e, distances to a genome A are related to simpler distances in the collapsed genome $K(A, L)$. This immediately translates to rank median instances. In other words, we have the following corollary.

Corollary 2. *For any instance $T = (A, B, C)$ of the rank median problem, any matching L, and any edge e compatible with L, we have:*

$$d(L \cup \{e\}; T) - d(L; T) = d(\{e\}; K(T, L)) - d(\emptyset; K(T, L)). \tag{5}$$

Proof. This is just a question of using the definitions of $d(X; T)$ and $K(T, L)$ and applying Lemma 5 to each of the genomes A, B, and C. $\qquad\square$

The relevant fact is that Lemma 5 can be extended to arbitrary, compatible matchings. However, we need auxiliary result to prove the extension. We begin with a few results on composite collapses.

Lemma 6. *If A and L are matchings, and e is an edge compatible with L, we have:*

$$K(A, L \cup \{e\}) = K(K(A, L), \{e\}).$$

Proof. See Appendix B.

The previous result can be extended to arbitrary, compatible matchings.

Lemma 7. *If A, L, and M are arbitrary matchings, with L and M compatible, we have:*

$$K(A, L \cup M) = K(K(A, L), M).$$

Proof. By induction on $|M|$. If $|M| = 0$, we have $M = \emptyset$ and the claim is true because $K(A, \emptyset) = A$.

If $|M| \geq 1$, let e be any edge in M, and consider $M' = M \setminus \{e\}$. Since M' is also compatible with L, the claim is true for M' by the induction hypothesis:

$$K(A, L \cup M') = K(K(A, L), M').$$

Then using Lemma 6 twice, plus the induction hypothesis, we end up with:

$$\begin{aligned}
K(A, L \cup M) &= K(A, L \cup M' \cup \{e\}) \\
&= K(K(A, L \cup M'), \{e\}) \\
&= K(K(K(A, L), M'), \{e\}) \\
&= K(K(A, L), M' \cup \{e\}) \\
&= K(K(A, L), M).
\end{aligned}$$

$\square$

We are now ready for the extension of Lemma 5 to arbitrary matchings.

Lemma 8. *If A, L, and M are arbitrary matchings, with L and M compatible, then*

$$d(L \cup M, A) - d(L, A) = d(M, K(A, L)) - d(\emptyset, K(A, L)).$$

Proof. By induction on $|M|$. If $|M| = 0$ then $M = \emptyset$, and both sides are equal to zero. If $|M| \geq 1$, take any edge $e \in M$ and let M' be $M \setminus \{e\}$. This matching M' is also compatible with L, so the induction hypothesis must hold for it, leading to:

$$d(L \cup M', A) - d(L, A) = d(M', K(L, A)) - d(\emptyset, K(A, L)). \tag{6}$$

Since M is a matching, we have that $\{e\}$ is compatible with M'. Therefore, we can apply Lemma 5, obtaining:

$$d(L \cup M' \cup \{e\}, A) - d(L \cup M', A) = d(\{e\}, K(A, L \cup M')) - d(\emptyset, K(A, L \cup M')). \tag{7}$$

Also, by Lemma 5 again, we have:

$$d(M' \cup \{e\}, A') - d(M', A') = d(\{e\}, K(A', M')) - d(\emptyset, K(A', M')), \tag{8}$$

where $A' = K(A, L)$. Now adding side by side Eqs. 6 and 7, and subtracting Eq. 8, we get:

$$d(L \cup M, A) - d(L, A) = d(M, K(A, L) - d(\emptyset, K(A, L)),$$

since $K(A, L \cup M') = K(A', M')$ by Lemma 7. $\square$

Corollary 3. *For any instance $T = (A, B, C)$ of the rank median problem, and any matchings L, M, with M compatible with L, we have:*

$$d(L \cup M; T) - d(L; T) = d(M; K(T, L)) - d(\emptyset; K(T, L)).$$

Proof. Again, this is just a question of using the definition of $d(M; T)$ and applying Lemma 8 to each of A, B, and C. $\square$

We are now ready for a result that relates medians in a multi-graph and their collapses.

Lemma 9. *Consider an instance $T = (A, B, C)$ of the rank median problem and let L be a partial median of T. Then, M is a median of $K(T, L)$ if and only if $L \cup M$ is a median of T.*

Proof. We start with the "only if" part. Assume $L \cup M$ is a median of T. Medians are matchings, so L and M are compatible. By definition, M is a median of $K(T, L)$ if it minimizes $d(M; K(T, L))$. By Corollary 3,

$$d(M; K(T, L)) = d(L \cup M; T) - d(L; T) + d(\emptyset; K(T, L)).$$

The only quantity dependent on M in the right hand side is the first term, and thus M is a median of $K(T, L)$ if it minimizes $d(L \cup M; T)$, which is trivially true since $L \cup M$ is a median of T.

For the "if" part, because L is a partial median, there is a matching M' compatible with L such that $L \cup M'$ is a median of T.

By Corollary 3, we have

$$d(L \cup M; T) - d(L; T) = d(M; K(T, L)) - d(\emptyset; K(T, L))$$

and

$$d(L \cup M'; T) - d(L; T) = d(M'; K(T, L)) - d(\emptyset; K(T, L)).$$

Now M is a median of $K(T, L)$, so $d(M; K(T, L)) \le d(M'; K(T, L))$, and thus

$$\begin{aligned}
d(L \cup M; T) &= d(L \cup M; T) - d(L; T) + d(L; T) \\
&= d(M; K(T, L) - d(\emptyset; K(T, L)) + d(L; T) \\
&= d(M; K(T, L)) + d(L \cup M'; T) - d(M'; K(T, L) \\
&\le d(M'; K(T, L)) + d(L \cup M'; T) - d(M'; K(T, L)) \\
&= d(L \cup M'; T).
\end{aligned}$$

But, since $L \cup M'$ is a median of T, this should be an equality, and so $L \cup M$ is also a median of T. $\square$

The result also partly applies to partial medians.

Corollary 4. *Consider an instance $T = (A, B, C)$ of the rank median problem and let L be a partial median of T. If M is a partial median of $K(T, L)$ then $L \cup M$ is a partial median of T.*

Proof. If M is a partial median of $K(T, L)$, then there exists a median M' in $K(G, L)$ with $M \subseteq M'$. By Lemma 9, $L \cup M'$ is a median of T and it contains $L \cup M$. $\square$

4   Relaxation Gap

As far as we know, the literature is lacking instances with a specific difference in score between relaxed medians and genomic medians. It is not even known if there is a limit on how large this difference can be. In this section, we show two classes of graphs witnessing arbitrary differences in score.

Define the *score gap* for an instance $T = (A, B, C)$ of the rank median problem as

$$g(T) = \min_{M\ genomic} d(M; T) - \min_{M\ any\ matrix} d(M; T).$$

Lemma 10. *If there exists a median M of $T = (A, B, C)$ such that $M \cap (A \cup B \cup C) = \emptyset$, then $\emptyset$ is a median of T.*

Proof. Let M be a matching with $M \cap (A \cup B \cup C) = \emptyset$. Denote by $\mathcal{Z}(X, Y)$ the set of cycles formed by matchings X and Y, so that $|\mathcal{Z}(X, Y)| = \mathcal{C}(X, Y)$, according to our previous definition of $\mathcal{C}(X, Y)$. Consider now the set of cycles $Z = \mathcal{Z}(M, A) \uplus \mathcal{Z}(M, B) \uplus \mathcal{Z}(M, C)$. We are using disjoint union because we want to consider an eventual cycle that appears, say, in $\mathcal{Z}(M, A)$ and also in $\mathcal{Z}(M, B)$ as two distinct entities. Now build a bipartite graph $\mathcal{B}$ with vertices in Z and M and edges linking a cycle $c \in Z$ to $e \in M$ exactly when $e \in c$. Let x_c be the degree of cycle $c \in Z$ in $\mathcal{B}$, and y_e the degree of $e \in M$ in $\mathcal{B}$.

Since M does not contain any edge in $A \cup B \cup C$, any cycle $c \in Z$ has degree at least two in $\mathcal{B}$, and thus

$$2|Z| \le \sum_{c \in Z} x_c.$$

We also know that each edge $e \in M$ can belong to at most three cycles in Z (at most one in each of $Z(A, M)$, $Z(B, M)$, and $Z(C, M)$) and thus

$$\sum_{e \in M} y_e \le 3|M|.$$

Since both sums equal the number of edges in $\mathcal{B}$, we have $2|Z| \le 3|M|$. With that we can calculate a lower bound on M's score:

$$
\begin{aligned}
d(M; A, B, C) &= d(M, A) + d(M, B) + d(M, C) \\
&= 3|M| + |A| + |B| + |C| - 2(\mathcal{C}(A, M) + \mathcal{C}(B, M) + \mathcal{C}(C, M)) \\
&= 3|M| + |A| + |B| + |C| - 2|Z| \\
&\ge 3|M| + |A| + |B| + |C| - 3|M| \\
&= |A| + |B| + |C| \\
&= d(\emptyset; A, B, C).
\end{aligned}
$$

Thus, if M is a median, so is $\emptyset$.   □

Lemma 11. *Let $G = G_1 \uplus G_2$, where $\uplus$ denotes disjoint union. Let A_1, B_1, and C_1 be matchings in G_1, and A_2, B_2, and C_2 be matchings in G_2. Finally, let M_1 be a median of A_1, B_1, and C_1, and M_2 be a median of A_2, B_2, and C_2. Then $M_1 \cup M_2$ is a median of $A = A_1 \cup A_2$, $B = B_1 \cup B_2$, and $C = C_1 \cup C_2$ in G.*

Proof. Let M be a median of A, B, and C. For $i = 1, 2$, take N_i as the set of edges of M that are incident to vertices in G_i. Since M is a median, and given that N_1 and N_2 are compatible, we have that $N = N_1 \cup N_2$ is a partial median, and by collapsing with it we get another instance $(K(A, N), K(B, N), K(C, N))$ of the rank median problem, with a solution $M' = M \setminus N$ by Lemma 9.

Now, since N contains all edges in M with both endpoints in the same subgraph, the matching M' can only contain edges with one end in G_1 and another in G_2. Because G_1 and G_2 are disconnected, and recalling that a collapse never augments the set of saturated vertices, we conclude that $M' \cap (E[K(A, N)] \cup E[K(B, N)] \cup E[K(C, N)]) = \emptyset$, and then we can apply Lemma 10 to get $\emptyset$ as a median of $K(A, N)$, $K(B, N)$, and $K(C, N)$. By Lemma 9 again, N is a median of A, B, and C.

Note that, since both G and N have no edges between the subgraphs G_1 and G_2, then there are also no cycles between them, and thus N minimizes the score for G_1 and G_2 separately. This implies that N_i is a median of A_i, B_i, and C_i, for $i = 1, 2$. Since M_1 and M_2 are also medians, we have

$$d(N_1; A_1, B_1, C_1) = d(M_1; A_1, B_1, C_1),$$
$$d(N_2; A_2, B_2, C_2) = d(M_2; A_2, B_2, C_2).$$

Now

$$\begin{aligned}
d(M_1 \cup M_2; A, B, C) &= d(M_1; A_1, B_1, C_1) + d(M_2; A_2, B_2, C_2) \\
&= d(N_1; A_1, B_1, C_1) + d(N_2; A_2, B_2, C_2) \\
&= d(N; A, B, C),
\end{aligned}$$

showing that $M_1 \cup M_2$ is a median of A, B, and C, as it equals the score of another median, namely, N. $\square$

Define an *AB 4-cycle* as a cycle with 4 edges, two of which are in matching A and the other two in matching B. If we have an instance $T = (A, B, C)$ of the rank median problem with just an AB 4-cycle and an empty C, it has a score gap of 1. Lemma 11 implies that, to achieve a score gap of k, it suffices to build an instance with k disjoint AB 4-cycles and empty C.

That result by itself does not say anything about the score gap for connected graphs. For the rest of this section, we will show how to join the AB 4-cycles with C-edges into a connected graph, without decreasing the score gap. We start with a result on 4-cycles.

Lemma 12. *If there is an AB 4-cycle in an instance $T = (A, B, C)$, then there is at least one edge e of the 4-cycle such that $\{e\}$ is a partial median.*

Proof. Let M be a median and let $N \subseteq M$ be the set of edges in M incident to the vertices of the cycle. Since M is a median, $M \setminus N$ is a partial median, and thus $K(G, M \setminus N)$ is a graph where N is a median.

Note that, because the cycle is a 4-cycle, there exists a subset of at most 8 vertices that contains both the vertices of the 4-cycle and the vertices incident to N. We can verify that at least one edge of the 4-cycle is a partial median by brute-forcing through all possible 8-graph vertices that contain an AB 4-cycle. See Appendix C for a piece of code that performs this verification. It runs in about 10 min in most modern computers or laptops.

Choosing a median N' that contains one of those edges, we conclude that $M \setminus N \cup N'$ is a median of G by Lemma 9. $\qquad\square$

Corollary 5. *If there is an AB 4-cycle in G, then at least one pair of edges in that cycle is a partial median.*

Proof. By collapsing the partial median edge e given by Lemma 12, we get a 2-cycle on the opposite edge f of the cycle. By Lemma 4, that edge f is a partial median in $K(G, \{e\})$. We conclude that $\{e, f\}$ is a partial median in G by Corollary 4. $\qquad\square$

Theorem 2. *If the genomes A, B of $T = (A, B, C)$ are perfect matchings, with AB cycles being either 2-cycles or 4-cycles, then there exists a median given by choosing either the edges in A or B for each AB cycle.*

Proof. This can be done through a proof by induction on the number n of AB cycles. If $n = 1$, Lemma 4 and Corollary 5 show that either A or B is a partial median, which is in fact a median since both A and B are perfect matchings.

For $n \geq 2$, let M be a partial median in one of the AB cycles. Now $K(T, M)$ is a graph with $K(A, M)$ and $K(B, M)$ as perfect matchings, again with each AB-cycle being of size 2 or 4, so the induction hypothesis applies, yielding a median M' in $K(T, M)$. By Lemma 9, the matching $M \cup M'$ satisfies the theorem. $\qquad\square$

Now, let a **path of cycles** be an instance $T = (A, B, C)$ constructed by joining k disjoint AB 4-cycles, with C being composed of $k - 1$ edges joining the ith cycle with the $(i + 1)$th cycle, for $1 \leq i < k$.

By Theorem 2, for any path of cycles there exists a median M in this graph that is given by choosing either A or B for each 4-cycle. Now, independently of whether the A or B edges are chosen, we have that the number of AM-cycles plus the number of BM-cycles is equal to 3 for each AB cycle, and we can also verify that there are no cycles in $C + M$, so

$$\mathcal{C}(A, M) + \mathcal{C}(B, M) = 3k,$$
$$\mathcal{C}(C, M) = 0.$$

Then, the score of the genomic median is

$$
\begin{aligned}
d(M; A, B, C) &= d(A, M) + d(B, M) + d(C, M) \\
&= 3|M| + |A| + |B| + |C| - 2\mathcal{C}(A, M) - 2\mathcal{C}(B, M) - 2\mathcal{C}(C, M) \\
&= 6k + 2k + 2k + (k - 1) - 6k - 0 \\
&= 5k - 1.
\end{aligned}
$$

The relaxed median R always reaches the lower bound, so

$$
\begin{aligned}
d(R; A, B, C) &= \frac{1}{2} \left(d(A, B) + d(B, C) + d(C, A) \right) \\
&= \frac{1}{2} \left(2\,A| + 2|B| + 2\,C| - 2\mathcal{C}(A, B) - 2\mathcal{C}(B, C) - 2\mathcal{C}(C, A) \right) \\
&= \frac{1}{2} (4k + 4k + (2k - 2) - 2k - 0 - 0) \\
&= 4k - 1.
\end{aligned}
$$

Thus their difference is

$$
d(M; A, B, C) - d(R; A, B, C) = k
$$

that is, we have a class of connected graphs in which the score gap grows with $1/4$ of the number of vertices in the graph.

5 Complexity of the Rank Median Problem

Although relaxed medians can be calculated in polynomial time, the fact that a connected graph can have a genomic median with a score arbitrarily distant from the score of a relaxed median suggests that the rank median problem does not have a polynomial time algorithm.

As it turns out, the rank median problem can be proven to be NP-hard building upon arguments that worked for the DCJ median problem. Tannier et al. proved that the DCJ median problem is NP-hard [8, Theorem 5], even for circular genomes, which correspond to perfect matchings. The proof was done through a reduction from the **Breakpoint Graph Decomposition** (BGD) problem. Looking at the reduction to the DCJ problem, a stronger statement can be made.

Theorem 3. *The DCJ median problem is NP-hard even for instances $T = (A, B, C)$ where A, B, and C are perfect matchings and $A + B$ is a disjoint union of 4-cycles.*

Proof. In the reduction performed by Tannier et al. [8] only genomes of this kind appear when an instance of BGD is transformed into an instance of DCJ median. $\square$

Now, we are ready for the result on rank medians.

Theorem 4. *The rank median problem for three genomes is NP-hard.*

Proof. We use a reduction from DCJ median, where A and B are perfect matchings and $A + B$ is a disjoint union of 4-cycles. If there was a polynomial time program P that solved rank medians, we could feed such an instance T to it and, by Theorem 2, we would get back a rank median M that is also a perfect matching. Since DCJ and rank distances differ by a factor of 2 for circular genomes (by Theorem 1), we have:

$$d_{DCJ}(M;T) = \frac{d(M;T)}{2}.$$

We claim that M is also a DCJ median. Indeed, if there was a genome M' with smaller DCJ score, it would also have smaller rank score, because

$$\frac{d(M';T)}{2} \leq d_{DCJ}(M';T) < d_{DCJ}(M;T) = \frac{d(M;T)}{2},$$

by Theorem 1. But this is impossible, given that M is a rank median. We conclude that M is indeed a DCJ median.

Therefore, a solver for rank medians would be a DCJ median solver as well, for instances of the kind specified above. Since DCJ median is NP-hard for these instances, so is the rank median problem for three genomes. $\qquad\square$

6   Conclusions

We have given an example of a class of graphs such that the error of their medians differs from the lower bound given by Eq. 2 by one fourth of the number of vertices.

We have also proven that the rank median problem is NP-hard, and given some tools that can help with the problem of finding a median as well as proving statements about medians on a certain class of instances. More specifically, Lemma 9 shows how, if we know that some edge set belongs to at least one median, to find a median for the entire instance by finding the median on a smaller instance.

We have also given a class of instances, in Theorem 2, that contains both the class of instances mentioned above as well as the instances used in the reductions for the NP-hardness proof. These instances have the special property of having a guaranteed median that belongs to a relatively smaller search space, and with the score of the matchings in that search space changing only with the number of cycles between the matching and one of the genomes.

6.1   Future Work

Although we have shown ways to simplify the problem of finding a median in an instance if it contains a 2-cycle or an alternate 4-cycle, the NP-hardness result implies that, in general, we cannot do much better than brute-forcing over all possible matchings in the corresponding complete graph to find a median.

The most straightforward way of finding all possible matchings is through a recursive function that chooses an edge in a graph, and for all matchings in the subgraph with that edge removed, create one matching that contains the edge

---

Algorithm 1 Straightforward function that returns all matchings in a graph

---

1: **function** MATCHINGS(G)
2:     $m \leftarrow \emptyset$
3:     $e \leftarrow$ an arbitrary edge of G
4:     **for** each matching M in MATCHINGS($G \setminus \{e\}$) **do**
5:         $m \leftarrow m \cup \{M\}$
6:         $m \leftarrow m \cup \{M \cup \{e\}\}$
7:     **end for**
8:     **return** m
9: **end function**

---

and another that does not, as in Algorithm 1. This algorithm finds all possible matchings, and by keeping track of all their scores, it can be used to find all medians of an input instance.

By Lemma 10, to find a median in $T = (A, B, C)$, it suffices to be able to look for matchings that are subsets of $A \cup B \cup C$; together with Lemma 9, this would give a brute-force algorithm for finding a median: we iterate over all edges in $A \cup B \cup C$, and for all medians found in the instance resulting from the collapse of the current edge, construct a matching with the current edge and the current median of the collapsed instance, and choose the ones with the lowest score (comparing it with the score of the empty matching too). Algorithm 2 implements this idea.

Algorithm 2 Median finding by collapse

1: **function** MEDIANS($T = (A, B, C)$)
2:     $m \leftarrow \{\emptyset\}$
3:     $s \leftarrow |A| + |B| + |C|$
4:     **for** $e \in A \cup B \cup C$ **do**
5:         $m' \leftarrow$ MEDIANS($K(T, e)$)
6:         $n \leftarrow \min\{d(M \cup \{e\}; T) \ : \ M \in m'\}$
7:         **if** $n = s$ **then**
8:             $m \leftarrow m \cup \{M \in m' : d(M \cup \{e\}; T) = n\}$
9:         **end if**
10:       **if** $n < s$ **then**
11:           $m \leftarrow \{M \in m' : d(M \cup \{e\}; T) = n\}$
12:           $s \leftarrow n$
13:       **end if**
14:     **end for**
15:     **return** m
16: **end function**

However, the number of matchings checked by this algorithm, without any optimizations, may be greater than the number of matchings of G in some situations, for example, when A, B, and C are perfect matchings, and so it is not necessarily more efficient. This shows that there is room for improvement in this algorithm.

One possible direction for future work would be to optimize this algorithm by avoiding the inclusion of the same edge in different recursive branches, and investigate whether or not that would result in a better algorithm for finding medians. Another direction would be to prove or find a counterexample for the hypothesis that all medians that do not intersect A, B, or C are necessarily empty. If that hypothesis is true, it would imply Algorithm 2 is capable of finding all possible medians, rather than a non-empty subset of them.

Acknowledgments. This study was financed, in part, by the São Paulo Research Foundation (FAPESP), Brazil. Process Numbers #2024/01200-8 and #2025/05185-6.

Disclosure of Interests. The authors have no competing interests to declare that are relevant to the content of this article.

A   Proof of Lemma 3

In this section, we need the following auxiliary lemma, that specifies exactly what a collapse by a single edge looks like.

Lemma 13. *If A is any matching and e is any edge, we have:*

$$
K(A, \{e\}) = \begin{cases} A \setminus \{e\} & if\ e \in A, \\ A & if\ e\ is\ compatible\ with\ A, \\ A \setminus \{f\} & if\ e \notin A\ is\ incident\ to\ exactly\ one\ edge\ f \in A, \\ A \setminus \{f, f'\} \cup \{e'\} & if\ e \notin A\ is\ incident\ to\ two\ edges\ f, f' \in A. \end{cases}
$$

where e' is the edge that forms a 4-cycle with e, f, and f'.

Proof. By Definition:

$$
K(A, \{e\}) = \{g\ compatible\ with\ \{e\} \mid \mathcal{C}(A, \{e, g\}) = \mathcal{C}(A, \{e\}) + 1\}.
$$

But:

$$
\mathcal{C}(A, \{e\}) = \begin{cases} 1\ if e \in A, \\ 0\ if e \notin A. \end{cases}
$$

Consider the case when $e \in A$. Then $\mathcal{C}(A, \{e\}) = 1$. To belong to $K(A, \{e\})$, an edge g needs to be part of another cycle in $A + \{e, g\}$. This happens if and only if g is an edge of A different from e. Hence, $K(A, \{e\}) = A \setminus \{e\}$ in this case.

Now assume that $e \notin A$. Then $\mathcal{C}(A, \{e\}) = 0$. To belong to $K(A, \{e\})$, an edge g needs to be independent from e and be part of cycle in $A + \{e, g\}$. Edges that form cycles in $A + \{e, g\}$ are the ones that connect nodes in the same connected component in $A + \{e\}$. All these connected components are paths. If e is compatible with A, all these components are single edges, and therefore $K(A, \{e\}) = A$, since we have to exclude e for not being compatible with itself. If there is only one edge $f \in A$ incident to e, its component is a 3-path containing both e and f, and there is no way to form a cycle with an edge compatible with e there. Then $K(A, \{e\}) = A \setminus \{f\}$ in this case, since the other edges of A form cycles.

Finally, if e is incident to two edges $f, f' \in A$, then the edge e' that forms a 4-cycle with e, f, and f' is compatible with e and forms a cycle, as do the edges of A except f and f'. Therefore, $K(A, \{e\}) = A \setminus \{f, f'\} \cup \{e'\}$ in this case. $\square$

Lemma 3. Let A and L be matchings, and e an edge not in L and incompatible with L. Then:
$$d(A, L \oplus e) = \begin{cases} d(A, L) - 2 \ if e \in A, \\ d(A, L) + s \ otherwise, \end{cases}$$
where $s \leq 2$.

Proof. By case analysis. We have two possibilities for edge e: either $e \in A$ or $e \notin A$.

1. **Edge $e \in A$:** Since $e \notin L$ and e is not compatible with L, we have only two subcases here: either e is incident to a single edge $f \in L$ or it is incident to two, $f, f' \in L$.

   (a) **Edge e is incident to a single edge $f \in L$:** In this first subcase, $L \oplus e = L \setminus \{f\} \cup \{e\}$, so $|L \oplus e| = |L|$. Then $A + (L \oplus e)$ will have a 2-cycle in e while $A + L$ does not have any cycles involving e or f, so $\mathcal{C}(A, L \oplus e) = \mathcal{C}(A, L) + 1$. Therefore:

   $$\begin{aligned} d(A, L \oplus e) &= |A| + |L \oplus e| - 2\mathcal{C}(A, L \oplus e) \\ &= |A| + |L| - 2\mathcal{C}(A, L) - 2 \\ &= d(A, L) - 2, \end{aligned}$$

   as claimed.

   (b) **Edge e is incident to two edges $f, f' \in L$:** In this second subcase for $e \in A$, we have e incident to two edges $f, f' \in L$. Then $L \oplus e = L \setminus \{f, f'\} \cup \{e, e'\}$, so $|L \oplus e| = |L|$ again and $A + (L \oplus e)$ will have a 2-cycle in e that does not exist in $A + L$. If the connected component of e, f, and f' in $A + L$ is a path (see Fig. 6, left), then e' will not be in a cycle in $A + (L \oplus e)$. But if this component is a cycle $ef'Pf$ in $A + L$ (Fig. 6, right), this cycle will be shortened to $e'P$ when f and f' are replaced by e and e'. In any case, $\mathcal{C}(A, L \oplus e) = \mathcal{C}(A, L) + 1$ and the same derivation as above shows that $d(A, L \oplus e) = d(A, L) - 2$.

2. **Edge $e \notin A$:** Here again we have that e is incident to either one or two edges of L. As in the previous block, in any case we have $|L \oplus e| = |L|$. All we have to do here is to show that $\mathcal{C}(A, L \oplus e) \geq \mathcal{C}(A, L) - 1$, because this leads to:

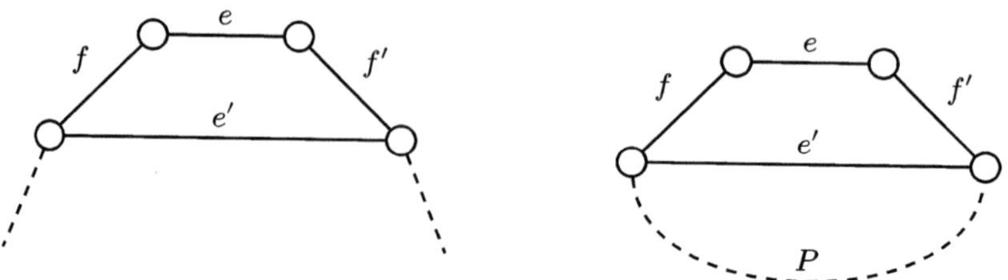

Fig. 6. Two subcases where $e \in A$. In the left picture, e, f, and f' belong to a path in $A + L$. In the picture to the right, e, f, and f' belong to a cycle in $A + L$. The dots in the pictures represent arbitrarily long, alternating A-L paths.

$$
\begin{aligned}
s &= d(A, L \oplus e) - d(A, L) \\
&= |A| + |L \oplus e| - 2\mathcal{C}(A, L \oplus e) - |A| - |L| + 2\mathcal{C}(A, L) \\
&= -2\mathcal{C}(A, L \oplus e) + 2\mathcal{C}(A, L) \\
&\leq -2(\mathcal{C}(A, L) - 1) + 2\mathcal{C}(A, L) \\
&= 2.
\end{aligned}
$$

In other words, all we have to do is convince ourselves that the multigraph $A + (L \oplus e)$ has at most one less cycle than $A + L$. However, edge addition does not lead to cycle loss, and edge removal accounts for at most one cycle lost per edge removed. Consequently, we do not have to worry about the case where e is incident to a single edge $f \in L$, since in this case a single edge is removed, causing the loss of at most one cycle.

The real concern is the case where e is incident to two edges $f, f' \in L$, because then there is the potential of losing two cycles as we remove f and f' from L and add e and e'. But, for this potential to materialize, both f and f' must belong to cycles in $A + L$, and distinct ones for that matter, as depicted in Fig. 7. However, in this situation, notice that we also create a new cycle $ePe'P'$ in $A + (L \oplus e)$. Therefore, the net loss is at most one cycle, since we lose two but create one cycle.

$\square$

B   Proof of Lemma 6

Lemma 6. If A and L are matchings, and e is an edge compatible with L, we have:

$$
K(A, L \cup \{e\}) = K(K(A, L), \{e\}).
$$

Proof. Consider a connected component X of $A + (L \cup \{e\})$ that does not contain e. This component will be present in $A + L$ as well. This component will contribute one edge to $K(A, L \cup \{e\})$ if it is a path with both vertices of degree

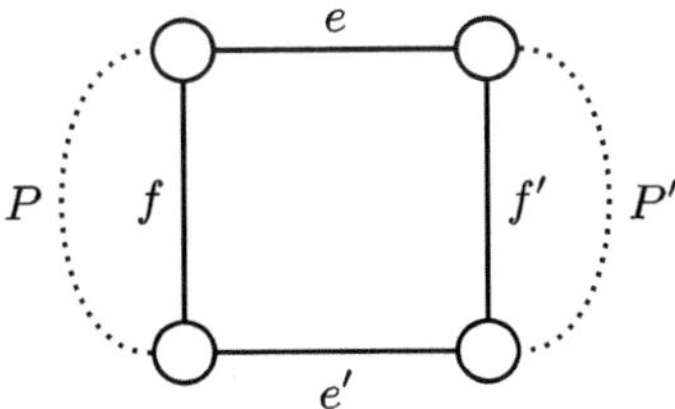

Fig. 7. Case 2: The removal of f and f' leads to the loss of two cycles. However, the inclusion of e and e' causes a new cycle to appear: $ePe'P'$. The dotted lines P and P' represent arbitrarily long, alternating A-L paths.

one A-saturated, and no edge at all otherwise. The component's contribution to $K(A, L)$ will be exactly the same, since $e \notin X$. Therefore, $K(A, L)$ and $K(A, L \cup \{e\})$ receive identical contributions from all connected components not containing e.

Therefore, we focus on the connected component of e in the multigraph $A + (L \cup \{e\})$. This component can be a cycle or a path. If it is a path, we focus on the A-saturation status of its extreme points, that is, the degree one nodes of this path. They may be both A-unsaturated, one A-unsaturated an the other A-saturated, or both A-saturated. We have therefore 4 cases, analyzed below.

1. **Edge e belongs to a cycle in $A+(L\cup\{e\})$:** Then $e \in K(A, L)$ by definition. In this case, $K(A, L \cup \{e\}) = K(A, L) \setminus \{e\}$, because e is removed from the collapse, no other edge in this cycle belong to either set, and the sets agree on the contributions from other connected components. On the other hand, given that $e \in K(A, L)$, Lemma 13 results in $K(K(A, L), \{e\}) = K(A, L)\setminus\{e\}$ as well.

2. **Edge e belongs to a path in $A + (L \cup \{e\})$:** Here we have three subcases, according to the A-saturation status of the path extremities.

   (a) **Both extremities A-unsaturated:** In this case, $e \notin K(A, L)$ and the entire path is absent from this set as well. Adding e to L will not change this, so $K(A, L \cup \{e\}) = K(A, L)$, as they agree in the contributions from other components. On the other hand, e is compatible with $K(A, L)$, because the entire path is absent, and Lemma 13 leads to $K(K(A, L), \{e\}) = K(A, L)$ as well.

   (b) **One extremity A-unsaturated and one A-saturated:** In this case, $e \notin K(A, L)$, the entire path is absent from this set as well, but there is an extra edge f in $K(A, L)$ linking an endpoint of e with the A-saturated extremity of the path (see Fig. 8). Adding e to L will remove this edge, so $K(A, L \cup \{e\}) = K(A, L) \setminus \{f\}$, as they agree in the contributions from other components. On the other hand, e is now not compatible with $K(A, L)$ and is incident to a single edge f in this set. Lemma 13 will imply that $K(K(A, L), \{e\}) = K(A, L) \setminus \{f\}$ as well.

   (c) **Both extremities A-saturated:** In this case, $e \notin K(A, L)$, the entire path is absent from this set as well, but there are two extra edges f and f'

in $K(A, L)$ linking the endpoints of e with the A-saturated extremities of the path (see Fig. 9). Adding e to L will remove these edges, but will also include a third edge e' linking both extremities of the path, so $K(A, L \cup \{e\}) = K(A, L) \setminus \{f, f'\} \cup \{e'\}$, as they agree in the contributions from other components. On the other hand, e is again not compatible with $K(A, L)$ and is indicent to f and f' in this set. Lemma 13 then implies that $K(K(A, L), \{e\}) = K(A, L) \setminus \{f, f'\} \cup \{e'\}$ as well.

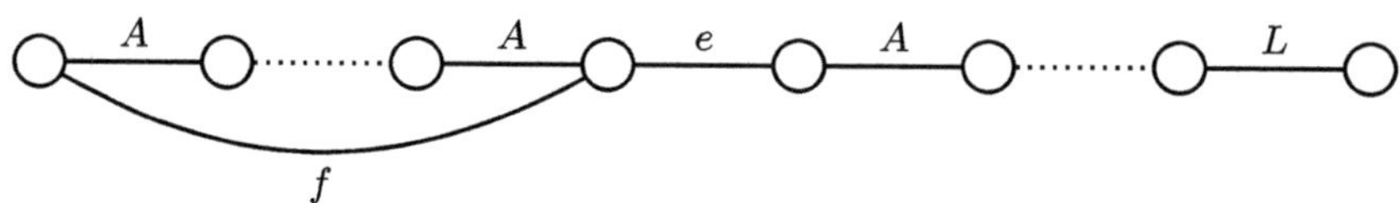

Fig. 8. Case 2b: One extremity A-unsaturated and one A-saturated in the e-path. The dots in the picture represent arbitrarily long, alternating A-L paths.

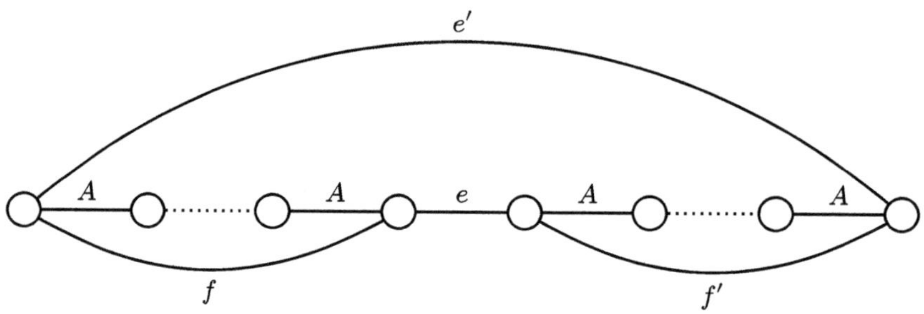

Fig. 9. Case 2c: Both extremities of the e-path are A-saturated. The dots in the picture represent arbitrarily long, alternating A-L paths.

We see that in all cases we have $K(A, L \cup \{e\}) = K(K(A, L), \{e\})$, proving the initial claim. $\qquad\square$

C   Code

To verify Lemma 12, copy the following code to a file `test.py` and run it with `python test.py`. The word 'problem' should not appear in the output.

```
#!/usr/bin/env python3


#################################################################
### Function that returns all matchings for a list
```

```python
### of vertices.

def matchings(l):
    """
    Returns all matchings of elements of list l
    """
    if len(l) <= 1:
        return [[]]
    else:
        m = matchings(l[1:])
        for i in l[1:]:
            l1 = l[1:]
            l1.remove(i)
            m += [m1 + [(l[0],i)] for m1 in matchings(l1)]
        return m

##################################################
### Node class for union-find structure

class Node:
    def __init__(self, value):
        self.value = value
        self.parent = self

##################################################
### Main class for union-find structure

class UnionFind:
    def __init__(self):
        self.nodes = {}

    def find(self, value):
        if value not in self.nodes:
            self.nodes[value] = Node(value)
            return value

        current_node = self.nodes[value]
        while current_node.parent != current_node:
            current_node = current_node.parent

        return current_node.value

    ### This is a special form of the union operation.
    ### It returns 1 if the arguments already belong
    ### to the same component, and 0 otherwise.
    ### In the context of two matchings, returning 1
    ### means closing a cycle.

    def union(self, value1, value2):
        root1 = self.find(value1)
```

```python
        root2 = self.find(value2)

        if root1 != root2:
            self.nodes[root2].parent = self.nodes[root1]
            return 0
        else:
            return 1


##########################################################
### Distance computation using union-find structure to
    determine
### the connected components of the union of two matchings.

def dist(x, y):
    """

    Compute distance between matchings.

    Method: use union-find structire to record the connected
    components of the union of two matchings.
    For each edge in either matching, add 1 to distance.
    For each cycle closed (union returns 1), subtract 2
    from distance.
    """

    d = 0
    uf = UnionFind()
    for u, v in x:
        d += 1
        uf.union(u, v)
    for u, v in y:
        d += 1
        d -= 2 * uf.union(u, v)
    return d


def score(m, a, b, c):
    """

    Computes the score d(m; a, b, c).
    """

    return dist(m, a) + dist(m, b) + dist(m, c)


def medians(a, b, c, candidates):
    """

    Returns all medians of the first three parameters
    among all matchings given by the fourth parameter.
    """

    minimum = -1
    for m in candidates:
        sc = score(m, a, b, c)
        if minimum == -1 or sc < minimum:
            minimum = sc
            medians = [m]
```

```python
        elif sc == minimum:
            medians.append(m)
    return medians

def occur(edges, lists):
    """
    Returns true when at least one of the edges
    belongs to at least one of the medians.
    """
    for e in edges:
        for l in lists:
            if e in l:
                return True
    return False


#####################################################
### Tests whether there is always an edge that is a
### partial median in a 4-cycle involving A and
### B in a graph with 8 vertices.

def main():
    ## Edges making up the 4-cycle
    ## It is important that all edges are increasing
    ### e.g., write (1,2), not (2,1)
    initial_a = [(0,1), (2,3)]
    initial_b = [(1,2), (0,3)]

    ## Complement: possible matchings on the
    ## vertices not involved in the initial edges
    compl = matchings(list(range(4,8)))

    ## All 8-vertex matchings: to loop over for
    ## the C matching and for median candidates
    all = matchings(list(range(8)))

    for ca in compl:
        ## ca is the complement for A
        a = initial_a + ca
        print("A", a)
        for cb in compl:
            ## cb is the complement for A
            b = initial_b + cb
            print("B", b)
            for c in all:
                print("C", c)
                ## Collects all medians, see if some edge
    belongs
                mds = medians(a, b, c, all)
                print("MS", mds)
                if not occur(initial_a + initial_b, mds):
```

```
                         print('problem', a, b, c, mds)

if __name__ == "__main__":
    main()
```

References

1. Chindelevitch, L., La, S., Meidanis, J.: A cubic algorithm for the generalized rank median of three genomes. Alg. Mol. Biol. **14**(1), 16 (2019). https://doi.org/10.1186/s13015-019-0150-y
2. Chindelevitch, L., Pereira Zanetti, J.P., Meidanis, J.: On the rank-distance median of 3 permutations. BMC Bioinform. **19**(6), 142 (2018). https://doi.org/10.1186/s12859-018-2131-4
3. Dias, Z., Meidanis, J.: Genome rearrangements distance by fusion, fission, and transposition is easy. In: Proceedings Eighth Symposium on String Processing and Information Retrieval, pp. 250–253 (2001). https://doi.org/10.1109/SPIRE.2001.989776
4. Feijão, P., Meidanis, J.: Extending the algebraic formalism for genome rearrangements to include linear chromosomes. IEEE/ACM Trans. Comput. Biol. Bioinform. **10**, 819–831 (2012). https://api.semanticscholar.org/CorpusID:2303303
5. Feijão, P.C.: On genome rearrangement models. Ph.D. thesis, University of Campinas (UNICAMP), Campinas, Brazil (2012). https://repositorio.unicamp.br/acervo/detalhe/896588
6. Meidanis, J., Chindelevitch, L.: Fast median computation for symmetric, orthogonal matrices under the rank distance. Linear Algebra Appl. **614**, 394–414 (2021). https://doi.org/10.1016/j.laa.2020.10.030
7. Meidanis, J., Dias, Z.: An alternative algebraic formalism for genome rearrangements. In: Sankoff, D., Nadeau, J.H. (eds.) Comparative Genomics: Empirical and Analytical Approaches to Gene Order Dynamics, Map Alignment and the Evolution of Gene Families, pp. 213–223. Springer, Dordrecht (2000). https://doi.org/10.1007/978-94-011-4309-7_20
8. Tannier, E., Zheng, C., Sankoff, D.: Multichromosomal median and halving problems under different genomic distances. BMC Bioinform. **10**(1), 120 (2009). https://doi.org/10.1186/1471-2105-10-120
9. West, D.: Introduction to Graph Theory. Prentice-Hall (2001)
10. Yancopoulos, S., Attie, O., Friedberg, R.: Efficient sorting of genomic permutations by translocation, inversion and block interchange. Bioinformatics **21**(16), 3340–3346 (2005). https://doi.org/10.1093/bioinformatics/bti535
11. Zanetti, J.P.P., Biller, P., Meidanis, J.: Median approximations for genomes modeled as matrices. Bull. Math. Biol. **78**(4), 786–814 (2016). https://doi.org/10.1007/s11538-016-0162-4

Quantifying the Rearrangement Complexity of Pangenomes

Leonard Bohnenkämper[iD] and Jens Stoye[(✉)][iD]

Faculty of Technology and Center for Biotechnology (CeBiTec), Bielefeld University, Bielefeld, Germany
`lbohnenkaemper@techfak.uni-bielefeld.de`, `jens.stoye@uni-bielefeld.de`

Abstract. The study of evolution between species (phylogenetics) and the study of evolution within a species (population genetics) are highly related, as the same biological mechanisms are fundamental to both fields. Although both have been studied for a long time, their joint study in a unified setting has been prevented by the different time scales they consider and the different data types they employ. A similar discrepancy holds for their whole-genome specializations, comparative genomics and pangenomics. Two active areas in these fields are genome rearrangement studies and graphical pangenomics, respectively. Since the emergence of graphical pangenomics, these have existed as separate fields, despite observations that central data structures representing genomic variants in both fields are highly similar.

While there exists a wealth of theoretical results for various rearrangement models in comparative genomics, the application to pangenomic data is hampered by the limitations of rearrangement problem formulations. On the practical side, pangenomes typically contain too many individual genomes for classical problems, such as the often NP-hard parsimony problems, to be solved, or for all-vs-all comparisons using rearrangement distances to be performed. On the theoretical side, some assumptions in the formulation of rearrangement problems, such as the assumption of an underlying tree, are inadequate for many pangenomes. In this work, we propose the *Complete Ancestral Reconstruction for Pangenomes (CARP)* problem, which overcomes these limitations while retaining intuitive relationships to both classical rearrangement problems and pangenome graphs.

Keywords: Pangenomics · Rearrangement Analysis · Single Cut or Join

1 Introduction

Rearrangements of genetic material have been described even before the molecular structure of DNA was known [4,15,44]. These rearrangements and other

---

Supplementary Information The online version contains supplementary material available at https://doi.org/10.1007/978-3-032-26891-4_2.

structural variations have been discovered not only to be ubiquitous, but also to play a significant role in evolution [16,19,49], as well as in genetic diseases like cancer [3,7,29,43]. Four years before the first whole eukaryotic genome was sequenced, David Sankoff proposed a framework to computationally quantify rearrangements between two genomes, the rearrangement distance problem [42]. Thirty years later, this basic distance problem has been solved for many different types of rearrangement models, some with simple results, like the Single-Cut-or-Join (SCJ), Double-Cut-and-Join (DCJ) and block interchange models [8,18,51], and some with more involved results, like the Reversal, Hannenhalli-Pevzner (HP) and DCJ-indel models [6,21,22], to name a few.

In the intervening decades, advances in sequencing technology have greatly enhanced the available material for analysis. This enabled the simultaneous study of several genomes in a clade – the field of pangenomics emerged. With advances in assembly quality, pangenomics itself moved from studying collections of genes [46] to collections of sequences [5] to studying entire genomes [47]. These sets of genomes are typically compressed for analysis, either via lossless text indices such as the Burrows Wheeler Transformation [41,52], or via lossy graphs such as colored de Bruijn [10,26,34] or variation graphs [17,20]. There exist many different ways to quantitatively characterize pangenomes, such as openness [37,46] or core percentage [25]. Many of these are already possible using gene-based pangenomes. One truly novel possibility, now that pangenomes are constructed from complete chromosomes, is the quantification of their structural complexity.

Examining the graphical representations of pangenomes, quantifying this complexity is intuitively possible. For example, in Fig. 1, one can hardly deny that Graph B is structurally more complex than Graph A. Of course, an objective, numerical measure is more desirable, especially in less obvious instances.

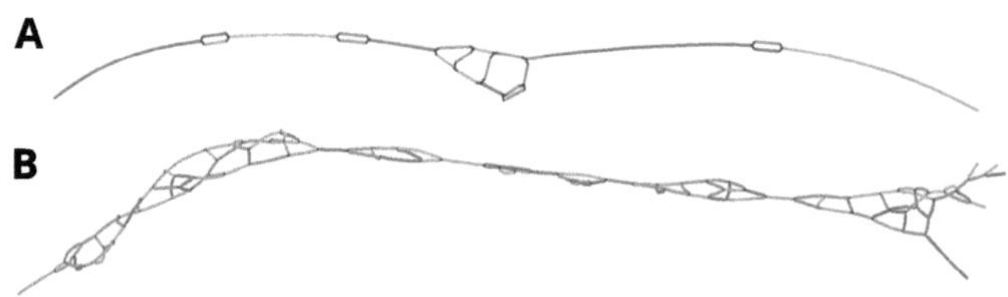

Fig. 1. 1000 BP regions in two compacted de Bruijn graphs ($k = 31$) built with Bifrost [26] and visualized with Bandage [48]. A - built from 56 *Yersinia pestis* genomes. B - built from 48 *Escherichia coli* genomes.

If one were to consider a "pangenome" of exactly 2 genomes, classical rearrangement analyses already give an answer to this quantification, namely the aforementioned rearrangement distance under a certain model. Notably, the rearrangement distance across various models can be quantified using the *breakpoint graph* [9,21,22,51]. This graph bears remarkable similarities to pangenome graphs. Specifically the close relationship between breakpoint and de-Bruijn graphs has been examined in [1] and [32].

Interestingly, the breakpoint graph of two identical genomes, whose distance is therefore 0 under classical rearrangement models, corresponds to a graph without any branching nodes, as seen in the bottom graph in Fig. 3. A pangenome with a graph like this has arguably minimal rearrangement complexity. Therefore, we say such a graph and any pangenome with such a graph is *simple*.

In this work, we propose the *Complete Ancestral Reconstruction for Pangenomes (CARP)* problem (Sect. 2), a new rearrangement problem designed to quantify the structural complexity of a pangenome based on how many graph altering operations, under a given model, it takes to transform the pangenome into a simple pangenome (see also Fig. 3). We then show how this problem relates to classical rearrangement problems (Sect. 3). We give a solution to this problem for the SCJ model (Sect. 4) and demonstrate in a number of experiments (Sect. 5) how even under this simple model, CARP can be used to derive valuable insights about pangenomes and pangenome graphs. This includes comparative analyses of the rearrangement complexities between different pangenomes, different pangenome graphs constructed from the same sequence set and different regions in the same pangenome graph. We further discuss the potential of the CARP problem formulation (Sect. 6), particularly once solutions for more advanced rearrangement models become available.

2   A New Rearrangement Problem

In a pangenome graph, each node represents a sequence segment. We call these segments *markers*. An edge connecting two nodes represents an *adjacency*, meaning that these two markers are immediate neighbors on a chromosome. To that end, an edge can either connect to the beginning or to the end of a marker. Each chromosome is then represented as a path or a cycle in this graph, depending on whether it is linear or circular.

A notable difference between pangenome and breakpoint graphs is that the latter aim to represent only rearrangements, and not local variations such as SNVs or small indels. To capture only structural variations, one can use genes or *collinear blocks* as markers. Collinear blocks can be generated as a by-product of whole genome alignment, such as in Mummer [14] or Mauve [11,12], but some pangenome graph construction methods, such as Minigraph-Cactus [23] or Pangraph [35], are also able to output collinear blocks. Additionally, there is dedicated software for collinear block detection, such as SibeliaZ [33].

What one considers a rearrangement, is to some degree dependent on a size threshold – note that one could even consider the deletion of a single base a "structural" variation. We will therefore presume that, using one of the aforementioned methods, a suitable level of abstraction that defines markers has been found.

Formally, a marker is a symbol from a large alphabet Σ. A *chromosome* is then associated with a string of oriented markers $S \in (\Sigma \cup \overline{\Sigma})^+$ where $\overline{\Sigma} := \{\overline{m} \mid$

$m \in \Sigma\}$ and $\overline{m}$ denotes marker m in reverse orientation. We call the string S a *chromosome string*. A convenient way of writing a marker m in this context is by referring to its *extremities*, that is, to its beginning m^t (*tail*) and to its end m^h (*head*). Then $m = m^t m^h$ and $\overline{m} = m^h m^t$. Two neighboring markers m and n in a string induce a pair of neighboring extremities, an *adjacency*. Depending on their relative orientation, adjacencies are $m^h n^t$, $m^t n^t$, $m^t n^h$ and $m^h n^h$ for mn, $\overline{m}n$, $\overline{m}\,\overline{n}$ and $m\overline{n}$, respectively. The identity of an adjacency is purely determined by the involved extremities, not their order, i.e. $m^h n^t$ is the same adjacency as $n^t m^h$.

A chromosome may be *linear*, in which case we denote it by $[S]$, or circular, in which case we denote it by (S). A circular chromosome $(s_1 \ldots s_l)$ in addition to the adjacencies of $s_1 \ldots s_l$ has the additional adjacency induced by $s_l s_1$. For a linear chromosome $[S]$ it is helpful to define the additional adjacencies $\varnothing u$ and $v\varnothing$ for the extremities u and v with which the chromosome starts and ends. We call $\varnothing$ a *telomere*. Note that we regard chromosomes as equal even when denoted in reverse orientation, i.e. $[\bar{s}_n \ldots \bar{s}_1] = [s_1 \ldots s_n]$ and circular chromosomes as equal even when denoted in different orientation $(s_k \ldots s_n s_1 \ldots s_{k-1}) = (s_1 \ldots s_{k-1} s_k \ldots s_n)$. A *genome* is a multiset of chromosomes, and a *pangenome* is a multiset of genomes.

We can now define the pangenome graph that represents the order of these markers as a bi-edged graph [38]. For ease of notation, we write the (undirected) edge between vertices u and v as uv or vu.

Definition 1. *The* Multiple Breakpoint Graph (MBG) $\mathcal{G}(\mathbb{P})$ *of a pangenome* $\mathbb{P}$ *is an undirected graph* $(V, E_{adj} \cup E_{mark})$ *where* V *represents the extremities of* $\mathbb{P}$ *including the telomere node* $\varnothing$ *and* E_{adj}, E_{mark} *are two types of edges:*

- E_{adj} *is a set of undirected edges, which is the union of all adjacencies in* $\mathbb{P}$. *We call* E_{adj} *adjacency edges.*
- E_{mark} *is a set of undirected edges that contains the edge* $m^t m^h$ *for each marker* m *in addition to a self edge* $\varnothing\varnothing$ *of the telomere* $\varnothing$. *We call* E_{mark} *marker edges.*

In principle, any graph in `gfa` format can be represented as a MBG if telomeres are added to those nodes at the end of linear paths. However, note that the theory presented here assumes all variants represented in the graph are captured by the rearrangement model used. An example of a MBG is found in Fig. 2.

Recall the earlier motivation of a *simple* pangenome without any branching paths in its pangenome graph as having minimal structural complexity. We can now formulate this more precisely. A pangenome and its MBG are *simple* if for each marker m (excluding $\varnothing$) at most one adjacency edge connects to its beginning m^t and at most one adjacency edge connects to its end m^h. We then wish to quantify the structural complexity of a pangenome by the number of rearrangement steps necessary to transform it into a simple pangenome. The operations used for this transformation are most meaningful when modeled after biological events, such as reversals or recombinations that occur in chromosomes of the pangenome. They must thus fundamentally operate on chromosomes, i.e.

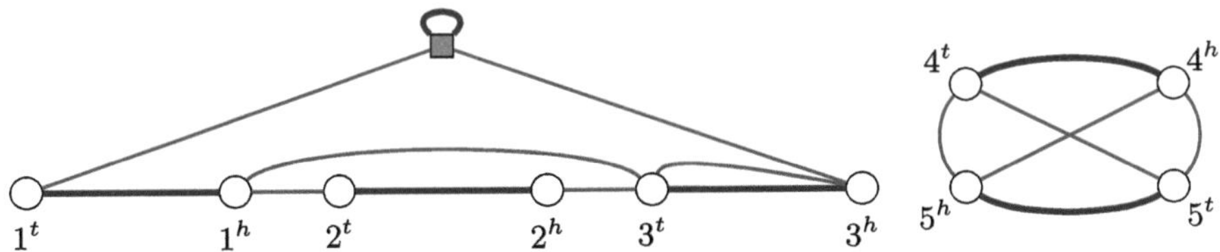

Fig. 2.   Multiple   Breakpoint   Graph   (MBG)   for   the   pangenome $\{\{[123],(45)\},\{[133],(45)\},\{[133],(4\bar{5})\}\}$. Blue edges are marker edges, red edges are adjacency edges, and the quadratic grey node is the telomere $\varnothing$.

the paths in the graph, not the graph itself. However, not all operations have the same effect on the MBG. There are operations that never affect the structure of the MBG, such as the duplication of a chromosome or the recombination of two chromosomes, as shown in the left box of Fig. 3. Other operations, typically ones that are associated with breaking a chromosome, may have an effect on the graph. Such an operation is shown in the right box of Fig. 3. We call these types of operations Path Altering Operations and Graph Altering Operations, respectively. Formally, an operation o is a *Path Altering Operation (PAO)* if, after applying it to a pangenome $\mathbb{P}$, denoted $\mathbb{P} \xrightarrow{o} \mathbb{P}'$, the resulting pangenome $\mathbb{P}'$ has the same MBG as $\mathbb{P}$, i.e. $\mathcal{G}(\mathbb{P}') = \mathcal{G}(\mathbb{P})$. Otherwise it is a *Graph Altering Operation (GAO)*.

Classical rearrangement operations, such as reversal, translocation, block interchange and DCJ, are all GAOs. In fact, one can reformulate many classical problems as transforming a set of input genomes into the closest set of genomes with a simple MBG by using the model operation and a limited set of supplementary PAOs while minimizing the number of model operations used. We examine this relationship for the distance, median, large parsimony and halving problems in Sect. 3.2.

However, in these formulations the interaction between lineages is strictly limited – owing to the fact that most classical rearrangement problems (implicitly) assume an underlying tree. In fact, the only PAOs necessary to model these classical problems are coalescence (unifying equal genomes) and some form of chromosome de-duplication (unifying equal chromosomes in the same genome). In a pangenome, however, the interactions between lineages can be much more frequent and complex, such as the transfer of plasmids between bacteria or the sexual reproduction of many eukaryotes. We therefore propose to use a "maximally powerful" set of PAOs to quantify the rearrangement complexity of a pangenome. We call such a set *complete*. Formally, we say a set of PAOs is *complete* if it is able to transform any two pangenomes with the same MBG into each other. Such a set exists; we give it in Theorem 1 (Sect. 3.3) and we prove that it is complete in Appendix B.

The definition for the *Complete Ancestral Reconstruction for Pangenomes (CARP)* problem is then as follows: Given a pangenome, a set of GAOs and a complete set of PAOs, find a simple pangenome into which it can be transformed

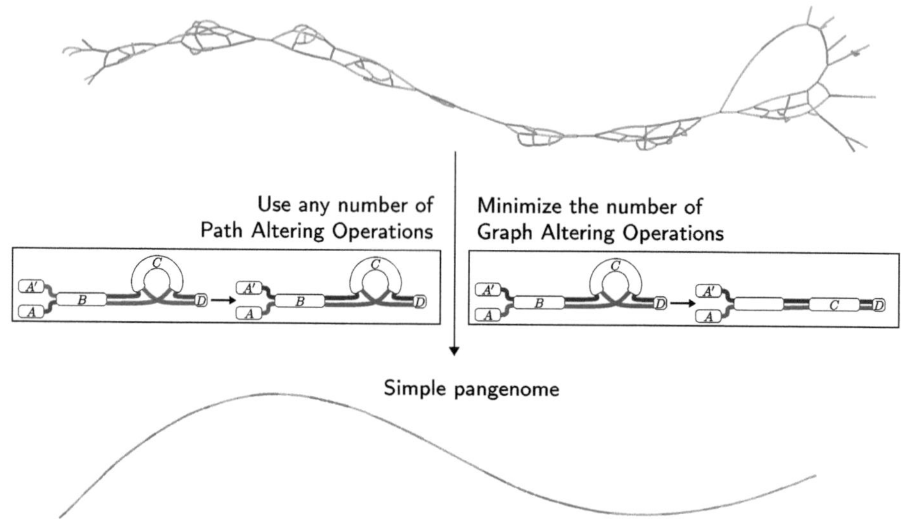

Fig. 3. Schematic representation of the CARP rearrangement problem: A pangenome is transformed into the closest simple pangenome using the minimum number of Graph Altering Operations while utilizing arbitrarily many operations from a complete set of Path Altering Operations. An example of a Path Altering Operation (a homologous recombination between the chromosome $A'B\bar{C}D$ (pink) and the chromosome $ABCD$ (purple) at marker B, transforming them into $A'BCD$ (blue) and $AB\bar{C}D$ (red)) is shown in the left box. An example of a Graph Altering Operation (a reversal of marker C in the chromosome $AB\bar{C}D$ (red)) is shown in the right box. Note that both types of operations change paths directly, but only Graph Altering Operations change edges in the graph. (Color figure online)

using any number of PAOs from a complete set and a minimal number of GAOs. A visual summary is found in Fig. 3. Formally, we define the problem as follows.

Problem 1. (Complete Ancestral Reconstruction for Pangenomes (CARP)). Given a pangenome $\mathbb{P}$, a PAOs N and a set of GAOs A, find the shortest sequence of GAOs $o_1 \ldots o_n \in A^*$, such that there is a sequence of operations $s = O_0 o_1 O_1 \ldots o_n O_n$ transforming $\mathbb{P}$ into a simple pangenome $\mathbb{A}$ where each O_i is a (possibly empty) sequence of PAOs, $O_i \in N^*$. We call the simple pangenome $\mathbb{A}$ a *CARP simplification*, s a *CARP scenario* and n the *CARP measure* of $\mathbb{P}$.

3 Relationship to Classical Rearrangement Problems

CARP is related to classical rearrangement problems, such as the distance, median and large parsimony problem. In fact, all of these can be expressed as specializations of a common superproblem, the *General Ancestral Reconstruction Problem*, which we examine in Sect. 3.2. In essence, these problems differ only by which path altering operations they permit. We thus first examine a few path altering operations in Sect. 3.1. Finally, we discuss a complete set of path altering operations in Sect. 3.3.

3.1 Path Altering Operations Relevant to Classical Problem Formulations

In this section, we give examples of PAOs that are useful to understand the relationship between CARP and classical rearrangement problems (see Sect. 3.2). One of the most fundamental PAOs is the emergence of new lineages, which when regarded backwards in time is known as *coalescence*. In the following, we use the multi-set addition $\oplus$ and multi-set subtraction $\ominus$.

Definition 2. (Coalescence/Divergence). *Given a pangenome $\mathbb{P}$ and genomes $G_1, G_2, \ldots, G_k \in \mathbb{P}$ with $G_1 = G_2 = \ldots = G_k$, a k-coalescence on genomes $G_1, G_2, \ldots, G_k$ creates the pangenome $\mathbb{P}' = \mathbb{P} \ominus \{G_2, \ldots, G_k\}$. Given a pangenome $\mathbb{P}$ and a genome $G_1 \in \mathbb{P}$, a k-divergence on genome G_1 creates the pangenome $\mathbb{P}' = \mathbb{P} \oplus \{G_2, \ldots, G_k\}$ with $G_2 = \ldots = G_k = G_1$.*

As we examine in Sect. 3.2, many classical problems examine only tree-like evolution without duplicated markers and can thus be formulated directly using only coalescence. However, some classical problems, such as double distance or halving problems, require the modeling of whole genome duplications. To that end, we first define the duplication and de-duplication of a single chromosome.

Definition 3. *A k-multiplication $(\rightarrow)$ transforms a single chromosome in one of the following ways and a k-de-multiplication $(\leftarrow)$ transforms either k identical or one repetitive circular chromosome in one of the following ways:*

$$[S_1] \rightleftarrows [S_1], [S_2], \ldots, [S_k] \quad (S_1) \rightleftarrows (S_1), (S_2), \ldots, (S_k) \quad (S_1) \rightleftarrows (S_1 S_2 \ldots S_k)$$

where $S_2 = \ldots = S_k = S_1$.

A whole genome duplication is then characterized by duplicating all of a genome's chromosomes. Again, there is an inverse when viewing time "backwards".

Definition 4. *A* genome duplication operation *transforms a genome by 2-multiplying all of its chromosomes. A* genome halving operation *is the inverse of a genome duplication operation.*

These definitions suffice to characterize many classical rearrangement problems as we examine in the following section.

3.2 The General Ancestral Reconstruction Problem (GARP)

Under the assumption that PAOs do not create new structural complexity while a simple set of genomes possess minimal structural complexity, we can quantify the complexity of a set of genomes by measuring how many GAOs it takes to transform it into a simple pangenome while using some supplementary set of PAOs. This formulation (see Problem 2) is then a common framework for CARP and classical rearrangement problems.

Problem 2. (General Ancestral Reconstruction Problem (GARP). Given a pangenome $\mathbb{P}$, a set of PAOs N and a set of GAOs A, find the shortest sequence of GAOs $o_1 \ldots o_n \in A^*$, such that there is a sequence of operations $s = O_0 o_1 O_1 \ldots \ldots o_n O_n$ transforming $\mathbb{P}$ into a simple pangenome $\mathbb{A}$ where each O_i is a (possibly empty) sequence of PAOs, $O_i \in N^*$.

Note that CARP (Problem 1) is a specialization of Problem 2, where the set of PAOs is required to be *complete*. In classical rearrangement problems, this is not the case.

For a simple example from classical rearrangement literature, consider the genomic distance problem [42]. In this problem, two genomes X and Y are given that contain the same set of markers, each exactly once. The goal is to find the minimum number of rearrangements to transform X into Y under a given model M. If each rearrangement operation in M has an inverse, which is typically the case, we can model this in the GARP framework.

We leave the set of PAOs N empty and define the set of GAOs A as the operations of M. Then for each scenario transforming X into Y, $X = X_0 \overset{o_1}{\to} X_1 \overset{o_2}{\to} \ldots \overset{o_n}{\to} X_n = Y$, there is a GARP solution with the same number of GAOs, namely $o_1 o_2 \ldots o_n$. Note that two genomes without duplicate markers are equal if and only if their adjacencies are equal [2], meaning the MBG constructed from both genomes is simple. Since $X_n = Y$, the final MBG is simple. Thus, the number of GAOs required in the GARP formulation is a lower bound for the distance.

Conversely, the distance also bounds the number of GAOs in the GARP scenario. To see why, we can find a classical rearrangement scenario for any GARP scenario. In this case, the scenario can modify both X and Y with $X_0 \overset{x_1}{\to} \ldots \overset{x_k}{\to} X_k$ and $Y_0 \overset{y_1}{\to} \ldots \overset{y_j}{\to} Y_j$. Because $\mathcal{G}(\{X_k, Y_j\})$ is simple, $X_k = Y_j$. Thus, we can use the inverse operations $y'_1, \ldots, y'_j$ of $y_1, \ldots, y_j$ respectively to build the classical scenario $X = X_0 \overset{x_1}{\to} \ldots \overset{x_k}{\to} X_k = Y_j \overset{y'_j}{\to} \ldots \overset{y'_1}{\to} Y_0 = Y$. The problem formulations are thus equivalent.

A slightly more involved example is the large parsimony problem [2] (called "Multiple Genome Rearrangement Problem" there). In this problem, extant genomes $X_1, \ldots, X_n$ are given and one is asked to find a binary tree T with $X_1, \ldots, X_n$ at the leaves and genomes $Y_1, \ldots, Y_{n-1}$ at the internal nodes, such that the total distance along the edges $E(T)$, $\sum_{(U,V) \in E(T)} d(U, V)$, is minimized. Under the assumption that the operations of M are symmetric, the following GARP formulation is equivalent: In Problem 2, let $\mathbb{P} = \{X_1, \ldots, X_n\}$, let N contain only 2-coalescence operations, and let A be the operations of M. We show how to obtain a GARP solution from the large parsimony solution. Assume the children V, W of an internal node U exist in the current pangenome. From the large parsimony solution we can deduce operations $v_1 \ldots v_k$ transforming U into V and operations $w_1 \ldots w_j$ transforming U into W. Since operations have an inverse, there are operations $v'_k \ldots v'_1$ and $w'_j \ldots w'_1$ transforming V and W into U, which we can use in the GARP scenario and finally use a 2-coalescence to unify the genomes into U. We can apply this procedure bottom up to arrive at the corresponding GARP scenario. Conversely, one can find the corresponding large parsimony problem solution for a GARP scenario by using the genomes unified by the 2-coalescence as inner nodes and reversing the scenarios to be top-down. In fact, this is how Alekseyev and Pevzner constructed the MGRA heuristic for the large parsimony problem for 2-breaks (DCJ) in [2], although they do not explicitly use coalescence, but instead rely on the concept of "strict transformations", i.e. such transformations that conform to a phylogeny.

In the same way one can find a GARP formulation for many classical rearrangement problems. We give a selection in Table 1. We note that coalescence is the only interaction between genomes in these problems. As mentioned earlier, due to the phylogenetic closeness of the genomes in a pangenome, the interactions in a pangenomic context may well be more frequent and direct. For our pangenome focused specialization of GARP, CARP, we permit much more interaction by using a complete set of PAOs. We give one example of such a complete set in the following section.

3.3   A Complete Set of Path Altering Operations

We will now show that there is a set of operations, which is a complete set of PAOs and describes meaningful biological interactions. In fact, we do so by extending the PAOs already described in Sect. 3.1. We only add a direct way by which genomes interact and an operation modeling chromosomal recombination.

While there are many ways how genomes can interact, the most direct one is by transferring chromosomal material.

Definition 5 (Chromosome Transfer). *Given a pangenome* $\mathbb{P}$ *that includes genomes* G_1, G_2 *and a multiset of chromosomes* $C \subseteq G_1$, *a chromosome transfer between* G_1 *and* G_2 *creates* $\mathbb{P}' = \mathbb{P} \ominus \{G_1, G_2\} \oplus \{G_1', G_2'\}$ *where* $G_1' := G_1 \ominus C$ *and* $G_2' := G_2 \oplus C$.

If multiple copies of the same chromosome exist in the same genome, new chromosome structures can arise by recombinations, such as shown in Fig. 4. These homologous recombinations are one mechanism of how rearrangements occur [40].

Fig. 4. A homologous recombination may create new chromosome structures not previously present in the pangenome.

We can define a homologous recombination in our context roughly as taking two (sub-)strings SmT and UmV, $m \in \Sigma$, of one or two chromosomes and transforming them into SmV and UmT. Depending on the connectivity of these substrings, this then has different effects. More formally:

Table 1. Classical genome rearrangement problems on a model M with symmetric operations and an equivalent GARP problem (see Problem 2). A dash ($-$) represents that no PAOs are needed while brackets indicate those PAOs that one can include into the problem formulation without losing the equivalence. In each case the set of GAOs A equals the set of operations of M and $\mathbb{P}$ is the multiset containing all input genomes.

Problem name	Ref.	Given genomes	Classical Formulation	PAOs (N)
Distance problem	[42]	Genomes X, Y on the same marker set (single copy)	Find the minimum number of operations in M to transform X to Y	$-$ (2-coalescence)
Median problem	[45]	Genomes $X_1, \ldots, X_k$ on the same marker set (single copy)	Find a genome Y that minimizes the total distance to $X_1, \ldots, X_k$	$-$ (k-coalescence)
Large parsimony problem	[2]	Genomes $X_1, \ldots, X_k$ on the same marker set (single copy)	Find a binary tree with leaves $X_1, \ldots, X_k$ and genomes $Y_1, \ldots Y_{k-1}$ as inner nodes with minimal total distance between the genomes at each edge	2-coalescence
Genome halving problem	[45]	Genome Y with each marker occurring twice	Find a genome X in which each marker occurs once, such that the genome obtained by doubling X minimizes the distance to Y	genome halving
Guided genome halving problem	[45]	Genome Y with each marker occurring twice and genome Z with each marker occurring once	Find a genome X in which each marker occurs once, such that the distance of the genome obtained by doubling X to Y plus the distance of X to Z is minimized	genome halving (2-coalescence)

Definition 6. *A homologous recombination transforms chromosomes of the same genome in one of the following ways*

$$\begin{aligned}
[SmT], [UmV] &\rightleftarrows [SmV], [UmT] &&\textit{(Translocation)} \\
[SmT\overline{V}\overline{m}\overline{U}] &\rightleftarrows [SmV\overline{T}\overline{m}\overline{U}] &&\textit{(Inversion)} \\
(SmT\overline{V}\overline{m}\overline{U}) &\rightleftarrows (SmV\overline{T}\overline{m}\overline{U}) &&\textit{(Inversion)} \\
[SmT], (UmV) &\rightleftarrows [SmVUmT] &&\textit{(Circular inclusion/excision)} \\
(SmT), (UmV) &\rightleftarrows (SmVUmT) &&\textit{(Circular fusion/fission)}
\end{aligned}$$

where $m \in \Sigma$ and $S, T, U, V \in (\Sigma \cup \overline{\Sigma})^*$.

Note that a homologous recombination is a PAO, because while the chromosome connectivity is transformed, the adjacencies (particularly at marker m) remain the same and therefore the graph remains unchanged.

Transfer and recombination in conjunction with chromosome (de-) multiplication, coalescence and divergence are then a complete set of PAOs, as we show in Appendix B.

Theorem 1. *The set of operations $\tilde{N}$, which consists of (de-) multiply, homologous recombination, coalescence, divergence and chromosome transfer operations is a PAOs.*

Note that we only discussed a small subset of PAOs here. There are many more possible PAOs. Some rearrangement operations which have been studied recently, such as flanked transpositions [50] or flanked block interchanges [30], are in fact PAOs, although none of them form a complete set by themselves. Nonetheless, this means the complete set of PAOs $\tilde{N}$ discussed here is not the only such set. However, since PAOs are used without a cost in Problem 1, the concrete set of PAOs does not change the CARP measure or CARP simplification as long as the set is complete.

4   Solution Under the SCJ Model

The *Single Cut or Join (SCJ)* model, established by Feijão and Meidanis [18], is a simple rearrangement model closely related to the breakpoint distance. The model allows to cut a chromosome in one place (single cut) or to join two ends from one or two linear chromosomes together (join). These two operations have analogues in biology, namely double strand breaks and non-homologous end joining. However, care needs to be taken in interpreting genomes reconstructed by SCJ since events with breakpoint reuse are not reflected in SCJ scenarios.

Nonetheless, it is a useful model as many otherwise NP-hard problems are tractable under SCJ. However, reconstruction without a tree (large parsimony problem), remains NP-hard [18]. One might thus expect that CARP is NP-hard as well under SCJ. However, we show here that the problem has a surprisingly simple linear time solution.

Formally, an SCJ is either a *Single Cut* or a *Single Join*. A Single Cut transforms a circular chromosome (S) into a linear chromosome $[S]$ or a linear chromosome $[ST]$ into two linear chromosomes $[S]$ and $[T]$. A *Single Join* transforms a linear chromosome $[S]$ into a circular chromosome (S) or two linear chromosomes $[S]$ and $[T]$ into one linear chromosome $[ST]$. S and T here represent arbitrary strings of oriented markers.

In the MBG, cutting between two markers m, n, i.e. $[SmnT] \to [Sm], [nT]$, creates the telomeric adjacencies $m^h\varnothing, n^t\varnothing$ and removes the adjacency edge $m^h n^t$ if $[SmnT]$ was the only chromosome with that adjacency. Conversely, the join $[Sm], [nT] \to [SmnT]$ creates the adjacency edge $m^h n^t$ and removes $m^h\varnothing, n^t\varnothing$ if $[Sm], [nT]$ are the only linear chromosomes with these (telomeric) adjacencies. The effects for other combinations of orientations and chromosome types are analogous. Note that if uv and $u'v'$ are joined to uu', then $v = v' = \varnothing$.

From this follows that, to arrive at a simple pangenome, we need to expend at least one SCJ operation per non-telomeric adjacency that involves a branching node. We call these adjacencies *contested*. We give an example graph where contested edges are marked in Fig. 5.

Lemma 1. *Transforming a pangenome* $\mathbb{P}$ *with MBG* $(\mathbb{P}) = (V, E_{adj} \cup E_{mark})$ *into a simple pangenome* $\mathbb{A}$ *requires at least* $|E_{cnt}|$ *SCJ operations where* $E_{cnt} := \{uv \in E_{adj} \mid u = v\} \cup \{uv \in E_{adj} \mid u, v \neq \varnothing, \exists uv' \in E_{adj} \text{ with } v' \neq v\}$. *We call* E_{cnt} *the set of* contested *adjacencies.*

Proof. We prove this lemma using an injection f between edges in E_{cnt} and SCJ operations in the CARP scenario. Consider an edge $uv \in E_{cnt}$. We distinguish two cases.

(I) Let uv not be an adjacency in the simple MBG of $\mathbb{A}$. Then there must be a cut c_{uv} removing uv from the graph before $\mathbb{A}$ is reached, i.e. $uv \xrightarrow{c_{uv}} u\varnothing, v\varnothing$. We thus map uv to c_{uv}, i.e. $f(uv) = c_{uv}$.

(II) Let uv be an adjacency in the simple MBG of $\mathbb{A}$. Note that then $u \neq v$, otherwise the final MBG would not be simple. Since uv is contested, there is another adjacency edge at u or at v. Let w.l.o.g. $uv' \in E_{adj}$ with $v' \neq v$. In the CARP scenario, we may then have a series of alternating cuts and joins exchanging v' for other nodes: $uv' = uv_0 \xrightarrow{c_1} u\varnothing \xrightarrow{j_1} uv_1 \xrightarrow{c_2} u\varnothing \xrightarrow{j_2} uv_2 \ldots \xrightarrow{x_n} uv_n$. The MBG of $\mathbb{A}$, which contains uv, is simple only if $v_n = v \neq \varnothing$. Therefore the last operation in the series must be a join, $x_n = j_n$. We thus map uv to j_n, i.e. $f(uv) = j_n$. $\qquad\square$

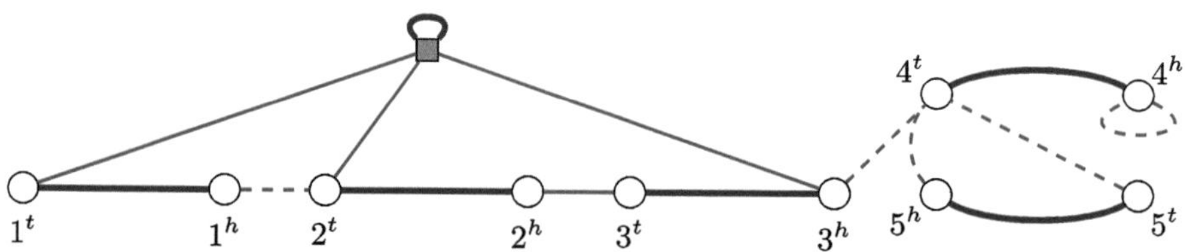

Fig. 5. MBG for the pangenome $\{\{[23], (4\bar{4}5)\}\}, \{[1234\bar{4}\bar{3}\bar{2}]\}\}$ with contested edges shown dashed. Blue edges are marker edges, red edges are adjacency edges, and the quadratic grey node is the telomere $\varnothing$.

We thus need to expend at least one SCJ operation per contested edge. In Appendix C we show that this lower bound is tight. Specifically, one can construct a simplification pangenome by simply cutting each edge in E_{cnt}, which then immediately implies the following:

Theorem 2. *The CARP measure under the SCJ model for a pangenome* $\mathbb{P}$ *with MBG* $(\mathbb{P}) = (V, E_{adj} \cup E_{mark})$ *is* $|E_{cnt}|$. *One possible SCJ-CARP simplification* $\mathbb{A}$ *is defined by the adjacencies* $E_{\mathbb{A}} := E_{adj} \setminus E_{cnt}$.

CARP under SCJ is thus solvable in time linear with respect to the graph size. Notably, the CARP measure corresponds directly to an easy to count graph metric, namely the edges connecting to branching nodes E_{cnt}. Indeed, this mirrors the results of classical rearrangement theory, where distances often correspond to simple graph structures (e.g. cycles (DCJ distance) or 2-cycles (breakpoint distance)) in specialized graphs, such as the breakpoint graph. The difference here is that the MBG can capture arbitrary (pangenome) graphs.

5   Evaluation

We implemented the SCJ-CARP reconstruction as well as utility functions to filter graph nodes based on size and extracting fixed size regions surrounding nodes from the graph. Relevant for the following experiments are the programs `carp`, which calculates the SCJ-CARP measure optionally filtering small nodes below a threshold and `carp-scan`, which calculates the SCJ-CARP measures for a fixed size region around each node. The implementation as well as snakemake [28] workflows for all following experiments are published on github[1]. The implementation is also available on bioconda under package name `carp`.

We evaluate our method for simulated datasets with regards to its ability to track the number of rearrangement events and to reconstruct ancestral adjacencies, and for its practical implications on biological datasets.

5.1   Simulated Experiments

In order to evaluate how well SCJ-CARP quantifies rearragements, we used ZOMBI [13] to simulate phylogenies of genomes from an initial genome with 1000 markers. We used default parameters unless otherwise specified. By default, ZOMBI generates phylogenies with 20 extant genomes.

In order to simulate genomes of varying complexity, we scaled rates in ZOMBI for gene originations, duplications, losses, inversions and transpositions from 0.1 to 1.0 of the default rate. A comparison between this rate scale and the SCJ-CARP measure calculated is shown in Fig. 6 (A). We see that the SCJ-CARP measure tracks the number of simulated rearrangement operations well, particularly for low rates of rearrangements. In fact the Pearson and Spearman correlation coefficients for the entire experiment are 0.89 and 0.91 respectively, indicating a strong correlation between the number of simulated operations and the SCJ-CARP measure.

We then proceeded to evaluate in which way the marker adjacencies in the SCJ-CARP simplification correspond to the ancestral root simulated by ZOMBI. In this analysis, we limited ourselves to non-telomeric adjacencies, since in the SCJ model telomeres are best interpreted as uncertainty about which adjacency should be reconstructed, not as actual ends of chromosomes. We evaluated precision and recall using the adjacencies of the root genome as the ground truth.

---

[1] https://github.com/gi-bielefeld/scj-carp.

The results are shown in Fig. 6 (B). We see that the precision is high across all samples with the majority of samples in each step reaching 100% precision. However, recall degenerates quickly with increasing numbers of rearrangements. This is not suprising as the simplifications reconstructed by `carp` are extremely conservative – like in other problems under SCJ, ancestral genomes tend to fragment unless adjacencies are perfectly conserved.

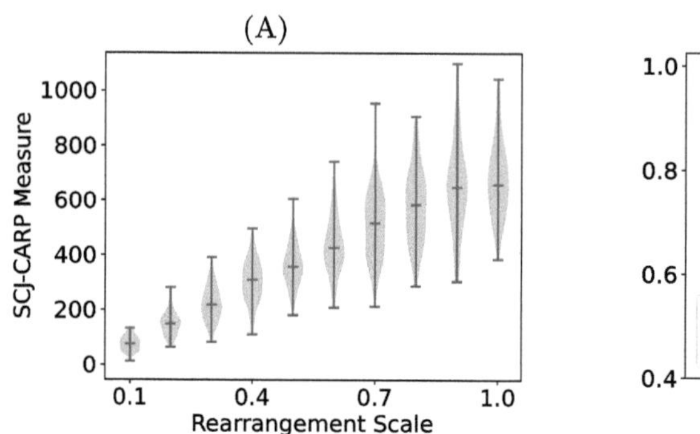
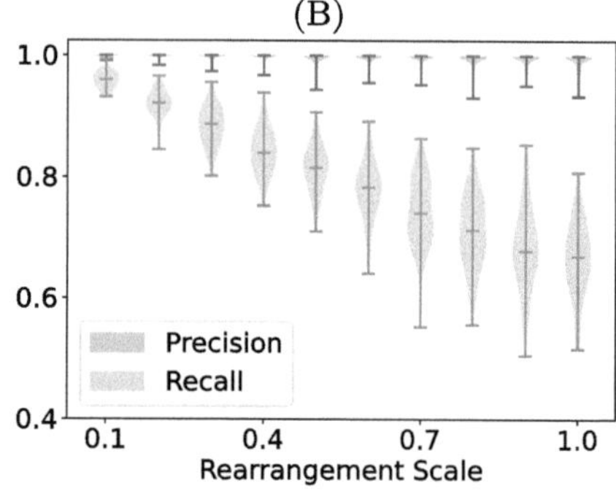

Fig. 6. SCJ-CARP measure (A) and reconstruction precision and recall (B) with increasing numbers of rearrangements.

5.2   Evaluating the SCJ-CARP Measure of Multiple Pangenomes

We downloaded multiple genomes for 8 different taxa using NCBI datasets [36] to compare their structural complexity with SCJ-CARP. We aimed for 10 genomes with NCBI assembly level "complete" when possible and opted for "chromosome+" and a slightly reduced number of genomes when necessary. Details are found in Appendix D. We ran Minigraph-Cactus [23] and SibeliaZ [33] with two different settings to determine collinear blocks. We then ran `maf2synteny` [27] with default settings on the output of both tools.

We evaluated the SCJ-CARP measure on the resulting genome permutation files. Each run of `carp` took less than a second using four threads on a cluster machine. The calculated SCJ-CARP measures are visualized in Fig. 7.

We see that the SCJ-CARP measures rank the relative pangenome complexity consistently even when using different block construction methods. The only disagreements are with respect to the order of *Y. pestis*, *E. coli* and *K. pneumoniae*, which are not well separated using any of the three block types, and the placement of *P. vivax*. Nonetheless, overall correlation between the CARP measures under different block definitions in this experiment is high (see Appendix Fig. 12).

Notably there are vast differences in the absolute SCJ-CARP measures between SibeliaZ and Minigraph-Cactus. Most interesting is the difference for *P. vivax*, for which both SibeliaZ runs evaluate to a much higher SCJ-CARP

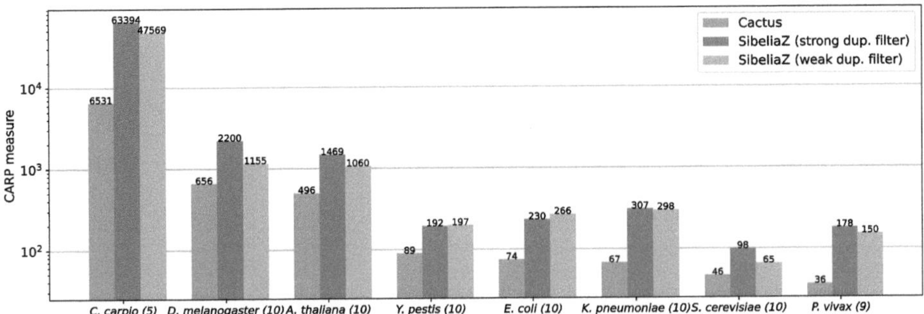

Fig. 7. SCJ-CARP measures on pangenomes with collinear blocks detected by Minigraph-Cactus and SibeliaZ. Two different settings were used for SibeliaZ, a strong duplication filter ($a = 2 \times$ #genomes) and a weak duplication filter ($a = 20 \times$ #genomes). Shown in parentheses behind the taxon name is the number of genomes used as input to the block detection.

measure than Minigraph-Cactus, which yields the lowest SCJ-CARP measure overall. We conjecture that this is due to the differing definitions of blocks used by the two tools. It might also be the case that the SCJ model struggles to capture the rearrangements displayed by SibeliaZ. Another model may score the Minigraph-Cactus and SibeliaZ blocks more similarly. Thus, once solutions to CARP under more complex models are available, the differences between these segmentation approaches may become better explainable.

The discrepancy also emphasizes how crucial the initial block definition is for the comparison of genomes, a known issue in rearrangement analyses. Especially note that the absolute CARP measures on differently determined blocks are usually not comparable. Nonetheless, using blocks determined by the same method, SCJ-CARP gives a relatively coarse but quickly computable measure for the rearrangement complexity of a pangenome.

5.3   Comparing Human Pangenome Graphs Using CARP

Given the effect of different marker definitions on the CARP measures observed in the previous section, it stands to reason that the discrepancy between CARP measures of pangenome graphs constructed by different methods can give insights about how these graphs differ structurally. Notably, such structural differences have been observed between Minigraph, Minigraph-Cactus and PGGB during the construction of the first human pangenome reference [31]. There the authors note that Minigraph-Cactus includes smaller variants than Minigraph while still constructing similar nodes for duplicate regions, while PGGB's all-vs-all alignment strategy creates a graph that collapses duplications resulting in an overall more complex graph (see Extended Data Fig. 3 in [31] for an example).

We downloaded the three respective graphs from the human-pangenomics S3 bucket using the CHM13-version of the Minigraph and Minigraph-Cactus graphs. We proceeded to analyze the graphs using `carp`, filtering out nodes below a threshold of progressively larger size. We measured the number of nodes in the graph and the SCJ-CARP measure. We show these results as well as the SCJ-CARP measure relative to the number of nodes in Fig. 8.

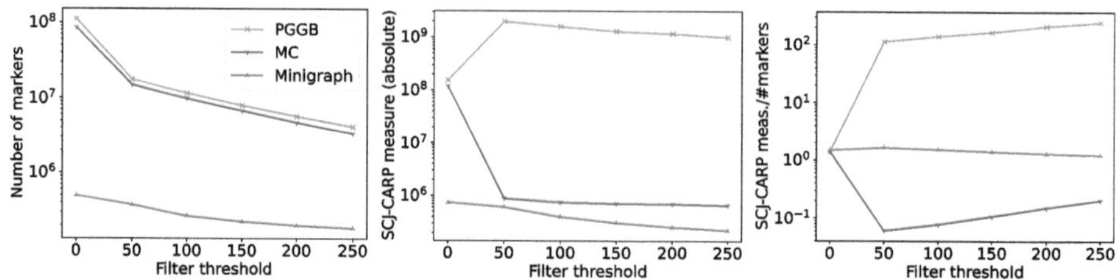

Fig. 8. CARP statistics for various levels of node filtering on human pangenome graphs built by PGGB, Minigraph-Cactus (MC) and Minigraph from [31]. Nodes of a size smaller than the filter threshold were removed and each path bridging two nodes that pass the threshold consisting of only filtered nodes was replaced with a single edge.

Without filtering the nodes (Filter threshold 0), we see that Minigraph is an outlier, while Minigraph-Cactus and PGGB are close together, both in the number of nodes and in the absolute SCJ-CARP measure. Nonetheless, the relative SCJ-CARP measure of all three tools is similar. Progressively filtering nodes, differences between Minigraph, Minigraph-Cactus and PGGB become more apparent. While the number of nodes decreases for all three graphs, again with a similar trajectory for Minigraph-Cactus and PGGB, the SCJ-CARP measure shows that PGGB's complexity increases, while the complexity of Minigraph-Cactus and Minigraph is now much more similar.

From these results one can conclude that while the graphs of PGGB and Minigraph-Cactus have similar numbers of nodes, the way variants are represented in Minigraph and Minigraph-Cactus is much more similar to each other than to PGGB. PGGB in turn tends to produce highly connected small nodes that lead to an increase in connectedness and consequently higher complexity once they are filtered.

These observations are of course consistent with the knowledge that Minigraph-Cactus graphs are constructed by adding smaller variants to Minigraph graphs while PGGB follows a different, alignment-based paradigm. However, note that this knowledge of the graph construction methods is not necessary to interpret the results computed by SCJ-CARP.

On 16 threads, calculations for Minigraph were completed within seconds while calculations for Minigraph-Cactus finished in a matter of minutes. The longest calculation for PGGB completed after about four hours. Overall, the entire analysis was completed within 20 hours. We give all runtimes of SCJ-CARP as Appendix Fig. 13.

We conclude that SCJ-CARP allows an analysis of pangenome graphs that highlights their differing structural qualities without prior knowledge of how these graphs were constructed. Additionally, the time frame for such an analysis is reasonable, taking about one day for human sized pangenomes.

5.4 Scoring Region Complexities in a *Bacillus Subtilis* de Bruijn graph

Typically the fastest way to build a pangenome graph is to construct a de Bruijn graph of all genome sequences in the pangenome of interest. However, de Bruijn graphs tend to be large and hard to interpret, especially in regions with a high complexity ("hair balls"), which typically arise due to nodes that appear very frequently in the pangenome. The SCJ-CARP measure can be used to score the complexity of the subgraph surrounding a node.

To demonstrate this possibility, we downloaded all 1508 available *Bacillus subtilis* genomes in genbank using NCBI datasets [36] and constructed a de Bruijn graph using Bifrost [26] for $k = 31$. The resulting graph had a total number of $3,665,906$ nodes. We then used the `carp-scan` program to extract regions of 300 BP around a node and to score this environment using the SCJ-CARP measure. On 8 threads on a cluster machine, `carp-scan` finished after 8 minutes and 4 seconds. We show the histogram of SCJ-CARP measures of the $3,629,054$ least complex nodes and examples of node environments in Fig. 9.

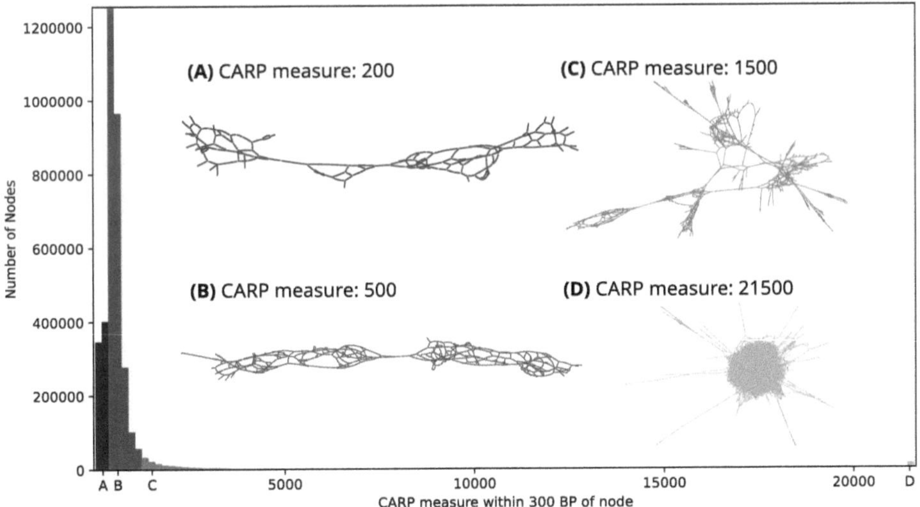

Fig. 9. Histogram of SCJ-CARP measures in the 300 BP environment of nodes in a compacted de Bruijn graph built from 1508 *Bacillus subtilis* genomes. For a better visualization, the histogram is artificially cut after 21500; the highest SCJ-CARP measure of a node environment was 175702. (A) to (D) visualize example environments from distinct regions of the distribution. These visualizations were created with Bandage [48].

We see that the majority of nodes are surrounded by a subgraph with a CARP measure below 1500, which appear to be interpretable (see Examples (A), (B) and (C) in Fig. 9), although ease of interpretation decreases as the complexity identified by the SCJ-CARP measure increases. True "hair ball" subgraphs, such as shown as Example (D) in Fig. 9, seem to be the exception in this graph, but there is a cluster of these nodes around SCJ-CARP measure 21500.

We also performed additional analyses varying the environment size. These histograms can be found in Appendix Fig. 14. Moreover, the `carp-scan` program can color graph nodes based on their surrounding SCJ-CARP measures, which enables to easily visualize where in the graph complexity arises. We give an example in Appendix Fig. 15. Additionally, one can map these complexities back to a reference genome and visualize it with a genome browser. We give an example in Fig. 10 using the UCSC Genome Browser [39].

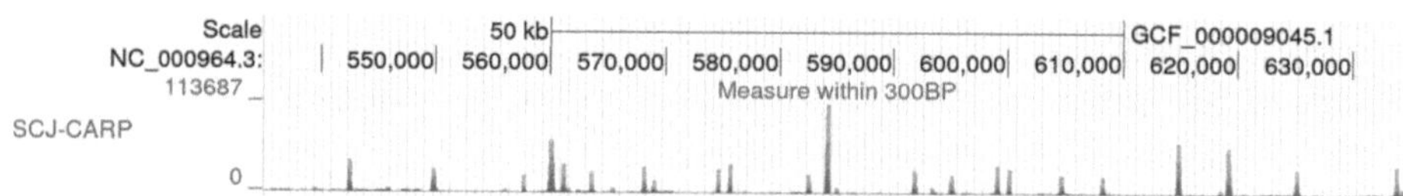

Fig. 10. 300 BP environment complexities in the de Bruijn graph ($k = 31$) mapped to the *Bacillus subtilis* reference `GCF_000009045.1`. Shown is the region `NC_000964.3:534,995-635,004` in the UCSC genome browser [39].

One can thus use SCJ-CARP to identify regions in a pangenome graph that are already interpretable and regions that still need refinement, which could be a useful tool or evaluation metric for pangenome graph construction methods. Moreover, this allows in principle to compare pangenomes and examine at which level structural complexity arises and how it is distributed throughout the pangenome graph.

6   Discussion

We developed CARP, a new rearrangement problem designed to quantify the rearrangement complexity of pangenomes. This problem relates intuitively to both pangenome graphs and classical genome rearrangement, thus forming a first step towards a joint modeling of these two, highly connected areas. We presented a solution to CARP under the simple SCJ model. This solution already enables fast analyses of pangenomes and their graphs, which yield valuable insights about construction methods and structural complexity. We expect this potential to further unfold once solutions under other models become available. For example, one could characterize a pangenome by a vector of CARP measures of different models (i.e. inversions, block interchanges, etc.), indicating which model best explains its complexity. This would also allow a high level comparison between pangenomes that do not share any sequence material. A prerequisite to perform such a comparison is some normalization of the CARP measure. Like other rearrangement-based methods, CARP yields an absolute number of rearrangements, which of course depends on the size and number of the input genomes. Future research may explore normalizing the CARP measure as has been done with other pangenome metrics (see for example [37]).

We have also seen that the CARP measure under the SCJ-model corresponds to a simple to evaluate graph metric in a general pangenome graph, which mirrors the fact that this is the case for many models in classical rearrangement analyses in the specialized breakpoint graph. It thus stands to reason that the CARP measure for other rearrangement models may be related to other known graph theoretic quantities. This would allow the study of abstract graph properties under the more biologically meaningful lens of rearrangement models and, vice versa, the mathematical understanding of a rearrangement model would be enhanced by connecting it to graph theory in general.

Furthermore, it is worth investigating the implications that Path Altering Operations used in a given scenario make about the phylogenetic relationship between genomes of the pangenome. This may well lead to an easily computable rearrangement-based tree reconstruction.

Acknowledgements. This work was funded the by the Federal Ministry of Research, Technology and Space in the frame of de.NBI & ELIXIR-DE (W-de.NBI-004).

LB has received funding from the European Union's Horizon 2020 research and innovation programme under the Marie Skłodowska-Curie grant agreement PANGAIA No. 872539.

This work was supported by the de.NBI Cloud within the German Network for Bioinformatics Infrastructure (de.NBI) and ELIXIR-DE (Forschungszentrum Jülich and W-de.NBI-001, W-de.NBI-004, W-de.NBI-008, W-de.NBI-010, W-de.NBI-013, W-de.NBI-014, W-de.NBI-016, W-de.NBI-022).

References

1. Alekseyev, M.A., Pevzner, P.A.: Colored de Bruijn graphs and the genome halving problem. IEEE/ACM Trans. Comput. Biol. Bioinf. **4**(1), 98–107 (2007). https://doi.org/10.1109/TCBB.2007.1002
2. Alekseyev, M.A., Pevzner, P.A.: Breakpoint graphs and ancestral genome reconstructions. Genome Res. **19**(5), 943–957 (2009). https://doi.org/10.1101/gr.082784.108
3. Baker, T.M., Waise, S., Tarabichi, M., Van Loo, P.: Aneuploidy and complex genomic rearrangements in cancer evolution. Nat. Cancer **5**(2), 228–239 (2024). https://doi.org/10.1038/s43018-023-00711-y
4. Belling, J.: Crossing over and gene rearrangement in flowering plants. Genetics **18**(4), 388 (1933). https://doi.org/10.1093/genetics/18.4.388
5. Bentley, S.: Sequencing the species pan-genome. Nat. Rev. Microbiol. **7**(4), 258–259 (2009). https://doi.org/10.1038/nrmicro2123

6. Braga, M.D., Willing, E., Stoye, J.: Double cut and join with insertions and deletions. J. Comput. Biol. **18**(9), 1167–1184 (2011). https://doi.org/10.1089/cmb.2011.0118

7. Chen, J.M., Cooper, D.N., Férec, C., Kehrer-Sawatzki, H., Patrinos, G.P.: Genomic rearrangements in inherited disease and cancer. Semin. Cancer Biol. **20**(4), 222–233 (2010). https://doi.org/10.1016/j.semcancer.2010.05.007

8. Christie, D.A.: Sorting permutations by block-interchanges. Inf. Process. Lett. **60**(4), 165–169 (1996). https://doi.org/10.1016/S0020-0190(96)00155-X

9. Compeau, P.E.: DCJ-indel sorting revisited. Algorithms Molecular Biol. **8**(1), 6–6 (2013). https://doi.org/10.1186/1748-7188-8-6

10. Cracco, A., Tomescu, A.I.: Extremely fast construction and querying of compacted and colored de Bruijn graphs with GGCAT. Genome Res. **33**(7), 1198–1207 (2023). https://doi.org/10.1101/gr.277615.122

11. Darling, A.C., Mau, B., Blattner, F.R., Perna, N.T.: Mauve: multiple alignment of conserved genomic sequence with rearrangements. Genome Res. **14**(7), 1394–1403 (2004). https://doi.org/10.1101/gr.2289704

12. Darling, A.E., Mau, B., Perna, N.T.: progressiveMauve: multiple genome alignment with gene gain, loss and rearrangement. PLoS ONE **5**(6), 1–17 (2010). https://doi.org/10.1371/journal.pone.0011147

13. Davín, A.A., Tricou, T., Tannier, E., de Vienne, D.M., Szöllősi, G.J.: Zombi: a phylogenetic simulator of trees, genomes and sequences that accounts for dead linages. Bioinformatics **36**(4), 1286–1288 (2019). https://doi.org/10.1093/bioinformatics/btz710

14. Delcher, A.L., Salzberg, S.L., Phillippy, A.M.: Using MUMmer to identify similar regions in large sequence sets. Curr. Protocols Bioinf. **00**, 10.3.1–10.3.18 (2003). https://doi.org/10.1002/0471250953.bi1003s00

15. Dobzhansky, T., Sturtevant, A.H.: Inversions in the chromosomes of Drosophila pseudoobscura. Genetics **23**(1), 28 (1938). https://doi.org/10.1093/genetics/23.1.28

16. Dunham, M.J., et al.: Characteristic genome rearrangements in experimental evolution of Saccharomyces cerevisiae. Proc. National Academy Sci. U.S.A. **99**(25), 16144–16149 (2002). https://doi.org/10.1073/pnas.242624799

17. Eizenga, J.M., et al.: Pangenome graphs. Annu. Rev. Genomics Hum. Genet. **21**, 139–162 (2020). https://doi.org/10.1146/annurev-genom-120219-080406

18. Feijao, P., Meidanis, J.: SCJ: a breakpoint-like distance that simplifies several rearrangement problems. IEEE/ACM Trans. Comput. Biol. Bioinf. **8**(5), 1318–1329 (2011). https://doi.org/10.1109/TCBB.2011.34

19. Feulner, P.G.D., De-Kayne, R.: Genome evolution, structural rearrangements and speciation. J. Evol. Biol. **30**(8), 1488–1490 (2017). https://doi.org/10.1111/jeb.13101

20. Garrison, E., et al.: Variation graph toolkit improves read mapping by representing genetic variation in the reference. Nat. Biotechnol. **36**(9), 875–879 (2018). https://doi.org/10.1038/nbt.4227

21. Hannenhalli, S., Pevzner, P.A.: Transforming men into mice (polynomial algorithm for genomic distance problem). In: Proceedings of IEEE 36th Annual Foundations of Computer Science, pp. 581–592. IEEE Press, Milwaukee, WI, USA (1995). https://doi.org/10.1109/SFCS.1995.492588

22. Hannenhalli, S., Pevzner, P.A.: Transforming cabbage into turnip: polynomial algorithm for sorting signed permutations by reversals. J. ACM **46**(1), 1–27 (1999). https://doi.org/10.1145/300515.300516

23. Hickey, G., et al.: The human pangenome reference consortium: pangenome graph construction from genome alignments with minigraph-cactus. Nat. Biotechnol. **42**(4), 663–673 (2024). https://doi.org/10.1038/s41587-023-01793-w
24. Hierholzer, C., Wiener, C.: Über die Möglichkeit, einen linienzug ohne wiederholung und ohne unterbrechung zu umfahren. Math. Ann. **6**(1), 30–32 (1873). https://doi.org/10.1007/BF01442866
25. Hiller, N.L., et al.: Comparative genomic analyses of seventeen Streptococcus pneumoniae strains: insights into the pneumococcal supragenome. J. Bacteriol. **189**(22), 8186–8195 (2007). https://doi.org/10.1128/jb.00690-07
26. Holley, G., Melsted, P.: Bifrost: highly parallel construction and indexing of colored and compacted de Bruijn graphs. Genome Biol. **21**(1), 249 (2020). https://doi.org/10.1186/s13059-020-02135-8
27. Kolmogorov, M., et al.: Chromosome assembly of large and complex genomes using multiple references. Genome Res. **28**(11), 1720–1732 (2018). https://doi.org/10.1101/gr.236273.118
28. Köster, J., Rahmann, S.: Snakemake—a scalable bioinformatics workflow engine. Bioinformatics **28**(19), 2520–2522 (2012). https://doi.org/10.1093/bioinformatics/bts480
29. Krupina, K., Goginashvili, A., Cleveland, D.W.: Scrambling the genome in cancer: causes and consequences of complex chromosome rearrangements. Nat. Rev. Genet. **25**(3), 196–210 (2024). https://doi.org/10.1038/s41576-023-00663-0
30. Li, T., Jiang, H., Zhu, B., Wang, L., Zhu, D.: Flanked block-interchange distance on strings. IEEE/ACM Trans. Comput. Biol. Bioinf. **21**(2), 301–311 (2024). https://doi.org/10.1109/TCBB.2024.3351440
31. Liao, W.W., et al.: A draft human pangenome reference. Nature **617**(7960), 312–324 (2023). https://doi.org/10.1038/s41586-023-05896-x
32. Lin, Y., Nurk, S., Pevzner, P.A.: What is the difference between the breakpoint graph and the de Bruijn graph? BMC Genomics **15**(6), S6 (2014). https://doi.org/10.1186/1471-2164-15-S6-S6
33. Minkin, I., Medvedev, P.: Scalable multiple whole-genome alignment and locally collinear block construction with SibeliaZ. Nat. Commun. **11**(1), 6327 (2020). https://doi.org/10.1038/s41467-020-19777-8
34. Muggli, M.D., Alipanahi, B., Boucher, C.: Building large updatable colored de Bruijn graphs via merging. Bioinformatics **35**(14), i51–i60 (2019). https://doi.org/10.1093/bioinformatics/btz350
35. Noll, N., Molari, M., Shaw, L.P., Neher, R.A.: Pangraph: scalable bacterial pangenome graph construction. Microbial Genomics **9**(6), 001034 (2023). https://doi.org/10.1099/mgen.0.001034
36. O'Leary, N.A., et al.: Exploring and retrieving sequence and metadata for species across the tree of life with NCBI datasets. Sci. Data **11**(1), 732 (2024). https://doi.org/10.1038/s41597-024-03571-y
37. Parmigiani, L., Garrison, E., Stoye, J., Marschall, T., Doerr, D.: Panacus: fast and exact pangenome growth and core size estimation. Bioinformatics **40**(12), btae720 (2024). https://doi.org/10.1093/bioinformatics/btae720
38. Paten, B., Eizenga, J.M., Rosen, Y.M., Novak, A.M., Garrison, E., Hickey, G.: Superbubbles, ultrabubbles, and cacti. J. Comput. Biol. **25**(7), 649–663 (2018). https://doi.org/10.1089/cmb.2017.0251
39. Perez, G., et al.: The UCSC genome browser database: 2025 update. Nucleic Acids Res. **53**(D1), D1243–D1249 (2025). https://doi.org/10.1093/nar/gkae974

40. Porubsky, D., et al.: Recurrent inversion polymorphisms in humans associate with genetic instability and genomic disorders. Cell **185**(11), 1986-2005.e26 (2022). https://doi.org/10.1016/j.cell.2022.04.017
41. Rossi, M., Oliva, M., Langmead, B., Gagie, T., Boucher, C.: Moni: a pangenomic index for finding maximal exact matches. J. Comput. Biol. **29**(2), 169–187 (2022). https://doi.org/10.1089/cmb.2021.0290
42. Sankoff, D.: Edit distance for genome comparison based on non-local operations. In: Proceedings of the Third Annual Symposium on Combinatorial Pattern Matching (CPM), pp. 121–135. Springer Berlin Heidelberg (1992). https://doi.org/10.1007/3-540-56024-6_10
43. Schuy, J., Grochowski, C.M., Carvalho, C.M., Lindstrand, A.: Complex genomic rearrangements: an underestimated cause of rare diseases. Trends Genet. **38**(11), 113–1146 (2022). https://doi.org/10.1016/j.tig.2022.06.003
44. Sturtevant, A.H.: A case of rearrangement of genes in Drosophila. Proc. National Academy Sci. U.S.A. **7**(8), 235–237 (1921). https://doi.org/10.1073/pnas.7.8.235
45. Tannier, E., Zheng, C., Sankoff, D.: Multichromosomal median and halving problems under different genomic distances. BMC Bioinf. **10**(1), 120 (2009). https://doi.org/10.1186/1471-2105-10-120
46. Tettelin, H., et al.: Genome analysis of multiple pathogenic isolates of Streptococcus agalactiae: Implications for the microbial "pan-genome". Proceedings of the National Academy of Sciences of the U.S.A. **102**(39), 13950–13955 (2005). https://doi.org/10.1073/pnas.0506758102
47. The computational pan-genomics consortium: computational pan-genomics: status, promises and challenges. Brief. Bioinform. **19**(1), 118–135 (2018). https://doi.org/10.1093/bib/bbw089
48. Wick, R.R., Schultz, M.B., Zobel, J., Holt, K.E.: Bandage: interactive visualization of de novo genome assemblies. Bioinformatics **31**(20), 3350–3352 (2015). https://doi.org/10.1093/bioinformatics/btv383
49. Wilson, A.C., Sarich, V.M., Maxson, L.R.: The importance of gene rearrangement in evolution: evidence from studies on rates of chromosomal, protein, and anatomical evolution. Proc. National Academy Sci. U.S.A. **71**(8), 3028–3030 (1974). https://doi.org/10.1073/pnas.71.8.3028
50. Xu, H., Tong, X., Jiang, H., Wang, L., Zhu, B., Zhu, D.: On sorting by flanked transpositions. In: Guo, X., Mangul, S., Patterson, M., Zelikovsky, A. (eds.) Bioinformatics Research and Applications, pp. 292–311. Springer Nature Singapore, Singapore (2023). https://doi.org/10.1007/978-981-99-7074-2_23
51. Yancopoulos, S., Attie, O., Friedberg, R.: Efficient sorting of genomic permutations by translocation, inversion and block interchange. Bioinformatics **21**(16), 3340–3346 (2005). https://doi.org/10.1093/bioinformatics/bti535
52. Zakeri, M., Brown, N.K., Ahmed, O.Y., Gagie, T., Langmead, B.: Movi: a fast and cache-efficient full-text pangenome index. iScience **27**(12) (2024). https://doi.org/10.1016/j.isci.2024.111464

Breakpoint Medians and Anti-medians for Signed Genomes

Poly H. da Silva[1], Arash Jamshidpey[2(✉)], and David Sankoff[3]

[1] Independent Researcher, Berkeley, CA, USA
[2] Department of Statistics, University of California at Berkeley, Berkeley, CA, USA
arash.jamshidpey@berkeley.edu
[3] Department of Mathematics and Statistics, University of Ottawa, Ottawa, ON, Canada
sankoff@uottawa.ca

Abstract. The breakpoint distance is a fundamental measure for comparing gene orders and plays a central role in genome rearrangement phylogenetics. For more than two genomes, breakpoint medians are widely used to infer ancestral gene orders, yet their typical behavior is difficult to characterize due to the complex, non-geodesic geometry of genome space. A conjecture of Haghighi and Sankoff states that for random genomes, breakpoint medians tend to lie near one of the input genomes rather than constituting genuinely intermediate solutions. In this paper, we study breakpoint medians of independently and uniformly sampled signed multichromosomal genomes. We prove the Haghighi–Sankoff conjecture in this setting for the first time. We show that with high probability, any breakpoint median must remain close to a single input genome, rather than drawing adjacencies from multiple genomes. Our analysis relies on probabilistic bounds for usable adjacencies in random signed genomes and structural constraints of the signed multichromosomal model. We further provide a game-theoretic interpretation that explains why compromise strategies lead to increased total distance and the emergence of anti-medians.

Keywords: Genomes · Breakpoint distance · Medians · Anti-medians

1 Introduction

Breakpoint distance, introduced in [10] as a combinatorial measure of gene-order dissimilarity, has proved to be a powerful tool for studying similarities between genomes. For more than two genomes, the notion of a median genome plays a central role in comparative genomics, particularly in the reconstruction of gene-order phylogenies. In the breakpoint framework, the median problem was introduced by Sankoff *et al.* [11], who showed that computing a breakpoint median reduces to the Traveling Salesman Problem (TSP) making it NP-hard for unichromosomal unsigned genomes. Since then, the breakpoint median problem has been extensively studied [1,2,4,5,15,16,18].

M. Lafond (Ed.): RECOMB-CG 2026, LNBI 16569, pp. 51–74, 2026.
https://doi.org/10.1007/978-3-032-26891-4_3

The difficulty of finding breakpoint medians is rooted in the complex geometry of genome space endowed with the breakpoint distance. More precisely, the breakpoint distance is not geodesic, meaning that midpoints between genomes may fail to exist. As a result, shortest paths need not behave continuously, and standard geometric intuition breaks down (see [7] for an accessible discussion). This lack of geodesic structure significantly complicates the median problem: algorithms that attempt to approach a median iteratively, starting from an input genome, often fail. Although many heuristic algorithms for finding breakpoint medians have been proposed, they are insufficient for exploring the global geometry of genome space or for understanding the typical positions and density of breakpoint medians of sampled genomes.

Using heuristic algorithms, Haghighi and Sankoff [6] observed an unexpected phenomenon: for k randomly sampled unichromosomal unsigned genomes with n genes, the breakpoint median value is approximately $(k-1)n$. Even more surprisingly, their simulation studies suggested that medians tend to lie near the "corners" of genome space, meaning that most heuristic medians were positioned very close to one of the input genomes. Nevertheless, a small proportion of medians appeared far from the corners, with some even roughly equidistant from all inputs. Identifying and understanding such "midpoint" medians is of particular interest, as these genomes potentially carry balanced genomic information from all inputs and may be relevant for ancestral genome reconstruction.

In subsequent work, Jamshidpey *et al.* [7] partially proved the Haghighi–Sankoff conjecture by establishing the median value $(k-1)n$. That work also introduced the notions of partial geodesics (or geodesic patches) as a means of reaching non-trivial medians far from the corners, termed "accessible genomes". A key result was that almost all adjacencies of a median must already appear among the adjacencies of at least one input genome. In other words, medians are not allowed to introduce many new adjacencies outside the union of adjacencies of the input genomes. This observation is crucial: it opens a principled pathway to constructing medians far from corners by selecting adjacencies from all input genomes, rather than predominantly from a single one.

Motivated by this observation, Larlee *et al.* [8,9] attempted to construct medians by compromising the number of adjacencies taken from k randomly sampled signed genomes. Specifically, for $x \in (0, 1)$, they sought to construct a genome G that takes a proportion x of adjacencies from each of the k input genomes, yielding a total of kx adjacencies, and to maximize x subject to the constraint that kx be as close as possible to 1. Unexpectedly, this approach failed and produced the opposite effect. Larlee *et al.* [8,9] observed that the more one attempts to compromise, the smaller the maximal achievable value of x becomes, and hence the greater the total distance of G from the input genomes. They introduced the notion of anti-medians to describe genomes that are equidistant from all inputs yet as far as possible from being medians.

Random sampling of adjacencies from input genomes [8] offers little control over the structural compatibility of adjacencies in signed multichromosomal genomes with circular chromosomes. The original Haghighi–Sankoff conjecture, however, concerned unichromosomal unsigned linear genomes, and additional constraints arising from unichromosomal structure must be handled carefully. Motivated by the work of Larlee *et al.* [8], da Silva *et al.* [12] introduced necessary conditions and constraints governing how adjacencies can be taken from each input genome. We believe that the framework developed in [12] will play a key role in answering whether the Haghighi–Sankoff conjecture holds or fails for unichromosomal unsigned genomes. By contrast, the case of multichromosomal signed genomes is more tractable, and in this setting, we are able to fully prove the HaghighiâĂŞSankoff conjecture in this work.

As discussed above, understanding how far medians can lie from the corners is deeply tied to the geometry of genome space. Moreover, the technical tools required for signed versus unsigned, and for unichromosomal versus multichromosomal genomes, differ substantially. In particular, the constraints involved in constructing signed genomes, which are the focus of this paper, are significantly stronger than those arising in the unsigned setting. This added structure makes it possible to prove versions of the Haghighi–Sankoff conjecture for signed genomes that remain out of reach for unsigned ones at this moment. Nonetheless, the large body of heuristic algorithms for breakpoint medians (many based on TSP heuristics) remains insufficient for understanding the geometry of the breakpoint distance and the relative position of medians relative to input genomes.

A recent breakthrough in this direction was achieved in our earlier work [13], where for the first time all exact (non-heuristic) medians of randomly sampled unichromosomal genomes were enumerated (the method applies to multichromosomal genomes as well). This exhaustive enumeration enabled a much more precise understanding of the geometry of genome space and of the positions of medians relative to the sampled input genomes.

In this paper, we study the relative position of breakpoint medians for k independently and uniformly sampled signed multichromosomal genomes. For simplicity, we assume that all chromosomes are circular; the methods extend naturally to linear chromosomes via a standard telomere-capping construction. For the first time, we prove the Haghighi–Sankoff conjecture for signed multichromosomal genomes. More precisely, we show that with high probability, for any median genome $\mathcal{G}$ of k independently and randomly sampled genomes $\mathcal{G}_1, \ldots, \mathcal{G}_k$, if x_i denotes the proportion of adjacencies shared between $\mathcal{G}$ and $\mathcal{G}_i$, with $x_1 + \cdots + x_k = 1$, then there exists an index j such that $x_j = 1$ and $x_i = 0$ for all $i \neq j$. Equivalently, any median must lie very close to one of the input genomes. Our results continue the line of work initiated by [8,9], but differ fundamentally from attempts to prove analogous statements for unsigned unichromosomal genomes, such as the approach in [12]. The stricter constraints in signed multichromosomal genome models play a decisive role in enabling our proofs.

More precisely, motivated by similar ideas for signed genomes [8], da Silva *et al.* [12] introduced a classification of gene adjacencies of a random genome $\mathcal{G}'$ relative to a given reference set of adjacencies $\mathcal{I}$. This classification allows one to count the number of usable adjacencies of $\mathcal{G}'$ with respect to $\mathcal{I}$.

More generally, in the context of constructing a median genome for k independently and randomly sampled genomes $\mathcal{G}_1, \ldots, \mathcal{G}_k$, this implies that any median $\mathcal{G}$ must draw almost all of its adjacencies from at least one of the input genomes. Suppose that a set of adjacencies $\mathcal{I}_r$ has already been taken from $\mathcal{G}_1, \cdots, \mathcal{G}_r$, for $1 \leq r < k$. One then seeks the largest possible set of usable adjacencies $\mathcal{J}_{r+1}$ from $\mathcal{G}_{r+1}$ such that $\mathcal{I}_{r+1} = \mathcal{I}_r \cup \mathcal{J}_{r+1}$ can still be extended to a median genome. The goal is to optimize this sampling process so that $\mathcal{I}_k$ constitutes a complete genome $\mathcal{G}$. Finding such a $\mathcal{G}$ guarantees that it is a median of $\mathcal{G}_1, \cdots, \mathcal{G}_k$.

This is not always possible, and depending on the underlying restrictions, the size difference between a full genome and $\mathcal{I}_k$ may be substantial. These restrictions depend on the genome model, including whether genomes are signed or unsigned, and multichromosomal or unichromosomal. Comparing signed and unsigned cases, once a set of adjacencies $\mathcal{I}$ has been sampled, the average number of usable adjacencies of a random genome is significantly smaller in the signed case. This observation is the key mechanism that allows us to prove the Haghighi–Sankoff conjecture for signed multichromosomal genomes in this paper.

Finally, we use game-theoretic ideas to explain why compromising, as in [8], leads to increased distance rather than improved median quality. We model each genome as a player in a cooperative general-sum game with transferable utility. Each player $\mathcal{G}_i$ gains a payoff x_i, the proportion of its adjacencies used in the constructed genome $\mathcal{G}$, and the players jointly aim to maximize $x_1 + \cdots + x_k \leq 1$. We determine the Pareto-optimal boundary of this game and show that the maximum total payoff is achieved at the extreme points $e_i = (0, \ldots, 1, \ldots, 0)$. As one moves along the Pareto boundary toward more balanced compromises, the total payoff decreases, reaching its minimum at a symmetric point $x = (a, \ldots, a)$ with $a < 1$. This fully explains why increased compromise in the number of gene adjacencies taken from input genomes leads to genomes that are farther from being medians and clarifies the emergence of anti-medians.

The remainder of the paper is organized as follows. In Sect. 2, we review genome representations, breakpoint distance for signed genomes, and the behavior of breakpoint distance for two random genomes. In this section, we also establish probabilistic results for medians of k independently sampled genomes. In Sect. 3, we prove the Haghighi–Sankoff conjecture in the signed multichromosomal setting and show that, with high probability, all approximate (and hence exact) medians lie at the corners. Finally, in Sect. 4, we develop the game-theoretic framework explaining why compromise in the number of gene adjacencies taken from input genomes increases the total distance and leads to the appearance of anti-medians.

2 Probabilistic Structure of Common Adjacencies and Breakpoint Medians

Although the methods in this paper can be easily extended to multichromosomal genomes with linear and circular chromosomes, for the sake of simplicity we assume that all chromosomes are circular. Therefore, having no telomere in our model, a signed multichromosomal genome with circular chromosomes can be represented as a perfect matching on the set of vertices $\pm[n] :=\{\pm 1, \pm 2, \ldots, \pm n\}$.

More precisely, for a set of genes labeled by $1, 2, \ldots, n$, we denote by $-i$ and $+i$ the two extremities of gene i. In fact the signs may be interpreted as encoding gene polarity (strandedness). A genome with gene content $1, 2, \ldots, n$ is then represented by a perfect matching on $\pm[n]$, where each edge of the matching determines a gene adjacency, linking two neighbouring genes. Note that allowing chromosomes in arbitrary number and of arbitrary sizes places no restrictions on the associated perfect matching. For instance, an edge (gene adjacency) $\{-i, +i\}$ means that the genome contains a circular chromosome of size 1, with gene content i. We denote by $\mathbb{G}_n$ the set of all perfect matchings on $\pm[n]$.

In general, the gene content of a genome G may be any finite subset $V(G) \subseteq [n]$. In this case, the gene extremities of G are elements of $\pm V(G) := V(G) \cup (-V(G))$, and the genome G is a perfect matching on $\pm V(G)$. We represent such a genome as $G = (\pm V(G), E(G))$, where $\pm V(G)$ is the set of vertices and $E(G)$ is its set of edges. Note that for $G \in \mathbb{G}_n$, $\pm V(G) = \pm[n]$.

Introduced in [10], breakpoint distance (bp-distance) is often used to measure the similarity between two genomes. More specifically, for a pair of genomes G, G', the breakpoint distance $d(G, G')$ can be defined by

$$d(G, G') = \frac{1}{2} |E(G) \triangle E(G')|,$$

where $\triangle$ denotes the symmetric difference between sets and $|\cdot|$ denotes the cardinality of the set. In particular, for $G, G' \in \mathbb{G}_n$, this formula reduces to

$$d(G, G') = n - |E(G) \cap E(G')|.$$

Letting $\mathbb{G}$ be the space of all perfect matchings on $\pm S := S \cup (-S)$, for finite subsets $S \subset \mathbb{N}$, $(\mathbb{G}, d)$ is a metric space. In addition, for any $n \in \mathbb{N}$, $(\mathbb{G}_n, d)$ is also a metric space. Let $\mathbf{S}_{\pm n}$ be the group of permutations on $\pm[n]$ with the function composition as the group action. We can define the action of $\mathbf{S}_{\pm n}$ on $\mathbb{G}_n$ by $\sigma \cdot G \in \mathbb{G}_n$ for $\sigma \in \mathbf{S}_{\pm n}$ and $G \in \mathbb{G}_n$, where

$$\pm V(\sigma \cdot G) = \pm[n], \qquad E(\sigma \cdot G) = \{\{\sigma(i), \sigma(j)\} : \{i, j\} \in E(G)\}.$$

That is, the action relabels the gene extremities via the permutation $\sigma \in \mathbf{S}_{\pm n}$.

Indeed, the action of $\mathbf{S}_{\pm n}$ on $\mathbb{G}_n$ defined above is distance preserving, i.e., for any $G, G' \in \mathbb{G}_n$ and $\sigma \in \mathbf{S}_{\pm n}$, $d(\sigma \cdot G, \sigma \cdot G') = d(G, G')$. We say that the bp-distance is invariant under the action of $\mathbf{S}_{\pm n}$ on $\mathbb{G}_n$.

We begin by analyzing the distribution of common adjacencies between independent random genomes. This provides a probabilistic baseline for genome comparison and plays a key role in understanding the structure and typical behavior of breakpoint medians in later sections.

Let $\mathcal{G}$ and $\mathcal{G}'$ be two independent genomes chosen uniformly at random from $\mathbb{G}_n$. Let $o_n \in \mathbb{G}_n$ be the perfect matching on $\pm[n]$ defined by $E(o_n) = \{\{-i, +i\} :$

$i \in [n]\}$. Let $(\mathcal{G}')^{-1} \subset \mathbf{S}_{\pm n}$ denote the set of all permutations σ on $\pm[n]$ such that $\sigma \cdot \mathcal{G}' = o_n$. Let $\pi_{\mathcal{G}'}$ be selected uniformly at random from $(\mathcal{G}')^{-1}$.

Since $\mathcal{G}'$ is a uniformly random perfect matching in $\mathbb{G}_n$, the mapping π : $\mathbb{G}_n \to \mathbf{S}_{\pm n}$ defined by $\mathcal{G}' \mapsto \pi_{\mathcal{G}'}$ induces the uniform probability on $\mathbf{S}_{\pm n}$. In other words, $\pi_{\mathcal{G}'}$ is in fact a uniformly random permutation chosen from $\mathbf{S}_{\pm n}$. Furthermore, $\pi_{\mathcal{G}'}$ is independent of $\mathcal{G}$, and $\pi_{\mathcal{G}'} \cdot \mathcal{G}$ is a uniformly random perfect matching in $\mathbb{G}_n$. As a result, since d is invariant under the action of $\mathbf{S}_{\pm n}$ on $\mathbb{G}_n$, we deduce

$$d(\mathcal{G}', \mathcal{G}) \stackrel{d}{=} d(o_n, \pi_{\mathcal{G}'} \cdot \mathcal{G}) \stackrel{d}{=} d(o_n, \mathcal{G}),$$

where "$\stackrel{d}{=}$" indicates equivalence in distribution. This reduction allows us to replace a two-genome comparison with a one-genome problem relative to a fixed reference, greatly simplifying the analysis. More precisely, to study the bp-distance of two independent random genomes, it suffices to study the bp-distance of a uniformly random genome $\mathcal{G} \in \mathbb{G}_n$ and the reference genome $o_n \in \mathbb{G}_n$. Note that the choice of o_n is arbitrary and any other fixed perfect matching in $\mathbb{G}_n$ can serve as the reference genome.

To simplify notation, for the set of genomes $V = \{G_1, \ldots, G_m\}$, let

$$\mathcal{A}_V = \mathcal{A}_{G_1, G_2, \ldots, G_m} = E(G_1) \cap E(G_2) \cap \cdots \cap E(G_m).$$

Intuitively, in large random genomes the probability that a specific adjacency appears in both genomes is small, and distinct adjacencies behave almost independently. It is well-known that for two independent random genomes $\mathcal{G}_n, \mathcal{G}'_n \in \mathbb{G}_n$, as $n \to \infty$, the number of common adjacencies $|\mathcal{A}_{\mathcal{G}_n, \mathcal{G}'_n}|$ converges in distribution to Poisson$(1/2)$, i.e. to a Poisson random variable with parameter $1/2$ (see for example [17]). We give a full treatment of this statement in Appendix A, where we show that a stronger mode of convergence—namely, convergence in total variation—holds (cf. Appendix A).

In the sequel, we use the Poisson asymptotics obtained in Appendix A to analyze the typical behavior of breakpoint medians for multiple random genomes. In particular, we study the median of $k \in \mathbb{N}$ perfect matchings $\mathcal{G}_1, \mathcal{G}_2, \ldots, \mathcal{G}_k$, picked independently and uniformly at random from $\mathbb{G}_n$.

The geometric medians were used in comparative genomics for the first time by Sankoff $et\ al.$ [11]. They have extensive applications in gene-order phylogenetics. A breakpoint (bp) median of a set of genomes $A = \{G_1, G_2, \ldots, G_k\} \subseteq \mathbb{G}_n$ is a genome $G \in \mathbb{G}_n$ that minimizes the total breakpoint distance function

$$\tilde{G} \mapsto \sum_{i=1}^{k} d(\tilde{G}, G_i).$$

The minimum value of the function is called the bp-median value of A. We denote by $M_n(A) = M(A) = M(G_1, \ldots, G_k)$ the set of all bp-medians of A, and by $\mu_n(A) = \mu(A) = \mu(G_1, \ldots, G_k)$ the median value of A. The median of a single genome G is itself. Also, G is a median of two genomes G_1, G_2 if and only if $\mathcal{A}_{G_1, G_2} \subseteq \mathcal{A}_G \subseteq \mathcal{A}_{G_1} \cup \mathcal{A}_{G_2}$ (cf. [7] for unsigned genomes). In addition, for k genomes $A = \{G_1, G_2, \ldots, G_k\} \subseteq \mathbb{G}_n$,

$$\mu(A) \leq \sum_{i=1}^{k} d(G, G_i) \leq (k-1)n.$$

We continue by recalling that for k perfect matchings $\mathcal{G}_1, \mathcal{G}_2, \ldots, \mathcal{G}_k$, chosen independently and uniformly at random from $\mathbb{G}_n$, the pairwise distances satisfy $d(\mathcal{G}_i, \mathcal{G}_j) = n - O_{\mathbb{P}}(1)$, with Poisson$(1/2)$-type fluctuations in the number of common adjacencies. Consequently, studying the medians and median value of $\mathcal{G}_1, \ldots, \mathcal{G}_k$ is similar to studying those of deterministic genomes $G_1, \ldots, G_k$ with pairwise maximum distance $d(G_i, G_j) = n$ for $i \neq j$. The proof of the following proposition is given in Appendix B.

Proposition 1. *For* $k, n \geq 1$, *let* $A = \{G_1, \ldots, G_k\} \subset \mathbb{G}_n$ *such that* $d(G_i, G_j) = n$ *for all* $i \neq j$. *Then* $\mu(A) = (k-1)n$, *and for any* $G \in M(A)$,

$$\mathcal{A}_G \subseteq \bigcup_{i=1}^{k} \mathcal{A}_{G_i}.$$

Remark 1. Under the assumption of Proposition 1, $A \subseteq M(A)$.

Proposition 1 shows that when genomes are pairwise at maximum bp-distance, any median is forced to draw all its adjacencies from the input genomes. In particular, no median can introduce new adjacencies beyond those already present in the data. We now extend this deterministic picture to the case of k independent random genomes. Since random genomes are pairwise almost maximally distant, Proposition 1 suggests that their medians should behave similarly, up to small random fluctuations. We make this intuition precise by extending Proposition 1 to k independent random genomes in $\mathbb{G}_n$.

To this end, for $A = \{G_1, \ldots, G_k\} \subseteq \mathbb{G}_n$, let $\mathcal{B}_A^A = \mathcal{B}_{G_1, \ldots, G_k}^A := \mathcal{A}_{G_1, \ldots, G_k}$. Then, for any $j = 1, \cdots, k$, define

$$\mathcal{B}_{G_1, \ldots, G_{j-1}, G_{j+1}, \ldots G_k}^A := \mathcal{A}_{G_1, \ldots, G_{j-1}, G_{j+1}, \ldots G_k} \setminus \mathcal{B}_{G_1, \ldots, G_k}^A$$

Continuing inductively, for any $I = \{G_{i_1}, \ldots, G_{i_r}\} \subset A$, we set

$$\mathcal{B}_I^A = \mathcal{B}_{G_{i_1}, \ldots, G_{i_r}}^A := \mathcal{A}_I \setminus \left(\bigcup_{I \subsetneq V} \mathcal{B}_V^A \right).$$

In other words, $\mathcal{B}_I^A$ consists of all adjacencies that are present in every genome in I, but absent from every genome in $A \setminus I$. The sets $\mathcal{B}_I^A$ decompose shared adjacencies according to the subsets of genomes in which they appear.

Let $U_{n,k}, L_{n,k} : (\mathbb{G}_n)^k \to \mathbb{Z}$ be defined by

$$U_{n,k}(G_1, \ldots, G_k) = \sum_{\emptyset \neq I \subseteq [k]} (|I| - 1)|\mathcal{B}_I|,$$

and

$$L_{n,k}(G_1, \ldots, G_k) = \max_{i \in [k]} \sum_{j \neq i} |\mathcal{A}_{G_i, G_j}|,$$

where in the latter, for each fixed $i \in [k]$, the inner sum runs over all $j \in [k]$ with $j \neq i$, and computes

$$\sum_{j \neq i} |\mathcal{A}_{G_i, G_j}|.$$

The outer maximum then selects the largest value of this sum over all choices of i. Thus, for each genome G_i, we sum the quantities $|\mathcal{A}_{G_i, G_j}|$ over all $j \neq i$, i.e., we add up the sizes of its pairwise adjacency intersections with the other genomes. An adjacency that is shared between G_i and multiple genomes G_j contributes once for each such j, and may therefore be counted multiple times in this sum.

The next result provides explicit upper and lower bounds on how far the median value can fall below the maximal value $(k-1)n$. These bounds depend only on how adjacencies are shared among subsets of the input genomes. The proof is given in Appendix B.

Theorem 1. *For any $k \geq 3$, and any set $A = \{G_1, \ldots, G_k\}$ of perfect matchings in $\mathbb{G}_n$, we have*

$$L_{n,k}(G_1, \ldots, G_k) \leq (k-1)n - \mu(A) \leq U_{n,k}(G_1, \ldots, G_k). \tag{1}$$

In other words, Theorem 1 shows that the deviation of the median value from its maximum is entirely governed by the shared adjacencies among the input genomes. For random genomes, such shared structures occur only sparsely, leading to tight probabilistic control over the median value.

Now let us apply this result on a set of k independent random perfect matchings $A = \{\mathcal{G}_1, \ldots, \mathcal{G}_k\} \subset \mathbb{G}_n$. As mentioned, this is very similar to the case of k non-random perfect matchings with pairwise distance n.

More precisely, for any $I \subseteq [k]$ with $|I| \geq 3$, $\mathcal{A}_I \Rightarrow 0$ as $n \to \infty$, and $\mathcal{A}_{\mathcal{G}_i, \mathcal{G}_j}$ and $\mathcal{A}_{\mathcal{G}_r, \mathcal{G}_s}$ are asymptotically independent if $\{i, j\} \neq \{r, s\}$. This is immediate when $\{i, j\} \cap \{r, s\} = \emptyset$, since the perfect matchings are independent. To see the case $\{i, j\}, \{i, k\}$, let $\xi_{lr} = \mathbb{1}\{\{-r, +r\} \in E(\mathcal{G}_l)\}$. On the other hand, from the invariance of the action of the group of permutations on $\mathbb{G}_n$, we can assume that $\mathcal{G}_i = o_n$. Then

$$\mathcal{A}_{\mathcal{G}_i, \mathcal{G}_j} = \sum_{r=1}^{n} \xi_{jr}, \qquad \mathcal{A}_{\mathcal{G}_i, \mathcal{G}_k} = \sum_{r=1}^{n} \xi_{kr}.$$

But since $\mathcal{G}_j$ and $\mathcal{G}_k$ are independent, so is ξ_{jr} and ξ_{ks} for $1 \leq r, s \leq n$.

Then from the Poisson limit for common adjacencies discussed from Appendix A, we deduce, for any $i \in [k]$,

$$\sum_{j \in [k] \setminus \{i\}} |\mathcal{A}_{\mathcal{G}_i, \mathcal{G}_j}| \Rightarrow \text{Poisson}\left(\frac{k-1}{2}\right), \tag{2}$$

and

$$U_{n,k} \Rightarrow \mathrm{Poisson}\left(\frac{k(k-1)}{4}\right). \tag{3}$$

We have readily proved the following result. Let α and β be two independent Poisson random variables, i.e.

$$\alpha \sim \mathrm{Poisson}\left(\frac{k-1}{2}\right), \qquad \beta \sim \mathrm{Poisson}\left(\frac{(k-1)(k-2)}{4}\right).$$

Note that $\alpha + \beta$ has the same distributional limit as $U_{n,k}$ in (3). We say a nonnegative integer-valued random variable Z is stochastically dominated by Y, $Z \preceq_{st} Y$, if for every $k \in Z$, $\mathbb{P}(Z > k) \le \mathbb{P}(Y > k)$.

The following theorem provides stochastic bounds on the deviation of the median value of k independent random genomes from $(k-1)n$. The proof is given in Appendix B.

Theorem 2. *For a given $k \ge 3$, let $A_n = \{\mathcal{G}_1^{(n)}, \ldots, \mathcal{G}_k^{(n)}\} \subseteq \mathbb{G}_n$ be k perfect matchings chosen independently and uniformly at random from $\mathbb{G}_n$. Then for every $k \in \mathbb{Z}_+$,*

$$\mathbb{P}(\alpha > k) \le \liminf_{n \to \infty} \mathbb{P}((k-1)n - \mu(A_n) > k)$$
$$\le \limsup_{n \to \infty} \mathbb{P}(((k-1)n - \mu(A_n)) > k) \le \mathbb{P}(\alpha + \beta > k).$$

Furthermore, $((k-1)n - \mu(A_n))_{n \in \mathbb{Z}_+}$ is tight and hence every sequence of that has a further subsequence that converges in distribution. For every subsequence limit X^*, we have

$$\alpha \preceq_{st} X^* \preceq_{st} \alpha + \beta.$$

The previous theorem shows that the deviation $(k-1)n - \mu(A_n)$ remains stochastically bounded as $n \to \infty$. In particular, the median value differs from its maximal possible value $(k-1)n$ only by a random quantity of order $O(1)$. The next result formalizes this observation by showing that this deviation becomes negligible under any diverging normalization. The proof is deferred to Appendix B.

Theorem 3. *For any diverging sequence $(a_n)_{n \ge 1}$ of strictly positive numbers, $a_n \to \infty$, as $n \to \infty$, we have*

$$\frac{(k-1)n - \mu(A_n)}{a_n} \to 0, \quad \text{in probability.}$$

While the previous result concerns the median value, it does not yet describe the structure of the median genome itself. The following theorem addresses this question by showing that, with high probability, a median genome introduces only a vanishing proportion of adjacencies not present in the input genomes. The proof is given in Appendix B.

Theorem 4. *Let $k \geq 3$, and let $A_n = \{\mathcal{G}_1^{(n)}, \ldots, \mathcal{G}_k^{(n)}\}$ be k perfect matchings chosen independently and uniformly at random from $\mathbb{G}_n$. For any $G \in M(A_n)$, then as $n \to \infty$,*

$$\frac{|\mathcal{A}_G \setminus \bigcup_{i=1}^{k} \mathcal{A}_{\mathcal{G}_i}|}{n} \xrightarrow{p} 0.$$

Taken together, these results indicate that the bp-median of random genomes is both value-wise and structurally constrained.

3   Position of the Median of Random Genomes

In this section, we study the position of the median genome (perfect matching) of k independent random perfect matchings in $\mathbb{G}_n$. As shown in the previous section, the median value is asymptotically $(k-1)n$, and the number of adjacencies of the median that do not belong to the union of the input genomes becomes negligible. More precisely, after normalization by any diverging sequence $(a_n)_{n \geq 1}$, this quantity converges to zero in probability.

Before proceeding, we state the guiding question of this section. While the previous section showed that the median value of random genomes is close to its maximum possible value, this does not by itself determine where the median genome is located in genome space. Here we ask whether a median can lie in the interior of the simplex spanned by the input genomes, or whether it must be concentrated near one of the inputs.

Let $A_n = \{\mathcal{G}_1^{(n)}, \ldots, \mathcal{G}_k^{(n)}\}$ be a collection of independent random genomes. Given nonnegative weights $x_1, \ldots, x_k$ with $x_1 + \cdots + x_k = 1$, our goal is to determine whether there exists a median $\mathcal{G}^{(n)} \in M(A_n)$, whose bp-distance to $\mathcal{G}_i^{(n)}$ is approximately $(1 - x_i)n$, that is $|\mathcal{A}_{\mathcal{G}^{(n)}, \mathcal{G}_i^{(n)}}|/n \to x_i$, as $n \to \infty$, in probability. We denote by Δ_k the $(k-1)$-dimensional probability simplex,

$$\Delta_k := \left\{ (x_1, \ldots, x_k) \in \mathbb{R}^k : x_i \geq 0 \text{ for all } i, \text{ and } x_1 + \cdots + x_k = 1 \right\}.$$

In fact Δ_k parametrizes all possible normalized overlap profiles between a candidate median and the k input genomes. Points in the interior of Δ_k correspond to medians that take their adjacencies across multiple genomes, while the vertices represent medians that coincide almost entirely with a single input genome.

We will show that this is impossible except in the trivial cases where $x_i = 1$ for exactly one index i and $x_j = 0$ for all $j \neq i$. This means that the approximate medians (and hence exact medians) of k independent random genomes cannot lie far from the "corners", the input genomes.

We say that a sequence of random variables (Z_n) converges to a random variable Z in L^n, denoted $Z_n \xrightarrow{L^n} Z$, if $\mathbb{E}(|Z_n - Z|^n) \to 0$, as $n \to \infty$. It is well known that convergence in L^2 implies convergence in probability, and the latter implies convergence in distribution. In what follows, we exploit this fact through a precise control of certain edge counts in random genomes. A key ingredient in our analysis is a quantitative understanding of how many edges of a random genome can avoid a prescribed set of gene extremities. The following theorem provides such a result and serves as the main probabilistic input for ruling out interior points of Δ_k. The proof is deferred to Appendix B.

Theorem 5. *Let* $m = m_n$ *be such that* $m/n \to x$, *as* $n \to \infty$. *For any given set* $R_n \subseteq \pm[n]$ *of size* $|R_n| = 2m$. *Let* $\mathcal{G}^{(n)}$ *be a perfect matching selected uniformly at random from* $\mathbb{G}_n$, *and let* $F_n = F_n(R_n)$ *be the number of edges in* $E(\mathcal{G}^{(n)})$ *with both vertices in* $\pm[n] \setminus R_n$. *Then*

$$\frac{F_n}{n} \xrightarrow{L^2} (1-x)^2, \qquad n \to \infty.$$

Theorem 5 shows that, asymptotically, the number of edges that remain available after fixing a fraction x of gene extremities concentrates sharply around a deterministic value. In particular, once a positive proportion of edges has been fixed, the pool of remaining free edges shrinks quadratically. This concentration phenomenon severely limits the number of edges that can be simultaneously borrowed from multiple random genomes without creating conflicts, and is the key mechanism behind the collapse of feasible overlap profiles to the boundary of the simplex.

Remark 2. As F_n does not depend on the specific choice of R_n but only on its cardinality, we may regard F_n as a function of m, writing $F_n = F_n(m)$.

From the definition of F_n, it is clear that it counts the number of edges in $\mathcal{G}^{(n)}$ whose both ends are free (not used) with respect to R_n. Motivated by this, for a genome $G \in \mathbb{G}_n$ and a set $R \subset \pm[n]$, we say an edge of G is *free* (or *with free ends*) if both vertices covered by that edge are in $\pm[n] \setminus R$. Theorem 5 counts the number of free edges in a random genome $\mathcal{G}^{(n)}$ with respect to any arbitrary set of gene extremities R_n, with $|R_n| = 2m$.

For a given $x = (x_1, \ldots, x_k)$ with $x_1 + \cdots + x_k = 1$, $x_i \geq 0$, for $i = 1, \ldots, k$ and a set $A_n = \{\mathcal{G}_1^{(n)}, \ldots, \mathcal{G}_k^{(n)}\}$ of perfect matchings chosen independently and uniformly at random from $\mathbb{G}_n$, we are to see whether it is possible to find a median $\mathcal{G}^{(n)} \in M(A_n)$ so that $|A_{G,G_i}| \approx \lfloor x_i n \rfloor$.

Let $x_1, \ldots, x_k$ be as above, and let $m_1 = m_1(n), \ldots, m_k = m_k(n)$ be such that $m_i/n \to x_i$, as $n \to \infty$. We are to sample m_1 edges from $\mathcal{G}_1^{(n)}$, m_2 edges from $\mathcal{G}_2^{(n)}$, ..., and m_k edges from $\mathcal{G}_k^{(n)}$, so that these $m_1 + \cdots + m_k$ edges together with the remaining $n - (m_1 + \cdots + m_k)$ edges that are not from $\bigcup_{i=1}^{k} A_{G_i^{(n)}}$ form a perfect matching (genome) $\mathcal{G} \in \mathbb{G}_n$.

The sampling can be performed on $k!$ different orders. More precisely, any permutation σ on $[k] = \{1, 2, \cdots, k\}$ induces an order $\sigma(1), \cdots, \sigma(k)$ on $[k]$. Then for any permutation σ on $1, 2, \cdots, k$, we sample the adjacencies as follows. We first sample $m_{\sigma(1)}$ edges from $F_n(0) = n$ free edges available in $\mathcal{G}_{\sigma(1)}^{(n)}$, $m_{\sigma(1)} \leq F_n(0) = n$. Denote by $I_{\sigma(1)}$ the set of these sampled edges. Then sample $m_{\sigma(2)}$ edges from $\mathcal{G}_{\sigma(2)}^{(n)}$ that are free with respect to $I_{\sigma(1)}$, therefore we must have $m_{\sigma(2)} \leq F_n(m_{\sigma(1)})$. Continuing in this way, denoting by $I_{\sigma(1)}, \ldots, I_{\sigma(r)}$ the edges sampled from $\mathcal{G}^{(n)}_{\sigma(1)}, \ldots, \mathcal{G}^{(n)}_{\sigma(r)}$, we sample $m_{\sigma(r+1)}$ edges from $\mathcal{G}^{(n)}_{\sigma(r+1)}$ which are free with respect to $I_{\sigma(1)} \cup \cdots \cup I_{\sigma(r)}$, hence we must have

$$m_{\sigma(r+1)} \leq F_n(\widetilde{s}^{\sigma}_r),$$

where $\widetilde{s}^{\sigma}_r = \widetilde{s}^{\sigma}_r(n) = m_{\sigma(1)} + \cdots + m_{\sigma(r)}$ for $r \geq 0$, and $\widetilde{s}^{\sigma}_0 = 0$. This yields the constraints, for $r = 0, \ldots, k - 1$

$$0 \leq m_{\sigma(r+1)} \leq F_n(\widetilde{s}^{\sigma}_r), \qquad \widetilde{s}^{\sigma}_r / n \to s^{\sigma}_r,$$

as $n \to \infty$, with $s^{\sigma}_0 = 0$, $s^{\sigma}_r = x_{\sigma(1)} + \cdots + x_{\sigma(r)}$. In particular, $s^{\sigma}_k = x_{\sigma(1)} + \cdots + x_{\sigma(k)} = 1$.

From Theorem 5, we know

$$\frac{F_n(\widetilde{s}^{\sigma}_r)}{n} \xrightarrow{L^2} (1 - s^{\sigma}_r)^2, \qquad n \to \infty.$$

Hence

$$0 \leq x_{\sigma(r+1)} = \lim_{n \to \infty} \frac{m_{\sigma(r+1)}}{n} \leq \lim_{n \to \infty} \mathbb{E}\left(\frac{F_n(\widetilde{s}^{\sigma}_r)}{n}\right) = (1 - s^{\sigma}_r)^2.$$

Passing to the limit $n \to \infty$, these sampling constraints translate into deterministic inequalities on the limiting overlap proportions $(x_1, \ldots, x_k)$. Therefore $(x_1, \ldots, x_k) \in \Delta_k$ must satisfy

$$x_{\sigma(r+1)} \leq (1 - s^{\sigma}_r)^2, \qquad r = 0, \ldots, k - 1, \tag{4}$$

for any permutation σ on $[k]$. Inequalities (4) impose strong constraints on the feasible weight vectors $(x_1, \ldots, x_k)$. In particular, they rule out any balanced allocation of overlap across multiple genomes, a fact that will force all feasible limits to lie at the vertices of Δ_k.

Note that By exchangeability of $\mathcal{G}^{(n)}_1, \ldots, \mathcal{G}^{(n)}_k$, the order of sampling does not matter. For any given permutation σ on $[k]$, system (4) admits no non-trivial solutions: the only feasible points in Δ_k are the vertices $e_1, \ldots, e_k$. Equivalently, no interior point of the simplex can arise as an asymptotic overlap structure of a median for random genomes That is, the only solutions of (4) are $x = (x_1, \ldots, x_k) = e_i$ for $i = 1, \ldots, k$, where e_i denotes the i-th standard basis vector in $\mathbb{R}^k$.

We say a genome $\mathcal{G}^{(n)}$ is an approximate median for $\mathcal{G}^{(n)}_1, \ldots, \mathcal{G}^{(n)}_k$ if, as $n \to \infty$,

$$\frac{(k-1)n - \sum_{i=1}^{k} d(\mathcal{G}^{(n)}, \mathcal{G}^{(n)}_i)}{n} \to 0,$$

in probability, or equivalently,

$$\sum_{i=1}^{k} \frac{|\mathcal{A}_{\mathcal{G}^{(n)}, \mathcal{G}^{(n)}_i}|}{n} \to 1, \qquad \text{in probability.}$$

The arguments above guarantee that with high probability there is no approximate median far from $\mathcal{G}_1^{(n)}, \ldots, \mathcal{G}_k^{(n)}$. This conclusion has an important conceptual implication. Although the bp-median is well defined for random genomes, it does not behave like an "average" genome combining features from multiple inputs. Instead, the medians are forced to collapse toward one of the input genomes, both in its total bp-distance and in its adjacency structure. The following theorem is readily obtained from the arguments above, showing that any approximate median must lie asymptotically at a corner of the simplex Δ_k.

Theorem 6. *Let $A_n = \{\mathcal{G}_1^{(n)}, \ldots, \mathcal{G}_k^{(n)}\}$ be k perfect matchings chosen independently and uniformly at random from $\mathbb{G}_n$, and suppose $\mathcal{G} = \mathcal{G}^{(n)} \in \mathbb{G}_n$ is their approximate median such that as $n \to \infty$,*

$$\frac{|\mathcal{A}_{\mathcal{G},\mathcal{G}_i}|}{n} \to x_i,$$

with $x = (x_1, \ldots, x_k) \in \Delta_k$. Then $x \in \{e_1, \ldots, e_k\}$.

Remark 3. Theorem 6 implies that approximate medians (and hence exact medians) of a set of random genomes concentrate near the "corners"(i.e. input genomes) with high probability.

4 Anti-medians and a Game-Theoretic Interpretation

In this section we study the midpoints of a set of genomes $A_n = \{\mathcal{G}_1^{(n)}, \ldots, \mathcal{G}_k^{(n)}\}$, sampled independently and uniformly at random from $\mathbb{G}_n$. The midpoints of $\mathcal{G}_1, \ldots, \mathcal{G}_k$ are these perfect matchings (genomes) which are equidistant from the input genomes $\mathcal{G}_1, \ldots, \mathcal{G}_k$. In particular we try to find the midpoints with the closest total distance to A_n. These are called anti-medians and are obtained in an attempt to compromize the number of adjacencies shared from the input genomes [8]. We interpret this problem as a game of gene-sharing that is played by k players $\mathcal{G}_1, \ldots, \mathcal{G}_k$. This reduces to an optimization problem that we solve and deduce that compromising results in an anti-median behavior. This fully explains the simulation observation regarding the appearance of anti-medians after compromising the gene adjacencies in [8].

Let $A_n = \{\mathcal{G}_1^{(n)}, \mathcal{G}_2^{(n)}, \ldots, \mathcal{G}_k^{(n)}\}$ be k perfect matchings picked uniformly and independently at random from $\mathbb{G}_n$. We frame this as a game in which each genome $\mathcal{G}_i$ plays the role of player i. The players $\mathcal{G}_1, \ldots, \mathcal{G}_k$ cooperate. Each player $\mathcal{G}_i$, for $i = 1, \ldots, k$, shares a proportion α_i of its adjacencies, namely I_i, with others. We assume $|I_i| \approx \alpha_i n$. The goal for players is to share their adjacencies such that $\bigcup_{i=1}^{k} I_i$ be a part of genome G, that is if $\bigcup_{i=1}^{k} I_i$ constitutes a perfect matching on a subset of $\pm[n]$. Since the perfect matchings are sampled independently and uniformly at random, with high probability their pairwise common adjacencies are few, so we may assume that the I_i are pairwise disjoint, i.e. $I_i \cap I_j = \varnothing$ for $i \neq j$. The payoff for player i is α_i. This is a multiplayer cooperative game with transferable utility, which means side payments among players are allowed. Therefore the goal for players $\mathcal{G}_1, \ldots, \mathcal{G}_k$ is to maximize the utility (payment) function $u(\alpha_1, \ldots, \alpha_k) := \alpha_1 + \cdots + \alpha_k$, so the problem is to find $(\alpha_1, \ldots, \alpha_k)$ such that u is maximized.

It is well known that the optimal solutions of a multiplayer cooperative game with transferable utility lie on the Pareto optimal boundary of the set (domain) of possible payoff vectors (called feasible set).

Definition 1. *A feasible payoff vector $\alpha = (\alpha_1, \ldots, \alpha_k)$ is Pareto optimal if no feasible vector $\alpha' = (\alpha'_1, \ldots, \alpha'_k)$ satisfies $\alpha'_i \geq \alpha_i$ for all i unless $\alpha' = \alpha$.*

In other words, no player can gain more without making at least one player worse off. The set of all Pareto-optimal feasible vectors is called the Pareto optimal boundary of the game.

In the sequel, we find the Pareto optimal boundary ϕ_k of our gene-sharing game and show that the minimum payoff on ϕ_k occurs on its intersection with the set $\{(\alpha_1, \ldots, \alpha_k) : \alpha_1 = \alpha_2 = \cdots = \alpha_k\}$, which is the compromiser of the payoff. We first study the case of two players (two random genomes) and then extend the results to general k.

For two players $\mathcal{G}_1^{(n)}$ and $\mathcal{G}_2^{(n)}$, we may have two sampling mechanisms. Either player 1 first shares $\lceil \alpha_1 n \rceil$ of its adjacencies I_1, and then player 2 can only share $\lceil \alpha_2 n \rceil$ of its free adjacencies with respect to I_1, that enforces $\alpha_2 \leq (1 - \alpha_1)^2$; or vice versa, player 2 shares first and player 1 shares next, enforcing $\alpha_1 \leq (1-\alpha_2)^2$. This means that the set of feasible payoff vectors is

$$P_2 := \{(\alpha_1, \alpha_2) \geq (0,0) : \alpha_1 \leq (1 - \alpha_2)^2 \text{ or } \alpha_2 \leq (1 - \alpha_1)^2\}$$
$$= \{(\alpha_1, \alpha_2) \geq (0,0) : \alpha_1 \leq (1 - \alpha_2)^2\} \cup \{(\alpha_1, \alpha_2) \geq (0,0) : \alpha_2 \leq (1 - \alpha_1)^2\}.$$

The first condition $\alpha_2 \leq (1 - \alpha_1)^2$ is equivalent to either

$$\alpha_1 \leq \frac{2 - \sqrt{4\alpha_2}}{2} = 1 - \sqrt{\alpha_2} \quad \text{or} \quad \alpha_1 \geq \frac{2 + \sqrt{4\alpha_2}}{2},$$

where the latter is impossible as it contradicts with $\alpha_1 \in [0,1]$. This implies $\alpha_1 + \sqrt{\alpha_2} \leq 1$. This and the same argument for $\alpha_1 \leq (1 - \alpha_2)^2$ concludes

$$P_2 = \{(\alpha_1, \alpha_2) \geq (0,0) : \alpha_1 + \sqrt{\alpha_2} \leq 1\} \cup \{(\alpha_1, \alpha_2) \geq (0,0) : \alpha_2 + \sqrt{\alpha_1} \leq 1\}. \tag{1}$$

This is because the intersection of $\alpha_1 + \sqrt{\alpha_2} = 1$ and $\alpha_2 + \sqrt{\alpha_1} = 1$ occurs at $(\alpha_1, \alpha_2) = \left(\frac{3-\sqrt{5}}{2}, \frac{3-\sqrt{5}}{2}\right)$ that comes as the intersection of both with the line $\alpha_1 = \alpha_2$, that is the solution to $\alpha_1 + \sqrt{\alpha_2} = 1 = \alpha_2 + \sqrt{\alpha_1}$ (see Fig. 1).

We have readily the following result.

Theorem 7. *For two-player games with players $\mathcal{G}_1^{(n)}, \mathcal{G}_2^{(n)}$, the Pareto optimal boundary of feasible set of payoff vectors is the function $\phi_2 : [0,1] \to [0,1]$, defined by*

$$\alpha_2 = \phi_2(\alpha_1) = \begin{cases} (1 - \alpha_1)^2, & \text{for } \alpha_1 \leq \dfrac{3 - \sqrt{5}}{2}, \\[2mm] 1 - \sqrt{\alpha_1}, & \text{for } \alpha_1 \geq \dfrac{3 - \sqrt{5}}{2}. \end{cases}$$

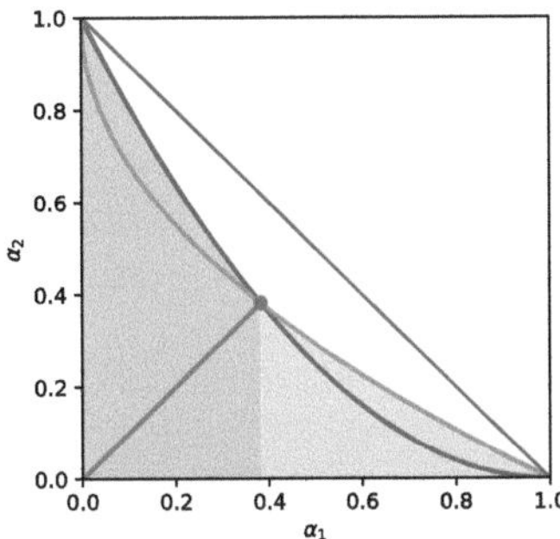

Fig. 1. Feasible payoff region P_2 (shaded) and Pareto boundary for the two-genome game. The boundary consists of the curves $\alpha_2 = (1 - \alpha_1)^2$ and $\alpha_2 = 1 - \sqrt{\alpha_1}$. The diagonal segment $\alpha_1 = \alpha_2$ connects $(0,0)$ to their intersection $\left(\frac{3-\sqrt{5}}{2}, \frac{3-\sqrt{5}}{2}\right)$, and the line $\alpha_1 + \alpha_2 = 1$ is shown for reference.

The set of feasible vectors P_2 is the area surrounded by (inside and including) $\alpha_1 = 0$, $\alpha_2 = 0$, and ϕ_2. As mentioned, it is well-known that the optimal solution of gene-sharing game are on ϕ_2. Otherwise, at least one player can increase its payoff without making the other player worse off. So the first step for players $\mathcal{G}_1$ and $\mathcal{G}_2$ is to agree to choose their payoff vector on ϕ_2. Therefore the optimization problem of finding the best and worst payoff $U(\alpha_1, \alpha_2) = \alpha_1 + \alpha_2$, reduces to:

Optimization Problem 1 (Median). Find $(\alpha_1, \phi_2(\alpha_1))$ with $\alpha_1 \in [0,1]$ such that the function $\alpha_1 \mapsto \alpha_1 + \phi_2(\alpha_1)$ is maximized.

Optimization Problem 2 (Anti-median). Find $(\alpha_1, \phi_2(\alpha_1))$ with $\alpha_1 \in [0,1]$ such that the function $\alpha_1 \mapsto \alpha_1 + \phi_2(\alpha_1)$ is minimized.

From the definition, it is clear that the function $\alpha_1 \mapsto \alpha_1 + \phi_2(\alpha_1)$ is strictly decreasing on $[0, \frac{3-\sqrt{5}}{2}]$ and strictly increasing on $[\frac{3-\sqrt{5}}{2}, 1]$. This implies that the solutions to the optimization problems (1) are $(\alpha_1, \alpha_2) = (0, 1)$ and $(\alpha_1, \alpha_2) = (1, 0)$. In other words the function $\alpha_1 \mapsto \alpha_1 + \phi_2(\alpha_1)$, attains its maximum at $\alpha_1 = 0$ (i.e. $(\alpha_1, \alpha_2) = (0, 1)$) and $\alpha_1 = 1$ (i.e. $(\alpha_1, \alpha_2) = (1, 0)$) with maximum value $U(0, 1) = U(1, 0) = 1$ that corresponds to the approximate medians tending to the corners (input genomes $\mathcal{G}_1^{(n)}$ and $\mathcal{G}_2^{(n)}$).

On the other hand, the unique solution to the optimization problem (2) is $(\alpha_1, \alpha_2) = \left(\frac{3-\sqrt{5}}{2}, \frac{3-\sqrt{5}}{2}\right)$. In other words, the function $\alpha_1 \mapsto \alpha_1 + \phi_2(\alpha_1)$ attains its minimum on $\alpha_1 = \frac{3-\sqrt{5}}{2}$ (i.e. $(\alpha_1, \alpha_2) = ((3 - \sqrt{5})/2, (3 - \sqrt{5})/2)$) which corresponds to full compromisation and anti-median.

Furthermore, as $\alpha_1 \mapsto \alpha_1 + \phi_2(\alpha_1)$ is decreasing on $\alpha_1 \leq (3 - \sqrt{5})/2$ and increasing on $\alpha_1 \geq (3 - \sqrt{5})/2$, we can conclude that the more $\mathcal{G}_1$ and $\mathcal{G}_2$ compromise, the less the genome G constructed from their shared adjacencies behaves like a median, i.e. the more G behaves like an anti-median. This fully explains the observations from the simulation studies in [8,9] for $k = 2$.

The same idea extends to k players, although the Pareto optimal boundary will become a hyper-surface ϕ_k on $\mathbb{R}_+^k$.

Recall that the total payoff is $u(\alpha_1,\ldots,\alpha_k) = \alpha_1 + \cdots + \alpha_k$. Denote by P_k the set of feasible payoff vectors for $\mathcal{G}_1^{(n)},\ldots,\mathcal{G}_k^{(n)}$. Then $P_k = \cup_\sigma P_k^\sigma$, where the union is over all permutations σ of $[k]$, and P_k^σ is the set of all $\alpha = (\alpha_1,\cdots,\alpha_k) \in \mathbb{R}_+^k$, with $\alpha_1,\cdots,\alpha_k \le 1$, for which

$$\alpha_{\sigma(r+1)} \le \Big(1 - \sum_{t=1}^{r} \alpha_{\sigma(t)}\Big)^2 \quad r = 1,\ldots,k-1.$$

As the optimal payoff vectors are located on ϕ_k, first $\mathcal{G}_1,\ldots,\mathcal{G}_k$ agree to take their payoff vector on ϕ_k, and then they try to optimize it. Having themselves restricted to ϕ_k, the maximum and minimum payoff on ϕ_k can be stated and solved similarly.

It concludes that the maximum value of u on ϕ_k occurs at payoff vectors $\alpha = (\alpha_1,\ldots,\alpha_k) = e_i$, for $i = 1,\ldots,k$, where as before $e_i \in \mathbb{R}^k$ denotes the i-th standard basis vector. This corresponds to the approximate median of $\mathcal{G}_1^{(n)},\ldots,\mathcal{G}_k^{(n)}$ that all tends to the corners.

On the other hand the minimum value of u on ϕ_k occurs at the unique point on the intersection of ϕ_k and the diagonal of $\mathbb{R}_+^k$, i.e.

$$\{v^*\} = \phi_k \cap \{(\alpha_1,\ldots,\alpha_k) \in \mathbb{R}_+^k : \alpha_1 = \alpha_2 = \cdots = \alpha_k\}.$$

Letting $v^* = (\theta(k),\cdots,\theta(k))$, equivalently $\alpha_1 = \alpha_2 = \cdots = \alpha_k = \theta(k)$, we have

$$\theta(k) \le (1 - r\theta(k))^2,$$

for $r = 0,\cdots,k-1$. Since the right-hand side decreases as r increases, the strongest condition is when $r = k-1$. Therefore, to have v^* on the Pareto optimal boundary, we must have equality, i.e.

$$\theta(k) = (1 - (k-1)\theta(k))^2.$$

Note that the larger root of the last equation is not admissible in the present setting, as it does not lie in the set of feasible payoff vectors. Therefore, $\theta(k)$ is the smaller root, which is given by

$$\theta(k) = \frac{2k - 1 - \sqrt{4k - 3}}{2(k-1)^2}.$$

The total payoff is

$$u(v^*) = k\theta(k) = \frac{k(2k - 1 - \sqrt{4k - 3})}{2(k-1)^2},$$

and the total distance of v^* to $\mathcal{G}_1^{(n)},\cdots,\mathcal{G}_k^{(n)}$ is approximately $kn(1 - \theta(k))$.

This solution corresponds to the anti-median of $\mathcal{G}_1^{(n)},\ldots,\mathcal{G}_k^{(n)}$. Between these two scenarios, one can see that the total payoff decreases as we move on lines between e_i to v^*, for $i = 1, \ldots, k$. This fully explains the appearance of anti-medians after compromising the number of gene adjacencies taken from $\mathcal{G}_1^{(n)}, \ldots, \mathcal{G}_k^{(n)}$ and shows why the total distance increases as we increase compromization.

5   Discussion

We have shown that for independently and uniformly sampled signed multichromosomal genomes, breakpoint medians asymptotically lie at the corners: with high probability, any median genome shares almost all of its adjacencies with a single input genome. This rigorously establishes the Haghighi–Sankoff conjecture in this setting and provides a unified explanation for earlier heuristic and simulation-based observations.

This phenomenon is driven by the strong constraints inherent in the signed multichromosomal model. Although breakpoint medians are well defined, the set of feasible medians is highly restricted: balanced overlap profiles are ruled out, and any approximate median is forced to draw almost all of its adjacencies from a single input genome. As a consequence, strategies that attempt to combine adjacencies evenly from multiple genomes necessarily increase the total distance.

The game-theoretic framework developed here makes this mechanism explicit, explaining why increasing compromise along the Pareto boundary leads to increased total bp-distance to the input genomes and to the emergence of anti-medians.

Disclosure of Interests. The authors have no competing interests to declare that are relevant to the content of this article.

A   Bp-Distance of Two Random Genomes

In this part, we will obtain a quantitative control on the approximation error for the distance of two independent random genomes (perfect matchings). Indeed, it is not hard to show a stronger convergence. To see this, recall that the total variation distance for two random variables Y and Z with a common state space S is defined by

$$\|\mathcal{L}(Y) - \mathcal{L}(Z)\|_{TV} = \sup_{A \subseteq S} |\mathbb{P}(Y \in A) - \mathbb{P}(Z \in A)|,$$

where $\mathcal{L}(Y)$ and $\mathcal{L}(Z)$ denote the distributions of random variables Y and Z, respectively. When $S \subseteq \mathbb{Z}_+$, the last formula reduces to

$$\|\mathcal{L}(Y) - \mathcal{L}(Z)\|_{TV} = \frac{1}{2} \sum_{i \in S} |\mathbb{P}(Y = i) - \mathbb{P}(Z = i)|.$$

We show that $|\mathcal{A}_{\mathcal{G}_n, \mathcal{G}'_n}|$ converges to Poisson$(1/2)$ in total variation topology, i.e., as $n \to \infty$,

$$\|\mathcal{L}(|\mathcal{A}_{\mathcal{G}_n, \mathcal{G}'_n}|) - \mathcal{L}(\text{Poisson}(1/2))\|_{TV} \to 0.$$

As we discussed earlier, this is equivalent to showing that, as $n \to \infty$,

$$\|\mathcal{L}(|\mathcal{A}_{\mathcal{G}_n, o_n}|) - \mathcal{L}(\text{Poisson}(1/2))\|_{TV} \to 0. \tag{5}$$

For simplicity, let $C_n := |\mathcal{A}_{\mathcal{G}_n, o_n}|$. We proceed as follows. Let $\mathcal{G}_n$ be a random perfect matching in $\mathbb{G}_n$, and let

$$\xi_{in} = \mathbb{1}_{\{-i, +i\} \in E(\mathcal{G}_n)},$$

where $\mathbb{1}$ is the indicator function, i.e. $\xi_{in} = 1$ if $\{-i, +i\} \in E(\mathcal{G}_n)$ and $\xi_{in} = 0$ otherwise. Each indicator ξ_{in} records whether the i-th adjacency of the reference genome is preserved in the random genome. We have

$$\mathbb{E}(\xi_{in}) = \frac{1}{2n - 1}, \qquad \mathbb{E}(\xi_{in}\xi_{jn}) = \frac{1}{(2n-1)(2n-3)}, \quad i \neq j.$$

Therefore,

$$\text{Var}(\xi_{in}) = \frac{1}{2n-1}\left(\frac{2n-2}{2n-1}\right), \qquad \text{Cov}(\xi_{in}, \xi_{jn}) = \frac{2}{(2n-1)^2(2n-3)}.$$

On the other hand,

$$C_n = \sum_{i=1}^{n} \xi_{in}.$$

Hence, as $n \to \infty$,

$$\mathbb{E}(C_n) = \frac{n}{2n-1} \to \frac{1}{2},$$

and

$$\text{Var}(C_n) = \frac{n(2n-2)}{(2n-1)^2} + 2(n^2 - n)\left(\frac{2}{(2n-1)^2(2n-3)}\right) \to \frac{1}{2}.$$

In addition, for any $n \in \mathbb{N}$, the random variables $\xi_{1n}, \ldots, \xi_{nn}$ are exchangeable. That is, for any permutation $\sigma \in S_n$, where S_n is the group of permutations on $\{1, \ldots, n\}$, and any $(j_{1n}, j_{2n} \ldots, j_{nn}) \in \{0, 1\}^n$, we have

$$\mathbb{P}(\xi_{1n} = j_{1n}, \ldots, \xi_{nn} = j_{nn}) = \mathbb{P}(\xi_{\sigma(1)n} = j_{1n}, \ldots, \xi_{\sigma(n)n} = j_{nn}).$$

Thus, by the Stein–Chen method [3, 14], we have

$$\|\mathcal{L}(C_n) - \mathcal{L}(\text{Poisson}(n/(2n-1)))\|_{TV} \le$$

$$\frac{1 - e^{-n/(2n-1)}}{n/(2n-1)}\left(\text{Var}(C_n) - \frac{n}{2n-1} + \frac{2n}{2n-3}\right) \xrightarrow[n \to \infty]{} 0. \tag{6}$$

Hence (5) follows from (2) and triangle inequality for the total variation distance. It is well-known that convergence in total variation implies convergence in distribution

$$C_n \Rightarrow \text{Poisson}(1/2), \quad n \to \infty. \tag{7}$$

That is,

$$\mathbb{P}(C_n = i) \to \frac{(1/2)^i e^{-1/2}}{i!}, \quad i \in \mathbb{Z}_+.$$

If we normalize C_n by a sequence of positive real numbers $(a_n)_{n\geq 1}$ such that $a_n \to 0$, then from (5) or (7) we obtain

$$\frac{C_n}{a_n} \to 0, \quad \text{in probability.} \tag{8}$$

Recall that, for a sequence of random variables Z_n, $n \in \mathbb{N}$, we say Z_n converges in probability to a random variable Z, denoted by $Z_n \xrightarrow{p} Z$, if for any $\varepsilon > 0$, as $n \to \infty$,

$$\mathbb{P}(|Z_n - Z| > \varepsilon) \to 0.$$

B   Proofs

B.1   Proof of Proposition 1

Proof We follow the proof from [7], for unsigned genomes. For $G \in M(A)$, suppose $\mathcal{A}_G \setminus \bigcup_{i=1}^k \mathcal{A}_{G_i} \neq \varnothing$. Let $e \in \mathcal{A}_G \setminus \bigcup_{i=1}^k \mathcal{A}_{G_i}$. Since $e \notin \mathcal{A}_{G_i}$ for every i, this edge contributes to none of the intersections $\mathcal{A}_{G,G_i}$, hence

$$\sum_{i=1}^k |\mathcal{A}_{G,G_i}| \leq n - 1.$$

Therefore

$$\sum_{i=1}^k d(G, G_i) = \sum_{i=1}^k (n - |\mathcal{A}_{G,G_i}|) = kn - \sum_{i=1}^k |\mathcal{A}_{G,G_i}| > kn - n = (k-1)n.$$

On the other hand, for each $i = 1, 2, \ldots, k$ we have

$$\sum_{j\neq i} d(G_i, G_j) = (k-1)n,$$

so G cannot minimize the total distance, a contradiction with the fact that $G \in M(A)$. This implies $\mathcal{A}_G \setminus \bigcup_{i=1}^k \mathcal{A}_{G_i} = \emptyset$. Therefore $\sum_i |\mathcal{A}_{G,G_i}| = n$, and from the above,

$$\sum_{i=1}^k d(G, G_i) = (k-1)n = \mu(A).$$

∎

B.2   Proof of Theorem 1

Proof. For a given set A of genomes, the best scenario for a median, i.e. the smallest possible median value is attained, happens if not only is every adjacency of every median $G \in M(A)$ an adjacency of a genome in A, but also

$$\bigcup_{j \neq i} \mathcal{A}_{G_i,G_j} \subseteq \mathcal{A}_G.$$

From definition, then $\mathcal{A}_G$ can be partitioned into $\{\mathcal{B}_{G_iG_j} : 1 \leq i < j \leq k\} \cup \{\mathcal{B}_{GG_i} : i \in [k]\}$. Hence

$$\sum_{i=1}^{k} d(G, G_i) = \sum_{i=1}^{k} \left(n - |\mathcal{B}_{GG_i}| - \sum_{\substack{i \in I \\ |I| \geq 2}} |I||\mathcal{B}_I| \right)$$

$$= kn - \left(\sum_{i=1}^{k} |\mathcal{B}_{GG_i}| + \sum_{\substack{i \in I \\ |I| \geq 2}} |\mathcal{B}_I| \right) - \sum_{\substack{i \in I \\ |I| \geq 2}} (|I| - 1)|\mathcal{B}_I|$$

$$= (k - 1)n - U_{n,k}(G_1, \ldots, G_k).$$

Thus the right hand side of (1). For the left side of (1), note that for any $G \in M(A)$,

$$\mu(A) = \sum_{j=1}^{k} d(G, G_j) \leq \sum_{j \neq i} d(G_i, G_j) \leq \min_i \sum_{j \neq i} (n - |\mathcal{A}_{G_i,G_j}|),$$

therefore

$$(k - 1)n - \mu(A) \geq L_{n,k}(G_1, \ldots, G_k).$$

∎

B.3   Proof of Theorem 2

Proof. The first part follows (2), (3) and applying Theorem 1. Note that $((k - 1)n - \mu(A_n))_{n \in \mathbb{Z}_+}$ is bounded between two sequences converging to α and $\alpha + \beta$, respectively. Therefore we can see that $((k-1)n - \mu(A_n))_{n \in \mathbb{Z}_+}$ is tight. Therefore, from Helly's selection theorem (Billingsley ...), every subsequence of that has a further subsequence that converges in distribution. Hence, for any subsequence limit X_n, namely X^*, we will have

$$\mathbb{P}(\alpha > k) \leq \mathbb{P}(X^* > k) \leq \mathbb{P}(\alpha + \beta > k).$$

∎

B.4   Proof of Theorem 3

Proof. Let $X_n := (k-1)n - \mu(A_n) \geq 0$. By the previous theorem, the sequence $(X_n)_{n\geq 1}$ is tight. Hence, for every $\varepsilon > 0$, there exists $M < \infty$ such that

$$\sup_{n\geq 1} \mathbb{P}(X_n > M) < \varepsilon.$$

Let $(a_n)_{n\geq 1}$ be any diverging sequence of positive numbers with $a_n \to \infty$. Fix $\delta > 0$. Choose N such that for all $n \geq N$ we have $\delta a_n \geq M$. Then, for $n \geq N$,

$$\mathbb{P}\left(\frac{X_n}{a_n} > \delta \right) = \mathbb{P}(X_n > \delta a_n) \leq \mathbb{P}(X_n > M) \leq \sup_{m\geq 1} \mathbb{P}(X_m > M) < \varepsilon.$$

Since $\varepsilon > 0$ is arbitrary, it follows that

$$\frac{X_n}{a_n} \xrightarrow{p} 0, \qquad n \to \infty.$$

∎

B.5   Proof of Theorem 4

Proof. Fix n and write $A_n = \{\mathcal{G}_1^{(n)}, \ldots, \mathcal{G}_k^{(n)}\}$. Let $G \in M(A_n)$ and set

$$t_n := \left| \mathcal{A}_G \setminus \bigcup_{i=1}^{k} \mathcal{A}_{\mathcal{G}_i^{(n)}} \right|.$$

For each adjacency (edge) $e \in \mathcal{A}_G$, define

$$c(e) := \left| \{\, i \in [k] : e \in \mathcal{A}_{\mathcal{G}_i^{(n)}} \,\} \right| \in \{0, 1, \ldots, k\}.$$

Then $c(e) = 0$ if and only if $e \notin \bigcup_{i=1}^{k} \mathcal{A}_{\mathcal{G}_i^{(n)}}$. Moreover,

$$\sum_{i=1}^{k} |\mathcal{A}_{G,\mathcal{G}_i^{(n)}}| = \sum_{e \in \mathcal{A}_G} c(e) = \sum_{e \in \mathcal{A}_G \cap \bigcup_{i=1}^{k} \mathcal{A}_{\mathcal{G}_i^{(n)}}} c(e).$$

Since $c(e) \geq 1$ on the intersection, we may write

$$\sum_{e \in \mathcal{A}_G \cap \bigcup_{i=1}^{k} \mathcal{A}_{\mathcal{G}_i^{(n)}}} c(e) = (n - t_n) + \sum_{e \in \mathcal{A}_G \cap \bigcup_{i=1}^{k} \mathcal{A}_{\mathcal{G}_i^{(n)}}} (c(e) - 1).$$

On the other hand,

$$(k-1)n - \mu(A_n) = \sum_{i=1}^{k} |\mathcal{A}_{G,\mathcal{G}_i^{(n)}}| - n,$$

since $\mu(A_n) = \sum_{i=1}^{k} d(G, \mathcal{G}_i^{(n)}) = kn - \sum_{i=1}^{k} |\mathcal{A}_{G,\mathcal{G}_i^{(n)}}|$. Combining the last two displays gives

$$(k-1)n - \mu(A_n) = -t_n + \sum_{e \in \mathcal{A}_G \cap \bigcup_{i=1}^{k} \mathcal{A}_{\mathcal{G}_i^{(n)}}} (c(e) - 1),$$

and hence

$$t_n = \sum_{e \in \mathcal{A}_G \cap \bigcup_{i=1}^{k} \mathcal{A}_{\mathcal{G}_i^{(n)}}} (c(e) - 1) - ((k-1)n - \mu(A_n)) \leq \sum_{e \in \bigcup_{i=1}^{k} \mathcal{A}_{\mathcal{G}_i^{(n)}}} (c(e) - 1).$$

The right-hand side equals $U_{n,k}(\mathcal{G}_1^{(n)}, \ldots, \mathcal{G}_k^{(n)})$ by the definition of $U_{n,k}$ (equivalently, by the decomposition into the $\mathcal{B}_I^A$ sets). Therefore,

$$0 \leq t_n \leq U_{n,k}(\mathcal{G}_1^{(n)}, \ldots, \mathcal{G}_k^{(n)}).$$

Finally, by (3) we have $U_{n,k} \Rightarrow \mathrm{Poisson}\left(\frac{k(k-1)}{4}\right)$, hence $(U_{n,k})_{n \geq 1}$ is tight. For any $\varepsilon > 0$ choose M such that $\sup_{n \geq 1} \mathbb{P}(U_{n,k} > M) < \varepsilon$. Then, for any $\delta > 0$ and all n large enough that $\delta n \geq M$,

$$\mathbb{P}\left(\frac{t_n}{n} > \delta\right) \leq \mathbb{P}\left(\frac{U_{n,k}}{n} > \delta\right) = \mathbb{P}(U_{n,k} > \delta n) \leq \mathbb{P}(U_{n,k} > M) < \varepsilon.$$

Since ε is arbitrary, this implies $t_n/n \xrightarrow{p} 0$. ∎

B.6   Proof of Theorem 5

Proof. Consider the n edges of $\mathcal{G}^{(n)}$ and order them in random $\tilde{e}_1, \ldots, \tilde{e}_n$. For an edge e, let $\mathrm{end}(e)$ be the two vertices incident to it. For $i = 1, \ldots, n$, let

$$\xi_i^{(n)} = \mathbb{1}_{\{\mathrm{end}(\tilde{e}_i) \cap R_n = \varnothing\}}.$$

Then $F_n = \sum_{i=1}^{n} \xi_i^{(n)}$. We can compute the expected value and variance of F_n. Write

$$\mathbb{E}(F_n) = \sum_{i=1}^{n} \mathbb{P}(\mathrm{end}(\tilde{e}_i) \cap R_n = \varnothing) = \sum_{i=1}^{n} \frac{(2n - 2m)(2n - 2m - 1)}{2n(2n-1)}.$$

So

$$\mathbb{E}\left(\frac{F_n}{n}\right) \to (1-x)^2.$$

Also,

$$\sum_{i=1}^{n} \mathrm{Var}(\xi_i^{(n)}/n) = \frac{n}{n^2}(\mathbb{E}(\xi_i) - (\mathbb{E}(\xi_i))^2) = \frac{1}{n}\mathbb{E}(\xi_i)(1 - \mathbb{E}(\xi_i))$$

$$= \frac{(2n - 2m)(2n - 2m - 1)}{n(2n(2n-1))}\left(1 - \frac{(2n - 2m)(2n - 2m - 1)}{2n(2n-1)}\right) \to 0,$$

and

$$2\sum_{i<j} \mathrm{Cov}(\xi_i^{(n)}/n, \xi_j^{(n)}/n) = \frac{(n^2-n)}{n^2} \times$$

$$\left(\frac{(2n-2\,m)(2n-2\,m-1)(2n-2\,m-2)(2n-2\,m-3)}{(2n)(2n-1)(2n-2)(2n-3)} - \frac{(2n-2\,m)^2(2n-2\,m-1)^2}{2n(2n-1)} \right)$$

$$= \frac{(n-1)(2n-2\,m)(2n-2\,m-1)}{n(2n)(2n-1)} \times$$

$$\left(\frac{(2n-2\,m-2)(2n-2\,m-3)}{(2n-2)(2n-3)} - \frac{(2n-2\,m)(2n-2\,m-1)}{2n(2n-1)} \right)$$

where the right-hand side converges to 0, as $n \to \infty$. Thus, $\lim_{n\to\infty} \mathrm{Var}(F_n) = 0$.

Since $\mathbb{E}(F_n) \to (1-x)^2$ and $\mathrm{Var}(F_n) \to 0$, as $n \to \infty$, from Chebyshev's inequality we conclude that F_n converges to $(1-x)^2$ in L^2. $\blacksquare$

Remark 4. One can obtain the asymptotics for the normalized number of edges in $\mathcal{G}^{(n)}$ with both ends in R_n, denoted by $\widetilde{F}_n = \widetilde{F}_n(R_n)$, or the normalized number of edges in $\mathcal{G}^{(n)}$ with exactly one end in R_n and the other end in $\pm[n]\backslash R_n$, denoted by $\widetilde{F}'_n = \widetilde{F}'_n(R_n)$. For $\widetilde{F}_n$, we apply Theorem 5 to the set $\pm[n] \backslash R_n$ instead of R_n, and obtain

$$\frac{\widetilde{F}_n}{n} \xrightarrow{L^2} x^2, \qquad n \to \infty.$$

For $\widetilde{F}'_n$, we note that $\widetilde{F}'_n = n - \widetilde{F}_n - F_n$, hence

$$\frac{\widetilde{F}'_n}{n} \xrightarrow{L^2} 1 - x^2 - (1-x)^2 = 2x(1-x).$$

References

1. Bryant, D.: The complexity of the breakpoint median problem. Centre de recherches mathematiques (1998)
2. Caprara, A.: The reversal median problem. INFORMS J. Comput. **15**(1), 93–113 (2003)
3. Chen, L.H.Y.: Poisson approximation for dependent trials. Ann. Probab. **3**(3), 534–545 (1975). https://doi.org/10.1214/aop/1176996303
4. Feijão, P., Meidanis, J.: SCJ: A Variant of Breakpoint Distance for Which Sorting, Genome Median and Genome Halving Problems Are Easy. In: Salzberg, S.L., Warnow, T. (eds.) WABI 2009. LNCS, vol. 5724, pp. 85–96. Springer, Heidelberg (2009). https://doi.org/10.1007/978-3-642-04241-6_8
5. Fertin, G., Labarre, A., Rusu, I., Tannier, E., Vialette, S.: Combinatorics of Genome Rearrangements. The MIT Press (2009)
6. Haghighi, M., Sankoff, D.: Medians seek the corners, and other conjectures. BMC Bioinf. **13**(19), S5 (2012)

7. Jamshidpey, A., Jamshidpey, A., Sankoff, D.: Sets of medians in the non-geodesic pseudometric space of unsigned genomes with breakpoints. BMC Genomics **15**(6), S3 (2014)
8. Larlee, C.A., Brandts, A., Sankoff, D.: Compromise or optimize? the breakpoint anti-median. BMC Bioinf. **17**(Suppl 18), 473 (2016)
9. Larlee, C.A., Zheng, C., Sankoff, D.: Near-medians that avoid the corners; a combinatorial probability approach. BMC Genomics **15**(6), S1 (2014)
10. Sankoff, D., Blanchette, M.: The median problem for breakpoints in comparative genomics. Comput. Comb. pp. 251–263 (1997)
11. Sankoff, D., Sundaram, G., Kececioglu, J.: Steiner points in the space of genome rearrangements. Int. J. Found. Comput. Sci. **7**(01), 1–9 (1996)
12. da Silva, P.H., Jamshidpey, A., Sankoff, D.: Sampling gene adjacencies and geodesic points of random genomes. In: Scornavacca, C., Hernández-Rosales, M. (eds.) RECOMB International Workshop on Comparative Genomics, pp. 189–210. Springer, Cham. https://doi.org/10.1007/978-3-031-58072-7_10(2024)
13. da Silva, P.H., Jamshidpey, A., Sankoff, D.: Identifying breakpoint median genomes: a branching algorithm approach. In: 25th International Conference on Algorithms for Bioinformatics (WABI 2025), pp. 18–1. Schloss Dagstuhl–Leibniz-Zentrum für Informatik (2025)
14. Stein, C.: A bound for the error in the normal approximation to the distribution of a sum of dependent random variables. In: Proceedings of the Sixth Berkeley Symposium on Mathematical Statistics and Probability, Volume II: Probability Theory, pp. 583–602. University of California Press (1972)
15. Tannier, E., Zheng, C., Sankoff, D.: Multichromosomal median and halving problems under different genomic distances. BMC Bioinf. **10**(1), 120 (2009)
16. Xu, A.W.: The median problems on linear multichromosomal genomes: graph representation and fast exact solutions. J. Comput. Biol. **17**(9), 1195–1211 (2010)
17. Xu, W., Alain, B., Sankoff, D.: Poisson adjacency distributions in genome comparison: multichromosomal, circular, signed and unsigned cases. Bioinformatics **24**(16), i146–i152 (2008)
18. Zanetti, J.P.P., Biller, P., Meidanis, J.: Median approximations for genomes modeled as matrices. Bull. Math. Biol. **78**, 786–814 (2016)

Genome Evolution and Reconstruction

Reconstructing the Constituent Genomes of the Ancestral Angiosperm Pangenome

David Sankoff[1], Jiazhen Leng[1], Pratheesh Soman[2], Qiaoji Xu[1],
Chunfang Zheng[3], Alex Liu[4], James H. Leebens-Mack[5],
and Lingling Jin[2(✉)]

[1] Department of Mathematics and Statistics, University of Ottawa, Ottawa, ON,
Canada
sankoff@uottawa.ca
[2] Department of Computer Science, University of Saskatchewan, Saskatoon, SK,
Canada
lingling.jin@cs.usask.ca
[3] Agriculture and Agri-Food Canada, Ottawa, ON, Canada
[4] School of Computer Science, University of Waterloo, Waterloo, ON, Canada

[5] Department of Plant Biology, University of Georgia, Athens, GA, USA

Abstract. To reconstruct the gene orders of the constituent genomes
in an ancestral pangenome, we propose an analysis of RACCROCHE maximum weight matching output of contigs based on adjacency pairs from
phylogenetically disparate genomes. The key idea is to use the multiple
solutions to the matching optimization as a sample of the set of constituent genomes. We identify those gene-order contigs present in all the
solutions as the "core" of the pangenome, and those absent in some of the
solutions as the pangenome "shell". Different cliques of mutually compatible shell contigs identified different constituent genomes. A significant
proportion of shell genes in each pangenome was inherited from the set
of shell genes in its ancestor pangenome. As a hint to the chromosomal
structure, we performed hierarchical clustering on the combined set of
contigs based on the number of solutions shared by each pair of contigs,
in the search for chromosomal fragments, large clusters present in many
ancestors, and some limited but clear results were obtained.

1 Introduction

Pangenomes aim to represent all the variation found in the genomes of a set
of related organisms [1]—populations, species, genera—which we call the constituent genomes. There are two main approaches to the formal study of the
gene complement of pangenomes. One is identification of the "core" genes (or
ortholog group) present in all constituent genomes, versus the "accessory" or
"shell" genes, constituent genomes, and the "unique" or "cloud" genes, present in
a single genome. (The meanings of terms like accessory and cloud vary in the
literature.) The core may contain fewer than 10% of the pangenome genes, as in
the case of some bacteria [2,3], from 30–70% for many plants and animals [4–6],
or over 95% for humans [7].

M. Lafond (Ed.): RECOMB-CG 2026, LNBI 16569, pp. 77–90, 2026.
https://doi.org/10.1007/978-3-032-26891-4_4

The second gene-centric approach to the pangenome is that of "pangenome graphs". Here, genes (or ortholog groups) are represented as vertices. Adjacent genes in a chromosome of a constituent genome are connected by an edge, often a directed edge. The massive redundancies and conflicts inherent in the resulting raw structure are then reduced by various algorithms to acyclic or locally acyclic graphs. Many types of graph are used to represent the output of these algorithms, but most of these focus on sequences, where full analysis of the gene content is secondary or absent, e.g., de Bruijn graphs [8], cactus graphs [9]. A number of primarily gene-centric pangenome-graph algorithms and packages are, however, available [10–12].

One topic almost never broached in the pangenome literature is the phylogeny of pangenomes. But in the context of the phylogenetics of a number of species or genera, each represented by a pangenome, why settle for simply reducing each pangenome to a linear, or at least locally acyclic, order and then proceed with a traditional phylogenetic analysis of these linearized genomes? After all, it is not a new idea that an ancestral population may be more or less heterogeneous with respect to the genomes of individuals or groups. This is explicit in the modern recognition of incomplete lineage sorting [13], but it was understood earlier, such as in the description of species as clouds or quasispecies of more or less closely related individuals [14]. In this paper, then, following previous suggestions [15], we develop a "small" phylogenetic analysis of pangenomes, where the inferred ancestors are also pangenomes. We apply this analysis to flowering plants, with representative genomes from each of the major angiosperm clades, as in Fig. 1.

Our analysis is based on an earlier gene-order inference of ancestral genomes: the RACCROCHE pipeline [16,17]. This analysis generates a non-unique optimal solution. In this paper, we capitalize on the non-uniqueness property to reconstruct the constituent genomes of the pangenome for each of the ancestors, as in Fig. 2. Note that this preservation and exploitation of the non-uniqueness of RACCROCHE solutions is *a complete antithesis* of the traditional goal of reducing the ambiguity inherent in multiple solutions to arrive at a single ancestral genome.

2   Data and Methods

Source Data. The original data were 16 high-quality genomes reported in [18] and depicted in Fig. 1. Indicating the ancestral node of each clade with boldface numbers, we use the nuclear genomes of

ANA grade ①
    Amborella trichopoda [19]
    Water lily (*Nymphaea colorata*) [20]
Chloranthus sessilifolius ② [21].
Magnoliids ⑦
    Cinnamon (*Cinnamomum chago*) [22]
    Tulip tree (*Liriodendron chinense*) [23,24]

    Aristolochia fimbriata [25]
Monocots ⑤
    Asparagus (*Asparagus officinalis*) [26]
    Pineapple (*Ananas comosus*) [27]
    Yam (*Dioscorea alata*) [28]
    Acorus americanus [29]
Ranunculales ⑧
    Poppy (*Papaver bracteatum*) [30]
    Columbine (*Aquilegia coerulea*) [31] (Eudicots),
Proteales ⑪
    Lotus (*Nelumbo nucifera*) [32]
Buxales ⑩
    Boxwood (*Buxus austroyunnanensis*) [33]
Core eudicots ⑨
    Grapevine (*Vitis vinifera*) (NCBI RefSeq assembly: GCF_030704535.1)
    Lindenbergia philippensis [34]

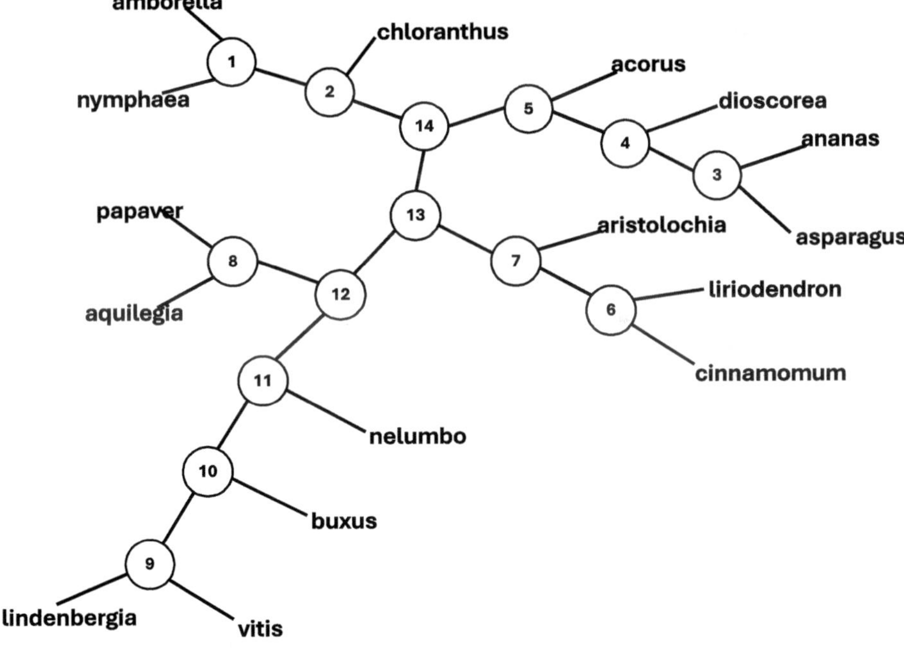

Fig. 1. Angiosperm phylogeny. The goal is to reconstruct the 14 ancestral pangenomes associated with the numbered nodes, based on the 16 extant genomes.

The Matching. The initial analysis was carried out using the RACCROCHE pipeline [16–18,35,36]. All adjacencies and near-adjacencies[1] identified by

---

[1] All pairs of genes in a window size 3 in the gene order of a chromosome. Because of frequent single-gene inversions, the orientation of genes was ignored.

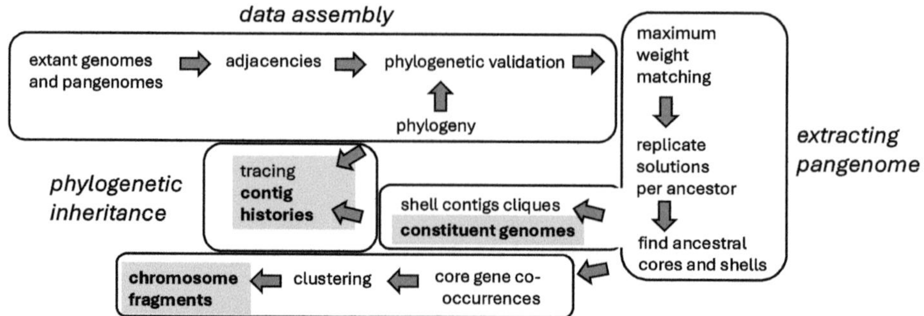

Fig. 2. Protocol for ancestral pangenome construction.

ortholog groups in all these genomes were assembled as an input graph to a maximum weight matching (MWM) algorithm, with specific "phylogenetic validation" restrictions as in Fig. 3 pertaining to each of the various ancestors. For an adjacency X to appear at an internal vertex associated with an ancestral pangenome of a binary branching phylogenetic tree, X must appear in at least two trees subtended by the internal vertex.

For each ancestor, 101 distinct replica solutions of the MWM algorithm were generated by varying the data input order. Since the matching algorithm ensured that each gene in an adjacency was matched to at most one gene in another adjacency, the output for each replicate solution was a set of disjoint "contigs" representing linearly ordered fragments of the chromosomes of the ancestor. Combining the results from all the replicates provided the summary under "contigs" in Table 1. Most of these contigs (from 85% to 92%) for each ancestor, containing from 90% to 95% of the genes, were in all replicates.

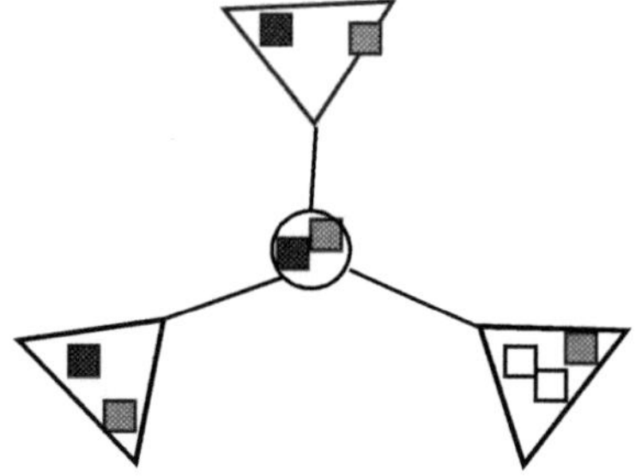

Fig. 3. Phylogenetic validation of adjacencies [15]. Necessary condition for adjacencies to appear at an internal vertex associated with an ancestral pangenome of a binary branching phylogenetic tree. Light shaded adjacency (small square) appears in all three trees (triangles) subtended by the internal vertex (circle). Dark shaded adjacency appears in only two of the trees. Unshaded adjacency appears in only one subtree so does not affect internal vertex. The shaded adjacencies are "phylogenetically validated" with respect to the internal vertex. The unshaded one is not validated.

Accessory Contigs Analysis. In order to examine contigs that are not part of the core (shared) genome, we constructed an overlap graph represented by a binary matrix in which each entry (i, j) is set to 0 if contigs i and j share no genes, and 1 otherwise. Under our evolutionary model, ancestral genomes are monoploid [36] and thus free of paralogy; consequently, any two contigs that share one or more genes are incompatible and cannot co-occur within the same constituent genome. In contrast, contigs with no shared genes are mutually compatible.

Algorithm 1 identifies compatible contig sets by iteratively extracting maximum subsets of pairwise compatible contigs. Specifically, it searches for a maximum clique in the complement of the overlap graph—that is, a largest set of vertices connected exclusively by 0-entries, corresponding to contigs that are mutually non-overlapping in gene content. Once such a clique is identified, the associated contigs are removed from the graph, and the procedure is repeated on the remaining contigs until the graph is exhausted. This iterative extraction yields a partition of the contigs into sets that are internally compatible and can therefore represent candidate constituents of accessory genomes.

Algorithm 1 first constructs the compatibility graph on the set of contigs, in which two contigs are adjacent if they do not overlap in gene content. A clique in this graph therefore represents a set of pairwise compatible contigs. At each iteration, the algorithm identifies a maximum clique, removes its vertices from the graph, and repeats the procedure on the remaining contigs until no vertices remain. This greedy sequential extraction yields an ordered partition of the contigs into internally compatible sets. Maximum cliques were computed exactly using a branch-and-bound–type algorithm with pruning, as implemented in the NetworkX library [37], following standard approaches to the maximum clique problem [38–40].

With this type of algorithm, no polynomial time guarantee is available. But for the moderate size data sets under study, the time complexity is no hindrance.

Inheritance of Shell Genes. Because each ancestor and its descendants are reconstructed independently by the MWM procedure, we examined whether the resulting reconstructions nonetheless exhibit signatures of inheritance across successive ancestral nodes. To this end, we focused on shell genes and quantified the extent to which the shell gene set of a given ancestor is inherited from its immediate predecessor. Specifically, we calculated the proportion of an ancestor's shell genes that are shared with the shell gene set of its parent ancestor.

Chromosomal Fragments. The core of the pangenomes contains, by definition, the set of genes that are present in all constituent genomes, excluding any genes that are absent from any of them. These core genes are distributed across a large number of contigs, which are not ordered relative to one another. However, because evolutionary breakpoints are far less frequent than genes, substantial fragments of chromosomes are expected to be conserved across multiple ancestral genomes.

Accordingly, as in Algorithm 2, we quantify the co-occurrence of pairs of genes within the same contig across all 14 reconstructed ancestors. We then

---

Algorithm 1: Sequential extraction of compatible contigs

Input: A set of contigs $C = \{1, 2, \ldots, n\}$ and an overlap matrix M, where

$$
M_{ij} = \begin{cases} 0, & \text{if contigs } i \text{ and } j \text{ share no genes (compatible)}, \\ 1, & \text{if contigs } i \text{ and } j \text{ share at least one gene (incompatible)}. \end{cases}
$$

Output: An ordered partition $\mathcal{P} = (K_1, K_2, \ldots, K_T)$ of C into pairwise compatible contig sets.

Construct an undirected graph $G = (V, E)$ with $V \leftarrow C$;

for $i < j$ *with* $i, j \in C$ **do**

    if $M_{ij} = 0$ **then**

        add edge (i, j) to E;

$\mathcal{P} \leftarrow \emptyset$;

while $V \neq \emptyset$ **do**

    $K \leftarrow$ a maximum clique of G;

    append K to $\mathcal{P}$;

    remove all vertices in K from G;

return $\mathcal{P}$;

---

apply a clustering analysis to these co-occurrence patterns to identify groups of genes that are likely to have resided on the same chromosome during the evolution of the pangenomes. Complete-link clustering was performed using MAT-LAB's `linkage` implementation, which uses optimized nearest-neighbor chain–type methods with quadratic time complexity.

3   Results

3.1   Replicates

One hundred and one replicate runs, produced 101 different ancestral genomes for each of intermediate ancestors.

3.2   The Core

Most of the 596–750 contigs for each ancestor,including all the longest contigs, containing an even greater proportion of the 3834-6073 genes, were part of every replicate solution, as is clear from the small numbers listed under "shell" in Table 1.

We hypothesize that genes in the great majority of these contigs form the core of the ancestral pangenome. Those counted under "shell" would form part of the accessory portion or a unique portion of the pangenome since they are present in at least one replicate, but not all.

Algorithm 2: Search for conserved chromosomal fragments

Input: Set A of ancestor pangenomes;
         for each $a \in A$: partition of gene set G into contigs

Output: Genes clustered by co-occurrence frequency in ancestral contigs

Initialize co-occurrence matrix $M \in \mathbb{R}^{|G| \times |G|}$ with zeros;

foreach *ancestor* $a \in A$ **do**
    foreach *contig* c *in* a **do**
        foreach *pair of genes* (g_i, g_j) *in* c *where* $i < j$ **do**
             $M[g_i][g_j] \leftarrow M[g_i][g_j] + 1$;
             $M[g_j][g_i] \leftarrow M[g_j][g_i] + 1$;

Convert to distance matrix: $D[i][j] \leftarrow |A| - M[i][j]$ for all i, j;

Perform complete-linkage hierarchical clustering on D;

foreach $k \in \{0, 1, \ldots, |A| - 2\}$ **do**
    Extract clusters at tree level k;
    `// Genes in each cluster co-occur in` $\geq (|A| - k)$ `ancestors`

return *All clusters for* $k \in \{0, 1, \ldots, |A| - 2\}$

3.3 How Many Constituent Genomes Make up a Pangenome?

Do the clique sizes reveal anything about the structure of the ancestral genomes in terms of these constituent genomes? Table 2 reports the maximum clique sizes and the results obtained by successively removing previously identified cliques. Because cliques are mutually incompatible, each maximum clique must correspond to a distinct constituent genome within the pangenome. This is a minimum, however, since each clique could potentially be subdivided into many smaller sets, each of which might determine a separate constituent genome. Nevertheless, these results provide the first evidence for the presence of distinct constituent genomes in our analysis.

3.4 Inheritance of Shell Genes

Table 1 shows that shell genes make up approximately 5% to 10% of the gene complement in each ancestral pangenome. Importantly, Table 3 shows that these shell genes are not randomly drawn from the ancestor's full gene set: up to 29% (average 16%) of them are inherited from the shell genes of the immediately preceding ancestor, clearly exceeding the 5%-10% that would be expected from random (Table 1). This indicates the presence of a detectable evolutionary signal from ancestors to descendants. This enrichment demonstrates that a distinct evolutionary signal is encoded within the pangenomes, reflecting non-random transmission of gene content from ancestor to descendant. Recall that each ancestor pangenome is reconstructed independently using different sets of phylogenetically validated adjacencies as input to the MWM algorithm, indicating that the

Table 1. Number of contigs and genes in ancestors, partitioned by core and shell membership.

ancestor	contigs			genes		
	core	shell	total	core	shell	total
2	557 (85%)	97 (15%)	654	4961 (90%)	557 (10%)	5518
1	593 (92%)	54 (8%)	647	4293 (95%)	216 (5%)	4509
14	549 (85%)	98 (15%)	647	5410 (91%)	557 (9%)	5967
5	627 (91%)	65 (9%)	692	4742 (95%)	264 (5%)	5006
4	633 (87%)	93 (13%)	726	4083 (92%)	362 (8%)	4445
3	638 (86%)	101 (14%)	739	3514 (92%)	320 (8%)	3834
13	518 (87%)	79 (13%)	597	5578 (92%)	495 (8%)	6073
7	594 (87%)	89 (13%)	683	5277 (93%)	377 (7%)	5654
6	624 (88%)	86 (12%)	710	5120 (93%)	394 (7%)	5514
12	570 (88%)	81 (12%)	651	5522 (92%)	468 (8%)	5990
8	593 (90%)	68 (10%)	661	4547 (94%)	292 (6%)	4839
11	577 (88%)	75 (12%)	652	5612 (94%)	357 (6%)	5969
10	618 (88%)	85 (12%)	703	5391 (93%)	386 (7%)	5777
9	688 (92%)	62 (8%)	750	4888 (95%)	247 (5%)	5135

Table 2. Sequentially identified maximum-weight cliques. For each ancestor, we show the number of cliques; then the ordered list of (clique size (contigs), number of genes).

ancestor	number of cliques	number of (contigs, genes)
1	3	(32,212),(21,192),(1,9)
2	3	(52,552),(44,533),(1,10)
3	4	(58,305),(40,278),(2,12),(1,5)
4	4	(51,353),(37,319),(3,19),(2,7)
5	3	(34,260),(30,248),(1,4)
6	4	(45,389),(35,368),(4,50),(2,29)
7	4	(49,373),(38,353),(1,15),(1,15)
8	2	(38,285),(30,269)
9	2	(36,241),(26,223)
10	9	(40,380),(34,363),(3,31),(2,30),(2,29), (1,20),(1,20),(1,8),(1,4)
11	4	(36,512),(36,512),(2,66),(2,66)
12	3	(37,565),(36,563),(3,33)
13	3	(29,366),(28,362),(1,9)
14	4	(38,219),(38,219),(1,7),(1,3)

observed evolutionary signal is inherently embedded in these adjacency patterns and is systematically propagated along evolutionary lineages.

Table 3. Number and proportion of genes inherited by ancestor j from ancestor i, partitioned by core and shell membership.

lineage		core i		shell i	
ancestor i	ancestor j	in core j	in shell j	in core j	in shell j
2	1	3635 (96%)	169 (4%)	398 (93%)	32 (7%)
14	5	4149 (96%)	189 (4%)	415 (87%)	61 (13%)
5	4	3422 (93%)	270 (7%)	186 (83%)	37 (17%)
4	3	3009 (93%)	224 (7%)	234 (77%)	71 (23%)
14	13	4749 (92%)	423 (8%)	497 (93%)	40 (7%)
13	7	4664 (94%)	290 (6%)	381 (84%)	72 (16%)
7	6	4539 (95%)	264 (5%)	250 (71%)	101 (29%)
13	12	4700 (94%)	309 (6%)	331 (73%)	121 (27%)
12	8	3993 (94%)	255 (6%)	364 (93%)	27 (7%)
12	11	4902 (95%)	268 (5%)	383 (86%)	62 (14%)
11	10	4858 (94%)	310 (6%)	264 (80%)	67 (20%)
10	9	4228 (96%)	191 (4%)	287 (88%)	39 (12%)

3.5   Chromosome Structure of the Core

All the constituent genomes of a reconstructed pangenome ancestor share the same core, containing the same genes organized within the same contigs (in our construction). However, there is currently no information or constraint that allows us to determine the relative order of these contigs within any constituent, and the same is true of the shell contigs.

Nevertheless, we can assume that each constituent genome of a pangenome possesses a chromosomal architecture, in which genes are partitioned and ordered along some number of chromosomes. During evolution, as a pangenome gives rise to its descendants, the chromosomes of its constituent genomes undergo various genome rearrangements. These include inversions of long or short segments, transpositions of segments to new positions along the same chromosome, translocations involving exchanges of segments between chromosomes, and segmental duplications.

Despite these rearrangements, the number of breakpoints in chromosomes remains approximately two orders of magnitude smaller than the total number of genes. Consequently, we can expect some chromosomal fragments to survive intact from one pangenome ancestor to its descendants, providing continuity in the genomic structure over evolutionary time.

To detect signals of gene retention across ancestors, we examined all pairs among the 8150 genes appearing in all the ancestors. We counted how many

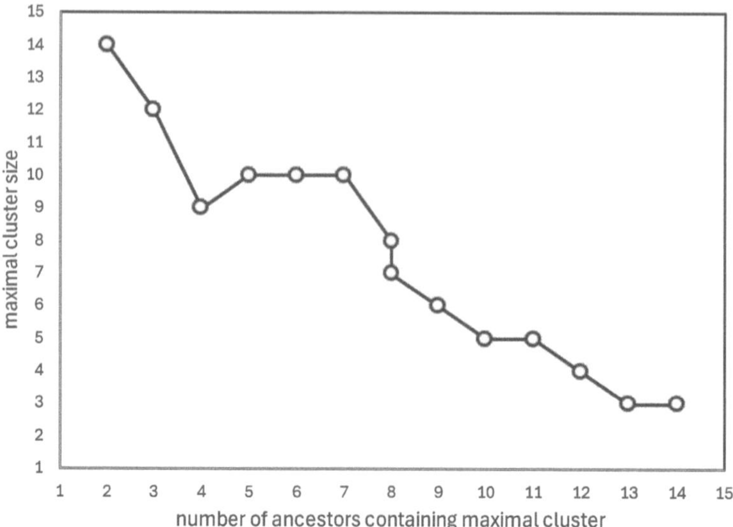

Fig. 4. Maximum number of occurrences of genes clusters of each size.

times they co-occurred in the same contig of an ancestor, from 0 co-occurrences to 14.

The co-occurrence matrix thus constructed was subjected to a complete-link clustering analysis, from which we could extract clusters of genes that consistently co-occurred with each other across contigs. Specifically, we could extract clusters that co-occurred in just one ancestor pangenome, in two ancestor pangenomes, and so on, thereby quantifying the degree of retention of the cluster across the ancestral pangenomes.

Figure 4 shows, for each number A of ancestor pangenomes, from 2 to 14, how large is the largest cluster of genes co-occurring in A ancestors. Thus for any group of seven ancestors, the maximum cluster of genes contained ten genes. For wider co-occurrence, such as 11 ancestors, the maximum cluster size is only 5.

4   Discussion

Our analysis demonstrates the first methodology for reconstructing gene orders of ancestral pangenomes by leveraging the multiple solutions generated by RAC-CROCHE. By distinguishing core contigs (those consistently present across all solutions) from shell contigs, which are variably present, we provide a systematic method to identify conserved versus variable genomic components. Grouping mutually compatible shell contigs into distinct constituent genomes reveals the modular architecture of ancestral pangenomes, whereas hierarchical clustering of core contigs highlights chromosomal fragments and large, evolutionarily conserved gene clusters.

As the first approach to the reconstruction of ancestral gene-order pangenomes, this work is based on a number of assumptions that could be explored or relaxed in further work. Perhaps the most contentious may be the identification of the alternate solutions to the MWM optimization output as potential constituent genomes of the pangenome. This is not, however, the arbitrary interpretation of combinatorial algorithmic choices. Signals present in the complete input genomes, reflecting differences among ancestral constituent genomes, would be carried through the adjacencies to the MWM analysis, and can be captured by the optimization process. This is supported by the observation that the proportion of inherited shell contigs and shell genes is markedly larger than that of the core, indicating a meaningful signal.

We have previously justified our working assumption of ancestral monoploidy in the context of the RACCROCHE gene-order reconstruction of unique ancestral genomes. This assumption is particularly important when partitioning shell genes among the constituent genomes of a pangenome.

The reliance on a single history of evolutionary divergence, in the context of the small phylogeny problem, could be controversial, given that the origin of the major angiosperm clades is not settled.

We have determined that there are generally at least two constituent genomes each containing approximately equal numbers of shell contigs and shell genes, with one to seven additional constituent genomes with fewer shell elements. These numbers should be considered lower bounds for each ancestor, as the data are also consistent with larger numbers of constituent genomes. Further research could refine these estimates, potentially by extending our heritability-based approach together with chromosome-level searches.

The discovery of a few sets of genes suggestive of conserved chromosomal fragments is encouraging, although it does not yet provide a detailed picture of the rate of angiosperm pangenome evolution. Several strategies could increase the sensitivity of this search: for example, using alternative clustering approaches such as average-link, k-means (our use of complete-link clustering may be overly stringent); examining whether smaller clusters are concentrated in specific clades, which could indicate genome rearrangement events at the founding of these clades; or identifying clusters of genes that are mutually exclusive in terms of co-occurrence patterns.

Together, this approach not only refines our understanding of ancestral pangenome organization but also offers a scalable methodology for studying pangenome evolution, structural variation, and lineage-specific innovations across larger phylogenies.

References

1. Matthews, C.A., Watson-Haigh, N.S., Burton, R.A., Sheppard, A.E.: A gentle introduction to pangenomics. Briefings Bioinf. **25**, bbae588 (2024)
2. Tantoso, E., Eisenhaber, B., Kirsch, M., et al.: To kill or to be killed: pangenome analysis of Escherichia coli strains reveals a tailocin specific for pandemic ST131. BMC Biol. **20**, 146 (2022)

3. Inglin, R.C., Meile, L., Stevens, M.J.A.: Clustering of pan- and core-genome of Lactobacillus provides novel evolutionary insights for differentiation. BMC Genomics **19**, 284 (2018)

4. Qin, P., et al.: Pan-genome analysis of 33 genetically diverse rice accessions reveals hidden genomic variations. Cell **184**(13), 3542–3558 (2021)

5. Gerdol, M., Moreira, R., Cruz, F., et al.: Massive gene presence-absence variation shapes an open pan-genome in the mediterranean mussel. Genome Biol. **21**, 275 (2020)

6. Gong, Y., Li, Y., Liu, X., et al.: A review of the pangenome: how it affects our understanding of genomic variation, selection and breeding in domestic animals. J. Animal Sci. Biotechnol. **14**, 73 (2023)

7. Liao, W.-W., Asri, M., Ebler, J., et al.: A draft human pangenome reference. Nature **617**, 312–324 (2023)

8. Depuydt, L., Renders, L., Abeel, T., et al.: Pan-genome de Bruijn graph using the bidirectional FM-index. BMC Bioinf. **24**, 400 (2023)

9. Hickey, G., et al.: Pangenome graph construction from genome alignments with Minigraph-Cactus. Nat. Biotechnol. **42**(4), 663–673 (2024)

10. Page, A.J., Cummins, C.A., Hunt, M., et al.: Roary: rapid large-scale prokaryote pan genome analysis. Bioinformatics **31**(22), 3691–3693 (2015)

11. Tonkin-Hill, G., MacAlasdair, N., Ruis, C., et al.: Producing polished prokaryotic pangenomes with the Panaroo pipeline. Genome Biol. **21**, 180 (2020)

12. Li, H., Marin, M., Farhat, M.R.: Exploring gene content with pangene graphs. Bioinformatics **40**(7), btae456 (2024)

13. Maddison, W.P., Knowles, L., Lacey, T.: Inferring phylogeny despite incomplete lineage sorting. Syst. Biol. **55**, 21–30 (2006)

14. Eigen, M., Schuster, P.: A principle of natural self-organization. Naturwissenschaften **64**, 541–565 (1977)

15. Zhou, X., Sankoff, D.: Ancestral pangenomes and their phylogenetic reconstruction. In: Song, G., (ed.) Comparative Genomics, vol. 15666, Lecture Notes in Computer Science. Springer, Cham (2026)

16. Xu, Q., Jin, L., Zheng, C., Leebens Mack, J.H., Sankoff, D.: RACCROCHE: Ancestral Flowering Plant Chromosomes and Gene Orders Based on Generalized Adjacencies and Chromosomal Gene Co-occurrences. In: Jha, S.K., Măndoiu, I., Rajasekaran, S., Skums, P., Zelikovsky, A. (eds.) ICCABS 2020. LNCS, vol. 12686, pp. 97–115. Springer, Cham (2021). https://doi.org/10.1007/978-3-030-79290-9_9

17. Xu, Q., Jin, L., Zheng, C., Zhang, X., Leebens-Mack, J., Sankoff, D.: From comparative gene content and gene order to ancestral contigs, chromosomes and karyotypes. Sci. Rep. **13**(1), 6095 (2023)

18. Soman, P., Sankoff, D., Jin, L.: Deciphering the angiosperm phylogeny using ancestral genome reconstruction. USRA Research Program, University of Saskatchewan, Poster presentation (2025)

19. Guo, Z., et al.: Near-gapless telomere-to-telomere reference nuclear genome and variable mitochondrial genome of Amborella trichopoda. J. Genet. Genomics (2025)

20. JGI. Jgi data portal (2020). https://data.jgi.doe.gov/phytozome?organism=Ncolorata&expanded=566. Accessed 22 July 2025

21. Ma, J., et al.: The Chloranthus sessilifolius genome provides insight into early diversification of angiosperms. Nat. Commun. **12**(1) (2021)
22. Tao, L., Guo, S., Xiong, Z., Zhang, R., Sun, W.: Chromosome-level genome assembly of the threatened resource plant Cinnamomum chago. Sci. Data **11**(1) (2024)
23. Chen, J., et al.: Liriodendron genome sheds light on angiosperm phylogeny and species–pair differentiation. Nat. Plants **5**(1), 18–25 (2018)
24. Wu, H., et al.: Re-annotation of the Liriodendron Chinense genome identifies novel genes and improves genome annotation quality. Tree Genet. Genomes **19**(4) (2023)
25. Qin, L., et al.: Insights into angiosperm evolution, floral development and chemical biosynthesis from the Aristolochia fimbriata genome. Nat. Plants **7**(9), 1239–1253 (2021)
26. PRJNA317340. Asparagus (ID 317340) - BioProject - NCBI — ncbi.nlm.nih.gov (2017). https://www.ncbi.nlm.nih.gov/bioproject/PRJNA317340/. Accessed 22 July 2025
27. Zhou, X., et al.: Metabolome and transcriptome profiling reveals anthocyanin contents and anthocyanin-related genes of chimeric leaves in Ananas comosus var. bracteatus. BMC Genomics **22**(1) (2021)
28. Zhang, Y.-M., et al.: A telomere-to-telomere genome assembly for greater yam (Dioscorea alata). Plant Commun. **6**(7), 101326 (2025)
29. JGI. Jgi data portal (2020). https://data.jgi.doe.gov/refine-download/phytozome?q=Acorus+americanus. Accessed 22 July 2025
30. NGDC. Browse - BioProject - CNCB-NGDC — ngdc.cncb.ac.cn (2024). https://ngdc.cncb.ac.cn/bioproject/browse/PRJCA018701. Accessed 22 July 2025
31. Filiault, D.L., et al.: The Aquilegia genome provides insight into adaptive radiation and reveals an extraordinarily polymorphic chromosome with a unique history. eLife **7** (2018)
32. TNGR. TNGR-Download — nelumbodb.cn (2022). http://www.nelumbodb.cn/Download-Sequence. Accessed 22 July 2025
33. Wang, Z., et al.: A high-quality Buxus austro-yunnanensis (Buxales) genome provides new insights into karyotype evolution in early eudicots. BMC Biol. **20**(1) (2022)
34. JGI. JGI Data Portal — data.jgi.doe.gov (2021). https://data.jgi.doe.gov/search?q=Lindenbergia+philippensis&expanded=Phytozome-689. Accessed 22 July 2025
35. Xu, Q., Jin, L., Leebens-Mack, J.H., Sankoff, D.: Validation of automated chromosome recovery in the reconstruction of ancestral gene order. Algorithms **14**(6), 160 (2021)
36. Xu, Q., et al.: The monoploid chromosome complement of reconstructed ancestral genomes in a phylogeny. J. Bioinf. Comput. Biol. **19**(6) (2021)
37. NetworkX Developers. Networkx: Network analysis in python (2023). https://networkx.org. Accessed 27 Mar 2026
38. Carraghan, R., Pardalos, P.M.: An exact algorithm for the maximum clique problem. Oper. Res. Lett. **9**(6), 375–382 (1990)
39. Östergård, P.R.J.: A fast algorithm for the maximum clique problem. Discr. Appl. Math. **120**(1–3), 197–207 (2002)
40. Tomita, E., Tanaka, A., Takahashi, H.: The worst-case time complexity for generating all maximal cliques. Theoret. Comput. Sci. **363**(1), 28–42 (2006)

A Multi-flow Approach for Binning Circular Plasmids from Short-Reads Assembly Graphs

Victor Epain[1], Aniket Mane[1], Gianluca Della Vedova[2],
Paola Bonizzoni[2], and Cedric Chauve[1(✉)]

[1] Department of Mathematics, Simon Fraser University, Burnaby, BC, Canada
{victor_epain,aniket_mane,cedric_chauve}@sfu.ca
[2] Università degli Studi di Milano-Bicocca, Milan, Italy
{gianluca.dellavedova,paola.bonizzoni}@unimib.it

Abstract. We address the problem of plasmid binning, that aims to group contigs—from a draft short-read assembly for a bacterial sample—into bins each expected to correspond to a plasmid present in the sequenced bacterial genome. We formulate the plasmid binning problem as a network multi-flow problem in the assembly graph and describe a Mixed-Integer Linear Program to solve it. We compare our new method, `PlasBin-HMF`, with state-of-the-art methods, `MOB-recon`, `gplasCC`, and `PlasBin-flow`, on a dataset of more than 500 bacterial samples, and show that `PlasBin-HMF` outperforms the other methods, by preserving the explainability.

Keywords: Plasmid binning · Mixed Integer Linear Programming · Flow network

1 Introduction

Plasmids are extra-chromosomal Mobile Genetic Elements (MGE) present in bacterial genomes that can easily transfer between bacterial cells, possibly of different bacterial species, and contribute widely to the propagation of genes responsible for Antimicrobial Resistance (AMR) [6,15]. Due to the importance of AMR in public health [18], an important aspect of epidemiological surveillance focuses on detecting plasmids in bacterial sequencing data [5]. Short-read sequencing is still the most commonly used data acquisition technology for epidemiological surveillance. As a consequence, the development of bioinformatics methods for the detection and characterization of plasmids in bacterial draft assemblies obtained from short-read sequencing data is an active research area [1,3,10,13,14,16,17,19,20,23].

Supplementary Information The online version contains supplementary material available at https://doi.org/10.1007/978-3-032-26891-4_5.

Detecting plasmids from a short-read draft assembly can be addressed at three different levels: contigs classification, plasmid binning and plasmid assembly. *Contigs classification* aims to separate contigs between plasmidic contigs and chromosomal contigs, although some contigs can also be classified as ambiguous if they are repeats present both in the chromosome and in some plasmid. There is a large corpus of classification methods, most of them based on machine-learning approaches [22]. As a bacterial genome can contain several plasmids, contigs classification does not provide a precise view of its plasmid content. This is addressed by *plasmid binning*, that computes groups of contigs (*plasmid bins*) with the aim that each plasmid bin contains the contigs constituting a single plasmid. *Plasmid assembly* can be seen as a refinement of plasmid binning where contigs in each plasmid bin are ordered and oriented.

`MOB-recon` [16] is one of the most popular plasmid binning methods, based on aligning contigs to a database of known plasmid-specific genes and a curated database of complete plasmid sequences. To the best of our knowledge, all other plasmid binning methods rely on a different approach that makes use the of *assembly graph*, a graph provided together with contigs by assemblers widely used for bacterial data, `SPAdes` [4], `Unicycler` [24] and `SKESA` [21]. The first methods of this kind were `plasmidSPAdes` [1] and `Recycler` [17]. Both rely on the expectation that contigs from the same plasmid (forming a true plasmid bin) should be co-located in the assembly graph and should have read coverage that is different from the expected coverage of chromosomal contigs and uniform within a plasmid bin. Unlike `MOB-recon`, `plasmidSPAdes` and `Recycler` are *de novo* methods that do not rely on homology with known plasmids. `HyAsP` [13] introduced the notion of scoring plasmids bins based on several features including the coverage of contigs by known plasmid genes, the uniformity of read coverage and of the GC content, and integrated this scoring scheme in an iterative greedy heuristic searching for high-score walks in the assembly graph, each such path forming a plasmid bin. `gplas` [2], similarly to `HyAsP`, is a greedy walk detection heuristic but it relies on a different scoring function and introduced the idea of using plasmid contigs classification results as an important signal to detect plasmid bins; the latest development of this method is `gplasCC` [14]. The `HyAsP` scoring function served as a basis for `PlasBin` [9], the first approach that introduced a rigorous combinatorial optimization approach to plasmid binning; `PlasBin` iteratively detected the plasmid bin of maximum score using a Mixed-Integer Linear Program (MILP), peeled the corresponding subgraph out of the assembly graph, and repeated until no good plasmid bin could be found. `PlasBin-flow` [10] improved upon `PlasBin` by using the concept of network flow to define plasmid bins as connected subgraphs that would optimize a function combining the features used in `PlasBin` and the flow value, considered as a proxy for the copy number of the hypothetically detected plasmid; however, similarly all other graph-based methods discussed above, `PlasBin-flow` is an iterative method, searching for one plasmid bin at a time.

Here we propose a novel plasmid binning method, `PlasBin-HMF` (standing for Plasmid Binning Hierarchical Multi-Flow), that builds upon the high-level principles of `PlasBin-flow` but does not search for plasmid bins one at a time and, instead, relies on the concept of multi-flow to detect a set of plasmid bins at once. To the best of our knowledge, `PlasBin-HMF` is the first plasmid binning method that defines the problem rigorously from a combinatorial optimization point of view and solves it exactly using an MILP. In Sect. 2, we formulate the plasmid binning problem as a multi-flow combinatorial optimization problem and describes how it can be solved using an MILP. In Sect. 3 we apply `MOB-recon`, `gplasCC`, `PlasBin-flow` and `PlasBin-HMF` to a dataset of more than 500 bacterial samples, and we show that `PlasBin-HMF` outperforms the other methods. We conclude by discussing avenues for improving the multi-flow model that serves as a basis for `PlasBin-HMF`.

2   Methods

Informally, given sequencing data from a bacterial sample and its assembly, plasmid binning aims to determine the contig content (a set of contigs) of all plasmids in the sample, without prior knowledge about the number of plasmids or their genomic sequences. From now on, we refer to a set of contigs as a *bin*.

In this section we describe a combinatorial optimization model to detect plasmid bins; we focus on describing the mathematical model, and postpone the description of the MILP formulation to the Supplementary Material. In the description that follows, we assume that plasmids are circular molecules, which is the case in the vast majority of cases.

2.1   Input Data

Assembly Graph. The assembly of short-read sequencing data results in a (unoriented) contig set $\mathcal{C}$ and a link set $\mathcal{L}$, a link being an ordered pairs of oriented contigs, $\mathcal{L} \subset (\mathcal{C} \times \{+, -\})^2$, where $+$ and $-$ respectively stand for the forward and the reverse orientations. Note that for each link $(c, d) \in \mathcal{L}$, its reverse $(\overleftarrow{d}, \overleftarrow{c})$ is also in $\mathcal{L}$, where $\overleftarrow{c}$ and $\overleftarrow{d}$ the reverse of the oriented contigs c and d. The link set naturally defines a directed graph, called the *assembly graph*, where every vertex corresponds to an oriented contig (see Figure S1 in Appendix A).

Contigs Length and Coverage. For a contig $c \in \mathcal{C}$, we denote by len_c its length and by cov_c its *normalized coverage*, i.e. its average read coverage divided by the mean read coverage over all contigs. The normalized coverage serves as a proxy for the expected number of copies of the contig sequence in the sequenced genome. The widely used assembler Unicycler provides this information by default.

Plasmidness Score and Plasmid Seeds. We also assume that each contig is provided with a *plasmidness score* $\mathsf{plm}_c \in [-1, 1]$ that estimates the nature of the contig as either chromosomal ($\mathsf{plm}_c < 0$) or plasmid-derived (plasmidic) ($\mathsf{plm}_c > 0$). Accurate plasmidness scores would assign a score of -1 to purely *chromosomal* contigs, 1 to purely *plasmidic* contigs, and 0 to contigs appearing both in the chromosome and in some plasmid(s) (*ambiguous* contigs). Finally, we assume that we are provided with a subset $\mathcal{C}_{seed} \subset \mathcal{C}$ of *seed* contigs, which are estimated to be plasmid-derived with high confidence. We describe in Sect. 3 how we use the plasmid contigs classifications tools RFPlasmid [23], plasmidCC [14] and Platon [19] to compute plasmidness scores and determine plasmid seeds.

2.2   Overview

Most plasmid binning methods that rely on exploring the assembly graph are based on the following principle: if we assume (1) uniform sequencing depth, (2) an accurate assembly graph, and (3) an accurate classification of the contigs as chromosomal, plasmidic or ambiguous, then a circular plasmid appears in the assembly graph as a circuit composed of plasmidic (and possibly ambiguous) contigs, with uniform normalized coverage that approximates well the copy number of the plasmid. The contig content of this circuit forms the (true) plasmid bin for the corresponding plasmid.

PlasBin-HMF is based on the same high-level principle, accounting for expected inaccuracies in the input data. Indeed, in practice, (1) sequencing depth is not uniform, (2) assembly graphs may exhibit spurious or missing links, and (3) contigs classification methods are prone to errors. PlasBin-HMF allows to define a (plasmid) bin as a connected subgraph of the assembly graph (instead of a circuit, to handle point (2)), whose contigs' normalized coverage is consistent with a flow through this subgraph (point (1)). PlasBin-HMF allows a bin to contain contigs classified as chromosomal but integrates in its scoring function the plasmidness score such that it assigns a better score to subgraphs that are denser in plasmidic contigs (point (3)). Unlike iterative methods, one of the main novelties of PlasBin-HMF is that it simultaneously identifies in the assembly graph multiple subgraphs and flows, each corresponding to a single plasmid bin.

2.3   Multi-flow Modelling of Plasmid Bins

We first describe the network derived from the assembly graph, on which we model plasmid bins as network flows (Definition 1). As we potentially search for several bins, we model one flow per bin (Definition 2, Fig. 1). We focus our search to circular bins in the network (Definition 3), and complete the list of bins with partially circular ones in the case some links are missing (Definition 4). We then model the plasmid binning as a combinatorial optimization problem on the network, using a multi-flow formulation framework, that we formalize as an MILP (Appendix B.2).

Definition 1 (Flow network). *Let $G = (V, A, \mathsf{s}, \mathsf{t})$ be the flow network defined as follows:*

- *the set V of vertices contains the vertices of the assembly graph (associated to the oriented contigs), the source s and the sink t: $V = V_\mathcal{C} \cup \{\mathsf{s},\mathsf{t}\}$, where $V_\mathcal{C} = \mathcal{C} \times \{+,-\}$;[1]*
- *the set A of arcs contains the link-arcs of the assembly graph and the arcs from the source and to the sink: $A = \mathcal{L} \cup \{(\mathsf{s},v) \mid v \in V_\mathcal{C}\} \cup \{(v,\mathsf{t}) \mid v \in V_\mathcal{C}\}$; [1]*

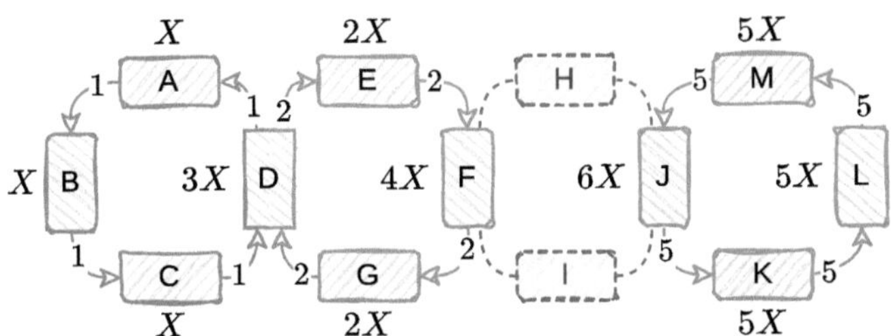

Fig. 1. Example of an assembly graph with three circular plasmid bins, $\{\mathsf{A},\mathsf{B},\mathsf{C},\mathsf{D}\}$, $\{\mathsf{D},\mathsf{E},\mathsf{F},\mathsf{G}\}$ and $\{\mathsf{J},\mathsf{K},\mathsf{L},\mathsf{M}\}$, with respective copy numbers 1, 2 and 5. The contig coverages are written as kX, $k \in \mathbb{N}$. The positive flows are numbers on the arcs. Contig D is shared between the blue and the orange plasmids. The total incoming flow in D equals its coverage $(1+2=3)$. The coverages of F and J are not fully explained by the flows, so one could assume they also belong to other elements of the genome than plasmids. (Color figure online)

Retrieving the contig from a vertex and conversely is enabled thanks to two functions $vtoc\colon V_\mathcal{C} \to \mathcal{C}$ and $ctov\colon \mathcal{C} \to {V_\mathcal{C}}^2$. For each vertex v, $V^-(v)$ and $V^+(v)$ denote respectively the set of predecessors and successors of v in the network. For a vertex $v \in V_\mathcal{C}$, $\overleftarrow{v}$ denotes its reverse (the same contig but with a reversed orientation, so $vtoc(v) = vtoc(\overleftarrow{v})$).

Definition 2 (Coverage-constrained multi-flow). *A multi-flow set of size $n \in \mathbb{N}_{\geq 1}$ is a set of n functions $\{f_1,\ldots,f_n\}$ such that:*

- $\forall 1 \leq k \leq n,\ f_k\colon A \to \mathbb{R}_{\geq 0}$
- $\forall c \in \mathcal{C}:$

$$\sum_{k=1}^{n} expcov(c,k) \leq \mathsf{cov}_c \tag{1}$$

where

$$expcov(c,k) = \sum_{u \in V^-(v)} f_k(u,v) + \sum_{u \in V^-(\overleftarrow{v})} f_k(u,\overleftarrow{v}), \quad ctov(c) = (v,\overleftarrow{v}) \tag{2}$$

---

[1] $V_\mathcal{C} = \mathcal{C} \times \{+,-\}$ is an abuse of notation. Rigorously, one should define $V_\mathcal{C}$ as a regular vertex set in bijection with $\mathcal{C} \times \{+,-\}$. The same remark holds for the arcs where one should define a set of link-arcs $A_\mathcal{L}$ in bijection with $\mathcal{L}$.

- $\forall v \in V_C, 1 \leq k \leq n:$

$$\sum_{u \in V^-(v)} f_k(u, v) = \sum_{w \in V^+(v)} f_k(v, w) \tag{3}$$

- $\forall 1 \leq k \leq n:$

$$\sum_{w \in V^+(\mathsf{s})} f_k(\mathsf{s}, w) = \sum_{u \in V^-(\mathsf{t})} f_k(u, \mathsf{t}) \tag{4}$$

Modelling Plasmids as Multi-flows. As discussed in Sect. 2.2, a bin is the set of contigs of a subgraph of the assembly graph. In `PlasBin-HMF`, each flow in a multi-flow set induces a non-empty subgraph, and thus a bin. We first search for a multi-flow composed only of circular flows (Definition 3), that models putative plasmids that appear as circuits in the assembly graph. Then we relax the circularity constraint (Definition 4) to account for possibly missing links in the assembly graph. Figure 2 illustrates the two kinds of structures we are looking for.

Definition 3 (Circular coverage-constrained flow). *A function f in the coverage constrained multi-flow set is circular if the set of strictly positive flow links $\{a \in \mathcal{L} \mid f(a) > 0\}$ is defining a circuit in G.*

Definition 4 (Partially circular coverage-constrained flow). *A function f in the coverage constrained multi-flow set is partially circular if:*

- *the set of strictly positive flow arcs augmented with an artificial arc from the sink to the source $\{a \in A \mid f(a) > 0\} \cup (\mathsf{t}, \mathsf{s})$ is defining a circuit in G;*
- *the subgraph without s and t is connected;*
- *each vertex connecting the source, respectively the sink, has no other positive-flow predecessor, resp. successor, i.e.*

$$\forall v \in V_C \mid f(\mathsf{s}, v) > 0, \qquad \{u \in V^-(v) \mid f(u, v) > 0\} = \{\mathsf{s}\} \tag{5}$$

$$\forall v \in V_C \mid f(v, \mathsf{t}) > 0, \qquad \{w \in V^+(v) \mid f(v, w) > 0\} = \{\mathsf{t}\} \tag{6}$$

Definition 4 ensures the contigs connected by the source and the sink are extremities of the "cut circuit". It prevents to artificially connecting contigs from other plasmids or misidentified non-plasmidic contigs to the extremities of the cut circuit.

Last, we describe how we make use of the plasmidness score of contigs in defining plasmid bins. Each flow function f_k in the multi-flow set is associated to a plasmidness score (Definition 5). The more the flow uses the coverage of a long plasmidic (non-plasmidic) contig, the better (worse) the score.

Definition 5 (Plasmidness score). *The plasmidness score of a flow function f_k in a multi-flow set is defined as:*

$$PlasmidnessScore_k = \sum_{c \in \mathcal{C}} \mathsf{len}_c \mathsf{plm}_c expcov(c, k) \tag{7}$$

In what follows, we require that any flow that would define a plasmid bin has a positive plasmidness score.

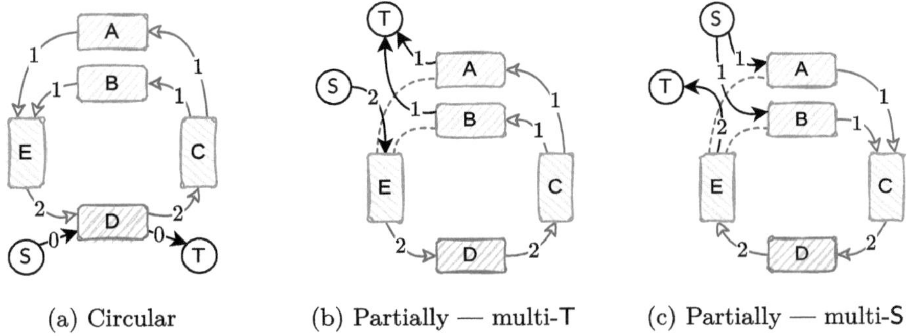

(a) Circular   (b) Partially — multi-T   (c) Partially — multi-S

Fig. 2. Fully and partially circular flows. In each subfigure, the red contig D is a seed, orange and red contigs have a coverage of $2X$ while the blue contigs A and B have coverage X. The numbers on the arcs are the flow values. For the ease of read, we do not show other links between other plasmids or chromosomes and suppose we cannot merge C, D and E in one unitig. The links are oriented according to the orientation of the positive flow. **(a)** No link is missing such that we can find a circuit using all the contig coverages. **(b)**—**(c)** Dashed links are missing links. The source S and the sink T simulate the circularity by connecting the extremities of the partial circuit. (Color figure online)

Scoring a Multi-flow Set. We describe how we associate with a multi-flow set a linear *explanation score* (Definition 6), where a higher scores implies a better explanation of the bins defined by the flows in terms of the input data. PlasBin-HMF aims to search for a multi-flow set maximizing the explanation score. On one hand, we favour multi-flow sets that maximizes the use of plasmidic contigs, avoiding the ones which are not (*PlasmidnessScores* term). On the other hand, we penalize multi-flow sets when they fail to fully explain the coverages of plasmidic contigs (*PosPlmCovPenalty* term). We model the search of the best multi-flow set of size n through an MILP that we describe formally in Appendix B.2.

Definition 6 (Explanation score). *Given a multi-flow set of size n, its explanation score is defined as follows:*

$$ExplanationScore = PlasmidnessScores + PosPlmCovPenalty \tag{8}$$

where:

$$PlasmidnessScores = \sum_{k=1}^{n} PlasmidnessScore_k \tag{9}$$

$$PosPlmCovPenalty = \sum_{\substack{c \in \mathcal{C} \\ \mathsf{plm}_c > 0}} \mathsf{len}_c \mathsf{plm}_c \left(\left(\sum_{k=1}^{n} expcov(c, k) \right) - \mathsf{cov}_c \right) \tag{10}$$

Deciding the Number of Flows. In what precedes, we always assumed that the number of flows in a multi-flow set was given. As the number of plasmid bins is exactly the number of flows in a multi-flow set, the question of choosing the size (number of flows) that is searched by `PlasBin-HMF` is an important question. Informally, we address this problem by computing optimal multi-flow sets of increasing size until a parsimony stopping criterion, based on the comparison among the explanation scores of successive multi-flow sets, is met.

Assume that during this process, at iteration m, we aim to find the best multi-flow set of size m. Each time we increment the number of flows, we penalize the next explanation score with an *additional flow penalty* (Definition 7). Given two parameters $\underline{L}$ (minimum length of a plasmid bin) and $\underline{cov}$ (minimum flow allowed to define a plasmid bin), the additional flow penalty is the plasmidness score of a flow defining a plasmid bin with a single contig, of length $\underline{L}$, plasmidness of 1 and normalized coverage $\underline{cov}$. We stop incrementing the number of flows if (1) either we reach the maximum number of flows $n \in \mathbb{N}_{\geq 1}$ we allowed to search for (see next paragraph) (2) the search is infeasible (3) or the new explanation score plus the additional flow penalty is not (strictly) greater than the previous explanation score (if $ExplanationScore_{m+1} - \mathsf{FlowPenalty} \leq ExplanationScore_m,\ 1 \leq m < n$).

Definition 7 (Additional flow penalty). *Given two parameters $\underline{L} \in \mathbb{N}$ and $\underline{cov} \in \mathbb{R}_{>0}$ the additional flow penalty is defined by:*

$$\mathsf{FlowPenalty} = \underline{L} \cdot \underline{cov} \tag{11}$$

Search Structure and Parameters. A multi-flow set is associated with a plasmid structure (fully or partially circular), which must be described by its flows. Additionally to the subgraph topology, the multi-flow set must respect a seed constraint: we first search for subgraphs containing at least one seed contig, and then relax the constraint. We first search for circular bins with seeds and then without seeds, and repeat the same procedure for partially circular bins, first with seeds and then without. The seed constraint provides us the maximum number of flows a multi-flow set can contain. When each bin must contain a seed, the maximum number of flows is the number of seeds ($n = |\mathcal{C}_{seed}|$). Otherwise, the upper bound equals to a user parameter $\overline{n_\mathsf{c}} \in \mathbb{N}$, respectively $\overline{n_\mathsf{pc}} \in \mathbb{N}$ when we search for circular flows, resp. for partially circular flows. Algorithms B.1 and B.2 detail the search prioritisation.

Finally, given a multi-flow set with a circularity and a seed constraints, we require that each flow function f_k (1) has a positive plasmidness score (Definition 5), (2) equals at least the normalized coverage parameter $\underline{cov}$ when it positively passes through an arc (i.e. $f_k \colon A \to \{0\} \cup [\underline{cov}, \infty]$), and (3) describes a bin such that the sum of its contigs lengths at least equals the parameter $\underline{L}$.

3   Results and Discussion

We implemented the method in `Python3` as a submodule of a larger package
https://github.com/AlgoLab/pangebin[2]. We evaluated `PlasBin-HMF` against
`MOB-recon`, `PlasBin-flow` and `gplasCC` on 581 bacterial samples for which
both short-read and long-read are publicly available on NCBI. In particular,
long-read data was used only to define the ground truth plasmid bins. We detail
in Appendix C how we obtained these 581 samples.

3.1   Data Preparation and Experiments

Input Data Preparation. For each sample, we assembled the Illumina short
reads with `Unicycler`, a read assembler aiming the completion of the circularity.
Contigs less than 100bp were removed from the assembly graph and transformed
into as many links required to connect their neighbours. The plasmidness of contigs was computed with the classification tools `RFPlasmid`. The plasmidness in
`PlasBin-flow` are between 0 and 1, while those of `PlasBin-HMF` are between
-1 and 1 (affine transformation). Internally, `PlasBin-HMF` raises the plasmidness of the seed contigs by computing the mean between one and the original
input plasmidness. The seeds for `PlasBin-flow` and `PlasBin-HMF` are the contigs
identified as plasmid contigs by `Platon` (binary classification).

Ground Truth. For each sample, we produced a hybrid assembly using both
the short and the (Oxford Nanopore) long reads. If a hybrid contig was longer
than 1Mbp, we annotated it as chromosomal, if it is circular and less than 1Mbp
we considered the contig to be a plasmid, otherwise we declared the contig unlabelled. Here, all the 581 samples have a ground truth with hybrid contigs labelled
as plasmids or chromosomal, and no contig is unlabelled. For each sample, we
mapped the short contigs against the hybrid ones with `minimap2` [8]. We labeled
a short contig as plasmidic if it mapped to a plasmidic hybrid contig, and as
non-plasmidic otherwise (i.e., if it mapped only to chromosomal hybrid contigs
or to no hybrid contig). To each plasmidic hybrid contig corresponds a ground
truth bin, containing the plasmidic short contig that mapped to the hybrid
contig. Note that, contrary to the recent plasmid binning method benchmark
study [22], a short contig can belong to multiple bins in our case.

Experiments. We use the default option values for `MOB-recon`, `PlasBin-flow`
and `gplasCC`. For `PlasBin-HMF`, we searched for fully and partially circular bins
with seeds, and complete the results with no more than two circular bins without seeds. We also fixed the minimum cumulative contig length a bin reach to
L = 1000, and a minimum of flow cov = 0.3. As `gplasCC` outputs bins that
exclusively contain contigs classified as plasmidic (even if chromosomal classified

---

[2] Pending the submodule becoming a stand-alone programme, `PlasBin-HMF` can be
launched via the command **pangebin asm-pbf hmf --help** to output a file in the
same format as `PlasBin-flow`.

contigs participate in the walks connecting the plasmidic ones) we add to the comparison a modified version of `PlasBin-flow` and of `PlasBin-HMF` that mimic the `gplasCC` filtering behaviour. Thus, for each bin, we only keep its seeds and its contigs with a plasmidness greater than 0.5 for `PlasBin-flow` and greater than 0 for `PlasBin-HMF`.

We ran the tools and `PlasEval` on the Digital Alliance Canada high-performance computing clusters Cedar and Fir. Each sample was processed using with 16 CPUs, 32 GB of RAM, with a time limit set to 10 h. `PlasBin-flow` and `PlasBin-HMF` ran with the MILP solver Gurobi.

3.2   Accuracy Measures

We evaluate the methods according to two accuracy measures. The first has been introduced in `PlasBin-flow` [10] and is an adaptation of the recall, precision and the F1 statistics, weighted by contig length. For each sample, they evaluate the best pairs of bins between a list of predicted bins P and a list of ground truth bins T. In Eqs. (12) and (13), $overlap(p, t)$ is the sum of length of contigs in both the predicted bin p and the ground truth one v, while $size(b)$ is the lengths sum of contigs in bin b. The F1 measure is the harmonic mean between $Prec$ and $Recall$.

$$Prec = \frac{\sum_{p \in P} \max_{t \in T} overlap(p, t)}{\sum_{p \in P} size(p)} \tag{12}$$

$$Recall = \frac{\sum_{t \in T} \max_{p \in P} overlap(p, t)}{\sum_{t \in T} size(t)} \tag{13}$$

The second evaluation measures have been introduced in the plasmid binning evaluation tool `PlasEval` [11]. For each sample, `PlasEval` provides the costs associated to the minimum number of cut and join operations required to transform the predicted bins into the ground truth ones. The join and cut costs are linearly composed with the costs of extra predicted contigs and missing ground truth contigs to define a dissimilarity score, normalized between 0 and 1. In the next sections, we set the alpha `PlasEval` parameter to 0.5. While we aim maximizing the F1 score, we aim to minimize the dissimilarity score.

These measures are relevant to evaluate a binning result. Indeed, the binning results in a list of bins that may share some contigs. Consequently, classical clustering or partitioning evaluation measures do not handle this list. Furthermore, the adapted F1 and the dissimilarity measures are not affected by true chromosomal contigs that do not belong to any predicted bins.

3.3   Results

For the best prediction-ground truth pairing evaluation, Fig. 3 shows that `PlasBin-HMF` is slightly better (mean 0.61, median of 0.78, before `MOB-recon`,

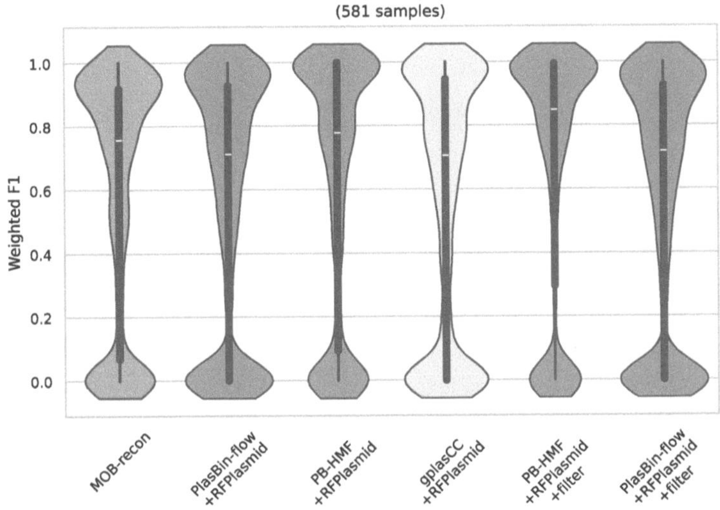

Fig. 3. Weighted F1 scores

with 0.58 and 0.76), however the filtered version is the best (respectively 0.66 and 0.85, with a higher Q1 of 0.30 versus 0.07 for `MOB-recon`).

Figure 4 shows the dissimilarity scores distributions. `PlasBin-HMF` and its filtered version are the best (mean and median equal to 0.3 and 0.27, versus 0.39 and 0.3 for `gplasCC`), with Q1 quartile that reaches 0 (0.05 for `gplasCC`) and Q3 0.44 (0.66 for `gplasCC`). However, the conclusions vary depending on the sequenced bacteria species (Fig. 5). `PlasBin-HMF` (especially filtered) keeps the best place for the *Staphylococcus aureus*, *Acinetobacter baumannii* and *Pseudomonas aeruginosa* species. While `MOB-recon` is not using an assembly graph, it is still efficient for *Escherichia coli* species for which it has been originally developed.

One of the main hypothesis on the encouraging results of `PlasBin-HMF` is its high recall (and low missing contigs costs, Figures S3 and S5) combined with a moderate raise of the cut costs (the cost of splitting a bin that corresponds to several ground truth bins, Figure S6). On the one hand, we penalize the plasmidic coverages not consumed by the flows (Eq. (10)). On the other hand, we search for circuits in the assembly graph, and due to the flow conservation constraints (Eq. (3)), if two plasmids share contigs and their circularity is preserved, our model artificially tends to merge the two circular flows in one bigger circular flow. In that case, what was a shared contig between two plasmids is now a repeat in the artificial plasmid containing the two true ones. Although we have this bias by design, the plasmid subgraphs, while still connected, appear sufficiently dispersed in the assembly graphs such that we do not merge them. In other words, the assembler `Unicycler` correctly untangled the shared subsequences between the plasmids.

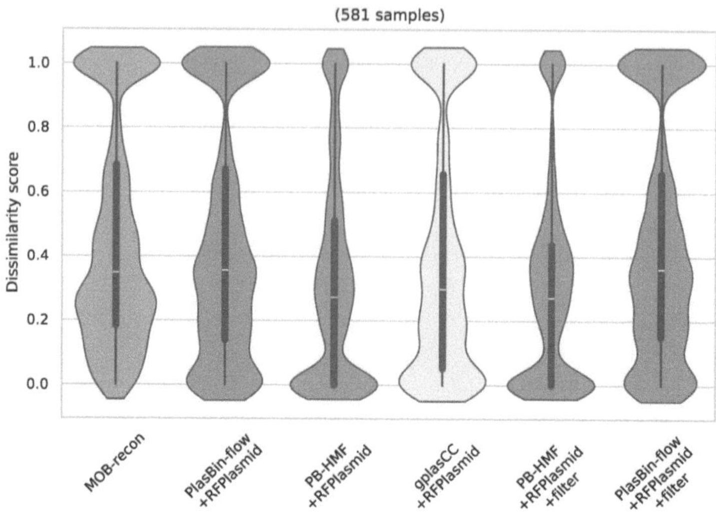

Fig. 4. Dissimilarity scores

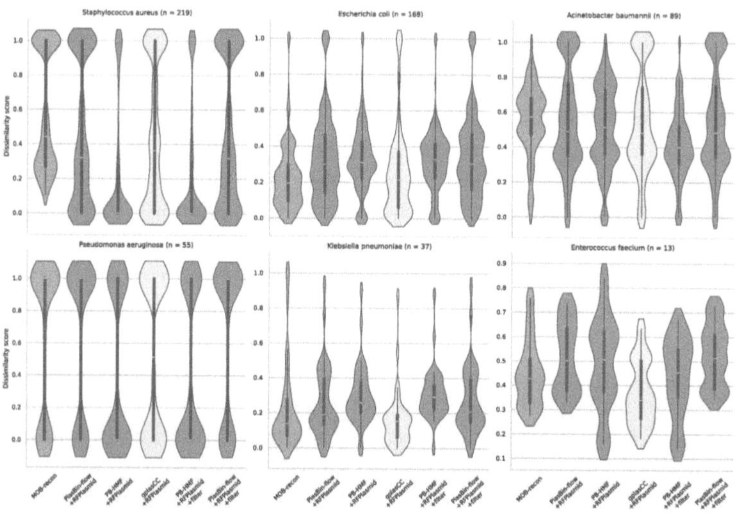

Fig. 5. Dissimilarity scores per species.

4   Conclusion

In this paper, we present the first hierarchical multi-flow approach to address the problem of recovering an unknown number of plasmid bins from a short-read assembly graph. Our methodology introduces a novel plasmid binning strategy that exploits the circular structure of most plasmids. We model the binning task as a combinatorial optimisation problem: we use a mixed-integer linear programming (MILP) formulation to search for a fixed, yet flexible, number of circular flows. Indeed, such number is increased during the process of computing flows. In a second step, we also consider partially circular flows to account for cases where missing links in the data disrupt plasmid circularity. To the best of our knowledge, `PlasBin-HMF` is the first method that simultaneously searches for multiple circular plasmids. We present experimental results comparing `PlasBin-HMF` with state-of-the-art binning tools, showing that our approach achieves the best performance on most bacterial samples.

Here we compared the exclusively binning tools (`gplasCC`, `PlasBin-flow` and `PlasBin-HMF`) with a classifier, `RFPlasmid`. Future comparisons would require us to use different classifiers, such as `plasmidCC`, the default classifier in `gplasCC`. Overall, the current results indicate that `PlasBin-HMF` has strong potential to substantially improve upon existing binning tools. However, several directions remain for further enhancing the model.

One of the main limitations of our approach is an inherent bias toward merging two circular plasmids when they share some contigs. One possible solution is to extend the MILP formulation with constraints on specific genetic content that should not be duplicated within a single plasmid. Such constraints must be carefully selected based on the plasmid genomics literature. Plasmid typing, which typically follows binning, could provide useful information for defining these constraints.

Alternatively, without introducing additional biological constraints and in order to keep the model minimalist, we could limit the amplitude of copy numbers (i.e., explained coverage variation) within each bin subgraph. This may help remove short, low-coverage chromosomal contigs that connect two plasmids and lead the model to incorrectly merge them. However, such a constraint may be less effective when separating plasmids with similar copy numbers. Similar challenges also arise in the haplotype assembly problem.

Another important direction concerns the number of flows in the multi-flow formulation, which corresponds to the unknown number of plasmids in the sample. Since this number is not known a priori, we currently implement a greedy iterative optimisation procedure to determine the appropriate number of flow functions. However, this strategy does not always scale well, as the branch-and-bound process may spend substantial time proving that the final multi-flow solution is not better than the previous one, leading to long runtimes before reaching an optimality certificate. In future work, rather than iteratively increasing the size of the multi-flow set, we aim to allow the MILP solver to directly decide the number of flows.

Acknowledgments. This study was funded by the Simon Fraser University Project N001347 Grant and V. E., P.B and G.D.V. have received funding from the European Union's Horizon 2020 Innovative Training Networks programme under the Marie Skłodowska-Curie grant agreement No. 956229. P.B and G.D.V have been funded by the Horizon Europe program under grant agreement No. 101160008 (FORGENOM II).

References

1. Antipov, D., Hartwick, N., Shen, M., Raiko, M., Lapidus, A., Pevzner, P.A.: plasmidSPAdes: assembling plasmids from whole genome sequencing data. Bioinformatics **32**(22), 3380–3387 (2016). https://doi.org/10.1093/bioinformatics/btw493

2. Arredondo-Alonso, S., et al.: Gplas: a comprehensive tool for plasmid analysis using short-read graphs. Bioinformatics **36**(12), 3874–3876 (2020). https://doi.org/10.1093/bioinformatics/btaa233

3. Arredondo-Alonso, S., et al.: Mlplasmids: a user-friendly tool to predict plasmid- and chromosome-derived sequences for single species. Microb. Genom. **4**(11), e000224 (2018). https://doi.org/10.1099/mgen.0.000224

4. Bankevich, A., et al.: SPAdes: a new genome assembly algorithm and its applications to single-cell sequencing. J. Comput. Biol. J. Comput. Mol. Cell Biol. **19**(5), 455–477 (2012). https://doi.org/10.1089/cmb.2012.0021

5. Castañeda-Barba, S., Top, E.M., Stalder, T.: Plasmids, a molecular cornerstone of antimicrobial resistance in the One Health era. Nat. Rev. Microbiol. **22**(1), 18–32 (2024). https://doi.org/10.1038/s41579-023-00926-x

6. De Oliveira, D.M.P., et al.: Antimicrobial resistance in ESKAPE pathogens. Clin. Microbiol. Rev. **33**(3) (2020). https://doi.org/10.1128/cmr.00181-19

7. Epain, V., Thuillier, K.: Mixed Integer Linear Programming Framework for SAT Variants of the Connected Flow Problem (2025)

8. Li, H.: Minimap2: pairwise alignment for nucleotide sequences. Bioinformatics **34**(18), 3094–3100 (2018). https://doi.org/10.1093/bioinformatics/bty191

9. Mane, A., Faizrahnemoon, M., Chauve, C.: A mixed integer linear programming algorithm for plasmid binning. In: Jin, L., Durand, D. (eds.) Comparative Genomics, pp. 279–292. Springer, Cham (2022). https://doi.org/10.1007/978-3-031-06220-9_16

10. Mane, A., Faizrahnemoon, M., Vinař, T., Brejová, B., Chauve, C.: PlasBinflow: a flow-based MILP algorithm for plasmid contigs binning. Bioinformatics **39**(Supplement_1), i288–i296 (2023). https://doi.org/10.1093/bioinformatics/btad250

11. Mane, A., Sanderson, H., White, A.P., Zaheer, R., Beiko, R., Chauve, C.: PlasEval: a framework for comparing and evaluating plasmid detection tools. BMC Bioinform. **25**(1), 365 (2024). https://doi.org/10.1186/s12859-024-05941-0

12. Mane, A.C.: Optimization-based methods for plasmid analysis. Ph.D. thesis, Simon Fraser University (2025)

13. Müller, R., Chauve, C.: HyAsP, a greedy tool for plasmids identification. Bioinformatics **35**(21), 4436–4439 (2019). https://doi.org/10.1093/bioinformatics/btz413

14. Paganini, J.A., et al.: gplasCC: classification and reconstruction of plasmids from short-read sequencing data for any bacterial species (2024). https://doi.org/10.1101/2024.11.28.625923

15. Partridge, S.R., Kwong, S.M., Firth, N., Jensen, S.O.: Mobile genetic elements associated with antimicrobial resistance. Clin. Microbiol. Rev. **31**(4) (2018). https://doi.org/10.1128/cmr.00088-17

16. Robertson, J., Nash, J.H.E.: MOB-suite: software tools for clustering, reconstruction and typing of plasmids from draft assemblies. Microb. Genom. **4**(8), e000206 (2018). https://doi.org/10.1099/mgen.0.000206

17. Rozov, R., et al.: Recycler: an algorithm for detecting plasmids from de novo assembly graphs. Bioinformatics **33**(4), 475–482 (2017). https://doi.org/10.1093/bioinformatics/btw651

18. Sanderson, H., et al.: Exploring the mobilome and resistome of Enterococcus faecium in a One Health context across two continents. Microb. Genom. **8**(9), 000880 (2022). https://doi.org/10.1099/mgen.0.000880
19. Schwengers, O., Barth, P., Falgenhauer, L., Hain, T., Chakraborty, T., Goesmann, A.: Platon: identification and characterization of bacterial plasmid contigs in short-read draft assemblies exploiting protein sequence-based replicon distribution scores. Microb. Genom. **6**(10), e000398 (2020). https://doi.org/10.1099/mgen. 0.000398
20. Sielemann, J., Sielemann, K., Brejová, B., Vinař, T., Chauve, C.: plASgraph2: using graph neural networks to detect plasmid contigs from an assembly graph. Front. Microbiol. **14** (2023). https://doi.org/10.3389/fmicb.2023.1267695
21. Souvorov, A., Agarwala, R., Lipman, D.J.: SKESA: strategic k-mer extension for scrupulous assemblies. Genome Biol. **19**(1), 153 (2018). https://doi.org/10.1186/ s13059-018-1540-z
22. Teixeira, M., et al.: Circling in on plasmids: benchmarking plasmid detection and reconstruction tools for short-read data from diverse species. Brief. Bioinform. **26**(6), bbaf589 (2025). https://doi.org/10.1093/bib/bbaf589
23. van der Graaf-van Bloois, L., Wagenaar, J.A., Zomer, A.L.: RFPlasmid: predicting plasmid sequences from short-read assembly data using machine learning. Microb. Genom. **7**(11), 000683 (2021). https://doi.org/10.1099/mgen.0.000683
24. Wick, R.R., Judd, L.M., Gorrie, C.L., Holt, K.E.: Unicycler: resolving bacterial genome assemblies from short and long sequencing reads. PLoS Comput. Biol. **13**(6), e1005595 (2017). https://doi.org/10.1371/journal.pcbi.1005595

Modeling the Mutational Dynamics of Very Short Tandem Repeats

Amos Onn[1,2(✉)] ⓘ, Tzipy Marx[3], Liming Tao[4] ⓘ, Tamir Biezuner[3] ⓘ, Ehud Shapiro[3], Christoph A. Klein[1,5] ⓘ, and Peter F. Stadler[2,6,7,8,9,10] ⓘ

[1] Chair of Experimental Medicine and Therapy Research, University of Regensburg, Universitätsstraße 31, 93053 Regensburg, Germany
`{christoph.klein,amos.onn}@klinik.uni-regensburg.de`
[2] Bioinformatics Group, Faculty of Mathematics and Computer Science, and Interdisciplinary Center for Bioinformatics, University of Leipzig, Härtelstrasse 16-18, 04107 Leipzig, Germany
`amos.onn@uni-leipzig.de`
[3] Department of Computer Science and Applied Mathematics, Weizmann Institute of Science, 234 Herzl Street, POB 26, Rehovot 7610001, Israel
`{tamir.biezuner,ehud.shapiro}@weizmann.ac.il`
[4] Cellular Tissue Genomics, Genentech, 94080 South San Francisco, CA, USA
[5] Fraunhofer Institute for Toxicology and Experimental Medicine Regensburg, Am BioPark 13, 93053 Regensburg, Germany
[6] Max Planck Institute for Mathematics in the Sciences, Inselstraße 22, 04103 Leipzig, Germany
[7] Institute for Theoretical Chemistry, University of Vienna, Währingerstraße 17, 1090 Vienna, Austria
[8] Facultad de Ciencias, Universidad Nacional de Colombia, Bogotá, Colombia
`studla@bioinf.uni-leipzig.de`
[9] Center for Non-coding RNA in Technology and Health, University of Copenhagen, Ridebanevej 9, 1870 Copenhagen, Denmark
[10] Santa Fe Institute, 1399 Hyde Park Road, 87501 Santa Fe, New Mexico, USA

Abstract. Short tandem repeats (STRs) are low-entropy regions in the genome, consisting of a short (1–6 bp) unit that is consecutively repeated multiple times. They are known for high mutational instability, due to so-called stutter-mutations, in which the number of units in the run increases or decreases. In particular, STRs with repeat unit length of 1–2 bp are prone to mutate even within several cell divisions. The extremely rapid accumulation of variation makes them interesting phylogenetic markers for retrospective single-cell lineage reconstruction. Here we model their mutational dynamics at the level of individual repeat unit motif and then aggregate length variations over many STR loci with the aim of obtaining a very fast "molecular clock". We calibrate our model based on several datasets with known lineage structure prepared from cultured cells. We find that the mutational dynamics of STRs are reasonably consistent for a given cell-line, but vary among different ones. This suggests that the dynamics are not entirely explained by mutations in caretaker genes, rather, various other factors play a role—possibly tissue

© The Author(s), under exclusive license to Springer Nature Switzerland AG 2026
M. Lafond (Ed.): RECOMB-CG 2026, LNBI 16569, pp. 106–118, 2026.
https://doi.org/10.1007/978-3-032-26891-4_6

origin and differentiation state. Further data and research is necessary to asses their relative effects.

Keywords: short tandem repeats · microsatellites · STR · MS · lineage · single cell

1   Introduction

Reconstructing the lineage of single cells is a problem of ongoing interest, with implications in various fields. Most methods typically employed for this purpose, however, require intensive intervention during the cell division process—either manipulation of the cells themselves [13], or some modification of their genetic material [9–11]. Hence they are not applicable to retrospective lineage reconstruction of human individual samples, where data are by definition limited to *a posteriori* observations of naturally-occuring somatic mutations, and intervention *in vivo* is not possible. While it is possible in principle to use single nucleotide polymorphisms (SNPs) for this purpose, the resolution of such data is limited by the comparably low mutations rates, around 10^{-8} per site per cell division [16], implying that even genome-wide sequencing with high coverage—and the associated high effort and cost—will capture only a very small number of somatic mutations [6,18]. A possibly cost-saving alternative is to focus on "hot-spot" SNPs; however, this results in a panel of SNPs partially driven by selective pressure, thus invalidating the assumption of independence of mutations and skewing the results [7]. This is particularly relevant in the case of lineage-tracing of cancer, which is one of the major use-cases [5].

To overcome some of these problems, we have developed an assay for reconstructing lineage based on sequencing particular regions of the genome known as short tandem repeats (STRs, also known as microsatellites) [1,12,14]. These are low-entropy regions, where a short (1–6 bp) motif of nucleotides, repeats itself in direct consecution, with numerous (≥ 5) copies of the repeat unit following each other. Due to the low entropy, during DNA duplication, the strands may detach and reattach at an offset of one or several repeat units, leading to a so-called stutter mutation. As a consequence the number of repeat units, also referred to as the length of the STR, increases or decreases. The rate for these mutations is several orders of magnitude higher than that of random single point mutations: estimates range from 10^{-3}–10^{-5} per site per cell division for di-repeats [17]. In addition, STRs are considered evolutionary neutral [3]. The extremely rapid accumulation of differences in repeat length makes STRs, in particular those with very short units, excellent candidates for tracing single cell lineage. The high mutability, however, also creates difficulties for the analysis of such data:

(1) The STR length observable in sequencing data differ from the *in vivo* state because the unavoidable amplification steps during library preparation introduce additional stutter mutations.

(2) Mutation rates differ substantially between loci and cell types.

The first point was addressed in a previous work [12]; our goal is to devise a continuous-time Markov model describing *in vivo* temporal evolution of STR length, focusing on second point. This model may be then used as a fast molecular clock, enabling estimations of distance in cell-divisions between single-cell samples. This distance can then be used for reconstructing lineage trees of these cells; the Markov model can also be used directly to asses and maximize the likelihood of the tree topologies. Such trees, which, in addition to a topological structure, also have reliable edge lengths, contribute toward a better understanding of the evolution of the analysed tissues.

Our aim is to allow for different rates of evolution at each locus. The size of our panel, consisting of thousands or tens of thousands of loci, however, makes it infeasible to optimise all parameters simultaneously. Therefore, we proceed in two steps. First, we estimate a separate base rate for the evolution of STR length at each locus, and then we aggregate across groups of loci of the same repeat unit motif. Then we iteratively repeat these two steps until the estimated parameters converge. We use eight artificial lineage trees constructed from four different cell-lines to estimate the evolutionary dynamics of STR evolution. Although there is coherence between data from the cell types, we also observe significant variations of the model parameters.

2   STR Data

STR data are produced by a specialized assay comprising the following steps:

(1) Following DNA extraction from the sample, whole genome amplification is performed. Different protocols were used for different datasets: the Repli-G protocol in the DU145 dataset, the Ampli1 protocol [2] in the HESC dataset, and an in-house phi29-based protocol [15] in the two HCT116 datasets.
(2) The amplified DNA is hybridized with a panel of duplex molecular inversion probes (MIP) that target genomic regions known to harbour a STR with unique flanking regions. For details about this panel see [14]. In the HESC dataset and the two HCT116 datasets, OM9 was used. In the DU145 dataset, OM6 was used.
(3) The selected DNA is then gap-filled, ligated, prepared as standard Illumina libraries and sequenced [14].
(4) The forward and reverse reads are merged using PEAR [19].
(5) The paired-end reads are aligned to a custom index comprising the same loci as the hybridization panel. For each locus, target sequences with different number of repetitive units are enclosed between unalterered flanking sequences. Each mapped read therefore specifies a locus and STR length.
(6) For each STR locus, a histogram of observed lengths is determined.

These STR length measurements cannot be used without corrections because the *in vitro* amplification steps during library preparation introduce further stutter mutations. The raw STR length data therefore follow a characteristic distribution. In previous work, a collection of reference distributions for the same panel of loci was generated for various lengths of the original alleles [12]. The distributions obtained from the sequencing data are then compared to the reference distributions taking 1 minus correlation as distance metric. In order to find the best match, we use k-d tree search [4]. A further difficulty arises for the fact that the cells of interest are derived of a diploid genome and hence for each locus there are (usually) two alleles. We therefore estimate a superposition of two reference distributions. The result of this procedure is the best estimate of the true *in vivo* STR length for each allele. In typical samples we assay 3000–8000 loci per sample, translating to 5000–10000 alleles.

Table 1. Overview of STR datasets used for model inference.

Tree	# of leaves	# of unique cells	Max. depth of leaf
DU145_A4	671	138	8
DU145_A4_deep	405	84	4
DU145_A4_rest	266	54	7
HCT116_MSI	80	80	3
HCT116_MSS	92	92	3
WIS_A8	90	35	1
WIS_D1	523	438	11
WIS_D11	33	15	1

In order to derive a model for converting STR length data into evolutionary distances we use a collection of datasets that were specifically prepared to investigate the use of STR length data. In brief, a culture is grown starting from a single cell. At predefined time points, single cells are picked from this culture. Some of these cells extracted and STR lengths data are measures, while other cells form the seed for new cultures. This process is then repeated for several generations to produce an artificial lineage tree. See [1] for more details about the preparation. For the present study we use the following eight datasets:

- a tree origined from a single DU145 cell (prostate cancer, displaying microsatellite instablity) [1] (DU145_A4) as well as deep sub-clade of this tree (DU145_A4_deep) and the complement of this sub-clade (DU145_A4_rest)
- a tree origined from a single HCT116 cell (colon cancer, displaying microsatellite instablity) [15] (HCT116_MSI)
- a tree origined from a single HCT116 cell with a functional version of the MMR gene hMLH1 reintroduced [8,15] (HCT116_MSS)
- three trees origined from three related (same ampule) single HESC cells (first published here) (WIS_A8, WIS_D1, WIS_D11)

Sizes of the datasets are detailed in Table 1. For a sketch of the structure of the trees, see Fig. 5.

3 Model of STR Length Evolution

We model the temporal evolution of STR length as a continuous-time Markov chain. A practical difficulty is that STR lengths show large variations across loci and hence we need to consider a fairly large number of states. For the data described in the previous section we consider $5 \leq j, k \leq 38$, i.e., lengths between 5 and 38. To avoid complicated treatment of "partial" repeat units, we only count complete repeat units and therefore take the STR length to be an integer. In order to reduce the number of parameters that need to be estimated independently for each locus we assume that we can decompose the rate matrix for a given locus ℓ and repeat unit motif τ in the following simple product form:

$$\mathbf{R}(\ell, \tau) = \mu(\ell)\mathbf{R}(\tau) \tag{1}$$

Here $\mu(\ell)$ is a single mutation rate specific for each locus ℓ. In contrast, $\mathbf{R}(\tau)$ is a matrix common to all loci of the same motif that describe the *relative* rates of length changes. Note that we allow both $\mu(\ell)$ and $\mathbf{R}(\tau)$ to depend on the cell-line. These parameters need to be therefore inferred independently for each dataset. We further constrain the model by two additional plausible assumptions:

(1) The Markov chain is reversible and hence has a stationary distribution;
(2) This stationary distribution coincides with the empirical distribution of STR lengths in the human genome.

The first assumption is clearly an approximation since STR that become too short will not behave like STRs any more. Since the data we actually measure are far away from this limit, it is nevertheless a safe assumption to make in our setting. The short time-scale of stutter mutations strongly suggests that their length distributions are equilibrated in the human genome.

In order to reduce the number of parameters that need to be estimated we use a simple model for the relative rates $\mathbf{R}_{jk}$ of a transition from length j to k, and assume that these values depend only on the length of the STR and the change in length. We first construct a symmetric, doubly-stochastic rate matrix as following:

$$\tilde{\mathbf{R}}_{jk} := \begin{cases} \exp(\gamma + \alpha(j + k) - \lambda|j - k|), & j \neq k \\ -\sum_{i \neq j} \tilde{\mathbf{R}}_{ji}, & j = k \end{cases} \tag{2}$$

The parameter γ is a scaling parameter. Such a symmetric rate matrix will lead to a uniform stationary distribution. It can be adjusted to enforce a given empirical stationary distribution by setting

$$\mathbf{R}_{jk}(\tau) := s_j^{-1}\tilde{\mathbf{R}}_{jk}(\tau) \tag{3}$$

where $s_j(\tau)$ is the proportion of STRs of repeat unit motif τ of length j in the genome. A short derivation of this correction can be found in Sect. 5. The empirical distributions for the human genome are shown in Fig. 1.

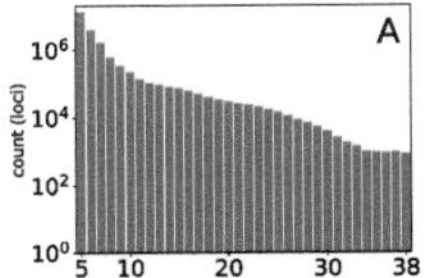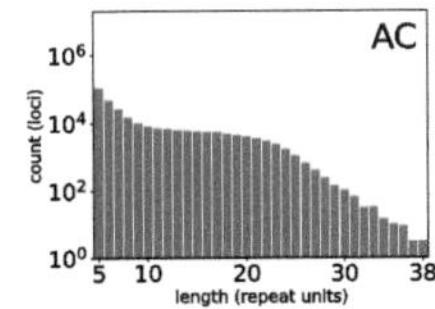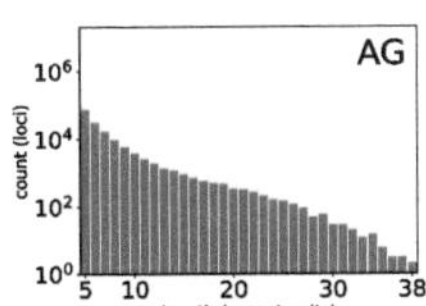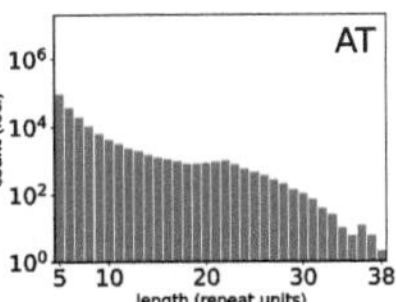

Fig. 1. Stationary distribution of STR length for the repeat unit motifs $\tau \in \{\texttt{A}, \texttt{AC}, \texttt{AG}, \texttt{AT}\}$ in the reference genome hg38.

In order to efficiently fit the model to the observed STR count data for a given sample, we proceed iteratively. We start with a uniform matrix of transition rates $\mathbf{R} = \mathbf{J} - N\mathbf{I}$, where $\mathbf{I}$ is the identity matrix, $\mathbf{J}$ is the matrix with entries 1, and N is the size of the matrices (in our case 34). For this transition rate matrix, we have the following closed-form solution for the probabilities $\mathbf{S}_{jk}$ that j is substituted by k after time t:

$$\mathbf{S}_{kk}(t) = \mathbf{S}_{=} = \frac{1}{N}\left(1 + (N-1)\exp(-Nt)\right)$$
$$\mathbf{S}_{jk}(t) = \mathbf{S}_{\neq} = \frac{1}{N}\left(1 - \exp(-Nt)\right) \quad \text{if } j \neq k \tag{4}$$

We aim to use the maximum-likelihood estimator to obtain $\mu(\ell)$ for each locus. To this end, we compare the STRs in each pair of leaves u and v in the known trees. We write $d(u,v)$ for their divergence in the tree measured as the number of generations, more precisely, passages of the cultures separating the two cell colonies, plus 2 for the divergence of each sample from the founder of its colony. For two cells within the same colony, this number is still taken as 2; a short derivation of this can be found in Appendix B. Now we substitute the time as $t = \mu(\ell)d(u,v)$ in the above solution. Moreover, we write $n_{=}(\ell, u, v)$ and $n_{\neq}(\ell, u, v)$ for the number of equal and distinct alleles between the two samples, respectively. Note that $n_{=}(\ell, u, v) + n_{\neq}(\ell, u, v) = 2$ if u and v have two alleles each for locus ℓ, and $n_{=}(\ell, u, v) + n_{\neq}(\ell, u, v) = 1$ otherwise. The likelihood $\mathcal{L}(\ell)$ of observing a set pairs of STRs at given locus ℓ thus takes the form

$$\log \mathcal{L}(\ell) = \sum_{u,v} n_{=}(\ell, u, v) \log \mathbf{S}_{=}(\mu(\ell)d(u,v)) + n_{\neq}(\ell, u, v) \log \mathbf{S}_{\neq}(\mu(\ell)d(u,v)) \tag{5}$$

A Newton-Rhapston search is sufficient to determine $\max \log \mathcal{L}$ as a function of μ.

Given the estimated values of $\mu(\ell)$, we proceed by refining the model for the rate matrices $\mathbf{R}(\tau)$. We estimate a single matrix for all the loci of each repeat unit motif $\ell \in L(\tau)$. Again this is done for each tree separately. For a given locus ℓ and two samples u, v, we count the number $a_{j,k}(\ell, u, v)$ of transitions $j \to k$ from allele length j in u to allele length k in v, as follows:

– If both u and v have a single allele, we count the single transition $u_1 \to v_1$.

- If both u and v have two distinct alleles, we order them as $u_1 < u_2, v_1 < v_2$, and count two transitions $u_1 \to v_1$ and $u_2 \to v_2$.
- If u has two alleles and v has one, we count a single transition of the allele of u closer to the one in v, i.e. denoting $i = minarg\{|u_2 - v_1|, |u_1 - v_1|\}$ we count $u_i \to v_1$.
- If u has one allele and v has two, we count a single transition of the allele of u to the one in v closer to it, i.e. denoting $i = minarg\{|u_1 - v_1|, |u_1 - v_2|\}$ we count $u_1 \to v_i$.

The likelihood function for a motif then takes the form

$$\log \mathcal{L}(\tau; \gamma, \alpha, \lambda) = \sum_{u,v} \sum_{\ell \in L(\tau)} \sum_{j,k} a_{j,k}(\ell, u, v) \log \mathbf{S}_{jk}(\mu(\ell)d(u,v)) \tag{6}$$
$$\text{with } \mathbf{S}(t) = \exp(t\mathbf{R}(\gamma, \alpha, \lambda))$$

For each value of the three parameters γ, α, and λ we can calculate the rate matrix, and from it the transition probability matrices for the required time points, in order to evaluate $\log \mathcal{L}$. Optimization of $\log \mathcal{L}$ over the parameter space is done using L-BFGS-B [20]; finding the maximal log-likelihood value, we obtain the optimal parameters and our estimation of the rate matrices $\mathbf{R}(\tau)$.

We now proceed to re-estimate the rate coefficients of $\mu(\ell)$ based on the newly estimated rate matrices. The expression for the likelihood

$$\log \mathcal{L}(\ell) = \sum_{u,v} \sum_{j,k} a_{j,k}(\ell, u, v) \log \mathbf{S}_{jk}(\mu(\ell)d(u,v)) \tag{7}$$
$$\text{with } \mathbf{S}(t) = \exp(t\mathbf{R}(\tau(\ell)))$$

is somewhat more involved than in (5), but still analytical and therefore a Newton-Rhapston search can again be used. After re-estimating μ, the rate matrices $\mathbf{R}(\tau)$ are re-estimated as well using (6) again. We observe that a third iteration of estimation of μ leads to no further noticable changes.

4   Results

In the top row of Fig. 2 we can see a density plot of the distributions of the locus-specific rates μ, for the eight datasets listed in Sect. 2. They are grouped by cell-line: the three DU145 trees (one tree and two subtrees), the two HCT116 trees, and the three HESC trees. We note that all curves have a similar form, suggesting some underlying common distribution, with varying parameters. We also note some skews of the center of this distribution, corresponding to a difference in the base rate of mutation.

For a more detailed comparison of the rates we computed linear regressions of shared loci for each pair of trees. We begin by taking the correlations (R-values) as a measure of similarity of the distributions (Fig. 3, left panel). Comparing rate estimations for trees of different cell-lines, we observe a substantial difference, reflected through low correlations. Comparing the rate estimates for the same cell-lines, however, we observe very high correlation for each pair DU145 trees and each pair of HESC trees, respectively. The locus-specific rates for two HCT116 trees exhibit a lower correlation than the DU145 and HESC data. This is consistent with the fact that these are two related cell-lines, but with a major genetic modification concerning mismatch repair (MMR) [8]. However the R-values are still higher than the R-values computed by comparing completely different cell-lines. This suggests that the common origin of the two variants of HCT116 still plays a significant role.

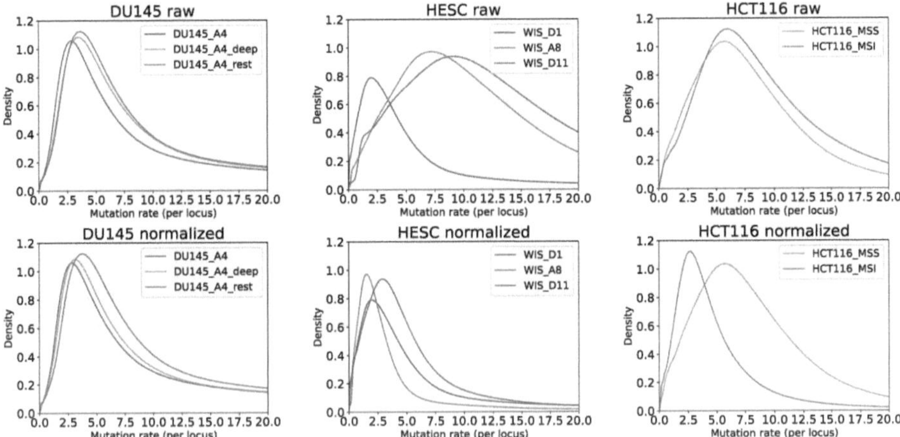

Fig. 2. Distribution of locus-specific rates μ estimated independently for the eight datasets, and grouped by cell-line: the three DU145 trees are subsets of the same tree; the three HESC trees are separately generated; the two HCT116 trees are grouped together, despite a genetic modification in the cell-line seeding HCT116-MSS. See also 1. Top row are the raw coefficients; bottom row they are scaled by the slope coefficients of linear regression within each group.

Using the slope of the linear regression as an estimation for overall scaling of rate (Fig. 3, right panel), we continue to the scaling of the rates estimated for trees of the same cell-line. We note that the rates for the DU145 trees have very little scaling, as would be expected since the three trees are essentially subsets of the same tree. A 2.2-fold scaling of the rates of HCT116_MSI as compared to the HCT116_MSS tree is also as expected, since the variants displaying microsatellite instability should indeed be more quickly mutating. The large factor between one of the HESC trees (WIS_D1) and the two others (WIS_A8, WIS_D11), of respectively 4.7 and 3.1, is somewhat surprising, but might be an artifact arising from the shallowness of the two latter trees, which allows proper calibration of only the faster-evolving loci. This might also explain the lower correlation compared to DU145.

Scaling the locus rate distributions by the these slopes, we obtain the bottom row of Fig. 2. For DU145 there is no large difference, but they are more closely aligned. For HESC this successfully corrects the shift seen in the top row, aligning the peaks of the density plot, and we can see the close relation of these distributions, despite the scaling on both axes. For HCT116 the peaks are not aligned, but the center of the distribution is. This highlights their lower correlation.

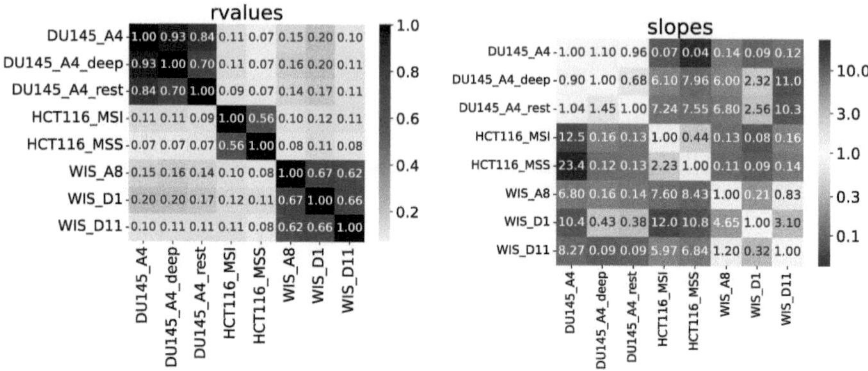

Fig. 3. Linear regressions of the locus-specific rate parameters $\mu(\ell)$ between pairs of samples: R-values (left) and estimated slopes (right). The slopes are column against row, so that a higher-than-1 slope means the tree of the column mutates more quickly than that of the row.

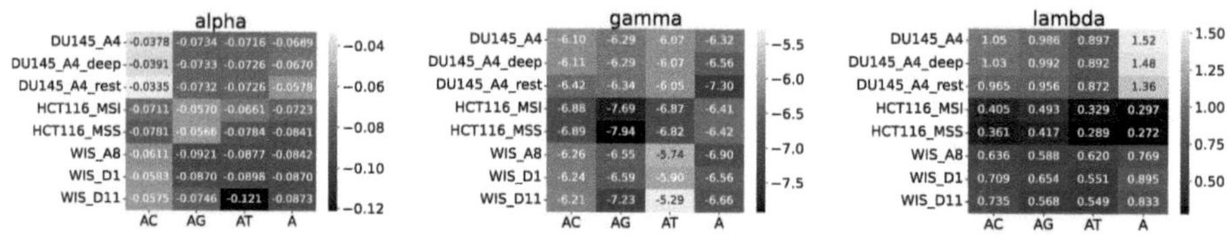

Fig. 4. Optimised rate-matrix model parameters, estimated separately for each tree and repeat unit motif.

Let us now turn to the motif-specific parameters γ, α, and λ, as seen in Fig. 4. These were independently estimated for each repeat unit motif, and we considered only data for the four motifs $\tau \in \{A, AC, AG, AT\}$. Although the parameters have a straightforward interpretation in in the symmetric rate model $\tilde{R}$, they are confounded by the stationary length distributions in Fig. 1. Even though the negative values of α appear to suggest that the mutation rate decreases with repeat length, we observe that R in fact contains higher values for large lengths.

Again, we see a similar pattern of coherence between cell types. The parameters for each given τ are very close among the DU145 trees, among the HESC trees, and also between the two HCT116 trees. Remarkably, the parameters for the `HCT116` trees are significantly more similar to each other than the locus rates are. This suggests that the common origin indeed plays a role in the length transition rates, and that the modification in MMR gene hMLH1 might affect certain loci more than others, but not so much the per-motif transition rates.

5 Concluding Remarks

The main application for estimating the rate models described in the previous section is retrospective single-cell lineage reconstruction, in particular of sample sets consisting of various healthy cells and cancer cells of various stages. The variations in mutational behaviour among cell-lines, in particular the low correlation of locus specific mutation rates, however, suggest that cells of different differentiation states might also vary in their mutational behaviour. This presents a problem when attempting to reconstruct a tree composed of heterogenous samples, which would require a unified model of mutation.

The datasets we analysed in this contribution provide us with limited insight as to the factors governing this difference in locus specific rates and of length transition rates. This is due to the fact, that they all belong to different individuals as well as different cell types. In order to further investigate those factors, we will require datasets controlling for these various factors—trees originating from the same cell type in different individuals, and trees originating from various cell types within a single individual. Such datasets may allow us to determine some commonalities and possibly develop a collection of models suitable for reconstructing lineage of various datasets. This will require in addition the development of a framework for estimating the likelihood of topologies which include shifts between mutational models.

In addition to the lineage reconstruction concerns, a better understanding of the factors determining the mutational behaviour of STRs might be of interest as of itself. Microsatellite instability (a marked increase in the mutation rate of STRs) is a well-studied phenomenon in cancer research, and further results of our method might contribute to this research.

Acknowledgments. A.O. and research in the Klein lab was supported by the Deutsche Forschungsgemeinschaft (DFG) (TRR-305/A01). Research in the Shapiro lab was supported by the European Union grants ERC-2014-AdG (project no. 670535) and EU-H2020-Health (project no. 874606). Research in the Stadler lab is supported by the German Federal Ministry of Research, Technology and Space (BMFTR) through DAAD project 57616814 (SECAI, School of Embedded Composite AI), the German Network for Bioinformatics Infrastructure, (de.NBI/RBC, grant W-de.NBI-018), and jointly be the BMFTR and the Sächsische Staatsministerium für Wissenschaft, Kultur und Tourismus in the programme Center of Excellence for AI-research *Center for Scalable Data Analytics and Artificial Intelligence Dresden/Leipzig*, project identification number: SCADS24B.

Author contributions. Conceptualization, A.O. and P.F.S.; formal analysis and software, A.O.; wet-lab methodology, single-cell isolation, and sample preparation,

T.M., L.T. and T.B.; funding acquistion, E.S. and C.A.K.; wet-lab supervision, E.S.; supervision, C.A.K and P.F.S.

Data Availability Statement. Data for the DU145 dataset is available as supplementary material for [14]. Additional data for that dataset, as well as data for the HESC dataset, will be gladly made available upon request by the corresponding author. Regarding the HCT116 datasets please contact the authors of [15] until that manuscript is published.

Appendix A

The following result seems to be well known although we are not aware of a convenient reference. We therefore include a short proof for completeness. Denote by $\mathbf{1}$ the vector with all entries 1 and write $\mathbf{0}$ for the zero vector (Appendix B). Note that any rate matrix $\tilde{\mathbf{R}}$ satisfies $\tilde{\mathbf{R}}\mathbf{1} = \mathbf{0}$ since $\mathbf{R}_{jj} = -\sum_{k \neq j} \mathbf{R}_{jk}$ by definition. Moreover, for a row vector $\mathbf{p}$, denote by $diag\,(\mathbf{p})^{-1}$ the diagonal matrix with entries $\mathbf{p}_j^{-1}$. One easily checks that $\mathbf{p}\,diag\,(\mathbf{p})^{-1} = \mathbf{1}$.

Lemma 1. *Let $\tilde{\mathbf{R}}$ be a symmetric rate matrix satisfying $\tilde{\mathbf{R}}\mathbf{1} = \mathbf{0}$, let $\mathbf{p}$ be a strictly positive (row) vector and set $\mathbf{R} := diag\,(\mathbf{p})^{-1}\tilde{\mathbf{R}}$. Then $\mathbf{p}$ is a (left) eigenvector of $\exp(t\mathbf{R})$ with eigenvalue $\mathbf{1}$ for all t.*
Proof. Symmetry of $\tilde{\mathbf{R}}$ implies $\mathbf{1}\tilde{\mathbf{R}} = \mathbf{0}$. The definition of $\mathbf{R}$ yields $\mathbf{p}\mathbf{R} = \mathbf{p}\,diag\,(\mathbf{p})^{-1}\tilde{\mathbf{R}} = \mathbf{1}\tilde{\mathbf{R}} = \mathbf{0}$. We compute

$$\mathbf{p}\exp(t\mathbf{R}) = \mathbf{p}\sum_{k=0}^{\infty}\frac{t^k}{k!}\mathbf{R}^k = \mathbf{p}\mathbf{I} + \underbrace{\mathbf{p}\mathbf{R}}_{\mathbf{0}}\sum_{k=1}^{\infty}\frac{t^k}{k!}\mathbf{R}^{k-1} = \mathbf{p}$$

i.e., $\mathbf{p}$ is indeed an eigenvector of $\exp(t\mathbf{R})$ with eigenvalue 1.   □

In particular, therefore, if $\mathbf{p}$ is a strictly positive probability distribution, it is a stationary distribution of the continuous time Markov chain with rate matrix $\mathbf{R} = diag\,(\mathbf{p})^{-1}\tilde{\mathbf{R}}$, where $\tilde{\mathbf{R}}$ is any symmetric rate matrix.

Appendix B

The expected cell division distance of two randomly selected cells in a culture can be estimated by assuming a constant duplication rate. Recall that each culture is started from a single cell in the experiments described in Sect. 2 (STR Data). Assuming all divisions are symmetric, the culture forms in the n-th generation (considering the seed cell a zeroth generation) a complete binary tree with $2^{n+1} - 1$ cells of which 2^n are leaves. In this case a given extant cell (leaf of the tree) has 2^h other extant cells within its clade of height h, and thus 2^{h-1} cells of distance exactly $2h$. Thus, denoting the depth of the entire colony as n, the sum of all distances is: $2\sum_{h=1}^{n} h2^{h-1} = (n-1)2^{n+1} + 2$, and hence the expected distance (divided by the number of other leaves, $2^n - 1$) is $2(n-1) + O(n2^{-n})$. Since we measure time in terms of passages of the culture, not replications, we obtain an average distance of $2\left(1 - \frac{1}{n}\right)$ up a finite size correction of order 2^{-n}. For estimated n values of 10–30 this is close enough to be taken as 2.

Appendix C

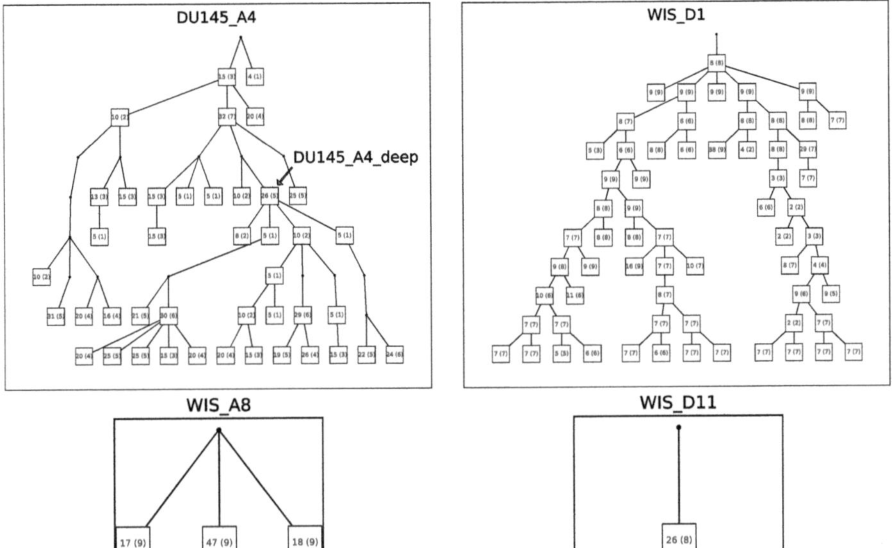

Fig. 5. Sketches of DU145 and HESC trees. Nodes represent colonies. The number of extracted samples is annotated in each node. The value in brackets gives the number of unique extracted cells. For DU145, the entire tree is DU145_A4, the marked clade is DU145_A4_deep, and the tree excluding that clade is DU145_A4_rest. See also Table 1 for an overview of the trees.

References

1. Biezuner, T., et al.: A generic, cost-effective, and scalable cell lineage analysis platform. Genome Res. **26**(11), 1588–1599 (2016). https://doi.org/10.1101/gr.202903. 115
2. Czyż, Z.T., Klein, C.A.: Deterministic whole-genome amplification of single cells. Methods Molecular Biol. (Clifton, N.J.) **1347**, 69–86 (2015). https://doi.org/10. 1007/978-1-4939-2990-0_5
3. Ellegren, H.: Microsatellites: simple sequences with complex evolution. Nat. Rev. Genetics **5**(6), 435–445 (2004). https://doi.org/10.1038/nrg1348
4. Freidman, J.H., Bentley, J.L., Finkel, R.A.: An algorithm for finding best matches in logarithmic expected time. ACM Trans. Math. Softw. **3**(3), 209–226 (1977). https://doi.org/10.1145/355744.355745
5. Hess, J.M., et al.: Passenger hotspot mutations in cancer. Cancer Cell **36**(3), 288–301.e14 (2019). https://doi.org/10.1016/j.ccell.2019.08.002
6. Hou, Y., et al.: Single-cell exome sequencing and monoclonal evolution of a JAK2-negative myeloproliferative neoplasm. Cell **148**(5), 873–885 (2012). https://doi. org/10.1016/j.cell.2012.02.028

7. Johnson, P.L.F., Hellmann, I.: Mutation rate distribution inferred from coincident snps and coincident substitutions. Genome Biol. Evolut. **3**, 842–850 (2011). https://doi.org/10.1093/gbe/evr044

8. Koi, M., et al.: Human chromosome 3 corrects mismatch repair deficiency and microsatellite instability and reduces N-methyl-N'-nitro-N-nitrosoguanidine tolerance in colon tumor cells with homozygous hMLH1 mutation. Can. Res. **54**(16), 4308–12 (1994)

9. Kretzschmar, K., Watt, F.: Lineage tracing. Cell **148**(1), 33–45 (2012). https://doi.org/10.1016/j.cell.2012.01.002

10. Lu, R., Neff, N.F., Quake, S.R., Weissman, I.L.: Tracking single hematopoietic stem cells in vivo using high-throughput sequencing in conjunction with viral genetic barcoding. Nat. Biotechnol. **29**(10), 928–933 (2011). https://doi.org/10.1038/nbt.1977

11. McKenna, A., Findlay, G.M., Gagnon, J.A., Horwitz, M.S., Schier, A.F., Shendure, J.: Whole-organism lineage tracing by combinatorial and cumulative genome editing. Science **353**(6298), aaf7907 (2016). https://doi.org/10.1126/science.aaf7907, https://www.science.org/doi/abs/10.1126/science.aaf7907

12. Raz, O., et al.: Short tandem repeat stutter model inferred from direct measurement of in vitro stutter noise. Nucleic Acids Res. **47**(5), 2436–2445 (2019). https://doi.org/10.1093/nar/gky1318

13. Sulston, J., Schierenberg, E., White, J., Thomson, J.: The embryonic cell lineage of the nematode Caenorhabditis elegans. Developm. Biol. **100**(1), 64–119 (1983). https://doi.org/10.1016/0012-1606(83)90201-4, https://www.sciencedirect.com/science/article/pii/0012160683902014

14. Tao, L., et al.: Retrospective cell lineage reconstruction in humans by using short tandem repeats. Cell Rep. Methods **1**(3), 100054 (2021). https://doi.org/10.1016/j.crmeth.2021.100054, https://www.sciencedirect.com/science/article/pii/S2667237521001028

15. Tao, L., Silverbush, D., Suva, M.: Decoding glioma plasticity via single-cell multiomics (unpublished)

16. Wang, J., Fan, H., Behr, B., Quake, S.: Genome-wide single-cell analysis of recombination activity and de novo mutation rates in human sperm. Cell **150**(2), 402–412 (2012). https://doi.org/10.1016/j.cell.2012.06.030

17. Willems, T., Gymrek, M., Poznik, G., Tyler-Smith, C., Erlich, Y.: Population-scale sequencing data enable precise estimates of Y-STR mutation rates. Am. J. Human Genet. **98**(5), 919–933 (2016). https://doi.org/10.1016/j.ajhg.2016.04.001

18. Xu, X., et al.: Single-cell exome sequencing reveals single-nucleotide mutation characteristics of a kidney tumor. Cell **148**(5), 886–895 (2012). https://doi.org/10.1016/j.cell.2012.02.025

19. Zhang, J., Kobert, K., Flouri, T., Stamatakis, A.: PEAR: a fast and accurate illumina paired-end read merger. Bioinformatics **30**(5), 614–620 (2013). https://doi.org/10.1093/bioinformatics/btt593

20. Zhu, C., Byrd, R.H., Lu, P., Nocedal, J.: Algorithm 778: L-BFGS-B: Fortran subroutines for large-scale bound-constrained optimization. ACM Trans. Math. Softw. **23**(4), 550–560 (1997). https://doi.org/10.1145/279232.279236

Phylogenetics I: Inferring Gene and Species Histories

Theoretical and Empirical Performance of Pseudo-likelihood-Based Bayesian Inference of Species Trees Under the Multispecies Coalescent

Nicolae Sapoval[1] , Zejian Liu[2] , Mehrdad Tamiji[1] , Meng Li[2] ,
and Luay Nakhleh[1,3(✉)]

[1] Department of Computer Science, Rice University, Houston, TX 77005, USA
nakhleh@rice.edu
[2] Department of Statistics, Rice University, Houston, TX 77005, USA
[3] Department of BioSciences, Rice University, Houston, TX 77005, USA

Abstract. Likelihood-based inference under the multispecies coalescent provides accurate estimates of species trees. However, maximum likelihood and Bayesian inference are both computationally very demanding. Pseudo-likelihood has been previously proposed as a computationally efficient alternative to full phylogenetic likelihood calculations in the context of maximum likelihood estimation. However, theoretical and practical aspects of pseudo-likelihood in the context of Bayesian inference have not been explored. In this work, we provide strong theoretical guarantees and an empirical evaluation of pseudo-likelihood-based Bayesian inference of species trees under the multispecies coalescent. Our contributions are threefold. First, we prove a bound on the convergence rate for species tree topology inference and a Bernstein-von Mises result for branch lengths under model misspecification. Second, we provide an empirical comparison of full- and pseudo-likelihood-based Bayesian inference on synthetic data. Finally, we demonstrate the practical scalability of pseudo-likelihood-based inference by analyzing two biological datasets.

Keywords: phylogenomics · Bayesian phylogenetic inference · pseudo-likelihood · multispecies coalescent

1 Introduction

Phylogenomic inference of the evolutionary history of a set of species from genomic data has become commonplace due to its ability to incorporate genome-wide signals that help resolve evolutionary relationships that are not easy to elucidate otherwise [5, 7]. However, this inference is very challenging owing in large

N. Sapoval, Z. Liu, M. Tamiji, M. Li and L. Nakhleh—Contributed equally.

Supplementary Information The online version contains supplementary material available at https://doi.org/10.1007/978-3-032-26891-4_7.

part to phylogenetic discordance—the phenomenon where evolutionary histories of different genomic regions disagree with each other as well as with the evolutionary history of the populations from which the genomes are obtained. Biological processes that could cause this phenomenon include incomplete lineage sorting (ILS), gene duplication and loss, and horizontal gene transfer [18,22].

The multispecies coalescent (MSC) [6,28] has emerged as a powerful mathematical model for inference of species trees from multi-locus data while accounting for ILS. Furthermore, significant progress has been made in developing species tree inference methods under the MSC [19]. In practice, likelihood-based approaches such as maximum likelihood [14,26,39] and Bayesian inference [3,24] provide high accuracy. Furthermore, Bayesian inference allows for incorporating prior knowledge and quantifying uncertainty in the inference results—two advantages over maximum likelihood method and the (much faster) summary statistic methods [19].

While MSC-based likelihood calculations using bi-allelic marker data can be done efficiently [3], that is not the case when the data consists of gene tree estimates, which has limited the applicability of likelihood-based inferences to datasets whose sizes do not exceed tens of species and hundreds of loci [25]. Thus, significant efforts have been made to accelerate Bayesian inference methods by introducing new sampling proposals [8,42] or restricting the space of explored trees [35]. In all of these cases, the full likelihood calculations still remain a bottleneck, since every iteration of the Markov chain Monte Carlo (MCMC) sampling algorithm requires evaluation of the full likelihood ratio for the proposal. To address this challenge, pseudo-likelihood was introduced as a computationally efficient alternative to full likelihood computation in the context of maximum likelihood inference [17]. Furthermore, the authors proved that the maximum pseudo-likelihood estimate is statistically consistent under the MSC [17] model.

Given the added advantages of Bayesian inference (which include incorporating prior information, estimating parameters beyond the tree topology, and quantifying uncertainty), it is important to investigate the question of whether the use of pseudo-likelihood instead of full likelihood in a Bayesian setting under the MSC improves scalability while achieving theoretical guarantees on the posterior distribution. To our knowledge, this question has not been explored and this study aims to address it.

The contributions of this work are as follows. We first prove theoretical guarantees for frequentist species tree topology estimation under pseudo-likelihood, which refines the known result of [17] with improved utility for practitioners. Unlike the non-Bayesian focus on point estimation, Bayesian inference requires characterizing the entire posterior distribution, which must account for the model misspecification inherent in pseudo-likelihood methods. Here, we address this challenge from a theoretical perspective to ensure broad applicability. In particular, we show that, under mild conditions, the pseudo-likelihood-based posterior distribution of branch lengths converges to a normal distribution with particular mean and covariance parameters. This characterizes the behavior of pseudo-likelihood-based posterior distributions and is achieved by proving a so-called Bernstein-von Mises (BvM) result under model misspecification. Our BvM result yields insights into the coverage of credible intervals and the consistency of the posterior distribution. In addition to our general theory, we provide empirical comparisons between full- and pseudo-likelihood-based Bayesian inference of species trees. We compare the two approaches on simulated datasets in terms of the accuracy of topology estimation and agreement in the inferred posteriors and species tree distributions. We also explore the computational speedup offered by pseudo-likelihood based inference. Finally, we explore scalability of pseudo-likelihood-based Bayesian analyses to biological data from spiders [7] and turtles [5].

It is important to emphasize here that summary statistic methods, such as ASTRAL [20], are much more scalable, yet provide less information, than Bayesian or maximum likelihood methods. The goal of this work is not to provide a method that outperforms methods like ASTRAL in terms of scalability of species tree topology inference. Our work, instead, is aimed at adding important theoretical and empirical results to the literature on Bayesian inference of species trees under the multispecies coalescent and providing practitioners with a new tool for phylogenomic analyses.

2   Methods and Theoretical Results

2.1   Phylogenetic Pseudo-likelihood Function

For the remainder of the manuscript, we will use N to denote the number of taxa and M to denote the number of gene trees. Let S be a rooted species tree and let $RT^S = \left\{ RT_j^S | j = 1, ... \binom{N}{3} \right\}$ be the set of rooted triples in S (i.e., rooted 3-taxon subtrees of S; see Fig. 1). Given a rooted triple RT_j^S, we associate with it a corresponding internal branch length b_j, and we will call the set of all such internal branch lengths $\mathbf{B} = \left\{ b_j | j = 1, ... \binom{N}{3} \right\}$ (e.g., in Fig. 1, we have $b_1 = x$, $b_2 = x + y$, $b_3 = b_4 = y$).

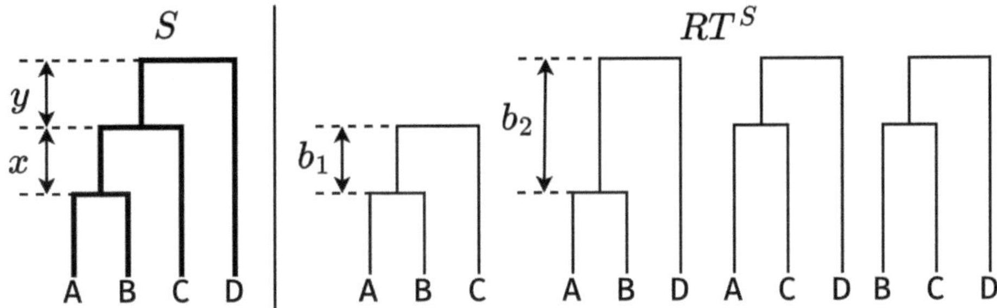

Fig. 1. A 4-taxon species tree S (left) and corresponding set of rooted triples RT^S (right). Each rooted triple has an associated internal branch length b_j.

If we consider an arbitrary rooted triple $RT_j^S = AB|C$ (i.e., A and B coalesce first; see leftmost 3-taxon tree in Fig. 1), then under the coalescent process the probability that $ab|c$ occurs in a gene tree generated from S is $1 - (2/3)e^{-b_j}$, and the probabilities of $ac|b$ and $bc|a$ are both equal to $(1/3)e^{-b_j}$. Let x_{j1}, x_{j2}, and x_{j3} denote the counts of triples $ab|c$, $ac|b$, and $bc|a$ observed in the gene trees (note that we have $x_{j1} + x_{j2} + x_{j3} = M$). Then it follows that the counts follow a multinomial distribution

$$f(x_{j1}, x_{j2}, x_{j3}|RT_j^S) = \frac{M!}{x_{j1}!x_{j2}!x_{j3}!}$$
$$\times (1 - (2/3)e^{-b_j})^{x_{j1}}((1/3)e^{-b_j})^{x_{j2}}((1/3)e^{-b_j})^{x_{j3}}.$$

Now, we can define the pseudo-likelihood of a species tree S given a collection of gene trees G as

$$\mathcal{L}(S|G) = \prod_{j=1}^{\binom{N}{3}} \frac{M!}{x_{j1}!x_{j2}!x_{j3}!} \tag{1}$$

$$\times \prod_{j=1}^{\binom{N}{3}} \left[(1 - (2/3)e^{-b_j})^{x_{j1}}((1/3)e^{-b_j})^{x_{j2}}((1/3)e^{-b_j})^{x_{j3}}\right]. \tag{2}$$

For example, consider the species tree S on four taxa and let $G = \{G_1, G_2, G_3\}$ be three gene tree topologies arising from S (Fig. 2).

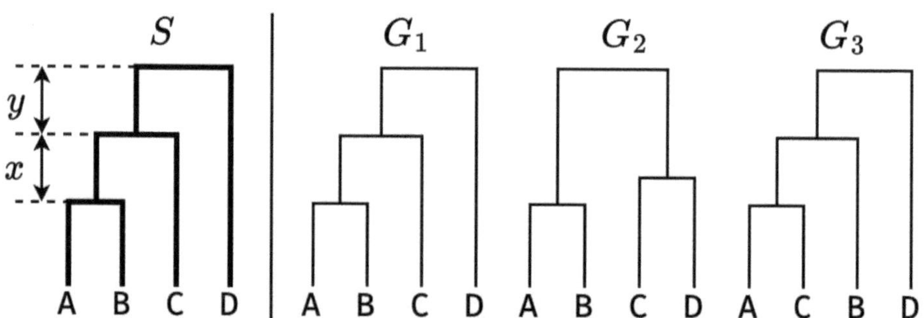

Fig. 2. A 4-taxon species tree S (left) and corresponding collection of gene tree topologies $G = \{G_1, G_2, G_3\}$ (right).

Then the full log-likelihood (ℓ_{full}) for observing G given S with branch lengths $x = 0.5$ and $y = 0.6$ is

$$\ell_{full} = \log\left[\frac{1}{3}e^{-y} - \frac{1}{6}e^{-(x+y)} - \frac{1}{18}e^{-(3y+x)}\right]$$

$$+ \log\left[1 - \frac{2}{3}(e^{-x} + e^{-y}) + \frac{1}{3}e^{-(x+y)} - \frac{1}{18}e^{-(3y+x)}\right]$$

$$+ \log\left[\frac{1}{3}e^{-x} - \frac{1}{3}e^{-(x+y)} + \frac{1}{18}e^{-(3y+x)}\right] \approx -6.35172.$$

In contrast, the log-pseudo-likelihood (ℓ_{pseudo}) in this case is equal to

$$\ell_{pseudo} = \log\left[3\left(1 - \frac{2}{3}e^{-x}\right)^2\left(\frac{1}{3}e^{-x}\right)\right] + \log\left[1 - \frac{2}{3}e^{-(x+y)}\right]^3$$

$$+ \log\left[3\left(1 - \frac{2}{3}e^{-y}\right)^2\left(\frac{1}{3}e^{-y}\right)\right]$$

$$+ \log\left[3\left(1 - \frac{2}{3}e^{-y}\right)^2\left(\frac{1}{3}e^{-y}\right)\right] \approx -5.311.$$

2.2 Pseudo-likelihood-Based Bayesian Inference

For the Bayesian inference under pseudo-likelihood we employ a standard Metropolis-Hastings algorithm where in the acceptance ratio calculation we substitute the value of the likelihood ratio of the proposal and current state by the corresponding pseudo-likelihood ratio. Note that since for a fixed collection of gene trees the value of $\prod_{j=1}^{\binom{N}{3}} \frac{M!}{x_{j1}!x_{j2}!x_{j3}!}$ in Eq. (1) is constant, it follows that the ratio can be simplified to

$$\frac{\tilde{\mathcal{L}}(S_{prop}|G)}{\tilde{\mathcal{L}}(S_t|G)} \frac{P(S_{prop})}{P(S_t)} \frac{Q(S_t|S_{prop})}{Q(S_{prop}|S_t)}, \tag{3}$$

where

$$\tilde{\mathcal{L}}(S|G) = \prod_{j=1}^{\binom{N}{3}}(1 - (2/3)e^{-b_j})^{x_{j1}}((1/3)e^{-b_j})^{x_{j2}}((1/3)e^{-b_j})^{x_{j3}},$$

$P(S)$ is the prior, and $Q(S|S_t)$ is the proposal probability. For the numerical stability of the calculation logarithms of all the ratios are used in the PhyloNet [33, 37] implementation.

2.3 Theoretical Results

Let the topology of a true species tree be S_0 and the corresponding set of branch lengths $\mathbf{B}_0$. We use $\hat{S}$ to denote the estimated species tree topology by maximizing the pseudo-likelihood function. For any set of branch lengths $\mathbf{B}$, we denote the minimum enclosed branch length as $\min \mathbf{B}$. Throughout this paper, the Euclidean norm is denoted by $\| \cdot \|$. Due to space limitations, we give the proofs in Appendix A.2.

We first refine the result of Liu *et al.* [17] by defining an explicit convergence rate for the topology estimation in terms of N, M, and a branch length parameter of the true species tree. This result holds in finite-sample settings, i.e., for any finite number of taxa N and number of gene trees M.

Theorem 1 (Convergence rate of topology estimation). *The probability that the tree topology $\hat{S}$ estimated from a set of gene trees G under the MSC, matches the true species tree topology S_0 is bounded by*

$$\mathbb{P}(\hat{S} = S_0) > 1 - \frac{25\binom{N}{3}}{36M\varepsilon^2},\tag{4}$$

where $0 < \varepsilon < (1 - e^{-\min \mathbf{B}_0})/2$.

Remark 1. Convergence rate of $\hat{S}$ has been studied in [17]. We obtain a refined error bound by revising the proof in (4) and additionally specify the required range of ε, as opposed to "a small positive real number" described in [17]. This explicit range in (4) leads to practical guidance on its optimal value for accurate topology estimation. For example, setting ε to its upper bound maximizes the error bound for the probability of $\hat{S}$ matching the true topology S_0. In particular, if we let p denote a target probability of this event that could be specified by users, then the minimum number of gene trees needed to achieve this probability is given by $M = \frac{25\binom{N}{3}}{36\bar{\varepsilon}^2(1-p)}$, where $\bar{\varepsilon} = (1 - e^{-\min \mathbf{B}_0})/2$. Also, for any N, letting $M \to \infty$ in (4) trivially leads to statistical consistency of $\hat{S}$, i.e., $\mathbb{P}(\hat{S} = S_0)$ converges to 1.

We demonstrate the application of the formula in Remark 1 through two examples. First, we fix the number of taxa N and illustrate the relationship between the probability of correct identification p and the necessary number of gene trees M. We select four different numbers of taxa, $N = 5, 10, 20, 50$, and set $\bar{\varepsilon} = 0.3$. The resulting curves of p against M for each N are depicted in Fig. 3.

We then fix the target probability p and visualize the relationship between the number of taxa N and the number of gene trees M. We choose four different target probabilities $p = 0.5, 0.8, 0.9, 0.95$, and fix the value of $\bar{\varepsilon} = 0.3$. The curves of N against M for each p are depicted in Fig. 4.

The finite-sample result in Theorem 1 concerns point estimation. Bayesian implementation based on the pseudo-likelihood would require the prior specification and posterior computation. Our next theorem studies the behavior of the pseudo-likelihood-based posterior distribution of branch lengths. Since the species tree topology can be estimated accurately with high probability, as established in Theorem 1, and our numerical experiments in Sect. 3 confirm the accuracy of this estimation using Bayesian methods, we fix the topology at S_0. This choice also simplifies interpretation, as branch lengths are closely linked

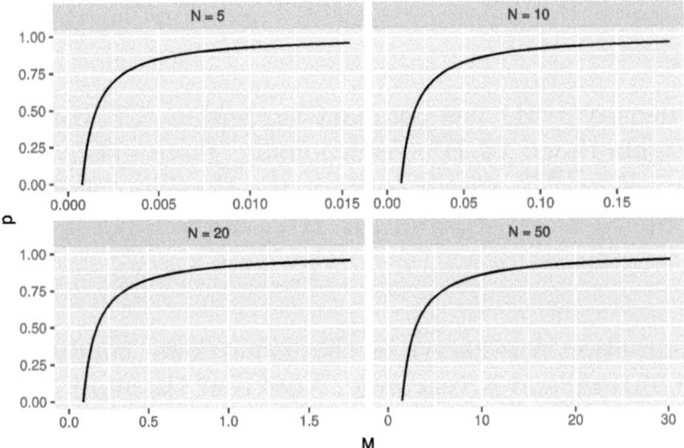

Fig. 3. Probability of correct identification p against the number of gene trees M ($\times 10^5$) for different numbers of taxa N and $\bar{\varepsilon} = 0.3$.

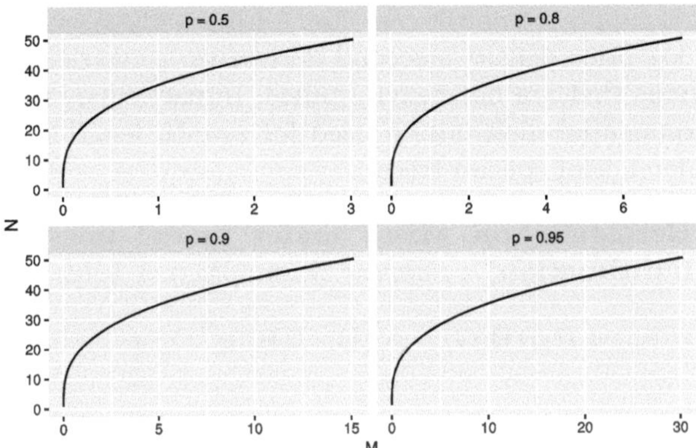

Fig. 4. The number of taxa N against the number of gene trees M ($\times 10^5$) for different probabilities of correct identification p for $\bar{\varepsilon} = 0.3$.

to the topology. Section 3 also presents empirical results that account for the uncertainty in inferring S_0.

Let $\Pi_M(\mathbf{B}|G)$ denote the posterior probability distribution of branch lengths given a collection of gene trees G. Our next main result shows that it converges to a normal distribution in total variation distance. Specifically, let $\mathbb{P}_0$ denote the data generating measure, p_0 denote its associated density relative to the Lebesgue measure, and the density according to the pseudo-likelihood as $p_{\mathbf{B}}$. Let $\mathbf{B}^* = (B_j^*)_{j=1}^{\binom{N}{3}}$ be the branch lengths that minimize the Kullback-Leibler divergence $D_{\mathrm{KL}}(p_{\mathbf{B}} \parallel p_0)$, assumed to exist and be unique, which is a key quantity to describe the behavior of $\Pi_M(\mathbf{B}|G)$ due to model misspecification. Let $\pi(\mathbf{B})$ denote the prior distribution assigned to $\mathbf{B}$ with respect to the Lebesgue measure. We assume the following general conditions on the prior, covering a broad class of priors:

(A) The prior density $\pi(\mathbf{B})$ has compact support and is continuous and positive in a neighborhood of $\mathbf{B}^*$. Additionally, there exists a constant c such that $\|H(\mathbf{B})\|_2 \leq ce^{\|\mathbf{B}\|^2}$, where $H(\mathbf{B})$ is the Hessian matrix of $\log\pi(\mathbf{B})$ (i.e., its second-order derivatives) and $\|H(\mathbf{B})\|_2$ is the 2-norm (spectral norm) of $H(\mathbf{B})$.

Remark 2. Assumption A is known as the "prior mass" condition in the statistical Bayesian literature, and is very mild. Note that the support of a distribution refers to the set of all possible values where the probability density is non-zero. Many prior distributions exist to ensure that the density functions are continuous and positive at any point within its support, including $\mathbf{B}^*$. The "compact support" condition means that the support of the prior is finite; for each dimension of the prior, one can choose a large enough but still finite interval to truncate a prior distribution supported on the real line. Additionally, Assumption A imposes a limit on the log prior density's second derivative, a restriction easily met by many distributions that are not heavily tailed. In particular, these assumptions hold for the exponential distribution used for branch lengths in PhyloNet [37] for Bayesian inference of species trees.

Theorem 2 (Bernstein-von Mises under model misspecification for branch length estimation). *Suppose Assumption A holds. Then,*

$$\mathbb{E}_{\mathbb{P}_0} \| \, \Pi_M(\mathbf{B}|G) - \mathcal{N}(\hat{\mathbf{B}}, (I(\mathbf{B}^*))^{-1}) \, \|_{TV} \to 0,$$

where

$$\hat{\mathbf{B}} = \left(-(\log(3(1 - x_{j1}/M)/2))\right)_{j=1}^{\binom{N}{3}},$$

$$I(\mathbf{B}^*) = \mathrm{diag}\left(\frac{2}{3}Me^{-B_j^*}/\{1 - \frac{2}{3}e^{-B_j^*}\}\right),$$

and $\| \cdot \|_{TV}$ is the total variation distance used for comparing two probability distributions.

A few remarks are in order.

Remark 3. Theorem 2 is referred to as a Bernstein-von Mises (BvM) result in the theoretical Bayesian literature. When there is no model misspecification, classical BvM asserts that, under certain conditions, the posterior distribution of a parameter becomes approximately normal and centered at the maximum likelihood estimator as the sample size grows; see, for example, Chap. 10.2 of [34]. However, pseudo-likelihood-based posterior distribution inherently misspecifies the model by replacing the full likelihood with the pseudo-likelihood. As such, we utilize techniques established in [13] to prove the BvM theorem under model misspecification.

Remark 4. The classical BvM phenomenon under the correct model reassures that Bayesian credible intervals are also frequentist confidence intervals. However, this does not hold for the pseudo-likelihood-based posterior distribution due to model misspecification. Let $\dot{\ell}(\mathbf{B}^*)$ denote the score of the log-pseudo-likelihood function evaluated at $\mathbf{B}^*$. Then the maximum likelihood estimator $\hat{\mathbf{B}}_M$ for the misspecified model has an asymptotic Gaussian approximation centered at $\mathbf{B}^*$, with covariance matrix given by? $\Sigma\mathbf{B}^* I(\mathbf{B})^{-1}\mathbb{P}_0(\dot{\ell}(\mathbf{B})\dot{\ell}(\mathbf{B})^T)I(\mathbf{B})^{-1}$. The covariance matrix $I(\mathbf{B}^*)^{-1}$ appearing in the Bernstein-von Mises theorem is generally not equal to the asymptotic covariance matrix $\Sigma_{\mathbf{B}^*}$, implying that Bayesian credible sets might not align with confidence sets at the nominal level. Such deviation from classical BvM has been observed and discussed in approximate Bayesian computation contexts, such as variational Bayes [36], and semiparametric settings with high-dimensional nuisance parameters [15]. Our numerical experiments in Sect. 3 also confirm this in finite samples.

An implication of Theorem 2 is that the posterior distribution is consistent at $\mathbf{B}^*$, instead of $\mathbf{B}_0$.

Corollary 1 (Consistency of posterior). *Under the conditions of Theorem 2, then for any $\epsilon > 0$,*

$$\Pi_M(\|\mathbf{B} - \mathbf{B}^*\| > \epsilon|G) \to 0 \tag{5}$$

in $\mathbb{P}_0$-probability.

3   Empirical Results

3.1   Simulation and Evaluation Setup

We simulated species trees using the `SiPhyNetwork` package [11] under the birth-death model with birth-rate λ sampled uniformly in the $[0.2, 0.5]$ interval, death-rate μ sampled uniformly in the $[0.1, \lambda]$ interval, and hybridization rate set to 0. We simulated a total of 10 species trees for N equal to 5 and 10 taxa. For each species tree, gene trees were simulated using `ms` [9] via the `SimGTinNetwork` command in `PhyloNet` [33,38]. For each species tree, we simulated $M \in \{100, 500, 1000, 2500, 5000, 10000\}$ gene trees. For each gene tree, gene DNA sequences of length 1,000 were simulated using Seq-Gen [27] under the Jukes-Cantor model [10] and branch length scaling of 0.01.

We sampled species trees from the posterior distribution using `PhyloNet` `MCMC_GT` command [37] with total chain length of 1,000,000 steps, burn-in length of 100,000 steps, and sampling frequency of one sample per 1,000 steps. Sampling from the posterior distribution under pseudo-likelihood is now implemented in PhyloNet as part of the `MCMC_GT` command using an additional `-pseudo` flag.

For inference from sequence data, gene trees were first inferred for individual genes using IQ-TREE [23] with ModelFinder Plus [12]. The average normalized Robinson-Foulds [29] distance for inferred trees was 0.025 ± 0.077 for 5-species trees and 0.044 ± 0.07 for 10-species trees. Then, species tree inference was performed analogously to the simulated true gene trees scenario described above.

Given that the theoretical analysis indicates that the Bayesian credible sets might not align with the confidence sets at the nominal levels, we assessed the frequentist coverage of the 95% credible sets following the standard methodology [4]. Hence, for each species tree S, we generated 100 independent draws of $M \in \{100, 500, 1000, 2500, 5000, 10000\}$ gene trees. Then, the pseudo-likelihood MCMC simulation was performed on each of the sets of gene trees yielding 900 samples $\hat{S}_i$ from the posterior distribution. Resulting samples were summarized via the `summarize_splits_on_tree` function in `dendropy` (v5.0.1) [21]. For each branch in S, we checked if the branch length θ fell within the 95% highest posterior density (HPD) credibility interval of the sampled trees $\hat{S}_i$. Finally, we computed the coverage of the 95% credibility interval as the ratio of the branch length values within the intervals to the total number of branch lengths.

In order to assess possible discrepancies between the posterior distributions inferred by pseudo- and full-likelihood-based MCMC samplers, we conducted an additional set of runs for $N = 5$ and the set of simulated gene trees of total length 10,100,000 steps with 100,000 steps used for burn-in. For each of the species tree candidates sampled from the posterior, we recorded its posterior probability and weighted Robinson-Foulds distance [30] to the ground-truth species tree.

3.2   Results

Species tree topology estimation First, we evaluated pseudo- and full-likelihood-based inference on a set of simulated ground-truth gene trees. The percentage of MCMC samples that had topology matching that of the ground-truth species tree increased as more gene trees were considered (Fig. 5). The pseudo-likelihood-based runs had slightly worse performance for 100 and 500 gene tree settings, but matched the full likelihood results for 5,000 and more gene trees (Fig. 5).

Next, we evaluated performance of both methods in the presence of potential gene tree estimation errors. After simulating sequences for each gene, the gene trees were inferred and the methods were evaluated with the inferred gene trees as the input. Similarly to the simulated gene trees scenario, as the number of gene trees increased the percentage of sampled trees with correct topology also increased (Fig. 6). Analogously, pseudo-likelihood-based inference had higher variance in the percentage of trees with matching topology for 100 gene trees input (Fig. 6). Both pseudo- and full-likelihood-based inference had a drop in performance when run on inputs consisting of 100 gene trees.

Branch length credible intervals Remark 4 implies that the frequentist coverage of the Bayesian credible intervals for the branch length estimates can deviate from its nominal value. We observe that in the empirical setting, the coverage of the 95% Bayesian credibility sets is lower than 95% with median coverage ranging from 0.762 to 0.818 for the 5-species case (Fig. 7, left) and from 0.654 to 0.668 for the 10-species case (Fig. 7, right). We also note that in the 10-species case, there is a reduced variance in the coverage of the credibility intervals.

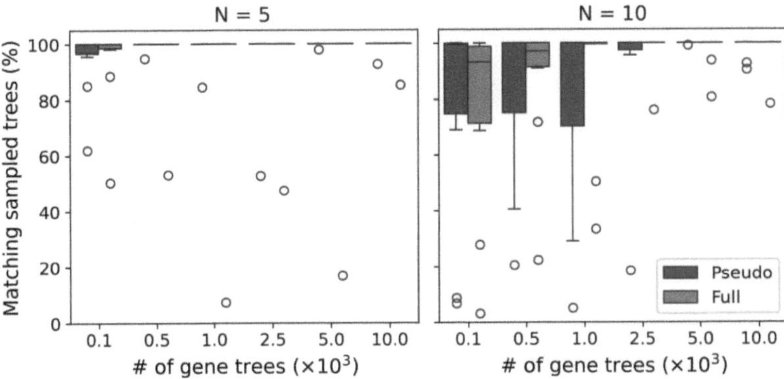

Fig. 5. Percentage of the species trees in the pseudo- and full-likelihood-based posterior distribution samples that have correct topology (y-axis) as a function of the number of gene trees (x-axis) and number of species (N). The input data consists of the true gene trees.

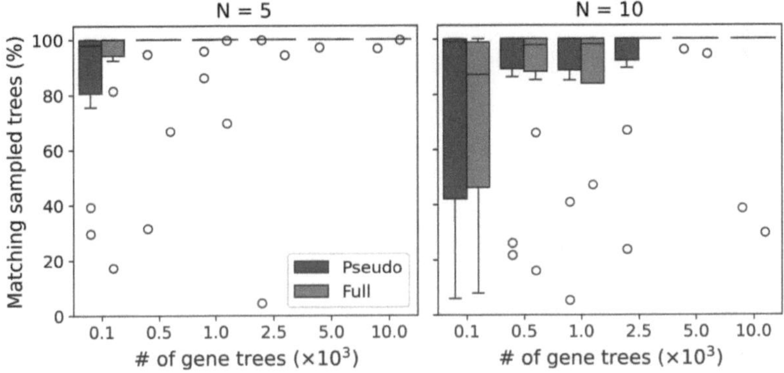

Fig. 6. Percentage of the species trees in the pseudo- and full-likelihood-based posterior distribution samples that have correct topology (y-axis) as a function of the number of gene trees (x-axis) and number of species (N). The input data consists of the estimated gene trees.

Posterior density estimation To further compare the behavior of pseudo- and full-likelihood-based Bayesian analysis, we compared the distributions of the log posteriors (Fig. 8). Since, the exact values of the log posteriors can differ between the pseudo- and full-likelihood-based analyses by more than an order of magnitude, we centered the distributions by subtracting their mean values.

We note that the centered distributions of log posteriors are similar for pseudo- and full-likelihood-based runs, with the L_1 norm of the histogram differences between the distributions ranging from 0.026 to 0.130 with median value of 0.061. We also compared distributions of weighted Robinson-Foulds distances for the inferred trees under pseudo- and full-likelihood-based inference (Supp.

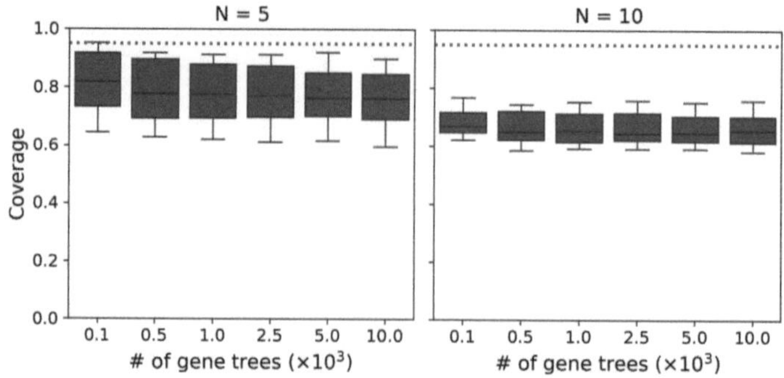

Fig. 7. Frequentist coverage of the 95% high posterior probability density intervals for the branch lengths of the species tree. The red dotted line indicates the nominal coverage. Box plots indicate coverage values across 10 replicate species trees. The input data consists of the true gene trees. (Color figure online)

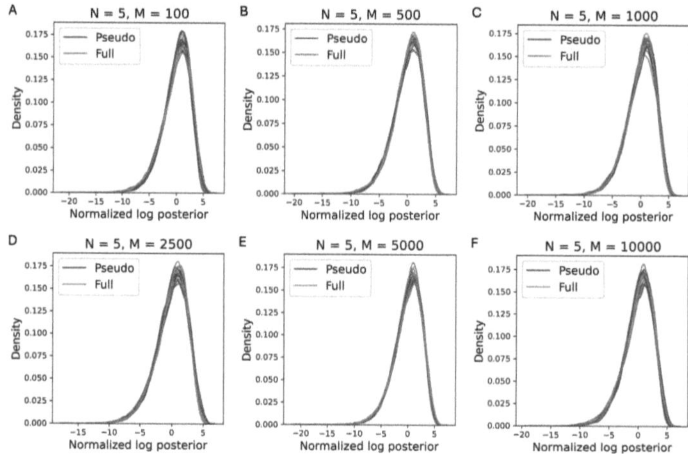

Fig. 8. Distribution of centered log posterior probabilities for pseudo- and full-likelihood-based MCMC runs. Each curve represents a single replicate species tree. The input data consists of the true gene trees.

Figure 13). Similarly to the distributions of the log posteriors these agree between the pseudo- and full-likelihood-based analyses.

Furthermore, we note that the acceptance rates of the pseudo- and full-likelihood-based MCMC runs are similar and confined to the $[0.4, 0.6]$ interval (Supp. Fig. 11, 12). As the number of gene trees increases, the acceptance rate decreases (Supp. Fig. 11, 12), matching the observation that as the number of gene trees increases the proportion of sampled topologies matching the true topology rapidly approaches 1 (Fig. 5, 6).

Computational performance We compared the CPU time required to complete 1,000,000 steps of the MCMC simulation under the pseudo- and full-likelihood functions. For both simulated and inferred gene trees, pseudo-likelihood-based MCMC outperforms full-likelihood-based computation by one to two orders of magnitude (Fig. 9A, B; Supp. Fig. 14). In particular, on the datasets with 10-species and > 1000 gene trees, pseudo-likelihood-based computation achieves more than 100x speedup compared to the full-likelihood-based analysis.

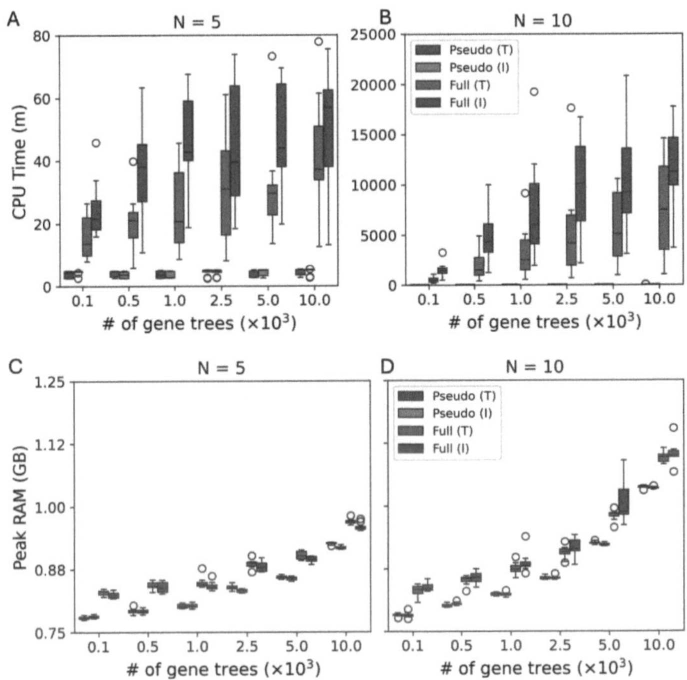

Fig. 9. (A, B) CPU time required (in minutes) to complete 1,000,000 steps of the MCMC simulation on ground-truth gene trees (T) and inferred gene trees (I). (C, D) Peak RAM usage (in GB) during the MCMC simulation.

The peak RAM usage of both methods is similar with full-likelihood-based analyses, requiring approximately 0.1 GB more RAM than pseudo-likelihood-based ones (Fig. 9C, D). All comparisons shown have been performed by running inference with a single thread on an Intel(R) Xeon(R) Gold 5220R CPU @ 2.20GHz.

3.3 Biological Data Analysis

We analyzed two biological phylogenomic datasets in this study. The first dataset comprised 233 alignments of ultra-conserved genetic elements (UCEs) from 32 taxa from order *Testudines* and 6 outgroup species (*H. sapiens, S. punctatus, A. carolinensis, P. molurus, C. porosus,* and *G. gallus*) [5]; we refer to this dataset as the **Turtles** dataset. The second dataset comprised 581 alignments from 83 taxa in the genus *Aphonopelma*; we refer to this dataset as the **Tarantulas** dataset.

For the **Turtles** dataset, all 900 sampled topologies were identical (Fig. 10) and had Robinson-Foulds distance of two with respect to the topology inferred by MrBayes [31] (from the concatenation of the 233 alignments) from the original study [5].

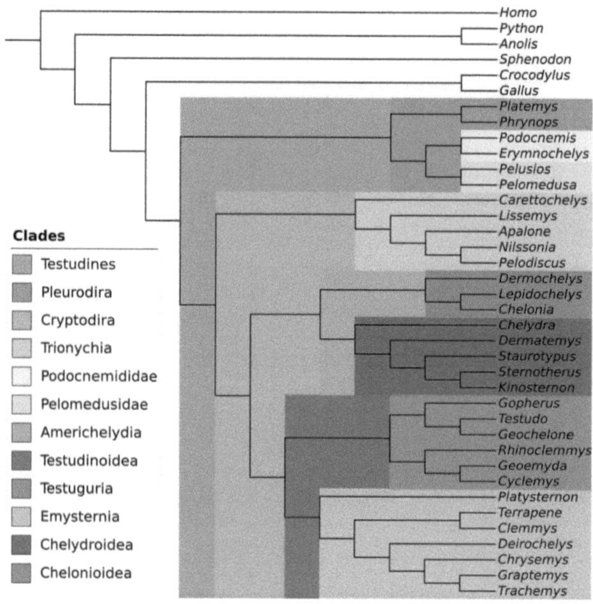

Fig. 10. A cladogram of the maximum a posteriori (MAP) estimate of the species tree for the **Turtles** dataset inferred from 233 UCE trees. Major clades in order *Testudines* are highlighted with colors.

The discrepancy is caused by the placement of the *Sphenodon* outgroup. The branch length estimates obtained from our analyses deviated from those inferred in the original study, with only 22 out of 72 branches that had support in our sampled topologies having the branch length inferred in the original study within the 95% Bayesian credible interval. It is important to note here that the original analysis of the authors using MrBayes does not account for ILS explicitly, as MrBayes does not employ the multispecies coalescent model, whereas the analyses using our method do.

Analysis of this dataset has required a total of 70.83 CPU hours utilizing a single thread on an AMD EPYC 7642 48-Core Processor @ 2.30 GHz.

For the **Tarantulas** dataset (Supp. Fig. 15), of the 2,235 sampled topologies, 1,199 had Robinson-Foulds distance of 14 and 1,036 had Robinson-Foulds distance of 16 to the ASTRAL tree inferred in the original study [7]. The topology for the paraphyletic species group *Aphonopelma iodius* matches the one previously reported [7]. Discrepancies between the inferred trees and the ones reported in the original study arise due to the placement of the outgroup species, resolution of polytomies present in the ASTRAL tree, and topologies for the individual species (collapsed in Supp. Fig. 15). Of the 154 branches in the ASTRAL inferred species tree that had support in the sampled topologies, 70 had their branch length values estimated by ASTRAL within the 95% credible intervals.

4   Discussion

Our theoretical and empirical analyses demonstrate the pseudo-likelihood provides a viable alternative to full likelihood for Bayesian phylogenomic inference tasks. It shows strong convergence guarantees with respect to the species tree topology supported by empirical evaluation in both idealized scenario where ground-truth gene trees are available and in a more realistic scenario where gene trees are estimated from the sequence data, although the impact of gene tree estimation error was only validated for low error rates ($<5\%$). Additionally, posterior distributions inferred based on pseudo-likelihood match those inferred based on full likelihood after appropriate centering. Furthermore, our observations on the required compute time indicate that the use of pseudo-likelihood can provide up to 100x speedup when compared with full-likelihood-based calculations. This is noteworthy given that the scale of biological datasets [5, 7] can completely prohibit Bayesian analyses under full likelihood. Additionally, since pseudo-likelihood-based inference speeds up individual proposal evaluations in the MCMC sampling procedure, it can in theory be paired with an improved proposal scheme [8, 42] to provide additional improvements to the computational cost.

However, as indicated by the theoretical analysis, branch length credible intervals do not match confidence intervals at the nominal level. While in theory the mismatch can result in both lower and higher empirical coverage, we note that in our experiments the coverage was always lower than the nominal value, that is, pseudo-likelihood-based Bayesian inference tends to be overconfident. Results on the biological data provide further evidence that the credible intervals estimated from pseudo-likelihood-based MCMC can be overconfident. These observations motivate a follow up investigation on whether there is a way to bound or correct this discrepancy, in particular by adjusting or reweighting MCMC samples or using coverage-oriented recalibration methods such as [16].

Finally, while species tree inference remains an important task, certain biological processes such as hybridization [1] and horizontal gene transfer can give rise to evolutionary histories that require inference of phylogenetic networks. The multispecies network coalescent (MSNC) has been previously proposed as a framework for statistical inference of phylogenetic species networks [37, 40].

However, computing full likelihood under the MSNC model is significantly more expensive when compared to species tree inference under the MSC model [43]. Thus, maximum pseudo-likelihood approaches have been proposed for the network inference problem under the MSNC [32,41,44]. However, suitability of pseudo-likelihood for Bayesian inference under the MSNC has not been investigated. Therefore, given the potential impact on the inference speed and promising results for species tree inference, we aim to expand current work to the network inference problem.

Acknowledgments. This work was in part supported by the NSF grants DMS/NIGMS-2153704 and DBI-2030604.

Code and Data Availability. Implementation of pseudo-likelihood-based MCMC is available in PhyloNet [33,38] (https://github.com/NakhlehLab/PhyloNet).

References

1. Barton, N.H., Hewitt, G.M.: Analysis of hybrid zones. Ann. Rev. Ecol. Systematics 113–148 (1985)
2. Bochkina, N.: Bernstein–von Mises theorem and misspecified models: a review. Found. Modern Stat. 355–380 (2019)
3. Bryant, D., Bouckaert, R., Felsenstein, J., Rosenberg, N.A., RoyChoudhury, A.: Inferring species trees directly from biallelic genetic markers: bypassing gene trees in a full coalescent analysis. Mol. Biol. Evol. **29**(8), 1917–1932 (2012)
4. Cook, S.R., Gelman, A., Rubin, D.B.: Validation of software for Bayesian models using posterior quantiles. J. Comput. Graph. Stat. **15**(3), 675–692 (2006)
5. Crawford, N.G., et al.: A phylogenomic analysis of turtles. Mol. Phylogenet. Evol. **83**, 250–257 (2015)
6. Degnan, J.H., Rosenberg, N.A.: Discordance of species trees with their most likely gene trees. PLoS Genet. **2**(5), e68 (2006)
7. Hamilton, C.A., Lemmon, A.R., Lemmon, E.M., Bond, J.E.: Expanding anchored hybrid enrichment to resolve both deep and shallow relationships within the spider tree of life. BMC Evol. Biol. **16**, 1–20 (2016)
8. Höhna, S., Drummond, A.J.: Guided tree topology proposals for Bayesian phylogenetic inference. Syst. Biol. **61**(1), 1–11 (2012)
9. Hudson, R.R.: Generating samples under a Wright-Fisher neutral model of genetic variation. Bioinformatics **18**(2), 337–338 (2002)
10. Jukes, T., Cantor, C.: Evolution of protein molecules. In: Munro, H. (ed.) Mammalian Protein Metabolism, pp. 21–132. Academic Press, NY (1969)
11. Justison, J.A., Solis-Lemus, C., Heath, T.A.: SiPhyNetwork: an R package for simulating phylogenetic networks. Methods Ecol. Evol. **14**(7), 1687–1698 (2023)
12. Kalyaanamoorthy, S., Minh, B.Q., Wong, T.K., Von Haeseler, A., Jermiin, L.S.: ModelFinder: fast model selection for accurate phylogenetic estimates. Nat. Methods **14**(6), 587–589 (2017)
13. Kleijn, B., van der Vaart, A.: The Bernstein-Von-Mises theorem under misspecification. Electronic J. Stat. **6**, 354–381 (2012)
14. Kubatko, L.S., Carstens, B.C., Knowles, L.L.: STEM: species tree estimation using maximum likelihood for gene trees under coalescence. Bioinformatics **25**(7), 971–973 (2009)

15. Li, M., Liu, Z., Yu, C.H., Vannucci, M.: Semiparametric Bayesian inference for local extrema of functions in the presence of noise. J. Am. Stat. Assoc. **119**(548), 3127–3140 (2024)
16. Liu, J., Li, M.: Bend to mend: Toward trustworthy variational Bayes with valid uncertainty quantification. arXiv preprint arXiv:2512.22655 (2025)
17. Liu, L., Yu, L., Edwards, S.V.: A maximum pseudo-likelihood approach for estimating species trees under the coalescent model. BMC Evol. Biol. **10**, 1–18 (2010)
18. Maddison, W.P.: Gene trees in species trees. Syst. Biol. **46**(3), 523–536 (1997)
19. Mirarab, S., Nakhleh, L., Warnow, T.: Multispecies coalescent: theory and applications in phylogenetics. Annu. Rev. Ecol. Evol. Syst. **52**(1), 247–268 (2021)
20. Mirarab, S., Reaz, R., Bayzid, M.S., Zimmermann, T., Swenson, M.S., Warnow, T.: Astral: genome-scale coalescent-based species tree estimation. Bioinformatics **30**(17), i541–i548 (2014)
21. Moreno, M.A., Sukumaran, J., Holder, M.T.: Dendropy 5: a mature python library for phylogenetic computing. arXiv preprint arXiv:2405.14120 (2024)
22. Nakhleh, L.: Computational approaches to species phylogeny inference and gene tree reconciliation. Trends Ecol. Evol. **28**(12), 719–728 (2013)
23. Nguyen, L.T., Schmidt, H.A., Von Haeseler, A., Minh, B.Q.: IQ-TREE: a fast and effective stochastic algorithm for estimating maximum-likelihood phylogenies. Mol. Biol. Evol. **32**(1), 268–274 (2015)
24. Ogilvie, H.A., Bouckaert, R.R., Drummond, A.J.: StarBEAST2 brings faster species tree inference and accurate estimates of substitution rates. Mol. Biol. Evol. **34**(8), 2101–2114 (2017)
25. Ogilvie, H.A., Heled, J., Xie, D., Drummond, A.J.: Computational performance and statistical accuracy of *BEAST and comparisons with other methods. Syst. Biol. **65**(3), 381–396 (2016)
26. Pei, J., Wu, Y.: STELLS2: fast and accurate coalescent-based maximum likelihood inference of species trees from gene tree topologies. Bioinformatics **33**(12), 1789–1797 (2017)
27. Rambaut, A., Grass, N.C.: Seq-Gen: an application for the Monte Carlo simulation of DNA sequence evolution along phylogenetic trees. Bioinformatics **13**(3), 235–238 (1997)
28. Rannala, B., Yang, Z.: Bayes estimation of species divergence times and ancestral population sizes using dna sequences from multiple loci. Genetics **164**(4), 1645–1656 (2003)
29. Robinson, D.F., Foulds, L.R.: Comparison of phylogenetic trees. Math. Biosci. **53**(1–2), 131–147 (1981)
30. Robinson, D.F., Foulds, L.R.: Comparison of weighted labelled trees. In: Combinatorial Mathematics VI: Proceedings of the Sixth Australian Conference on Combinatorial Mathematics, Armidale, Australia, August 1978, pp. 119–126. Springer (2006)
31. Ronquist, F., et al.: MrBayes 3.2: efficient Bayesian phylogenetic inference and model choice across a large model space. Systematic Biol. **61**(3), 539–542 (2012)
32. Solís-Lemus, C., Ané, C.: Inferring phylogenetic networks with maximum pseudo-likelihood under incomplete lineage sorting. PLoS Genet. **12**(3), e1005896 (2016)
33. Than, C., Ruths, D., Nakhleh, L.: PhyloNet: a software package for analyzing and reconstructing reticulate evolutionary relationships. BMC Bioinform. **9**, 1–16 (2008)
34. van der Vaart, A.: Asymptotic Statistics. Cambridge University Press (1998)

35. Wang, Y., Ogilvie, H.A., Nakhleh, L.: Practical speedup of Bayesian inference of species phylogenies by restricting the space of gene trees. Mol. Biol. Evol. **37**(6), 1809–1818 (2020)
36. Wang, Y., Blei, D.M.: Frequentist consistency of variational bayes. J. Am. Stat. Assoc. **114**(527), 1147–1161 (2019)
37. Wen, D., Yu, Y., Nakhleh, L.: Bayesian inference of reticulate phylogenies under the multispecies network coalescent. PLoS Genet. **12**(5), e1006006 (2016)
38. Wen, D., Yu, Y., Zhu, J., Nakhleh, L.: Inferring phylogenetic networks using PhyloNet. Syst. Biol. **67**(4), 735–740 (2018)
39. Wu, Y.: Coalescent-based species tree inference from gene tree topologies under incomplete lineage sorting by maximum likelihood. Evolution **66**(3), 763–775 (2012)
40. Yu, Y., Dong, J., Liu, K.J., Nakhleh, L.: Maximum likelihood inference of reticulate evolutionary histories. Proc. Natl. Acad. Sci. **111**(46), 16448–16453 (2014)
41. Yu, Y., Nakhleh, L.: A maximum pseudo-likelihood approach for phylogenetic networks. BMC Genomics **16**, 1–10 (2015)
42. Zhang, C., Huelsenbeck, J.P., Ronquist, F.: Using parsimony-guided tree proposals to accelerate convergence in Bayesian phylogenetic inference. Syst. Biol. **69**(5), 1016–1032 (2020)
43. Zhu, J., Liu, X., Ogilvie, H.A., Nakhleh, L.K.: A divide-and-conquer method for scalable phylogenetic network inference from multilocus data. Bioinformatics **35**(14), i370–i378 (2019)
44. Zhu, J., Nakhleh, L.: Inference of species phylogenies from bi-allelic markers using pseudo-likelihood. Bioinformatics **34**(13), i376–i385 (2018)

wQFM-GDL Enables Accurate Quartet-Based Genome-Scale Species Tree Inference Under Gene Duplication and Loss

Abdur Rafi[ID], Ahmed Mahir Sultan Rumi[ID], Sheikh Azizul Hakim[ID], and Md. Shamsuzzoha Bayzid[✉][ID]

Department of Computer Science and Engineering, Bangladesh University of Engineering and Technology, Dhaka, Bangladesh
`shams_bayzid@cse.buet.ac.bd`

Abstract. Species tree estimation from multi-copy gene family trees, including both paralogs and orthologs, is a challenging task due to the gene tree discordance caused by biological processes such as incomplete lineage sorting (ILS) and gene duplication and loss (GDL). Quartet-based species tree estimation methods, such as ASTRAL, Quartet Max-Cut (QMC), and Quartet Fiduccia–Mattheyses (QFM) frameworks have gained substantial popularity for their accuracy and statistical guarantee. However, most of these methods rely on single-copy gene trees and model only ILS, which limits their applicability to large genomic datasets. ASTRAL-Pro incorporates both orthology and paralogy for species tree inference under GDL by employing a refined quartet similarity measure based on the concept of species-driven quartets (SQs). In this study, we show that these SQ-based techniques can be effectively leveraged within the QFM framework. This required substantial algorithmic re-engineering, including the development of efficient techniques for computing the initial bipartition in QFM and novel combinatorial methods for computing refined quartet scores directly from gene family trees. We extensively evaluated our method, wQFM-GDL, on simulated and real biological datasets and compared it with leading methods ASTRAL-Pro3 and SpeciesRax. wQFM-GDL outperforms all other methods in 113 out of 124 model conditions considered in this study, with performance differences becoming more pronounced as dataset size increases. In particular, for larger datasets with 200 and 500 taxa, wQFM-GDL significantly outperforms all leading methods in all 72 out of 72 model conditions and achieves, on average, nearly a 25% reduction in reconstruction error compared with ASTRAL-Pro3. wQFM-GDL is freely available in open source form at https://github.com/abdur-rafi/wQFM-GDL.

Keywords: Species tree inference · Gene tree discordance · Gene duplication and loss · Orthologs and Paralogs

---

A. Rafi and A. M. S. Rumi—Contributed equally to this work.

M. Lafond (Ed.): RECOMB-CG 2026, LNBI 16569, pp. 139–160, 2026.
https://doi.org/10.1007/978-3-032-26891-4_8

1   Introduction

Species tree inference from genes sampled throughout the whole genome is complicated by the fact that gene trees can differ from each other (and from the true species tree) due to several biological processes, including gene duplication and loss (GDL), horizontal gene transfer, incomplete lineage sorting (ILS), and hybridization [20]. In particular, gene duplication and loss is one of the most important processes shaping gene family evolution and often results in multicopy gene trees [28].

Summary methods, which find a species tree by combining a set of gene trees, are becoming increasingly popular due to their accuracy and statistical guarantee under particular reasons for gene tree discordance [1,7,13–18,21,24,27]. A large family of summary methods considers GDL as a source of gene tree discordance. Most of these methods are gene tree parsimony methods that seek a species tree by minimizing the total number of duplications and losses required to explain the observed gene trees. DupTree [37], iGTP [6], and their recent improvement DupLoss-2 [29], DynaDup [2–4], and earlier similar dynamic programming based methods [10] are some of the well-known parsimony-based methods. However, there are other methods, such as PHYLDOG [5] and Guenomu [8], that are more agnostic about the reasons for gene tree discordance and do not necessarily rely on maximum parsimony reconciliation.

Quartet-based summary methods have been among the most widely used approaches for species tree inference over the last decade. These methods address the Maximum Quartet Support Species Tree (MQSST) problem, which seeks to infer a species tree that maximizes the number of quartets induced by the gene trees that are consistent with the species tree. Within this framework, methods can be broadly categorized based on how they solve the MQSST problem. ASTRAL, the most widely used summary method, employs a dynamic programming strategy to optimize quartet support [24]. In contrast, quartet amalgamation approaches, such as QMC (Quartet MaxCut) [1,35] and QFM (Quartet Fiduccia–Mattheyses) [22,34], adopt divide-and-conquer strategies to assemble quartets (with or without weights) induced by gene trees. While these approaches offer conceptual simplicity and flexibility–since they can operate directly on quartets, as in [7], without requiring explicit gene tree estimation–their reliance on explicit quartet enumeration poses significant computational challenges for large datasets. Recent methods, notably TREE-QMC [11] and wQFM-TREE [32], address this limitation by eliminating the need for explicit quartet enumeration, thereby substantially improving scalability without sacrificing accuracy. These methods were originally designed to handle gene tree heterogeneity due to ILS and are therefore applicable to single-copy gene trees.

ASTRAL-Pro [41] extends the ASTRAL framework to model GDL and remains the only quartet-based species tree inference method that explicitly accounts for duplication and loss. It introduces a new quartet similarity measure that accounts for both orthology and paralogy [41]. However, despite their effectiveness, no existing quartet amalgamation–based summary methods, such as wQFM and wQMC, explicitly model GDL.

In this study, we introduce wQFM-GDL, an extension of the QFM framework that explicitly accounts for GDL and is the first initiative to model GDL in a quartet amalgamation method. The proposed method includes two variants, wQFM-GDL-Q and wQFM-GDL-T, which operate on quartets and gene trees, respectively. In particular, we make the following key contributions. We extend the wQFM-TREE framework so it can account for speciation-driven quartets (SQs) by developing efficient graph-theoretic and combinatorial techniques to compute them directly from gene family trees without quartet enumeration – leading to wQFM-GDL-T. We also propose wQFM-GDL-Q where we enumerate the SQs from a given set of gene family trees and apply wQFM on this set of SQs. Notably, this set of SQs can be directly given as input to any quartet amalgamation methods (e.g., wQFM, wQMC). We found that, on small to moderate-sized datasets, where quartet enumeration is feasible, wQFM-GDL-Q is often more accurate than wQFM-GDL-T. Importantly, we also leveraged a locus-aware normalization scheme for improved accuracy.

Our extensive experimental analysis on established benchmark datasets shows that wQFM-GDL achieves the highest accuracy among the evaluated methods. Overall, across a wide range of model conditions with varying levels of duplication and loss, ILS, and gene tree estimation error, wQFM-GDL obtained the best performance in 113 out of 124 model conditions considered in this study. wQFM-GDL performs especially well on large datasets such as comprising 200 and 500 taxa, where wQFM-GDL significantly outperformed all other methods across all 72 model conditions with an average 25% reduction in tree error relative to ASTRAL-Pro - making it particularly suitable for large-scale phylogenomics datasets. We also reanalyzed the Plants83 [38] dataset, and the results provide evidence of wQFM-GDL's robustness and accuracy.

2   Materials and Methods

We first provide a brief overview of wQFM, wQFM-TREE, and the duplication-aware quartet similarity measure used in ASTRAL-Pro. Then, we describe the key algorithmic components introduced in wQFM-GDL in detail.

2.1   Overview of wQFM and wQFM-TREE

wQFM [22] is a quartet amalgamation method that takes as input a set of weighted quartets induced by a given set of gene trees and infers a species tree by maximizing quartet consistency. The algorithm follows a divide-and-conquer strategy (Fig. 1A): at each divide step, the current taxa set is partitioned into two disjoint subsets, defining two subproblems. For each divide step, wQFM starts from an initial bipartition on the current taxa set and iteratively refines it using a heuristic inspired by the Fiduccia–Mattheyses (FM) algorithm for hypergraph bipartitioning [9], with the objective of maximizing the difference between the number of satisfied and violated quartets. The procedure is applied recursively to the resulting subproblems until each subproblem contains at most three taxa, for which the solution is trivial. At each divide step, a dummy (artificial) taxon is introduced into each partition to represent the taxa in the complementary partition. During the conquer phase, the solutions to the subproblems (subproblem trees) are merged by connecting them through their corresponding dummy taxa, ultimately producing our final species tree.

Central to wQFM's divide-and-conquer process is the identification of a suitable bipartition of the taxa set at each divide step. Each candidate bipartition is scored with respect to the input quartet set. A quartet $q = ((A, B), (C, D))$ is *satisfied* with respect to a bipartition (P_a, P_b) if taxa A and B lie in one partition and taxa C and D lie in the other, and is *violated* if taxa A and C (or A and D) lie in one partition while taxa B and D (or B and C) lie in the other. Otherwise, the quartet is *deferred*. The *score of a bipartition* is defined as the difference between the number of satisfied and violated quartets.

The main computational bottleneck of wQFM is the explicit enumeration of all quartets induced by the input gene trees. In earlier work, we developed wQFM-TREE [32] to address this limitation by enabling the wQFM framework to operate directly on gene trees without explicit quartet enumeration. This was achieved through novel algorithmic techniques that combine a gene tree consensus–based heuristic for constructing initial bipartitions at each divide step with combinatorial and graph-theoretic methods for computing the scores of candidate bipartitions directly from gene trees. We now revisit the normalization schemes used in TREE-QMC [11] and later adopted by wQFM-TREE, as they play an important role in the accuracy of the methods and are essential for understanding the new locus-aware normalization approach we propose in this study (Fig. 1).

Normalization Scheme. Normalization is a critical component in the QMC and QFM frameworks because the divide-and-conquer process introduces artificial or dummy taxa into subproblems, which enables the subproblem trees to be merged in the conquer phase. Consequently, every non-root subproblem in the recursion contains one or more dummy taxa, whereas the input gene trees and the quartets induced contain only real taxa.

To compute satisfied and violated quartets with respect to a subproblem that contains dummy taxa, the leaves of each gene tree are relabeled appropriately by dummy taxa. A dummy taxon represents all real taxa in its sister subproblem, and the corresponding leaves are relabeled using that dummy taxon. Thus, the resulting gene trees only contain the taxa present in the subproblem, including both real and dummy taxa. An important consideration is that, in this relabeled representation, multiple leaves of a gene tree may be relabeled by the same dummy taxon, which can inflate the contribution of quartets involving dummy taxa. However, from the perspective of the current subproblem, a dummy taxon should have the same influence as any single real taxon. To address this, quartet weights are normalized so that every subset of four taxa in a subproblem, including those containing dummy taxa, contributes a single vote toward the tally of satisfied or violated quartets.

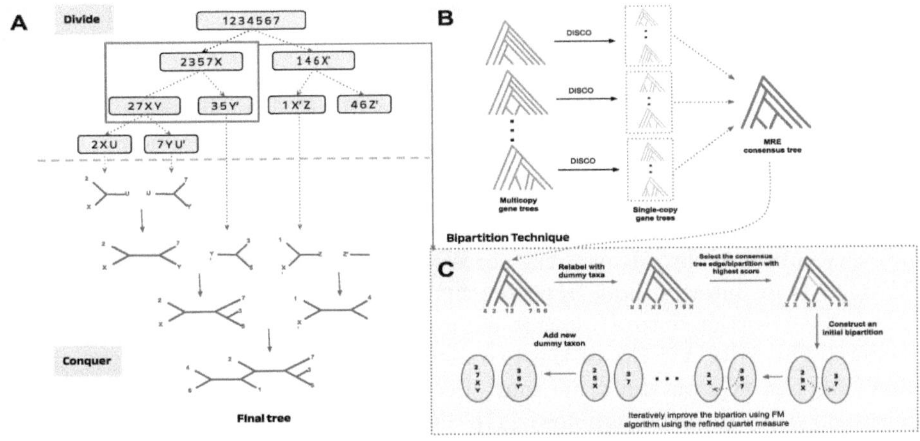

Fig. 1. Overview of our proposed method. (A) The divide-and-conquer framework used in all QFM methods. (B) An MRE consensus tree is constructed from the multicopy gene trees after decomposing them into single-copy gene trees using DISCO, which will be used to create an initial bipartition. (C) After generating the initial bipartition, it is iteratively improved using the FM heuristic, where the bipartitions are evaluated using the refined quartet measure addressing GDL.

Normalization was used in wQFM prior to TREE-QMC and wQFM-TREE, but it was fairly simple because the full quartet set is available as input. Recently, TREE-QMC [11] introduced an effective normalization strategy that directly works on gene trees, and wQFM-TREE adopts a similar approach. Under this strategy, each taxon in a gene tree is assigned a weight, and the weight of a quartet is defined as the product of the weights of its four taxa. Consequently, the total contribution of any quartet to the bipartition score reflects the effective representation of its constituent taxa in the current subproblem. The taxa that are absent from a subproblem but are represented through a dummy taxon receive appropriately reduced weights. This ensures that every subset of four taxa in a subproblem, including those containing dummy taxa, gets one vote per gene tree. The taxa are weighted non-uniformly depending on the subproblem decomposition in the recursion tree, and taxa that are more relevant to a particular subproblem are assigned higher weights. wQFM-GDL extends and improves the normalization scheme of wQFM-TREE to account for GDL (details in Sect. 2.3).

2.2 Overview of ASTRAL-Pro: Solving Maximum per-Locus Quartet-Score Species Tree (MLQST)

To account for orthology and paralogy, ASTRAL-Pro refines the quartet similarity measure of the original ASTRAL in mainly two ways.

- Firstly, it only considers "speciation-driven quartets" (SQs) and excludes duplication quartets (DQs), which contain no information regarding speciation events.
- Secondly, it aggregates the speciation quartets that mandatorily share the same topology and separately do not provide any new information. These are counted as one unit, forming a quartet equivalence class.

A quartet in a multicopy gene tree is classified as a speciation quartet (SQ) if, for every subset of three leaves, the corresponding most recent common ancestor (MRCA) is an internal node representing a speciation event. The subtree of a binary tree restricted to a quartet exhibits two degree-3 nodes referred to as anchors. The MRCA/LCA of the two anchors of a quartet is called the anchor LCA. All the SQs on the same four species that share the same anchor LCA must also share the same topology [41]. They are counted as one unit to prevent double-counting. Importantly, to classify quartets as SQs or DQs, the input trees must be rooted and tagged. ASTRAL-Pro defines the per-locus quartet score of a species tree with respect to a gene family tree with tagged internal nodes to be the number of quartet equivalence classes of the gene trees agreeing with the species tree. Thus, it tries to solve the Maximum per-Locus Quartet-score Species Tree (MLQST) problem by reconstructing the tree having the maximum total per-locus quartet score with respect to the multicopy gene trees.

2.3   wQFM-GDL-T: Extending wQFM-TREE for GDL

wQFM-GDL-T extends wQFM-TREE to accommodate GDL by solving the MLQST problem as in ASTRAL-Pro through three principal enhancements. First, we extend the consensus-tree-based heuristic to construct an appropriate initial bipartition of wQFM-TREE for multicopy gene trees. Second, we develop combinatorial techniques that enable efficient scoring of candidate bipartitions directly from multicopy gene trees without explicit quartet enumeration. Third, we enhance the normalization strategy of wQFM-TREE by proposing a locus-aware normalization scheme that explicitly accounts for GDL.

Initial Bipartition. At each divide step, we initiate the FM heuristic by generating an initial bipartition of the taxa set of the subproblem, which would then be refined iteratively, and the quality of this initial bipartition has a substantial impact on the quality of the final bipartition. In wQFM-TREE, an initial bipartition is obtained using a consensus-tree–based heuristic, in which an edge of a consensus tree constructed from gene trees is selected for initial bipartition. wQFM-TREE scores each bipartition according to its scoring scheme and selects the bipartition with the highest score. [32]. However, for multicopy gene trees, a consensus tree is not directly defined. To address this limitation, we employ the method DISCO [39] to decompose multicopy gene trees into single-copy gene trees. We then construct a greedy consensus tree using the Majority Rule Extended (MRE) method implemented in the PAUP package [36], which supports incomplete gene trees as well. Then, we score each bipartition of the consensus tree using the updated scoring method addressing GDL as described later in this section, and select the best one. The consensus tree is constructed once and used for all subproblems.

Scoring a Candidate Bipartition. After constructing the initial bipartition, we iteratively improve it. The taxa are transferred from one partition to the other iteratively, following the FM heuristic to obtain progressively improved bipartitions. Each bipartition is scored to assess its quality. As described earlier, the score has been defined in the QFM framework as the difference between the number of quartets from the gene trees that the bipartition satisfies and violates.

The scoring mechanism in wQFM-GDL differs substantially from wQFM-TREE, as it must account for the complexities introduced by GDL. In the original wQFM-TREE, all quartets from single-copy gene trees are considered when computing the score. In contrast, for multicopy gene family trees, following the GDL model of ASTRAL-Pro, we restrict the calculation of satisfied and violated quartets only to speciation-driven quartets (SQs) and treat each quartet equivalence class as a single unit to avoid double-counting. To integrate these duplication-aware measures while preserving the scalability of the original wQFM-TREE, we had to introduce substantial modifications to the underlying combinatorial and graph-theoretic techniques of wQFM-TREE to enable the exclusion of all the duplication-driven quartets and the aggregation of equivalent quartets within the scoring framework. The extended terminology and the updated scoring procedure are described below.

Let $\mathcal{G}$ be a set of rooted and tagged (each node labelled as either speciation or duplication) gene trees on the taxa set $\mathcal{X}$. We root and tag using the algorithm of ASTRAL-Pro. Let (A, B) be a candidate bipartition of the taxa set S in a divide step. We define R_A and D_A as the sets of real and dummy taxa in A, respectively. We define F_A as the set of all real taxa that are either in R_A or represented by a dummy taxon $X \in D_A$. We similarly define R_B, D_B, and F_B. Now, a quartet $ab|cd$ in a rooted and tagged gene tree is an SQ only if the MRCAs for all triplets in $\{a, b, c, d\}$ are speciation nodes in its respective gene tree. We categorize the SQs using the same traditional approach as in the QFM framework.

1. **Satisfied SQ:** SQ in which either $\{a, b\} \subseteq F_A, \{c, d\} \subseteq F_B$ or vice versa.
2. **Violated SQ:** SQ in which either $\{a, c\} \subseteq F_A, \{b, d\} \subseteq F_B$ or $\{a, d\} \subseteq F_B, \{b, c\} \subseteq F_A$ or vice versa.
3. **Deferred SQ:** Any other SQ.

Now, our algorithm assigns weight to each real taxon in $\mathcal{X}$ according to the normalization process. We consider a quartet as composed of two unordered pairs $\{a, b\}$ and $\{c, d\}$. The weight of a pair, $w(\{a, b\}) = w(a) \cdot w(b)$ and weight of a quartet $ab|cd$ is defined as $w(ab|cd) = w(\{a, b\}) \cdot w(\{c, d\}) = w(a) \cdot w(b) \cdot w(c) \cdot w(d)$. The weight of a set Q of quartets, $w(Q) = \sum_{q \in Q} w(q)$.

Let $S_{SQ}^{(g)}$ and $V_{SQ}^{(g)}$ denote the sets of satisfied and violated SQs, respectively, each containing only one representative from every equivalence class in a gene tree $g \in \mathcal{G}$. The score of a candidate bipartition (A, B) with respect to $\mathcal{G}$ is defined as: $\mathrm{Score}(A, B, \mathcal{G}) = \sum_{g \in \mathcal{G}} \left(w(S_{SQ}^{(g)}) - w(V_{SQ}^{(g)}) \right)$. In this current setting, the total weight of potential satisfied and violated quartets are dependent on the specific speciation and duplication event tagging within each gene family tree which necessitates the use of more complex combinatorial and graph-theoretic techniques to calculate $w(S_{SQ}^{(g)})$ and $w(V_{SQ}^{(g)})$ directly from the gene family trees. We provide a high-level overview of the process below.

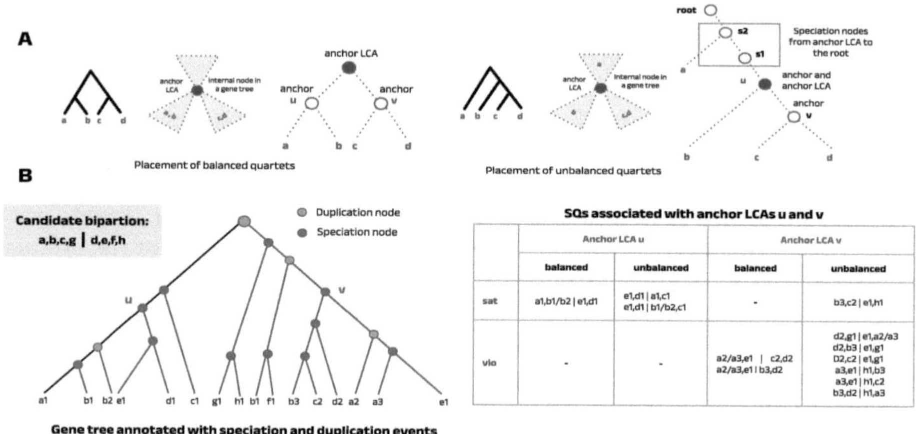

Fig. 2. **Scoring a candidate bipartition.** (A) Placement of a balanced and unbalanced quartet in a gene tree with respect to their anchor LCA. (B) SQs of different categories, illustrated using anchor LCAs u and v for an example bipartition $a, b, c, g | d, e, f, h$. Each speciation node may act as an anchor LCA for some SQs, and for scoring a bipartition, we compute the total weight of satisfied and violated SQs associated with each anchor LCA independently, treating balanced and unbalanced quartets separately, and then sum the results.

Computing $w(S_{SQ}^{(g)})$ for Multi-copy Trees. Our key observation is that the total weight of all satisfied SQs sharing the same anchor LCA (Least Common. Ancestor (LCA) of the two anchors of a quartet) can be computed efficiently and independently, and then aggregated for all anchor LCAs. Each internal node in the gene tree can serve as a potential anchor LCA for a set of SQs if and only if it is labeled as a speciation node, and the corresponding SQs associated with it can be computed as follows.

Rooted quartets exhibit two distinct forms: balanced and unbalanced, as shown in Fig. 2a. Our observation is that, for each particular anchor LCA, we can efficiently calculate the balanced and unbalanced satisfied SQs separately. Let u be an internal node tagged as speciation with child branches i and j. We define $PA_i^{(g,u)}$ and $PB_i^{(g,u)}$ as the weights of the unordered pairs of taxa in partitions A and B, respectively, within branch i. $PA_j^{(g,u)}$ and $PB_j^{(g,u)}$ are defined similarly. Importantly, multiple identical pairs may arise within the branches, but we count only one representative instance. This is crucial to ensure that only a single quartet from each equivalence class contributes to the score, as identical quartets formed from these identical pairs share the same anchor LCA and can be considered equivalent. Their topology is already determined by the speciation event at the anchor LCA. These quartets arise due to paralogy along branches i and j and do not provide additional speciation information. This represents an important distinction from the calculation procedure used in the original wQFM-TREE.

Balanced Structure. A rooted quartet is balanced if the root splits the taxa into two groups of two. A balanced SQ with anchor LCA u will be anchored by two internal nodes in branch i and j, each subtending exactly two leaves (Fig. 2a). The SQ is satisfied if the pairs of leaves from two branches belong to opposite partitions (A and B). Let $S_{1,SQ}^{(g,u)}$ denote the weight of balanced satisfied SQs with anchor LCA u. Thus, we compute

$$w(S_{1,SQ}^{(g,u)}) = w(PA_i^{(g,u)}) \cdot w(PB_j^{(g,u)}) + w(PA_j^{(g,u)}) \cdot w(PB_i^{(g,u)})$$

Unbalanced Structure. A rooted quartet is unbalanced if the root separates one taxon from the other three, creating a ladder-like (pectinate) structure. For an unbalanced SQ with anchor LCA u, one anchor lies in a child branch, while the second anchor is the node u itself. In this case, a pair of taxa from one child branch (e.g., branch i) is grouped with a taxon from a sibling branch j and a fourth taxon from the parent lineage k. Importantly, this fourth taxon cannot just be any arbitrary taxon from the parent branch. It must originate from one of the branches associated with speciation nodes along the path from u to the root. Otherwise, the defining condition of a speciation quartet, that the LCA of any selection of three leaves must be a speciation node, would be violated. Now, we define $PA_{i,k}^{(g,u)}$ as the weight of taxon pairs split between branch i and parent lineage k. $PB_{i,k}^{(g,u)}$ is defined similarly. Thus, the weight of unbalanced satisfied SQs with anchor LCA u,

$$w(S_{2,SQ}^{(g,u)}) = w(PA_i^{(g,u)}) \cdot w(PB_{j,k}^{(g,u)}) + w(PA_j^{(g,u)}) \cdot w(PB_{i,k}^{(g,u)})$$
$$+ w(PB_i^{(g,u)}) \cdot w(PA_{j,k}^{(g,u)}) + w(PB_j^{(g,u)}) \cdot w(PA_{i,k}^{(g,u)})$$

Finally, $w(S_{SQ}^{(g,u)}) = w(S_{1,SQ}^{(g,u)}) + w(S_{2,SQ}^{(g,u)})$. Now, we can simply aggregate the results for each anchor LCA. $w(S_{SQ}^{(g)}) = \sum_{u \in \text{SpeciationNodes}} w(S_{SQ}^{(g,u)})$. An example for this process is shown in Fig. 2.

Here, we observe that duplication-derived quartets in the gene trees are effectively excluded, as the LCA of at least one triplet among the four taxa of these quartets corresponds to a duplication node and therefore does not satisfy any of

148   A. Rafi et al.

the criteria mentioned for getting considered in both balanced and unbalanced quartet calculations.

Computing $w(V_{SQ}^{(g)})$ for Multi-copy Trees. The calculation of $w(V_{SQ}^{(g)})$ follows a similar approach. We compute the total weight of violated quartets independently for each anchor LCA, treating balanced and unbalanced structures separately, and then sum the results.

Let us consider the same scenario with anchor LCA u. For the balanced violated quartets, each branch will contain a pair, one from partition A and the other from B. Thus, the weight of violated quartets is as follows. $w(V_{1,SQ}^{(g,u)}) = w(PA_{i,j}^{(g,u)}) \cdot w(PB_{i,j}^{(g,u)})$. For the unbalanced ones, $w(V_{2,SQ}^{(g)}) = w(PA_{i,k}^{(g,u)}) \cdot w(PB_{i,j}^{(g,u)}) + w(PA_{i,j}^{(g,u)}) \cdot w(PB_{i,k}^{(g,u)}) + w(PA_{j,k}^{(g,u)}) \cdot w(PB_{j,i}^{(g,u)}) + w(PA_{j,i}^{(g,u)}) \cdot w(PB_{j,k}^{(g,u)})$. Then we can add the quantities as before.

The terms $w(PA_i^{(g,u)}), w(PB_i^{(g,u)}), w(PA_{i,j}^{(g,u)})$ are calculated efficiently and without actually enumerating the quartet sets using further combinatorial techniques and efficient graph traversal algorithms.

Locus-Aware Normalization Scheme. The prior normalization scheme used in wQFM-TREE could also be applied directly to wQFM-GDL. However, we identified an opportunity to improve normalization by introducing different normalization factors for distinct locus-specific regions in multicopy gene trees. Here, the concept of the locus tree plays a crucial role. Rasmussen and Kellis proposed the DLCoal model [33] for jointly modelling GDL and ILS, which introduces a third tree called the locus tree, alongside the gene tree and the species tree. The locus tree is derived from the species tree via a top-down duplication–loss process, with each internal node labeled as either a speciation or a duplication event. A duplication event creates a new daughter locus from the parent locus. Both the parent and daughter loci evolve independently thereafter and can undergo further duplications or losses in subsequent evolutionary steps. Finally, a gene family tree is constructed from the locus tree by applying a multi-locus coalescent process.

Unlike the case with single-copy gene trees, a particular set of four species (e.g., a, b, c, d) can appear multiple times within different locus-specific subtrees of a particular gene tree. Now, note that the set of taxa in the various locus-specific subtrees may not be identical because each locus undergoes a distinct evolutionary history, marked by different duplication and loss events. Importantly, the normalization factors used in our method depend on the taxa set. The normalization scheme used in QMC and QFM families of methods does not consider this, and thus any quartet on a, b, c, d in a gene tree is assigned the same weight. However, our observation is that the quartets on a particular set of four species across different locus-specific subtrees should be assigned different weights, as the taxa set of different locus-specific subtrees may vary. Recently, it has been shown that normalization of TREE-QMC can be affected in case of missing taxa in incomplete gene trees and thus needs correction [12]. However, in the case of gene family trees, identifying the taxa set associated with a particular locus within a multicopy gene tree is not straightforward. Our key step is that all quartets within an equivalence class descend from the same ancestral locus at the time of the speciation event corresponding to the anchor LCA, and the normalization factors for quartets associated with each anchor LCA are computed separately. This locus-aware normalization scheme provides a substantial accuracy improvement for wQFM-GDL while incurring only a minimal runtime overhead.

2.4   wQFM-GDL-Q: Extending wQFM for GDL

wQFM takes as input a set of weighted quartets where the weight is the frequency of the quartet in the input gene trees. We devise an algorithm to create an appropriate weighted quartet set that only contains SQs and only one representative from an equivalent class. Then, this quartet set is given input to the wQFM algorithm, which remains unchanged. We call this method wQFM-GDL-Q. For generating this quartet set, we adopt a similar algorithmic approach to the score calculation process in wQFM-GDL-T; however, in this case, we explicitly enumerate quartets with appropriate weights, rather than computing the sum of quartet weights. As expected, wQFM-GDL-Q is considerably slower than wQFM-GDL-T. However, its main advantage is that under smaller model conditions, where explicit quartet enumeration is feasible, we observe that it outperforms competing methods, including wQFM-GDL-T, in several settings. This may be because the heuristics used in wQFM-GDL-T, which allow it to operate directly on gene trees, can sometimes influence its accuracy. Further details of wQFM-GDL-Q are provided in Appendix A.

2.5   Complexity Analysis

The total time complexity for wQFM-GDL-T is $O(n^2k + \alpha n^2 Dd)$, where n is the number of taxa, k is the number of gene trees, D is the number of unique speciation-driven tripartitions, d is the number of dummy taxa, and α is the number of refinement iterations (consistently observed to be ≤ 5 in our studies). It has successfully analyzed 500 taxa datasets with a high duplication rate (more than 2,000 leaves) within 20 h and 16GB of memory. The running time and memory consumption statistics for the methods used in this study are reported in Appendix C.

3   Experimental Results and Discussions

We test the performance of wQFM-GDL-T and wQFM-GDL-Q on existing and our new simulated datasets and biological datasets against the leading methods: ASTRAL-Pro3 [40], and SpeciesRax [26].

3.1   Datasets

Simulated Datasets. We primarily use S25 from the ASTRAL-Pro study [41], which contains 25 species under diverse duplication, loss, and ILS conditions, and S100, simulated by Molloy and Warnow [25] based on a real fungal dataset. We note that the number of model conditions for large datasets was limited in the ASTRAL-Pro study. Therefore, to extensively evaluate the performance under GDL on large datasets, we generated two new large-scale simulated datasets comprising 200 and 500 taxa, which we call SIM200 and SIM500, respectively.

For both SIM200 and SIM500, species trees and gene family trees were simulated using SimPhy [23] under varying duplication rates, loss rates, and ILS levels. Sequence alignments were generated using AliSim [19], and maximum likelihood gene trees were estimated with FastTree [31] under the GTR+Gamma model. Duplication rates and haploid effective population sizes were determined empirically. Detailed parameter settings and the full simulation pipeline are described in Appendix B. The datasets are publicly available at https://zenodo.org/records/18605522.

Biological Datasets. We analyzed an important empirical dataset containing multicopy gene trees, Plants83 [38], which were inferred in the original study by Wicket et al., [38].

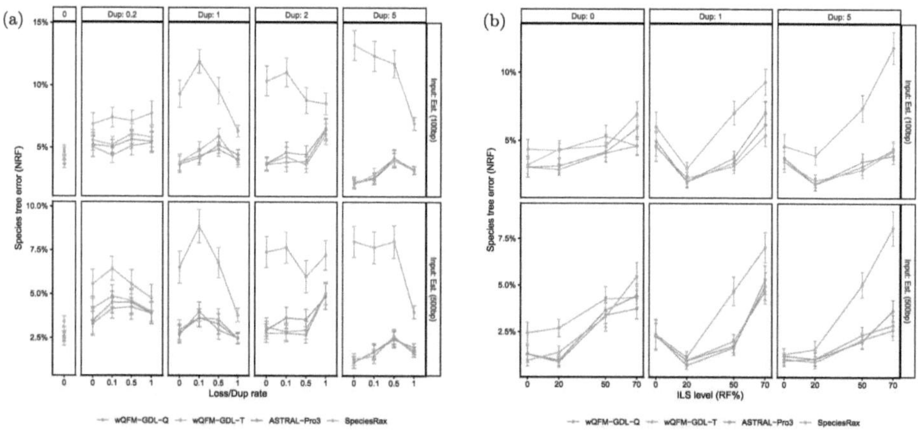

Fig. 3. Species tree error on the S25 dataset using 1,000 estimated gene trees from 100 and 500bp alignments. We show the average RF rates with standard error bars over 50 replicates. (a) Controlling duplication rate (box columns, labelled by mean number of copies per species minus one) and loss rate (labelled by ratio of loss and duplication rate). (b) Controlling the duplication rate and ILS level (RF rate between true gene trees and species tree).

3.2 Results on Simulated Datasets

Results for S25 Dataset. In S25, both versions of wQFM-GDL and ASTRAL-Pro3 substantially outperform SpeciesRax across nearly all model conditions (Fig. 3). While wQFM-GDL and ASTRAL-Pro3 perform similarly, wQFM-GDL-Q demonstrates better performance. It outperforms ASTRAL-Pro3 in 35 out of 52 model conditions, with statistically significant differences in 7 of these cases, and ties occurring in 8 conditions.

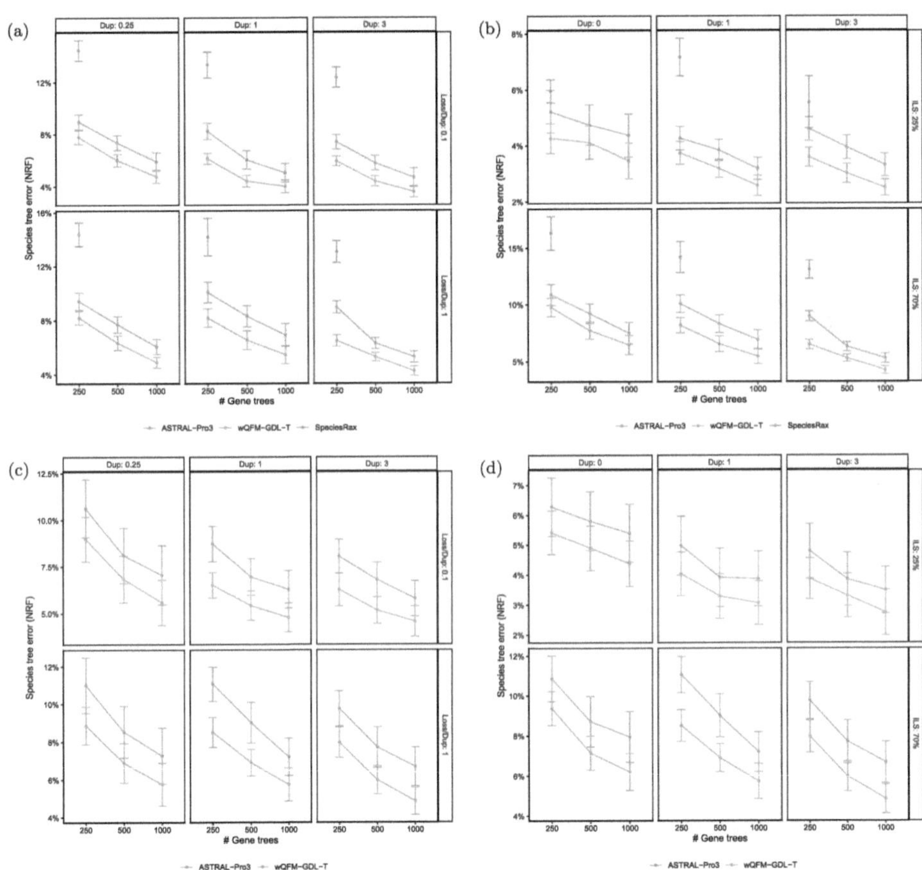

Fig. 4. Species tree error on the SIM200 (a-b) and SIM500 (c-d) dataset using estimated gene trees from 100 bp alignments. (a,c) Controlling duplication and loss rate. (b,d) Controlling the duplication rate and ILS level. SpeciesRax only completed the 250-gene model conditions of SIM200 within our resource limits.

Results for SIM200 and SIM500 Datasets. Figure 4 presents the performance of the methods on the SIM200 and SIM500 datasets. wQFM-GDL-Q could not be executed on these datasets due to its computationally intensive quartet generation. We could not finish running SpeciesRax on a single replicate with 500 or 1000 genes for some model conditions, even after running for more than two days. Therefore, within our resource limits (64 GB of CPU RAM and 48 h of computation per replicate), we could only analyze SpeciesRax on 250-gene model conditions of SIM200. Quite remarkably, wQFM-GDL-T convincingly outperformed all the methods in all 72 out of 72 model conditions, differences being statistically significant ($p < 0.05$). wQFM-GDL-T delivers almost 19% and 23% reduction in error compared to ASTRAL-Pro3 in SIM200 and SIM500, respectively. As expected, the performance of all methods decreases with fewer gene trees and higher levels of ILS. Interestingly, both wQFM-GDL and ASTRAL-Pro3 show improved accuracy with higher duplication rates, consistent with observations reported in the ASTRAL-Pro3 study. Thus, wQFM-GDL exhibits a more pronounced performance advantage in SIM200 and SIM500, i.e., under larger model conditions, compared to S25. The effectiveness of the inherent FM search strategy becomes more pronounced in larger search spaces, as observed in prior works [32], and this trend appears to carry over to the present setting. While it leads to higher accuracy, it may be slower than some alternative strategies.

3.3   Results on Biological Data

We analyze 9,237 multicopy gene trees of the Plants83 Dataset. The tree estimated by wQFM-GDL (Fig. 5) successfully recovers all major relationships and does not violate any known consensus among scientists. Following the most prominent recent methods, wQFM-GDL reconfirms the monophyly of Bryophytes (mosses, liverworts, and hornworts). Lycophytes are correctly recovered as the sister group of ferns and seed plants, and Gymnosperms are correctly recovered as the sister to flowering plants. It places Zygnematales (not Chara) as sister to all land plants and Amborella as sister to the rest of angiosperms. All of these relationships are well established.

However, there are three quite contentious areas. Alternative branching orders and quartet supports for those relationships are shown in Fig. 5. Firstly, wQFM-GDL, ASTRAL-Pro3, and SpeciesRax all support the Gnepine hypothesis for Gymnosperms (1.0 localPP), placing Gnetales and Pinaceae as sisters. However, in the 1kp plant analysis [38], ASTRAL with single-copy nuclear gene trees and plastome-based supermatrix analysis strongly suggested otherwise. Secondly, for the relative position of Coleochaetales and Chara, wQFM-GDL places Chara as the closest relative of Zygnematophyceae and Embryophyta (land plants), whereas ASTRAL-Pro3 and SpeciesRax place Coleochaetales algae. Both of the cases have almost exactly the same quartet support (36%). Finally, in the ASTRAL-Pro3 tree, Rosmarinus and Ipomoea are placed in the same clade. In contrast, wQFM-GDL recovers Ipomoea as sister to Catharanthus and Allamanda.

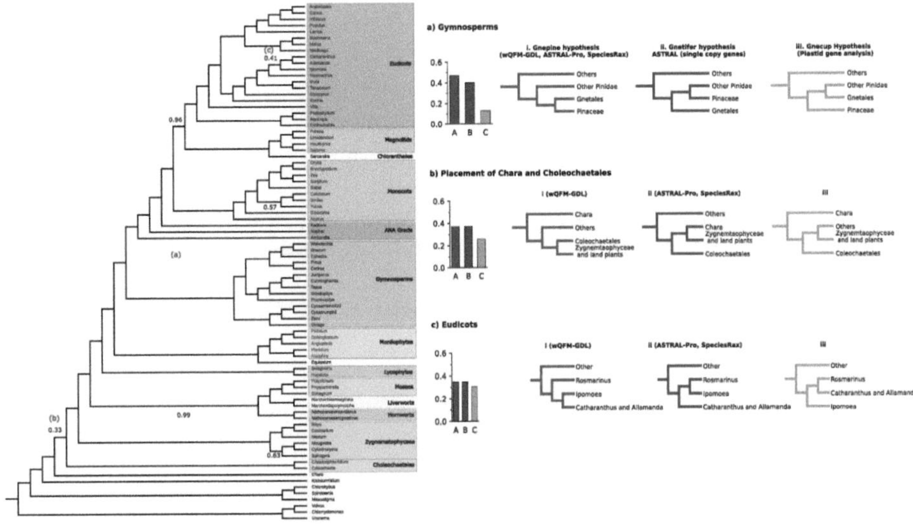

Fig. 5. The species tree inferred using wQFM-GDL for the Plants83 dataset and alternative branching orders for contentious relationships.

4 Conclusion

In this study, we proposed wQFM-GDL, a scalable and accurate quartet-based summary method, which accounts for both ILS and GDL. We have leveraged the concept of SQs within the QFM framework using appropriate algorithmic re-engineering and used a locus-aware normalization for improved accuracy. It demonstrates excellent performance and outperforms leading methods across most model conditions, particularly on large datasets. wQFM-GDL is capable of analyzing multicopy datasets containing thousands of taxa and gene trees with high duplication and loss rate within hours. Overall, these results position wQFM-GDL as a competitive and practical alternative to existing species tree inference methods for genome-scale data in the presence of GDL.

There are several possible future directions of this study. Currently, wQFM-GDL uses the tagging and rooting heuristic introduced by ASTRAL-Pro. Complex evolutionary scenarios involving ILS and GDL may cause this heuristic to produce suboptimal rooting and tagging, potentially reducing the accuracy of wQFM-GDL. A better tagging and rooting heuristic might enhance the performance of our framework. Furthermore, the divide-and-conquer structure of QFM naturally supports parallel execution, and a multi-processor-based implementation could significantly improve the scalability of QFM-based methods, including wQFM-GDL.

A   wQFM-GDL-Q: Details and Additional Results

Following an approach similar to the score calculation process of a candidate bipartition in wQFM-GDL-T, for each speciation node in the tagged and rooted gene trees, we independently generate the speciation-driven quartets with anchor LCA at that particular node and add them to the set. Similar to wQFM, the weight of a quartet is defined as its frequency across the gene trees. However, because the gene trees here are multicopy, multiple instances of the same quartet may occur within a single gene tree, and consequently, the total frequency (weight) of a quartet can be substantially larger than the number of gene trees, unlike in wQFM.

Now, we consider enumerating all SQs with a particular anchor LCA u with child branches i and j. For balanced quartets, we enumerate all possible taxon pairs formed within each child branch. Of course, we keep only one copy of a taxa pair as they form equivalent quartets, and we must consider one representative from them. Then, any pair from branch i can create a valid quartet with any pair from branch j. We generate all possible quartets by joining the pairs.

For the unbalanced quartets, one pair of taxa will be present in one of the child branches i or j. We generate all possible pairs for both child branches. For the second pair, one taxon will be present in the other child branch and one taxon in the parent branch. As discussed, not all taxa in the parent branch are considered here (Fig. 2 in the main paper). We list all taxa present in the child branches and the parent branch and merge them to create all valid second pairs. The pairs are then merged appropriately to enumerate quartets.

An important point to note is that, although the high-level idea resembles the score calculation procedure of wQFM-GDL-T, there is no notion of satisfied or violated quartets here because there is no bipartition to consider at this stage. Instead, our objective is to generate all speciation-driven quartets (SQs) from the gene trees so that wQFM can operate on them. Since this step focuses on explicit quartet enumeration rather than efficiently computing scores directly from the gene trees, the procedure is more straightforward compared to wQFM-GDL-T.

B   Simulation Details and Parameters

We simulated two large datasets, SIM200 and SIM500, containing model conditions of 200 and 500 taxa, respectively. In this section, we describe the simulation procedure in detail, including the commands used and the parameter settings. For both datasets, we varied three duplication rates. For each duplication rate, we considered two loss rates, two levels of ILS, and different numbers of gene trees (250, 500, and 1000). This resulted in a total of 36 model conditions per dataset, with 20 replicates per model condition for SIM200 and 10 replicates for SIM500. Simulation settings are summarized in Table 1.

Table 1. Simulation settings for SIM200 and SIM500 with varying parameters. For each of the duplication rates, we vary the loss rate, ILS level, and number of gene trees.

Condition	Parameter Ranges
No. of taxa	200,500
Varying no. of gene trees	250, 500, and 1000
Varying duplication rate	0.25, 1, and 3 (Mean number of extra copies per species)
	Corresponding duplication rates for S200: $\{0.8, 2.0, 4.1\} \times 10^{-10}$ events/generation
	Corresponding duplication rates for S500: $\{0.8, 2.1, 4.3\} \times 10^{-10}$ events/generation
Varying loss rate	0.1 and 1 (fraction of duplication rates)
Varying ILS	25% and 70% (mean RF distance between true gene trees and the species tree)
	Corresponding haploid effective population size for S200: $\{0.5, 1.3, 2.9\} \times 10^{8}$
	Corresponding haploid effective population size for S500: $\{0.5, 1.25, 2.7\} \times 10^{8}$
Sequence length	100bp

B.1 True Gene Family Tree Simulation

We used Simphy [23], which is a widely used program for simulation of gene family evolution under incomplete lineage sorting (ILS), gene duplication and loss (GDL), and horizontal gene transfer (HGT). The simulation parameters used in Simphy are presented in detail in Table 2.

For both SIM200 and SIM500, we varied the duplication rate to create model conditions with mean numbers of extra gene copies per species of 0.25, 1, and 3. That means, due to duplication and loss, the average numbers of gene copies per species are 1.25, 2, and 4, respectively. In SIM500, the model condition with an average of 3 extra copies per species contains gene trees with significant duplication, having more than 2,000 leaves on average. To determine the appropriate duplication rates in Simphy for generating these conditions, we systematically varied the duplication rate and, for each value, generated 1000 replicates for datasets with 200 taxa (SIM200) and 500 taxa (SIM500), using a loss rate of zero. As expected, the mean number of copies increased with higher duplication rates. We then identified the precise duplication rate that produced the desired mean copy number (0.25, 1, or 3) using a binary-search style procedure over the duplication parameter space. The loss rate is varied for each duplication rate simply as the ratio of 0.1 to 1 of the duplication rate. In the case of a 0.1 ratio, there is a small number of loss events, and the number of copies is high. In a ratio of 1, there is a significant number of loss events, and the mean number of copies is closer to the number of taxa, even with high duplication rates.

Similarly, for each duplication rate, we simulated two conditions varying ILS: 25% and 70% ILS, meaning the average RF rates of true gene trees and species tree are 25% and 70%, respectively. As the amount of ILS increases, the RF rate goes higher. We estimated the proper haploid effective population sizes for these two conditions in a procedure similar to the estimation of duplication rates.

Simphy Command for SIM200 with Standard Parameters:

```
simphy -sl f:200 -rs 20 -rl f:1000 -rg 1 -sb f:0.000000005 -
    sd f:0 -st ln:21.25,0.2 -so f:1 -si f:1 -sp f:435000000 -
```

```
 su ln: 21.9,0.1 -hh f:1 -hs ln:1.5,1 -hl ln: 1.551533,
  0.6931472 -hg ln:1.5,1 -cs 9644 -v 3 -o default -ot 0 -op
  1 -lb f:0.00000000041 -ld f:0.00000000041 -lt f:0
```

B.2 Simulating MSA

We use Alisim [19] to generate sequence alignments from the simulated true trees. The parameters are presented in Table 3.

AliSim is implemented in IQ-TREE version 2.4.0 or later. We use IQ-TREE to generate the sequence alignments.

IQ-TREE Command for SIM200 with Standard Parameters:

```
iqtree3 --alisim SimPhy_1.0.2/bin/default/1/MSA1 -t tree_file
    -m "GTR{1/3/0.6/1.2/3.2}+F
    {0.2950819672/0.2131147541/0.2295081967/0.2622950820}" --
    seqtype DNA --length 100 --seed 9644
```

B.3 Simulating Estimated Gene Trees from MSA

We use FastTree [30] to estimate maximum-likelihood gene trees from the sequence alignment under the general time-reversible model of nucleotide substitution with four discrete gamma rates (GTR+G4). We used the simple command:

```
FastTree -gtr -gamma -nt alignment_file > tree_file
```

C Running Time and Memory Consumption

The running times of all methods on the large datasets are reported in Table 4. All experiments were conducted on a single-core AMD Ryzen 9 7950X processor with 64 GB of RAM. In our machine, ASTRAL-Pro3 seems to be the fastest method on most model conditions, while SpeciesRax is the slowest. However, both ASTRAL-Pro3 and SpeciesRax can achieve good parallel efficiency and may run faster in environments that better support full parallel processing, which was not available in our setup. The running time of wQFM-GDL is comparable to ASTRAL-Pro3, though it is typically slower. However, it successfully analyzed 500-taxon model conditions with a high duplication rate (more than 2,000 leaves per gene family tree) in 20 h, showing its high scalability. The QFM framework is inherently parallelizable, and future upgrades can further improve its runtime performance.

wQFM-GDL successfully analyzed the large 200-taxon and 500-taxon datasets using no more than 8 GB and 16 GB of memory, respectively. In contrast, ASTRAL-Pro3 is more memory-efficient, requiring approximately 2 GB and 800 MB of memory to analyze both of these datasets.

Table 2. Simphy simulation Parameters for SIM200 and SIM500 dataset

Parameter name	Parameter value
Standard Parameters for SIM200/500	
No. of taxa	200/500 + outgroup
Speciation rate	5e-9
Extinction rate	0
Locus trees	1000
Gene trees	1
No. of replicates	20/10
Ingroup divergence to the ingroup ratio	1.0
Generations	LogN(21.25, 0.2)
Haploid effective population size	2.9e+8/2.7e+8
Global substitution rate	LogN(-21.9, 0.1)
Lineage specific rate gamma shape	LogN(1.5, 1)
Gene family specific rate gamma shape	LogN(1.551533, 0.6931472)
Gene tree branch specific rate gamma shape	LogN(1.5, 1)
Duplication rate	4.1e-10/4.35e-10
Loss rate to duplication rate ratio	1
Seed	9644
Varying Duplication and Loss Rates - SIM200	
Duplication rate	0.8e-10, 2e-10, 4.1e-10
Loss rate to duplication rate ratio	0.1, 1
Varying Duplication and Loss Rates - SIM500	
Duplication rate	0.8e-10, 2.1e-10, 4.35e-10
Loss rate to duplication rate ratio	0.1, 1
Controlling Duplication and ILS Rate - SIM200	
Duplication rate	0.8e-10, 2e-10, 4.1e-10
Haploid effective population size	0.5e+8, 2.9e+8
Controlling Duplication and ILS Rate - SIM500	
Duplication rate	0.8e-10, 2e-10, 4.1e-10
Haploid effective population size	0.5e+8, 2.7e+8

Table 3. Simulation Parameters for AliSim

Parameter name	Parameter value
Sequence length	100
Sequence base frequencies	Dirichlet(A=36, C=26, G=28, T=32)
Sequence transition rates	Dirichlet(TC=16, TA=3, TG=5, CA=5, CG=6, AG=15)
Seed	9644

Table 4. Running time (in minutes) of ASTRAL-Pro3, SpeciesRax, and wQFM-GDL-T under a few model conditions with low, moderate, and high duplication and loss across different numbers of gene trees (SpeciesRax could not complete all model conditions within our computational budget). All methods were executed on 5 replicates for each model condition to measure running time, and the results were averaged.

Model condition	Method	Number of gene trees		
		250	**500**	**1000**
sim200_dup1_loss1	ASTRAL-Pro3	2.86 ± 0.36	6.8 ± 0.73	9.73 ± 0.49
	SpeciesRax	279.27 ± 22.10	–	–
	wQFM-GDL-T	4.58 ± 0.75	11.92 ± 1.17	23.27 ± 1.98
sim200_dup0_ILS25	ASTRAL-Pro3	2.73 ± 0.25	6.4 ± 0.43	11.00 ± 0.82
	SpeciesRax	163.59 ± 15.69	–	–
	wQFM-GDL-T	0.85 ± 0.23	2.36 ± 0.33	4.39 ± 0.53
sim200_dup3_loss0.1	ASTRAL-Pro3	4.30 ± 0.76	11 ± 1.55	26.40 ± 3.31
	SpeciesRax	329.18 ± 20.58	–	–
	wQFM-GDL-T	13.15 ± 1.69	34.24 ± 2.84	68.44 ± 4.15
sim500_dup1_loss1	ASTRAL-Pro3	14.76 ± 1.53	31.87 ± 2.84	55.73 ± 4.97
	SpeciesRax	–	–	–
	wQFM-GDL-T	118.04 ± 16.29	277.73 ± 25.43	568.88 ± 45.12
sim500_dup0_ILS25	ASTRAL-Pro3	11.87 ± 0.55	31.13 ± 1.04	55.87 ± 1.73
	SpeciesRax	–	–	–
	wQFM-GDL-T	3.94 ± 1.25	12.98 ± 1.72	24.75 ± 3.51
sim500_dup3_loss0.1	ASTRAL-Pro3	33.95 ± 10.72	89 ± 22.48	234.25 ± 63.97
	SpeciesRax	–	–	–
	wQFM-GDL-T	264.17 ± 26.20	599.06 ± 43.28	1301.28 ± 80.55

References

1. Avni, E., Cohen, R., Snir, S.: Weighted quartets phylogenetics. Syst. Biol. **64**(2), 233–242 (2015)
2. Bayzid, M.S., Mirarab, S., Warnow, T.: Inferring optimal species trees under gene duplication and loss. In: Proceedings of Pacific Symposium on Biocomputing (PSB), vol. 18, pp. 250–261 (2013)
3. Bayzid, M.S., Warnow, T.: Gene tree parsimony for incomplete gene trees: addressing true biological loss. Algorith. Mol. Biol. **13**, 1 (2018)
4. Bayzid, M.S.: Inferring optimal species trees in the presence of gene duplication and loss: beyond rooted gene trees. J. Comput. Biol. **30**(2), 161–175 (2023)
5. Boussau, B., Szöllősi, G.J., Duret, L., Gouy, M., Tannier, E., Daubin, V.: Genome-scale coestimation of species and gene trees. Genome Res. **23**(2), 323–330 (2013)
6. Chaudhary, R., Bansal, M.S., Wehe, A., Fernández-Baca, D., Eulenstein, O.: iGTP: a software package for large-scale gene tree parsimony analysis. BMC Bioinform. 574–574 (2010)
7. Chifman, J., Kubatko, L.: Quartet from SNP data under the coalescent model. Bioinformatics **30**(23), 3317–3324 (2014)

8. De Oliveira Martins, L., Mallo, D., Posada, D.: A Bayesian supertree model for genome-wide species tree reconstruction. Syst. Biol. **65**(3), 397–416 (2016)

9. Fiduccia, C., Mattheyses, R.: A linear-time heuristic for improving network partitions. In: 19th Design Automation Conference, pp. 175–181 (June 1982). https://doi.org/10.1109/DAC.1982.1585498

10. Hallett, M.T., Lagergren, J.: New algorithms for the duplication-loss model. In: Proc RECOMB, pp. 138–146 (2000)

11. Han, Y., Molloy, E.K.: Improving quartet graph construction for scalable and accurate species tree estimation from gene trees. Genome Res. **gr.277629.122** (May 2023). https://doi.org/10.1101/gr.277629.122

12. Han, Y., Molloy, E.K.: Improved robustness to gene tree incompleteness, estimation errors, and systematic homology errors with weighted tree-qmc. Systematic Biology p. syaf009 (2025)

13. Islam, M., Sarker, K., Das, T., Reaz, R., Bayzid, M.S.: Stelar: a statistically consistent coalescent-based species tree estimation method by maximizing triplet consistency. BMC Genomics **21**(1), 1–13 (2020)

14. Kubatko, L.S., Carstens, B.C., Knowles, L.L.: Stem: species tree estimation using maximum likelihood for gene trees under coalescence. Bioinf **25**, 971–973 (2009)

15. Larget, B., Kotha, S.K., Dewey, C.N., Ané, C.: BUCKy: gene tree/species tree reconciliation with the Bayesian concordance analysis. Bioinf **26**(22), 2910–2911 (2010)

16. Liu, L., Yu, L.: Estimating species trees from unrooted gene trees. Syst. Biol. **60**(5), 661–667 (2011). https://doi.org/10.1093/sysbio/syr027

17. Liu, L., Yu, L., Edwards, S.V.: A maximum pseudo-likelihood approach for estimating species trees under the coalescent model. BMC Evol. Biol. **10**, 302 (2010)

18. Liu, L., Yu, L., Pearl, D.K., Edwards, S.V.: Estimating species phylogenies using coalescence times among sequences. Syst. Biol. **58**(5), 468–477 (2009)

19. Ly-Trong, N., Naser-Khdour, S., Lanfear, R., Minh, B.Q.: Alisim: a fast and versatile phylogenetic sequence simulator for the genomic era. Mol. Biol. Evol. **39**(5), msac092 (2022)

20. Maddison, W.P.: Gene trees in species trees. Syst. Biol. **46**, 523–536 (1997)

21. Mahbub, M., Wahab, Z., Reaz, R., Rahman, M.S., Bayzid, M.S.: wQFM: highly accurate genome-scale species tree estimation from weighted quartets. Bioinformatics **37**(21), 3734–3743 (2021). https://doi.org/10.1093/bioinformatics/btab428

22. Mahbub, M., Wahab, Z., Reaz, R., Rahman, M.S., Bayzid, M.S.: wqfm: highly accurate genome-scale species tree estimation from weighted quartets. Bioinformatics **37**(21), 3734–3743 (2021)

23. Mallo, D., de Oliveira Martins, L., Posada, D.: Simphy: phylogenomic simulation of gene, locus, and species trees. Syst. Biol. **65**(2), 334–344 (2016)

24. Mirarab, S., Reaz, R., Bayzid, M.S., Zimmermann, T., Swenson, M.S., Warnow, T.: ASTRAL: genome-scale coalescent-based species tree estimation. Bioinformatics **30**(17), i541–i548 (2014)

25. Molloy, E.K., Warnow, T.: FastMulRFS: fast and accurate species tree estimation under generic gene duplication and loss models. Bioinformatics **36**($Supplement_1$), i57–i65 (Jul 2020). https://doi.org/10.1093/bioinformatics/btaa444

26. Morel, B., Schade, P., Lutteropp, S., Williams, T.A., Szöllősi, G.J., Stamatakis, A.: Speciesrax: a tool for maximum likelihood species tree inference from gene family trees under duplication, transfer, and loss. Mol. Biol. Evol. **39**(2), msab365 (2022)

27. Mossel, E., Roch, S.: Incomplete lineage sorting: consistent phylogeny estimation from multiple loci. IEEE Comp. Biol. Bioinform. **7**(1), 166–171 (2011)

28. Ohno, S.: Evolution by gene duplication. Springer Science & Business Media (2013)
29. Parsons, R.A., Bansal, M.S.: Duploss-2: Improved phylogenomic species tree inference under gene duplication and loss. Syst. Biol. syaf073 (2025)
30. Price, M.N., Dehal, P.S., Arkin, A.P.: FastTree 2 - approximately maximum-likelihood trees for large alignments. PLoS ONE 5(3), e9490 (2010). https://doi.org/10.1371/journal.pone.0009490
31. Price, M.N., Dehal, P.S., Arkin, A.P.: Fasttree: computing large minimum evolution trees with profiles instead of a distance matrix. Mol. Biol. Evol. 26(7), 1641–1650 (2009)
32. Rafi, A., et al.: wQFM-TREE: highly accurate and scalable quartet-based species tree inference from gene trees. Bioinform. Adv. 5(1), vbaf053 (03 2025). https://doi.org/10.1093/bioadv/vbaf053
33. Rasmussen, M.D., Kellis, M.: Unified modeling of gene duplication, loss, and coalescence using a locus tree. Genome Res. 22(4), 755 (2012). https://doi.org/10.1101/gr.123901.111
34. Reaz, R., Bayzid, M.S., Rahman, M.S.: Accurate phylogenetic tree reconstruction from quartets: a heuristic approach. PLoS ONE 9(8), e104008 (2014)
35. Snir, S., Rao, S.: Quartet maxcut: a fast algorithm for amalgamating quartet trees. Mol. Phylogenet. Evol. 62(1), 1–8 (2012)
36. Swofford, D.: PAUP*: Phylogenetic analysis using parsimony (* and other methods). Ver. 4. Sinauer Associates, Sunderland, Massachusetts (2002)
37. Wehe, A., Bansal, M.S., Burleigh, J.G., Eulenstein, O.: Duptree: a program for large-scale phylogenetic analyses using gene tree parsimony. Am. J. Bot. 24(13), 1540–1541 (2008)
38. Wickett, N.J., et al.: Phylotranscriptomic analysis of the origin and early diversification of land plants. Proc. Natl. Acad. Sci. 111(45), E4859–E4868 (2014)
39. Willson, J., Roddur, M.S., Liu, B., Zaharias, P., Warnow, T.: DISCO: species tree inference using multicopy gene family tree decomposition. Syst. Biol. 71(3), 610–629 (08 2021). https://doi.org/10.1093/sysbio/syab070
40. Zhang, C., Nielsen, R., Mirarab, S.: Aster: a package for large-scale phylogenomic reconstructions. Mol. Biol. Evol. 42(8), msaf172 (2025)
41. Zhang, C., Scornavacca, C., Molloy, E.K., Mirarab, S.: ASTRAL-Pro: quartet-based species-tree inference despite paralogy. Mol. Biol. Evol. 37(11), 3292–3307 (2020). https://doi.org/10.1093/molbev/msaa139

On the Consistency of Duplication, Loss, and Deep Coalescence Gene Tree Parsimony Costs Under the Multispecies Coalescent

Nicolae Sapoval[1]([envelope]) [iD] and Luay Nakhleh[1,2]([envelope]) [iD]

[1] Department of Computer Science, Rice University, Houston, TX 77005, USA
{nsapoval,nakhleh}@rice.edu
[2] Department of BioSciences, Rice University, Houston, TX 77005, USA

Abstract. Gene tree parsimony (GTP) is a common approach for efficient reconciliation of multiple discordant gene tree phylogenies for the inference of a single species tree. However, despite the popularity of GTP methods due to their low computational costs, prior work has shown that some commonly employed parsimony costs are statistically inconsistent under the multispecies coalescent process. Furthermore, a fine-grained analysis of the inconsistency has indicated potentially complimentary behavior of duplication and deep coalescence costs for symmetric and asymmetric species trees. In this work, we prove inconsistency of GTP estimators for all linear combinations of duplication, loss and deep coalescence scores. We also explore empirical implications of this result evaluating inference results of several GTP cost schemes under varying levels of incomplete lineage sorting.

Keywords: Phylogenomics · Gene tree parsimony · Gene tree discordance · Gene duplication and loss · Multispecies coalescent

1 Introduction

Inference of the evolutionary history of a set of species—represented by a phylogenomic species tree—is challenging due to the discordance in the evolutionary histories of the individual genes within genomes [18,29]. Evolutionary histories of the individual homologous sequences, modeled as gene trees [14], are often discordant with the species tree tracing out the history of speciation events due to different biological processes. Gene duplication and loss [22,34] (GDL), and incomplete lineages sorting (ILS) [18] both contribute to this observed discordance. Statistical frameworks for modeling these biological phenomena have emerged, with multispecies coalescent (MSC) accounting for ILS [5,18,26], and DLCoal [27], MLMSC [10], MSC-DL [20] accounting for GDL and ILS at the same time. In order to address the underlying biological question of phylogenomic species tree inference multiple problem formulations and corresponding approaches have been proposed. In this work we will only focus on summary

M. Lafond (Ed.): RECOMB-CG 2026, LNBI 16569, pp. 161–179, 2026.
https://doi.org/10.1007/978-3-032-26891-4_9

methods that take as input a set of gene trees, and aim to reconstruct a species tree that optimizes some compatibility criterion.

Among the summary methods, there are multiple approaches that are proven to be statistically consistent under the MSC [1,9,11–13,19,21] Also, more recent work has shown that some of the proposed approaches are also statistically consistent under DLCoal or variants of it [8,16,23,25]. Conversely, some of the gene tree parsimony (GTP) methods that seek a species tree that minimizes the reconciliation cost with respect to the events of interest (duplications, losses [2,24], and deep coalescences [14,30,35] have been previously shown to be inconsistent under the MSC [31]. However, in practice, GTP methods are still widely used due to easily interpretable optimization criteria, and relative computational efficiency, especially when compared to fully statistical methods that use maximum likelihood estimation or Bayesian inference. Furthermore, while the inconsistency of the individual GTP costs has been previously established, the results for joint optimization under combined criteria have not been theoretically analyzed [28].

In this work, we prove that any linear combination of the gene duplication and deep coalescence costs yields a GTP estimator that is inconsistent under MSC. In particular, we prove that there exists a species tree topology and a set of branch length parameters (i.e. an anomaly zone) for which the GTP estimator converges to an incorrect species tree topology. This is particularly interesting, since the anomaly zone for duplications arises from a symmetric species tree topology, and the one for deep coalescence score from an asymmetric topology. Additionally, we provide an extensive simulation based comparison of different GTP scoring schemes which matches our theoretical results, and offers potential guidance for selecting optimal GTP scoring schemes in practical settings.

2   Preliminaries

Definition 1 (Gene Tree Parsimony Estimator). *Given a collection of gene trees $\mathcal{G} = \{G_i\}_{i=1}^m$ and a cost function $c(G, S)$ we call the species tree defined as*

$$\hat{S} = \arg\min_S \sum_{i=1}^m c(G_i, S),$$

the gene tree parsimony estimator of the species tree with cost function c.

Definition 2 (Consistent Estimator). *Let S_T denote the true species tree. Let $\mathcal{G} = \{G_i\}_{i=1}^m$ be a set of gene trees arising from a multispecies coalescent (MSC) process on S_T. Then, we say that a species tree estimator $\hat{S}_m$ is* consistent *if*

$$\mathbb{P}\left(\lim_{m\to\infty} \hat{S}_m = S_{GT}\right) = 1.$$

We define the gene duplication ($c_D(G, S)$), gene loss ($c_L(G, S)$), and deep coalescence ($c_X(G, S)$) costs analogously to [35]. Furthermore, analogously to [28] we define a generalized cost given by the linear combination of duplication, loss and deep coalescence costs

$$c_{wDLX}(G,S) = w_D c_D(G,S) + w_L c_L(G,S) + w_X c_X(G,S).$$

Observation 1 *Let $L(T)$ denote the multi-set of leaf labels of a tree T. If $L(G) = L(S)$ and all labels are unique, then from Theorem 3.1 in [35] it follows that*

$$c_X(G,S) = c_L(G,S) - 2c_D(G,S).$$

Thus, it follows that

$$\begin{aligned}
c_{wDLX}(G,S) &= w_D c_D(G,S) + w_L c_L(G,S) + w_X c_X(G,S) \\
&= w_X c_X(G,S) + w_D c_D(G,S) + w_L(c_X(G,S) + 2c_D(G,S)) \\
&= (w_X + w_L)c_X(G,S) + (w_D + 2w_L)c_D(G,S)
\end{aligned}$$

for G satisfying the condition above.

Thus, for the rest of the manuscript we will be concerned with proving statistical inconsistency of a cost given by $\alpha c_D(G,S) + \beta c_X(G,S)$ where $\alpha, \beta \in \mathbb{R}$.

Observation 2 *Let $\mathcal{G} = \{G_i\}_{i=1}^m$ be a set of gene trees arising from a MSC process on S_T, and let $\mathcal{T}_N$ denote the set of all trees on N taxa, with $S_T \in \mathcal{T}_N$. Let $\hat{S}_m$ be the gene tree parsimony estimator of the species tree with cost function c, and let $\mathbb{P}(T|S_T)$ denote the probability of observing gene tree T given species tree S_T under the MSC. From the strong law of large numbers it follows that*

$$\mathbb{P}\left(\lim_{m \to \infty} \hat{S}_m = \arg\min_S \sum_{T \in \mathcal{T}_N} c(T,S)\mathbb{P}(T|S_T) \right) = 1.$$

Hence, it follows that the consistency of a gene tree parsimony estimator with cost function c can be determined by checking if $S^* = \arg\min_S \sum_{T \in \mathcal{T}_N} c(T,S)\mathbb{P}(T|S_T) = \arg\min_S \mathbb{E}_{S_T}[c(S)]$ is unique and equal to S_T.

3   Theoretical Results

3.1   Statistical Inconsistency of GTP Under MSC

In prior work [31] it has been shown that the estimators under the deep coalescence cost are inconsistent when the true species tree topology is asymmetric. Furthermore, by explicitly analyzing all possible topologies for 4 species (Table 1), we note that the estimators under the duplication cost are inconsistent when the true species tree topology is symmetric.

This phenomenon is illustrated in Fig. 1. The parsimony estimator using duplication score is inconsistent on a symmetric (with respect to the short branch) topology (Fig. 1A,C) and the parsimony estimator using deep coalescence score is inconsistent on an asymmetric topology (Fig. 1B,D).

We can formalize and prove these observations as the following two lemmas.

Table 1. Probabilities, deep coalescence (c_X) and duplication (c_D) costs for each of the 15 rooted binary gene trees with leaf labels A, B, C, and D given the either the species tree $(S_1; \lambda) = ((a,b) : y, (c,d) : x)$ or $(S_4; \lambda) = (((a,b) : y, c) : x, d)$.

Gene tree T_i	$\mathbb{P}(T_i\lvert S_1; \lambda)$	c_X	c_D	$\mathbb{P}(T_i\lvert S_4; \lambda)$	c_X	c_D
$T_1 = ((a,b),(c,d))$	$1 - \frac{2}{3}(e^{-x} + e^{-y}) + \frac{4}{9}e^{-(x+y)}$	0	0	$\frac{1}{3}e^{-x} - \frac{1}{6}e^{-(x+y)} - \frac{1}{18}e^{-(3x+y)}$	1	1
$T_2 = ((a,c),(b,d))$	$\frac{1}{9}e^{-(x+y)}$	2	1	$\frac{1}{6}e^{-(x+y)} - \frac{1}{18}e^{-(3x+y)}$	2	1
$T_3 = ((a,d),(c,b))$	$\frac{1}{9}e^{-(x+y)}$	2	1	$\frac{1}{6}e^{-(x+y)} - \frac{1}{18}e^{-(3x+y)}$	2	1
$T_4 = (((a,b),c),d)$	$\frac{1}{3}e^{-x} - \frac{5}{18}e^{-(x+y)}$	1	1	$1 - \frac{2}{3}(e^{-x} + e^{-y}) + \frac{1}{3}e^{-(x+y)} + \frac{1}{18}e^{-(3x+y)}$	0	0
$T_5 = (((a,b),d),c)$	$\frac{1}{3}e^{-x} - \frac{5}{18}e^{-(x+y)}$	1	1	$\frac{1}{3}e^{-x} - \frac{1}{6}e^{-(x+y)} - \frac{1}{9}e^{-(3x+y)}$	1	1
$T_6 = (((c,d),b),a)$	$\frac{1}{3}e^{-y} - \frac{5}{18}e^{-(x+y)}$	1	1	$\frac{1}{18}e^{-(3x+y)}$	3	2
$T_7 = (((c,d),a),b)$	$\frac{1}{3}e^{-y} - \frac{5}{18}e^{-(x+y)}$	1	1	$\frac{1}{18}e^{-(3x+y)}$	3	2
$T_8 = (((a,c),b),d)$	$\frac{1}{18}e^{-(x+y)}$	2	2	$\frac{1}{3}e^{-y} - \frac{1}{3}e^{-(x+y)} + \frac{1}{18}e^{-(3x+y)}$	1	1
$T_9 = (((b,c),a),d)$	$\frac{1}{18}e^{-(x+y)}$	2	2	$\frac{1}{3}e^{-y} - \frac{1}{3}e^{-(x+y)} + \frac{1}{18}e^{-(3x+y)}$	1	1
$T_{10} = (((a,d),b),c)$	$\frac{1}{18}e^{-(x+y)}$	2	2	$\frac{1}{18}e^{-(3x+y)}$	3	2
$T_{11} = (((b,d),a),c)$	$\frac{1}{18}e^{-(x+y)}$	2	2	$\frac{1}{18}e^{-(3x+y)}$	3	2
$T_{12} = (((a,c),d),b)$	$\frac{1}{18}e^{-(x+y)}$	2	2	$\frac{1}{6}e^{-(x+y)} - \frac{1}{9}e^{-(3x+y)}$	2	1
$T_{13} = (((b,c),d),a)$	$\frac{1}{18}e^{-(x+y)}$	2	2	$\frac{1}{6}e^{-(x+y)} - \frac{1}{9}e^{-(3x+y)}$	2	1
$T_{14} = (((a,d),c),b)$	$\frac{1}{18}e^{-(x+y)}$	2	2	$\frac{1}{18}e^{-(3x+y)}$	3	2
$T_{15} = (((b,d),c),a)$	$\frac{1}{18}e^{-(x+y)}$	2	2	$\frac{1}{18}e^{-(3x+y)}$	3	2

Lemma 1. *For all $\alpha \in (0, \infty)$ the gene tree parsimony estimator with the cost function $\alpha c_D(G, S)$ is inconsistent for species trees with $N \geq 4$ taxa under the MSC model.*

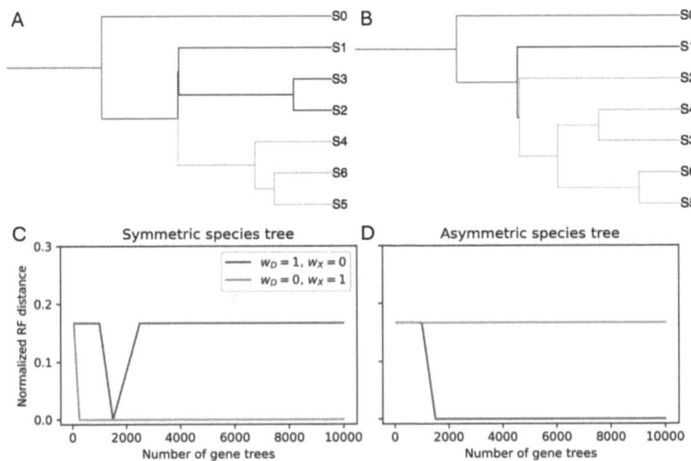

Fig. 1. Duplication and deep coalescence costs are inconsistent on symmetric and asymmetric topologies, respectively. Two example species tree topologies, with one having a symmetric subtree structure associated with the ILS-driver branch (A), and another one with asymmetric subtree structure associate with the ILS-driver branch (B). (C, D) Normalized Robinson-Foulds distances between species trees inferred by GTP and the ground truth species tree as the functions of number of gene trees used. Blue line shows the duplication cost, and orange line shows the deep coalescence cost.

Proof. Let $\mathbb{E}_S[c(S')] = \sum_{G \in \mathcal{T}_N} c(G, S')\mathbb{P}(G|S)$ denote the expected cost of a species tree S' for the ground truth species tree S. From Observation 2 it follows that a gene tree parsimony estimator with a cost function c is inconsistent under the MSC model if there exists $S' \in \mathcal{T}_N$ s.t. $\mathbb{E}_S[c(S')] \leq \mathbb{E}_S[c(S)]$.

Let T_1 and T_4 be the tree topologies as defined in the Table 1. We claim that for any $\beta \in \mathbb{R}^*$ there exist $x \in (0, \infty)$ and $y \in (0, \infty)$ s.t. for $S_T = (T_1; (x, y))$ the following inequality holds

$$\mathbb{E}_{S_T}[c(T_1)] - \mathbb{E}_{S_T}[c(T_4)] \geq 0. \tag{1}$$

Using the probabilities and cost values listed in Table 1, we have

$$\mathbb{E}_{S_T}[c(T_1)] - \mathbb{E}_{S_T}[c(T_4)] = \alpha \left(\frac{1}{18} e^{-(x+y)} + e^{-x} - 1 \right). \tag{2}$$

We seek values (x, y) s.t. the expression above is non-negative.

$$\frac{1}{18}e^{-(x+y)} + e^{-x} - 1 \geq 0$$

$$e^{-y} \geq \frac{18(1 - e^{-x})}{e^{-x}}$$

$$y \leq \log \frac{e^{-x}}{18(1 - e^{-x})}$$

We note that for $x < \log(19/18) \approx 0.054$ the right hand side of the inequality is strictly positive. Thus, it follows that for any $x \in (0, \log(19/18))$ and any $y \in (0, \log \frac{e^{-x}}{18(1-e^{-x})})$ the estimator is inconsistent.

Lemma 2. *For all $\beta \in (0, \infty)$ the gene tree parsimony estimator with the cost function $\beta c_X(G, S)$ is inconsistent for species trees with $N \geq 4$ taxa under the MSC model.*

Proof. Than and Rosenberg [31] showed that the gene tree parsimony estimator with the cost function $c_X(G, S)$ is inconsistent under the MSC model. By the Observation 2 it follows that for $\beta > 0$ the species tree estimated under the cost βc_X is the same as the tree estimated under the cost c_X.

In particular, Lemma 1 indicates that an estimator with the duplication cost will prefer asymmetric topologies in the branch length anomaly zone for symmetric species tree. While Lemma 2 implies that an estimator optimizing the deep coalescence cost will prefer symmetric topologies in the branch length anomaly zone for asymmetric species trees. Furthermore, we note that for any non-zero weight choice β for the deep coalescence cost the resulting combined cost will prefer symmetric topologies when the ground truth tree is asymmetric. More precisely the following lemma holds.

Lemma 3. *For any choice of weights $\alpha, \beta \in [0, \infty)$ the gene tree parsimony estimator with the cost function $c(G, S) = \alpha c_D(G, S) + \beta c_X(G, S)$ is statistically inconsistent for species trees with $N = 4$ taxa under the MSC model.*

Proof. The cases $\beta = 0$ and $\alpha = 0$ follow from Lemmas 1, 2 respectively. Hence, we can assume that $\beta \neq 0$.

Let $\mathbb{E}_S[c(S')] = \sum_{G \in \mathcal{T}_N} c(G, S')\mathbb{P}(G|S)$ denote the expected cost of a species tree S' for the ground truth species tree S. From Observation 2 it follows that a gene tree parsimony estimator with a cost function c is inconsistent under the MSC model if there exists $S' \in \mathcal{T}_N$ s.t. $\mathbb{E}_S[c(S')] \leq \mathbb{E}_S[c(S)]$.

Let T_1 and T_4 be the tree topologies as defined in the Table 1. We claim that for any $\alpha \in [0, \infty)$ and any $\beta \in (0, \infty)$ there exist $x \in (0, \infty)$ and $y \in (0, \infty)$ s.t. for $S_{\mathrm{GT}} = (T_4; (x, y))$ the following inequality holds

$$\mathbb{E}_{S_{\mathrm{GT}}}[c(T_4)] - \mathbb{E}_{S_{\mathrm{GT}}}[c(T_1)] \geq 0. \tag{3}$$

Using the probabilities and cost values listed in Table 1, we have

$$\mathbb{E}_{S_{GT}}[c(T_4)] - \mathbb{E}_{S_{GT}}[c(T_1)] = (\alpha+\beta)(e^{-x}-1) + \frac{1}{6}(\beta-\alpha)e^{-(x+y)} + \frac{2}{9}(\alpha+\beta)e^{-(3x+y)} \tag{4}$$

We seek values (x, y) s.t. the expression above is non-negative. We start by setting $y = -\log 0.9 \approx 0.105$ to obtain

$$(\alpha + \beta)(e^{-x} - 1) + \frac{3}{20}(\beta - \alpha)e^{-x} + \frac{1}{10}(\alpha + 2\beta)e^{-3x} \geq 0 \tag{5}$$

$$(4 + 2\gamma)e^{-3x} + (23 + 17\gamma)e^{-x} - 20\gamma - 20 \geq 0, \tag{6}$$

where $\gamma = \alpha/\beta$, which is well defined since $\beta \neq 0$.

Now, let $f(x) = (4+2\gamma)e^{-3x} + (23+17\gamma)e^{-x} - 20\gamma - 20$, clearly f is a continuous function with $\lim_{x\to\infty} f(x) = -20\gamma - 20$. Furthermore, we have that

$$f'(x) = -(23 + 17\gamma)e^{-x} - (6\gamma + 12)e^{-3x},$$

and hence the function $f(x)$ is decreasing for any value of $\gamma \in [0, \infty)$.

Thus, we know that $f(x)$ is decreasing on $[0, \infty)$ and $f(0) = 7 - \gamma > 0$ for $\gamma < 7$. Since $\lim_{x\to\infty} f(x) < 0$ by the intermediate value theorem it follows that there exists some $x_r \in (0, \infty)$ s.t. $f(x_r) = 0$, and subsequently for all $x \in (0, x_r)$ we have $f(x) > 0$. Therefore, we can conclude that for $y = -\log 0.9$ and $x \in (0, x_r)$ the inequality (3) holds for $\gamma = \alpha/\beta < 7$.

For the case $\gamma \geq 7$, we will consider T_1 and T_4 as before, and show that $\alpha \in [0, \infty)$ and any $\beta \in (0, \infty)$ s.t. $\alpha/\beta \geq 7$ there exist $x \in (0, \infty)$ and $y \in (0, \infty)$ s.t. for $S_T = (T_1; (x, y))$ the following inequality holds

$$\mathbb{E}_{S_T}[c(T_1)] - \mathbb{E}_{S_T}[c(T_4)] \geq 0. \tag{7}$$

Using the probabilities and cost values listed in Table 1, we have

$$\mathbb{E}_{S_T}[c(T_1)] - \mathbb{E}_{S_T}[c(T_4)] = (\alpha + \beta)(e^{-x} - 1) + \frac{5}{18}(7\alpha + \beta)e^{-(x+y)} - \frac{2}{3}\beta e^{-y}. \tag{8}$$

We seek values (x, y) s.t. the expression above is non-negative. We start by setting $y = -\log 0.9 \approx 0.105$ to obtain

$$(55\gamma + 25)e^{-x} - 20\gamma - 32 \geq 0, \tag{9}$$

where $\gamma = \alpha/\beta$. Proceeding the same way as in the previous case, we note that $f'(x) < 0$ and $\lim_{x\to\infty} f(x) < 0$. Furthermore, $f(0) = 35\gamma - 7 > 0$ for all $\gamma \geq 7$. Hence, once again by the intermediate value theorem it follows that there exists x for which the inequality holds.

Thus, combining these two results, we can conclude that for any choice of α, β there exists a 4-taxon species tree for which the parsimony estimator is inconsistent.

Than and Rosenberg [31] provided a general framework for recognizing an embedded 4-taxon tree within a larger species tree, thus extending proofs of statistical inconsistency of parsimony methods from 4-taxon trees to all $N \geq 4$.

Lemma 4. *Let S_T be a species tree with 5 or more leaves, and let S_1 and S_4 be the four-leaf tree as in Table 1. Let $S_4' = (((A_1, A_2), A_3), A_4)$ denote the embedded asymmetric structure in S_T, and let $S_1' = ((A_1, A_2), (A_3, A_4))$ denote an alternative symmetric structure (see [31], Section 3.3). Then the following inequalities hold:*

$$\rho \mathbb{E}_{S_4}[c(T_4)] \leq \mathbb{E}_{S_T}[c(S_4')] \tag{10}$$

$$\mathbb{E}_{S_T}[c(S_1')] \leq \mathbb{E}_{S_4}[c(T_1)] + (1 - \rho)(\alpha(N + 29) + \beta(N^2 + 30)), \tag{11}$$

where the coefficient $(\alpha(N + 29) + \beta(N^2 + 30))$ is determined by the maximum values of duplication [7] and deep coalescence [31] scores, respectively.

We note that the arguments used to derive these inequalities do not explicitly depend on the topology, and hence analogous bounds can be derived for an embedded symmetric structure and an alternative asymmetric structure. Hence, by leveraging this result, we can extend our observation to all trees with 4 or more taxa, proving that no linear combination of the duplication and deep coalescence costs (and by Observation 1 no linear combination of duplication, loss, and deep coalescence costs) is statistically consistent under the MSC model.

Theorem 1. *For any choice of weights $\alpha, \beta \in \mathbb{R}$ the gene tree parsimony estimator with the cost function $c(G, S) = \alpha c_D(G, S) + \beta c_X(G, S)$ is statistically inconsistent for species trees with $N \geq 4$ taxa under the MSC model.*

Proof. Combining the two inequalities from Lemma 4 it follows that the estimator is inconsistent as long as there exist ρ satisfying

$$\rho > \frac{\mathbb{E}_{S_4}[c(T_1)] + \alpha(N + 29) + \beta(N^2 + 30)}{\mathbb{E}_{S_4}[c(T_4)] + \alpha(N + 29) + \beta(N^2 + 30)}. \tag{12}$$

Now, from Lemma 3 we know that the right-hand side of the inequality above is less than 1 for $\alpha/\beta < 7$ the appropriate choice of parameters, and hence there exists ρ satisfying the above condition. Similarly, for $\alpha/\beta \geq 7$, we need to find ρ that is greater than $1/\text{RHS}$ in inequality 12, which is also satisfiable for appropriate parameter choice. Thus, we conclude that for any choice of weights α, β the gene tree parsimony estimator with the cost function $c(G, S) = \alpha c_D(G, S) + \beta c_X(G, S)$ is statistically inconsistent for species trees with $N \geq 4$ taxa under the MSC model.

Remark 1. The result above shows that given any linear combination of the scores, there exists an anomaly zone in which this combination is inconsistent. However, this does not prove that there exists an anomaly zone that simultaneously misleads both duplication and deep coalescence GTPs evaluated independently. While we do not have a rigorous proof of existence of such an anomaly zone, empirical evidence for it can be provided by a tree that simultaneously exhibits symmetric and asymmetric gadgets (Fig. 2).

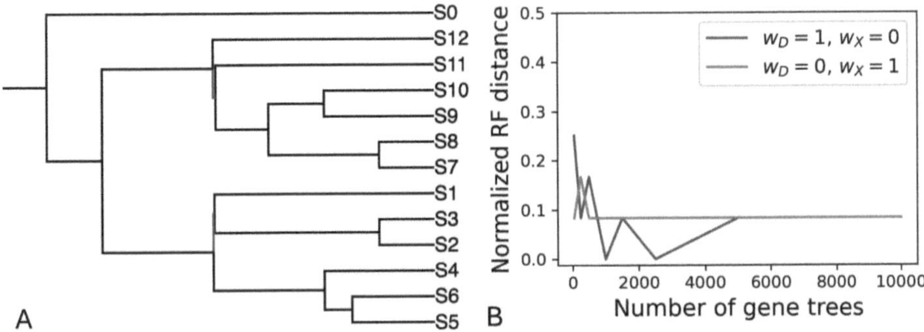

Fig. 2. (A) A phylogenetic tree exhibiting both kinds of problematic branches for GTP duplication and deep coalescence costs (highlighted in red). (B) Normalized Robinson-Foulds distance to the ground truth species tree as the function of the number of input gene trees. Neither the duplication-based nor the deep-coalescence-based GTP approach converges to the correct topology. Each approach fails to recover its corresponding anomaly branch.

4   Empirical Results

4.1   Experimental setup

In order to provide an empirical evaluation of GTP methods under MSC with gene duplication and loss, we have designed a set of simulated data experiments, largely following the setup from a prior study by Zhang *et al.* [33] and utilizing SimPhy [15]. We have considered a total of four simulation scenarios: (A) high ILS, duplication and loss rates; (B) low ILS, high duplication and loss rates; (C) low ILS, high duplication, and low loss rates; and (D) high ILS, and high duplication rates, and no loss rate (Table 2).

Table 2. Details of the simulation scenarios considered in this study.

Scenario	Eff. pop. size	Duplication rate	Loss rate
A	4.7×10^8	4.9×10^{-10}	4.9×10^{-10}
B	4.8×10^7	4.9×10^{-10}	4.9×10^{-10}
C	4.8×10^7	4.9×10^{-10}	4.9×10^{-11}
D	4.7×10^8	4.9×10^{-11}	0

We controlled the level of ILS via the effective population size parameter, and duplication and loss rates via respective duplication/loss rate parameters. For all scenarios, the number of independent replicates was kept at 50. The random seed was kept fixed at 1441 for all simulation scenarios. For scenario A, we simulated species trees with 10, 20, and 50 taxa. For all of the other scenarios, we only considered the 50-taxon case.

Species Tree Simulation. Species trees were simulated with SimPhy [15] under the pure birth model. The birth rate (-sb) was set to 5×10^{-9} for all simulations. Tree height (-st) was sampled from the log-normal distribution with the location parameter set to 21.25 and scale parameter set to 0.2. The ratio of ingroup tree height to the length of the branch to the ingroup (-so) was kept at 1 for all simulations. Tree-wide substitution rate (-su) was sampled from the log-normal distribution with the location parameter set to -21.9 and scale parameter set to 0.1. Finally, tree-wide effective population size (-sp) was set to 4.7×10^8 for scenarios A and D, and to 4.8×10^7 for scenarios B and C.

Gene Tree Simulation. For all scenarios 250 and 500 gene trees were simulated with SimPhy [15] with duplication and loss rates specified as indicated in Table 2. Substitution rate heterogeneity was modeled with: species-specific branch rate heterogeneity modifier (-hs) sampled from log-normal distribution (1.5, 1), gene-family-specific rate heterogeneity modifier (-hl) sampled from log-normal distribution (1.551533, 0.6931472), and gene-by-lineage-specific rate heterogeneity modifier (-hg) sampled from log-normal distribution (1.5, 1).

Sequence Simulation. For each of the four scenarios A, B, C, D we simulated multiple sequence alignments with INDELible [6] for the $N = 50$ case. We simulated three alignments with sequence lengths of 100, 250, and 500 respectively. For all simulated alignments, the same parameters have been used. Substitutions were modeled with the GTR model with rates sampled from a 6-dimensional Dirichlet distribution with parameters $(16, 3, 5, 5, 6, 15)$. Equilibrium probabilities were sampled from a 4-dimensional Dirichlet distribution with parameters $(36, 26, 28, 32)$. Indels were modeled using Zipfian distribution with parameter sampled uniformly from $[1.5, 2]$ and a maximum indel size of 10. The insertion rate was set equal to the deletion rate and sampled uniformly from $[0.001, 0.002]$.

Gene Tree Inference. For the simulated alignments, gene trees were inferred with IQ-TREE [17] under the $\text{GTR} + \Gamma$ substitution model with all other parameters set to their default values.

Species Tree Inference. Due to Observation 1 we only focused on weights for duplication and deep coalescence, and furthermore since $c_D(G, S) \leq c_X(G, S)$ for any G and S (Proof of Theorem 3.2 [35]), we only evaluated the cases where $w_D \geq w_X$. We used DynaDup v2.3.2 [3,28] to infer species tree from the collection of gene trees. We inferred species trees with all possible subsets of cost functions, as well as with duplication cost having weights $[2, 4, 8, 16, 32, 64]$ while the deep coalescence cost had a constant weight 1.

4.2   Characteristics of the Simulated Data

First, we quantified the gene tree estimation error (GTEE) measured as the normalized sum of the false positive and false negative splits in the inferred trees (Fig. 3). As expected, the GTEE decreased for longer alignments for all scenarios. The distribution of GTEE was highly similar among the four scenarios considered, with scenarios A and B showing slightly multimodal behavior of the error, absent from the scenarios C and D (Fig. 3).

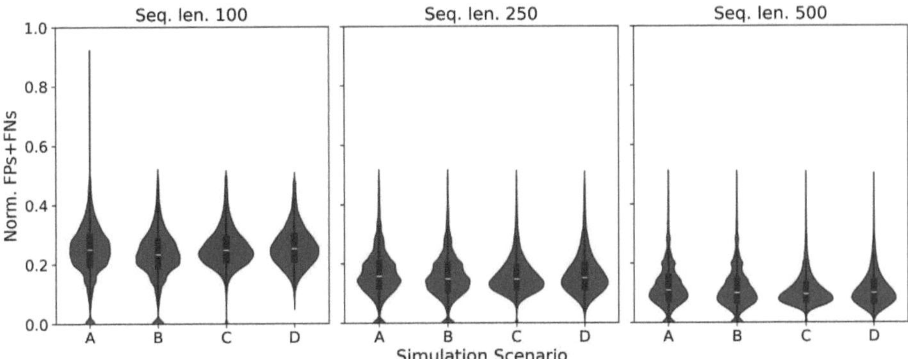

Fig. 3. Violin plots showing the distribution of the normalized sum of false positive (FP) and false negative (FN) splits in each of the inferred gene trees. Panels (left to right) show different sequence lengths for the DNA sequence in the simulation. x-axis indicates the simulation scenario. All reported data is for the trees on 50 taxa.

4.3   Performance on the Simulated Gene Trees

Next, we have evaluated how GTP methods based on duplication, loss, and deep coalescence costs, and an equal weight linear combination of all three costs perform on the simulated gene tree data in the presence of ILS, gene duplication and loss (Scenario A, Fig. 4). We also included ASTRAL-Pro 3, a method designed for handling paralogs while performing inference under MSC model, as a baseline in the evaluation.

We observe that as the number of species increases, the topological error in the inferred species trees increases for all methods (Fig. 4). With an increase in the number of gene trees used for inference, the topological error in the inferred species trees goes down for ASTRAL-Pro 3. However, for all GTP methods, the topological error does not consistently decrease as more gene trees are available, with the exception of duplication-based parsimony cost for 10 and 50 species cases (Fig. 4).

Next, we explored how these methods perform under varying duplication and loss rates, as well as at different levels of ILS.

In particular, across the four scenarios considered (Table 2) we observe that all methods have higher error when the ILS levels are higher (Scenarios A, D, Fig. 5). Notably, in all cases GTP method using only the duplication cost performed the best among GTP methods, and achieved topological error values comparable to or better than ASTRAL-Pro 3 (Fig. 5).

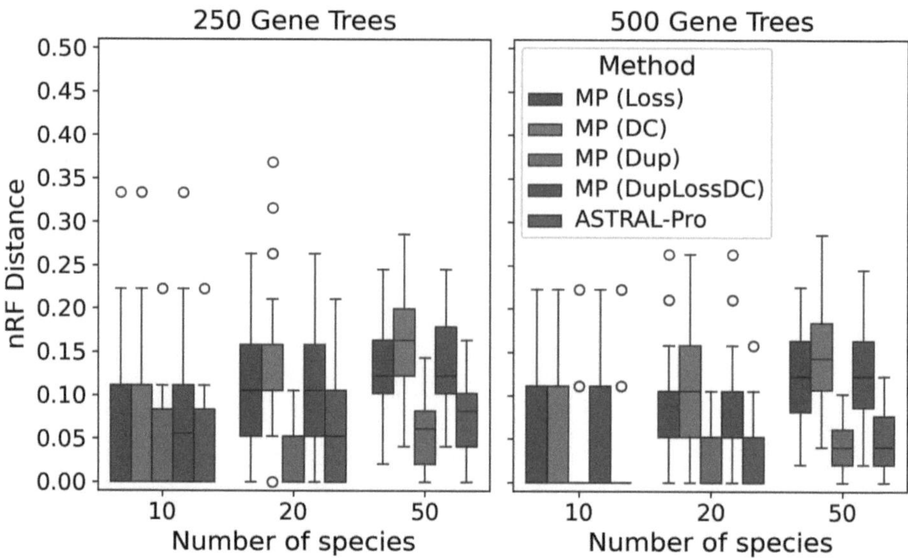

Fig. 4. Normalized Robinson-Foulds distance between inferred and true species trees across GTP scoring schemes in scenario A. y-axis shows normalized Robinson-Foulds distance between the true species tree and the one inferred from true gene trees by each of the methods (shown with different colors). x-axis shows the number of species in each instance of scenario A. All boxplots are based on 50 replicate runs.

To further explore the impact of various weight schemes when combining different GTP scores, we have evaluated how the ratio of the weight associated with the duplication cost (α) to the weight associated with the deep coalescence cost (β) impacts inference accuracy (Fig. 6).

We note that in all cases, as the weight given to duplication cost increases the topological error of the inferred species tree decreases (Fig. 6). This behavior is consistent across all scenarios, with the ratio of 32 achieving performance comparable to the use of duplication cost only (Fig. 6).

4.4   Performance on the Inferred Gene Trees

To further evaluate how GTP methods perform in practice, we have evaluated topological accuracy of the inferred species trees in the case when gene trees are inferred from the simulated multiple sequence alignment data (Figs. 7,8). We note that similarly to the experiments on simulated gene trees, ASTRAL-Pro 3 shows an improvement in performance across all scenarios with the increase in the number of gene trees available (Fig. 7, top vs bottom panels). Conversely, no such trend is observed for the GTP methods. Also, similarly to the case of simulated gene trees, in all scenarios we note that minimizing only the duplication cost (Fig. 7, green; Fig. 8) provides the highest accuracy among the GTP scoring schemes. Additionally, the ILS level is the biggest contributor to GTP

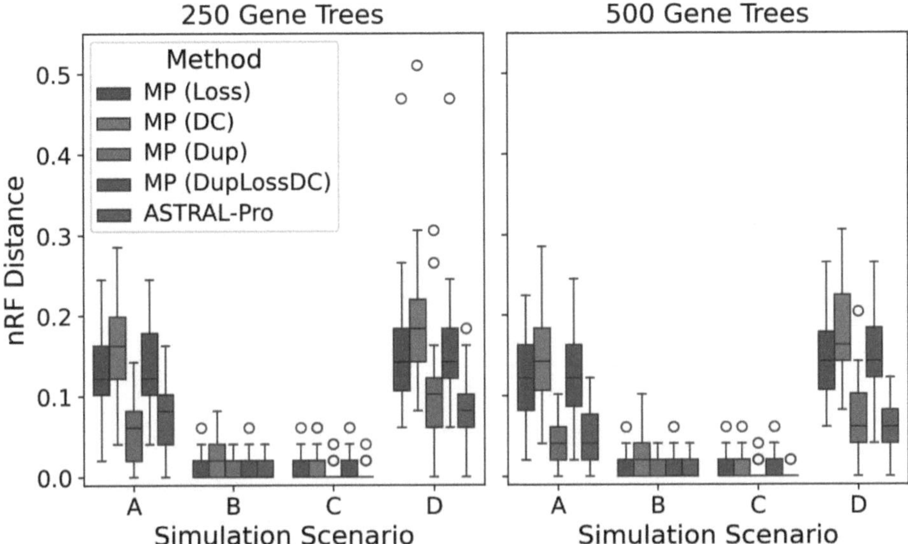

Fig. 5. Normalized Robinson-Foulds distance between inferred and true species trees for scenarios A-D. y-axis shows normalized Robinson-Foulds distance between the true species tree and the one inferred from true gene trees by each of the methods (shown with different colors). x-axis shows the simulation scenario defined by the tree-wide effective population size, and duplication and loss rates. All data shown are based on species trees with 50 taxa. All boxplots are based on 50 replicate runs.

errors leading to scenarios A and D having the highest topological distances to ground truth species trees.

The analysis of the varying ratios of duplication to deep coalescence costs on the inferred gene trees shows the same trends as the one performed on simulated gene trees (Fig. 8). We note that the higher weight given to duplication cost leads to better inference accuracy in all scenarios. Interestingly, unlike in the case of simulated gene trees, in scenario D using the ratio of weights equal to 32 yields better performance than using just the duplication costs alone (Fig. 8, middle and left panels).

4.5   Biological Fungi Data

Finally, we also evaluated performance of the GTP methods on a biological dataset of 16 species of fungi examined in [4,27]. We inferred species trees under various cost schemes for GTP, as well as a single ASTRAL-Pro 3 tree (Fig. 9). All inferred species trees had an identical topology (Fig. 9B) that differed by one split from the topology reported in the prior studies (Fig. 9A). The differing split has been consistently identified by multiple methods on this data in a prior study [32].

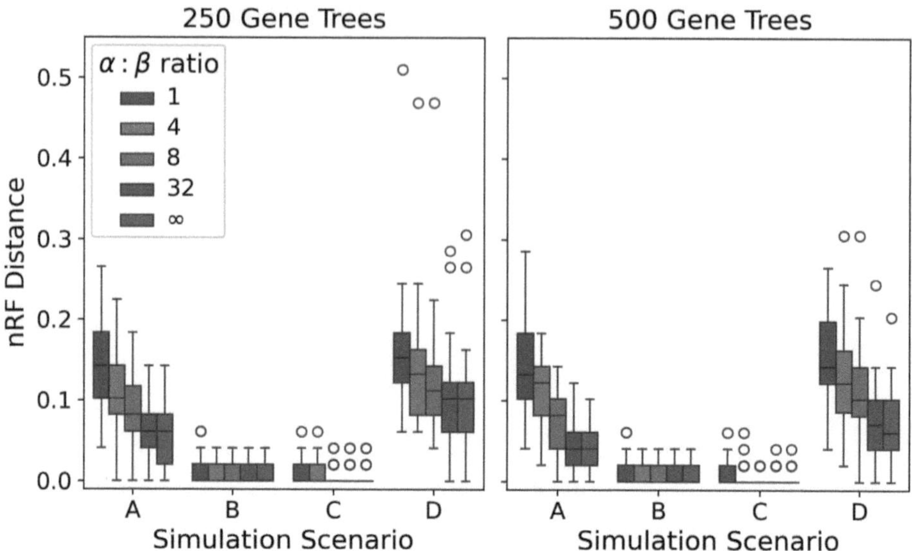

Fig. 6. Normalized Robinson-Foulds distance between inferred and true species trees with varying duplication-to-deep-coalescence cost ratios. y-axis shows normalized Robinson-Foulds distance between the true species tree and the one inferred from true gene trees under different ratios of duplication (α) to deep coalescence (β) costs shown with different colors. x-axis shows the simulation scenario defined by the tree-wide effective population size, and duplication and loss rates. All data shown are based on species trees with 50 taxa. All boxplots are based on 50 replicate runs.

5   Discussion

Theorem 1 proves that no linear combination of the gene duplication and deep coalescence costs yields a statistically consistent estimator. However, our empirical evaluation indicates that in the setting when the ILS rate is low, parsimony based estimators tend to perform well. This is also supported by work that investigates performance of parsimony methods in the absence of ILS [24]. Furthermore, we note that both theoretical analyses and empirical results suggest that the ratio of deep coalescence cost weight to that of duplication cost should be low in order to minimize the anomaly zone. This is in part due to deep coalescence cost always being greater or equal to the duplication cost for any given gene tree species tree pair [35].

We note that while our analysis focused on the inconsistency due to ILS under the MSC, additional work is required to properly investigate consistency of different approaches under unified duplication-loss-coalescence (DLCoal) model [27] and multiloucs-multispecies coalescent (MLMSC) [10]. Recent results show that quartet-based methods are statistically consistent under some formulations of the DLCoal [8, 16, 23]. However, theoretical and empirical results on sample com-

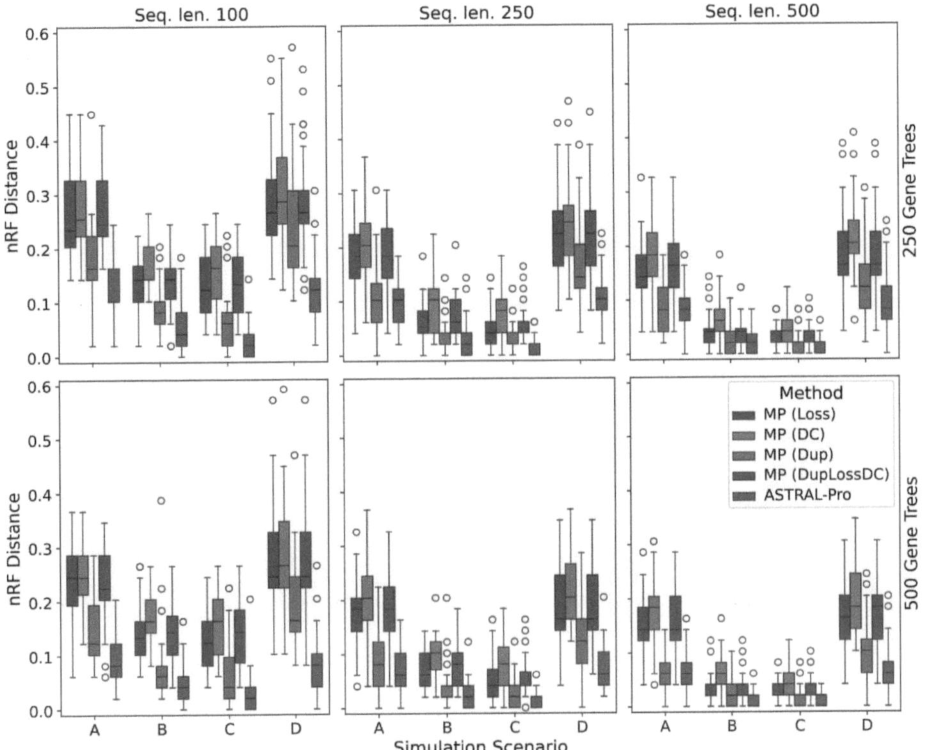

Fig. 7. Normalized Robinson-Foulds distance between inferred and true species trees for scenarios A-D. y-axis shows normalized Robinson-Foulds distance between the true species tree and the one inferred by each of the methods (shown with different colors). x-axis shows the simulation scenario defined by the tree-wide effective population size, and duplication and loss rates. All data shown are based on species trees with 50 taxa. Two rows correspond to 250 and 500 inferred gene trees, and each column corresponds to 100, 250, and 500 base pairs in the DNA sequence. All boxplots are based on 50 replicate runs.

plexity, as well as the investigation of certain practical concerns, such as impact of rooting errors, remain as open questions [23].

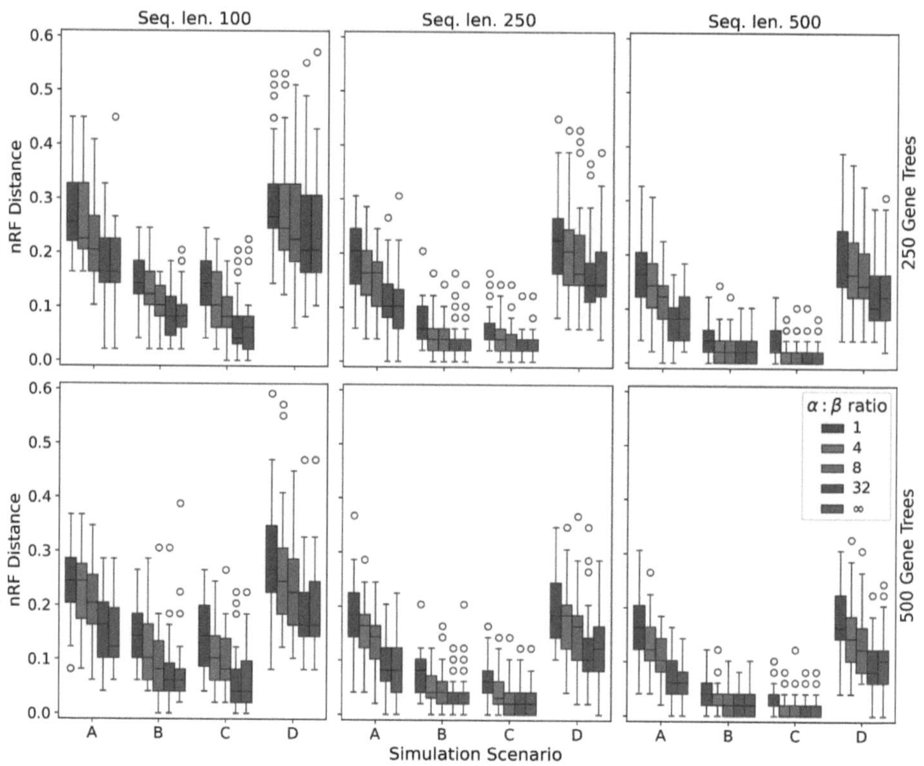

Fig. 8. Normalized Robinson-Foulds distance between inferred and true species trees with varying duplication-to-deep-coalescence cost ratios. y-axis shows normalized Robinson-Foulds distance between the true species tree and the one inferred under different ratios of duplication (α) to deep coalescence (β) costs shown with different colors. x-axis shows the simulation scenario defined by the tree-wide effective population size, and duplication and loss rates. All data shown are based on species trees with 50 taxa. Two rows correspond to 250 and 500 inferred gene trees, and each column corresponds to 100, 250, and 500 base pairs in the DNA sequence. All boxplots are based on 50 replicate runs.

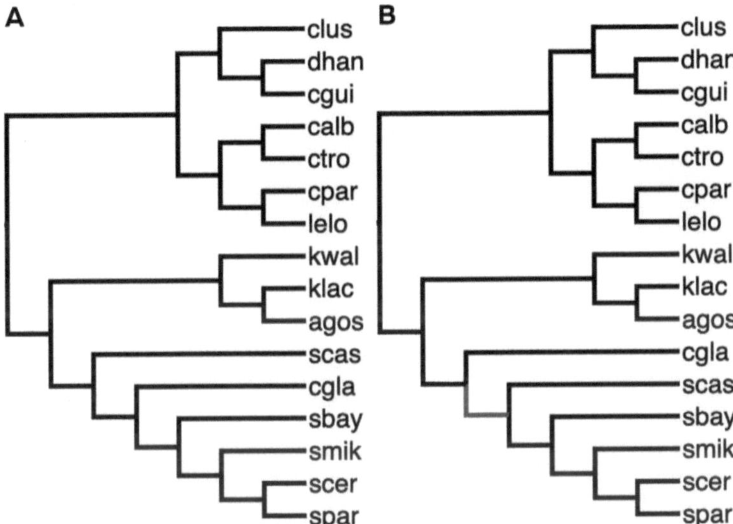

Fig. 9. Species tree topologies for 16 species of fungi: (A) the topology reported in the original publications, (B) topology inferred by ASTRAL-Pro 3 and DynaDup. The red edge highlights the single differing split between the topologies (A) and (B).

Acknowledgments. The authors would like to thank Zhi Yan for her feedback on the manuscript. This work was in part supported by the NSF grants DMS/NIGMS-2153704 and DBI-2030604.

Data Availability Statement. All code used to generate and process the data, as well as perform any plotting is available on GitHub: https://github.com/nsapoval/gt-parsimony.

References

1. Ané, C., Larget, B., Baum, D.A., Smith, S.D., Rokas, A.: Bayesian estimation of concordance among gene trees. Mol. Biol. Evol. **24**(2), 412–426 (2007)
2. Bayzid, M.S., Mirarab, S., Warnow, T.: Inferring optimal species trees under gene duplication and loss. In: Biocomputing 2013, pp. 250–261. World Scientific (2013)
3. Bayzid, M.S., Warnow, T.: Gene tree parsimony for incomplete gene trees: addressing true biological loss. Algorithms Molecular Biol. **13**, 1–12 (2018)
4. Butler, G., et al.: Evolution of pathogenicity and sexual reproduction in eight candida genomes. Nature **459**(7247), 657–662 (2009)
5. Degnan, J.H., Rosenberg, N.A.: Gene tree discordance, phylogenetic inference and the multispecies coalescent. Trends Ecol. Evolution **24**(6), 332–340 (2009)
6. Fletcher, W., Yang, Z.: Indelible: a flexible simulator of biological sequence evolution. Mol. Biol. Evol. **26**(8), 1879–1888 (2009)
7. Górecki, P., Mykowiecka, A., Paszek, J., Eulenstein, O.: Mathematical properties of the gene duplication cost. Discret. Appl. Math. **258**, 114–122 (2019)
8. Hill, M., Legried, B., Roch, S.: Species tree estimation under joint modeling of coalescence and duplication: sample complexity of quartet methods. Ann. Appl. Probab. **32**(6), 4681–4705 (2022)
9. Kubatko, L.S., Carstens, B.C., Knowles, L.L.: Stem: species tree estimation using maximum likelihood for gene trees under coalescence. Bioinformatics **25**(7), 971–973 (2009)
10. Li, Q., Scornavacca, C., Galtier, N., Chan, Y.B.: The multilocus multispecies coalescent: a flexible new model of gene family evolution. Syst. Biol. **70**(4), 822–837 (2021)
11. Liu, L., Yu, L.: Estimating species trees from unrooted gene trees. Syst. Biol. **60**(5), 661–667 (2011)
12. Liu, L., Yu, L., Edwards, S.V.: A maximum pseudo-likelihood approach for estimating species trees under the coalescent model. BMC Evol. Biol. **10**(1), 302 (2010)
13. Liu, L., Yu, L., Pearl, D.K., Edwards, S.V.: Estimating species phylogenies using coalescence times among sequences. Syst. Biol. **58**(5), 468–477 (2009)
14. Maddison, W.P.: Gene trees in species trees. Syst. Biol. **46**(3), 523–536 (1997)
15. Mallo, D., de Oliveira Martins, L., Posada, D.: Simphy: phylogenomic simulation of gene, locus, and species trees. Syst. Biol. **65**(2), 334–344 (2016)
16. Markin, A., Eulenstein, O.: Quartet-based inference is statistically consistent under the unified duplication-loss-coalescence model. Bioinformatics **37**(22), 4064–4074 (2021)
17. Minh, B.Q., et al.: Iq-tree 2: new models and efficient methods for phylogenetic inference in the genomic era. Mol. Biol. Evol. **37**(5), 1530–1534 (2020)
18. Mirarab, S., Nakhleh, L., Warnow, T.: Multispecies coalescent: theory and applications in phylogenetics. Annu. Rev. Ecol. Evol. Syst. **52**(1), 247–268 (2021)

19. Mirarab, S., Reaz, R., Bayzid, M.S., Zimmermann, T., Swenson, M.S., Warnow, T.: Astral: genome-scale coalescent-based species tree estimation. Bioinformatics **30**(17), i541–i548 (2014)
20. Mishra, S., Hahn, M.W.: Distribution of gene tree topologies with duplication, loss, and coalescence. bioRxiv, pp. 2026–01 (2026)
21. Mossel, E., Roch, S.: Incomplete lineage sorting: consistent phylogeny estimation from multiple loci. IEEE/ACM Trans. Comput. Biol. Bioinf. **7**(1), 166–171 (2008)
22. Ohno, S.: Evolution by gene duplication. Springer Science & Business Media (2013)
23. Parsons, R., Liu, Y., Dua, P., Markin, A., Molloy, E.: On the correctness of gene tree tagging and the consistency of astral-pro under a unified model of gene duplication, loss, and coalescence. bioRxiv pp. 2026–01 (2026)
24. Parsons, R.A., Bansal, M.S.: Duploss-2: Improved phylogenomic species tree inference under gene duplication and loss. Syst. Biol., syaf073 (2025)
25. Rabiee, M., Sayyari, E., Mirarab, S.: Multi-allele species reconstruction using astral. Mol. Phylogenet. Evol. **130**, 286–296 (2019)
26. Rannala, B., Yang, Z.: Bayes estimation of species divergence times and ancestral population sizes using DNA sequences from multiple loci. Genetics **164**(4), 1645–1656 (2003)
27. Rasmussen, M.D., Kellis, M.: Unified modeling of gene duplication, loss, and coalescence using a locus tree. Genome Res. **22**(4), 755–765 (2012)
28. Saha, P., et al.: Gene tree parsimony in the presence of gene duplication, loss, and incomplete lineage sorting. In: RECOMB International Workshop on Comparative Genomics, pp. 110–128. Springer (2024)
29. Steenwyk, J.L., Li, Y., Zhou, X., Shen, X.X., Rokas, A.: Incongruence in the phylogenomics era. Nat. Rev. Genet. **24**(12), 834–850 (2023)
30. Than, C., Nakhleh, L.: Species tree inference by minimizing deep coalescences. PLoS Comput. Biol. **5**(9), e1000501 (2009)
31. Than, C.V., Rosenberg, N.A.: Consistency properties of species tree inference by minimizing deep coalescences. J. Comput. Biol. **18**(1), 1–15 (2011)
32. Yan, Z., Smith, M.L., Du, P., Hahn, M.W., Nakhleh, L.: Species tree inference methods intended to deal with incomplete lineage sorting are robust to the presence of paralogs. Syst. Biol. **71**(2), 367–381 (2022)
33. Zhang, C., Scornavacca, C., Molloy, E.K., Mirarab, S.: Astral-pro: quartet-based species-tree inference despite paralogy. Mol. Biol. Evol. **37**(11), 3292–3307 (2020)
34. Zhang, J.: Evolution by gene duplication: an update. Trends Ecol. Evolution **18**(6), 292–298 (2003)
35. Zhang, L.: From gene trees to species trees ii: species tree inference by minimizing deep coalescence events. IEEE/ACM Trans. Comput. Biol. Bioinf. **8**(6), 1685–1691 (2011)

Gene Repertoire Evolution Minimizing Episodes of Gains and Losses

Mathieu Gascon, Mattéo Delabre, and Nadia El-Mabrouk(✉)

Département d'informatique et de recherche opérationnelle, Université de Montréal, Québec, Canada
{mathieu.gascon.1,matteo.delabre}@umontreal.ca, mabrouk@iro.umontreal.ca

Abstract. Given a tree topology and an assignment of character states to its leaves, the Small Parsimony Problem (SPP) consists in assigning character states to the internal nodes in a way maximizing a certain parsimony or probabilistic criterion. In the genome rearrangement field, tree leaves are permutations of gene sets, and the problem is to infer permutations at internal nodes minimizing a rearrangement distance. Almost all genome rearrangement models lead to intractable problems for the SPP. Considering only the numerical profiles on a phylogeny, the Count package (Csűrös, 2010) can be used to predict the size of gene families on internal nodes under a parsimony or probabilistic model. Here, we present a tractable version of the SPP: given a tree topology leaf-labeled by unordered gene sets, infer gene sets at internal nodes in a way minimizing the number of gain and loss episodes on the edges of the tree, while having a single gain point for each gene (i.e. under Dollo's law). We show that the entire solution space is covered by testing four possible cases on each internal node's content, leading to a linear-time dynamic programming algorithm for obtaining an optimal solution. We apply our *InOutParsimony* software to clusters of orthologous mitochondrial protein-coding genes (MitoCOGs) in both the mitochondrial and nuclear genomes of 11 land plant species. The results are discussed considering the Endosymbiotic Gene Transfer events shaping the mitochondria and nucleus contents and compared with Count's returned numerical profiles.

Keywords: Genome evolution · Gain and loss · Gene clusters · Synteny · Small parsimony · Dollo parsimony · Dynamic programming

Introduction

Understanding how a gene family, a set of genomic regions, or a set of related genomes have evolved from a common ancestor is an important step towards elucidating a large number of biological questions. If a rooted phylogenetic tree is already available for the considered biological units represented at the leaves of the tree, the problem reduces to the Small Parsimony Problem (SPP) which

© The Author(s) 2026
M. Lafond (Ed.): RECOMB-CG 2026, LNBI 16569, pp. 180–210, 2026.
https://doi.org/10.1007/978-3-032-26891-4_10

consists in inferring the content of ancestral nodes in a way maximizing a certain parsimony (or minimizing a certain cost) or probabilistic criterion.

Inferring ancestral gene sequences by minimizing the alignment score on a tree leaf-labeled by the sequences of extant genes is a well-known SPP handled by the Fitch [14] and Sankoff [21] algorithms. On a genomic level, an algorithm implemented in the software package Count [8] has been developed to predict the size of whole gene families on internal nodes of a phylogenetic tree, under the Dollo parsimony [10], Wagner parsimony [7], or under a probabilistic model [9]. On the other hand, in the genome rearrangement field, biological units are permutations of gene sets. In this case, minimization can be done on the number of order disruptions (number of breakpoints), on rearrangement events remodeling the gene order (inversions, transpositions, translocations, etc.), or on mathematical abstractions such as Double Cut-and-Join (DCJ) or Single-Cut-or-Join (SCJ) operations. Except for the latter case [13,19], almost all genome rearrangement models lead to intractable problems [5,12], even when restricted to the median problem [6,18,20,22]. The problem becomes even harder in the presence of gene duplicates [1,4].

Finally, if the biological units are homologous genomic regions containing genes, also called syntenies, evolving through content-modifying operations (such as gains, losses, duplications, transfers, etc.), the problem is to infer a history minimizing such operations. In a previous set of papers [2,11,15], we presented a generalization of the gene/species reconciliation [16] and developed *Synesth* taking as input a species tree and a synteny tree (tree topology leaf-labeled by unordered gene sets representing homologous genomic regions), and giving as output one or many possible histories (gene sets and events at the internal nodes of the tree) minimizing a certain function (minimum cost, Pareto-optimal vectors of event counts, etc.). However, to simplify the problem, gains were excluded from the optimization function. More precisely, contrary to losses, taking advantage of simultaneous gains of multiple genes to reduce the overall number of individual gain events was disregarded.

In this paper, we present a new and tractable version of the SPP restricting the evolutionary model to gains and losses of groups of genes. More precisely, the problem is to infer gene sets at internal nodes of a synteny tree in a way minimizing the number of gain and loss episodes on the edges of the tree, further considering an extension of Dollo's law, i.e. requesting to have a single gain point for each gene. This is the first attempt to address the above mentioned shortcoming of *Synesth* by accounting for the gain of multiple genes as a single event. We show that the entire solution space is covered by testing four cases on each possible internal node's content, leading to a linear-time dynamic programming algorithm for obtaining an optimal solution.

After setting our notations in the next section, the model is formally defined in Sect. 2 and the algorithm is described in Sect. 3. We then, in Sect. 4, apply the implemented software *InOutParsimony* to a dataset of mitochondrial-encoded protein families (called MitoCOGs) taken from [17], in order to study the exchange of those genes between the mitochondrial and nuclear eukaryotic genomes. For this purpose, we selected 11 land plant species containing a diversified MitoCOG's content in both their nuclear and mitochondrial genomes. The results are discussed in light of the Endosymbiotic Gene Transfer events shaping those genomes and numerical profiles of ancestral genomes are compared with those obtained using Count with both parsimony and probabilistic models. These results illustrate the potential of our method to account for the simultaneous loss of entire biological complexes, but also reflect its limitations related to overly restrictive cost models. We conclude by suggesting avenues for future extensions[1].

1   Notations

All trees considered in this paper are rooted. Given a tree T, we denote by $r(T)$ its root, by $V(T)$ its node set and by $L(T) \subseteq V(T)$ its leafset. A node v' is an *ancestor* of v if v' is on the (inclusive) path between v and the root, and we then call v a *descendant* of v'. The ancestor–descendant relation is denoted $\leq$ and forms a partial order on nodes, in which the root is minimal and the leaves are maximal. The node $v' = p(v)$ immediately preceding $v \neq r(T)$ on this path is the *parent* of v, and then v is a *child* of v'. The set of children of a node v is denoted by $ch(v)$; if $|ch(v)| = 1$, then v is said to be *unary*; if $|ch(v)| = 2$, then it is said to be *binary* and, unless specified otherwise, we denote its children by v_ℓ and v_r in an arbitrary order. A tree is *binary* if all its internal (non-leaf) nodes are binary.

For any node v of T, we denote by T_v the subtree of T rooted at v, i.e. obtained from T by removing all the nodes which are not descendant of v. We denote by $E(T)$ the edge set of T, where each edge is represented by a pair of nodes $(p(v), v)$.

The *lowest common ancestor* (lca) of a subset L' of $L(T)$, denoted $lca_T(L')$, is the ancestor common to all nodes in L' most distant from the root.

A *synteny* is a genomic region containing a set of genes from a set of gene families $\mathcal{F}$, where the genes of a given synteny all belong to different gene families (i.e. repeated gene copies inside a synteny are ignored). Therefore, from now on, a gene is simply identified by the family $g \in \mathcal{F}$ it belongs to. We call a *synteny tree* on $\mathcal{F}$ a tuple $\mathcal{T} = \langle T, \tilde{x} \rangle$ where T is a binary tree and $\tilde{x} : L(T) \rightarrow \mathcal{P}(\mathcal{F}) - \{\emptyset\}$ ($\mathcal{P}(\mathcal{F})$ is the powerset of $\mathcal{F}$) is the function mapping each leaf $v \in L(T)$ (element of $\mathcal{X}$) to its *synteny content*. We will say that v is *labeled* by $\tilde{x}(v)$.

Finally, the restriction of any function f to a subset A of its domain is denoted by $f|_A$.

2   The Small Gain-Loss Parsimony Problem

We consider a tree-like evolutionary model involving gain and loss episodes, starting with a unary root with an empty gene set, and giving rise to a set $\mathcal{F}$

---

[1] For the sake of clarity, the proofs are given in Appendix. The proof of Theorem 4 is omitted for space constraints. It will be included in a future journal version.

of gene families and to an extant set of syntenies on $\mathcal{F}$. Those syntenies may represent genomic regions as well as entire genomes, identified by their gene content (with no duplicates, i.e. with at most one copy per gene family).

Except for the root, all other internal nodes of a history are binary. Binary nodes may correspond to speciations, as well as duplications or transfers in the case of genomic regions. However, as the only events considered in our optimization function are gains and losses, we ignore the nature of those binary nodes.

Since we do not consider the order of genes inside the syntenies, the relative order of gain and loss events on a given edge of a history is irrelevant. Moreover, as we will aim to minimize the overall number of events, any sequence of consecutive gains or losses can be collapsed into a single event. Therefore at most one gain and/or one loss may occur on an edge, thus instead of representing gains and losses as unary nodes, we rather represent them as edge labels.

In the following, for a binary tree T, we will denote by $\dot{T}$ the tree obtained from T by rerooting it to a new unary root, i.e. by adding a new unary node u with a new edge $(u, \mathrm{r}(T))$. This is required in order to account for an initial gain event on an initial edge connecting a hypothetical ancestor with an empty content to the root of the input tree with a non-empty content.

Definition 1 (Explanatory history). *Let $\mathcal{T} = \langle T, \tilde{x} \rangle$ be a synteny tree on $\mathcal{F}$. A history $\mathcal{H}$ explaining $\mathcal{T}$ is a tuple $\langle \dot{T}, \epsilon, x \rangle$, where each node $v \in \mathrm{V}(\dot{T})$ is labeled with a synteny content $x(v) \in \mathcal{P}(\mathcal{F})$, and each edge $e \in \mathrm{E}(\dot{T})$ is labeled with an event subset $\epsilon(e) \subseteq \{\mathrm{Gain}, \mathrm{Loss}\}$ satisfying*

1. For each $l \in L(\dot{T})$, $x(l) = \tilde{x}(l)$
2. For each edge $(v, v_c) \in \mathrm{E}(\dot{T})$:
   (a) $\mathrm{Loss} \in \epsilon((v, v_c))$ iff $x(v) \not\subseteq x(v_c)$
   (b) $\mathrm{Gain} \in \epsilon((v, v_c))$ iff $x(v_c) \not\subseteq x(v)$
3. For each gene family $g \in \mathcal{F}$, there is a unique node $v \in \mathrm{V}(\dot{T}) - \{\mathrm{r}(\dot{T})\}$ such that $g \in x(v) - x(p(v))$. We say that v is the gain point for g and denote it as $\mathrm{Gain}_{\mathcal{H}}(g)$.
4. $x(\mathrm{r}(\dot{T})) = \emptyset$ and for all $v \in \mathrm{V}(\dot{T}) - \{\mathrm{r}(\dot{T})\}$, $x(v) \neq \emptyset$.

The set of all histories explaining a synteny tree $\mathcal{T}$ is denoted by $\mathbb{H}(\mathcal{T})$.

Given a history $\mathcal{H}$ and a node $v \in \mathrm{V}(T)$, we call subhistory of $\mathcal{H}$ rooted at v the history $\mathcal{H}_v = \langle T_v, \epsilon|_{\mathrm{E}(T_v)}, x|_{\mathrm{V}(T_v)} \rangle$, and hanging subhistory of $\mathcal{H}$ rooted at v the same subhistory but also including the edge of $\mathcal{H}$ from $\mathrm{p}(v)$ to v (or, said differently, the subhistory of $\mathcal{H}$ rooted at $\mathrm{p}(v)$ but only keeping the subhistory of its child v).

Finally, for a given history $\mathcal{H}$ explaining $\mathcal{T}$, let $c_{\mathrm{gain}}(\mathcal{H})$ (respec. $c_{\mathrm{loss}}(\mathcal{H})$) be the number of gains (respec. losses) of $\mathcal{H}$, and let $\delta_{\mathrm{gain}} > 0$ (respec. $\delta_{\mathrm{loss}} > 0$) be the gain (respec. loss) cost. Then the cost of $\mathcal{H}$ is $c(\mathcal{H}, \delta_{\mathrm{gain}}, \delta_{\mathrm{loss}}) = \delta_{\mathrm{gain}} \cdot c_{\mathrm{gain}}(\mathcal{H}) + \delta_{\mathrm{loss}} \cdot c_{\mathrm{loss}}(\mathcal{H})$. We are now ready to state the optimization problem. See Fig. 1 for an example.

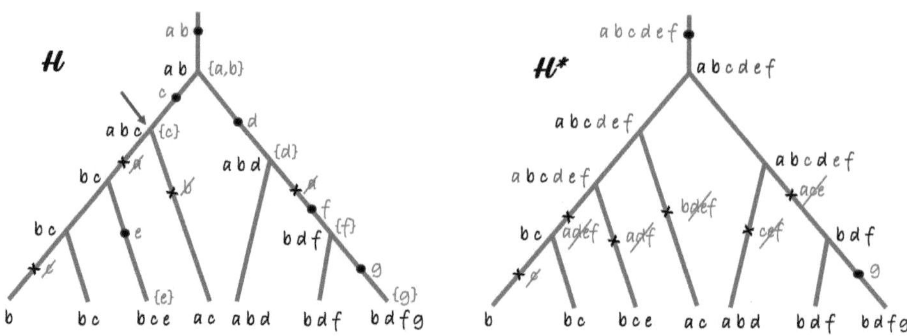

Fig. 1. Two different histories explaining the same synteny tree $\langle T, \tilde{x} \rangle$ on $\mathcal{F} = \{a, b, c, d, e, f, g\}$. In other words, T is simply the underlying binary tree (the same on the left and on the right), and $\tilde{x}$ is the synteny labeling of its leaves. In both histories, gains are represented by dots and losses by crosses. The gray characters represent the set of gained or lost characters. Although written as sequences, the syntenies are unordered sets of genes. In the left figure, the set of characters in curly brackets at a node v represents $x_{\text{lca}}(v)$ (see Definition 2). In both trees, each node v is labeled by $x(v)$ for the considered history. Three pairwise disjoint covering subsets of $x(v)$ are introduced in Definition 3; here, we color each gene $g \in x(v)$ black if $g \in x_{\min}(v)$, red if $g \in x_{\text{in}}(v)$ and blue if $g \in x_{\text{out}}(v)$. (Left) A suboptimal history of cost 10 (6 gains and 4 losses) for a unit cost obtained by assigning $x_{\min}(v)$ to each node v of T. (Right) An optimal history of cost 8 (2 gains and 6 losses), for the unit cost $\delta_{\text{loss}} = \delta_{\text{gain}} = 1$.

SMALL GAIN-LOSS PARSIMONY PROBLEM.

Input: A synteny tree $\mathcal{T} = \langle T, \tilde{x} \rangle$ on $\mathcal{F}$, δ_{gain} and δ_{loss}.

Output: $\mathcal{H}^* \in \text{argmin}_{\mathcal{H} \in \mathbb{H}(\mathcal{T})} \{c(\mathcal{H}, \delta_{\text{gain}}, \delta_{\text{loss}})\}$.

For conciseness, in the rest of this paper, we omit δ_{gain} and δ_{loss} in the input of the cost $c(\mathcal{H}, \delta_{\text{gain}}, \delta_{\text{loss}})$ and simply write $c(\mathcal{H})$.

3   Algorithm

In this section, we are given a synteny tree $\mathcal{T} = \langle T, \tilde{x} \rangle$ on a set $\mathcal{F}$ of gene families. By definition, the only difference between T and $\dot{T}$ is an additional unary root in $\dot{T}$. We will alternate between T and $\dot{T}$ depending on whether a unary root needs to be considered or not.

We first define the lowest possible gain point for each gene family.

Definition 2 (Set of LCAs at a node). *The least common ancestor of a gene family $g \in \mathcal{F}$ is defined as:*

$$\text{lca}(g) = \text{lca}_T(\{l \in L(T) \mid g \in \tilde{x}(l)\})$$

For any $v \in \text{V}(T)$, we define $x_{\text{lca}}(v) = \{g \in \mathcal{F} \mid \text{lca}(g) = v\}$.

We can then define the minimum and extra contents at a node v of T.

Definition 3 (Minimum and extra contents at a node). *Let v be a node of T. We define:*

$$x_{\min}(v) = \{g \in \mathcal{F} \mid \mathrm{lca}(g) \leq v \wedge g \in \cup_{l \in L(T_v)} \tilde{x}(l)\}$$
$$x_{\mathrm{out}}(v) = \{g \in \mathcal{F} \mid g \notin \cup_{l \in L(T_v)} \tilde{x}(l)\}$$
$$x_{\mathrm{in}}(v) = \{g \in \mathcal{F} \mid \mathrm{lca}(g) > v\}$$

For example, for the node v pointed by the blue arrow on the left tree of Fig. 1, $x_{\mathrm{lca}}(v) = \{c\}$, $x_{\min}(v) = \{a, b, c\}$, $x_{\mathrm{out}}(v) = \{d, f, g\}$ and $x_{\mathrm{in}}(v) = \{e\}$.

Notice that, for $v \in T$, $x_{\min}(v)$ corresponds to the set of gene families with lca above v and present in at least one leaf descendant of v, $x_{\mathrm{out}}(v)$ corresponds to the gene families that are not present in the leaves descendant of v and $x_{\mathrm{in}}(v)$ corresponds to the set of gene families with lca below v. Therefore, $x_{\min}(v) \cup x_{\mathrm{out}}(v) \cup x_{\mathrm{in}}(v) = \mathcal{F}$ and the three sets are pairwise disjoint. The next lemma trivially follows from this fact and from the fact that a gene cannot be gained twice in a history.

Lemma 1 (Content of a node). *Let $\mathcal{H} = \langle \dot{T}, \epsilon, x \rangle \in \mathbb{H}(\mathcal{T})$ and let $v \in \mathrm{V}(T)$. Then, $x(v) = x_{\min}(v) \cup O \cup I$ for some sets $O \subseteq x_{\mathrm{out}}(v)$ and $I \subseteq x_{\mathrm{in}}(v)$.*

According to Lemma 1, each node v of a history must be in one of the following four states: "min" if its content is exactly $x_{\min}(v)$, "out" if it additionally contains at least one gene from $x_{\mathrm{out}}(v)$ but none from $x_{\mathrm{in}}(v)$, "in" in the opposite case, or "in out" if it contains at least one gene from both $x_{\mathrm{out}}(v)$ and $x_{\mathrm{in}}(v)$. In the following, we will build recurrence equations to compute the minimum possible cost for each subhistory of an history $\mathcal{H}$ explaining $\mathcal{T} = \langle T, \tilde{x} \rangle$, distinguishing between these four possible states for each node.

Definition 4 (Cost of a subhistory). *Let v be a node of T. For a state $\sigma \in \{\min, \mathrm{out}, \mathrm{in}, \mathrm{in\,out}\}$, define $c_\sigma(v)$ as the minimum cost of an explanatory subhistory rooted at v such that v is in state σ, i.e.,*

- *for $c_{\min}(v)$, such that $x(v) = x_{\min}(v)$;*
- *for $c_{\mathrm{out}}(v)$, such that $x(v) = x_{\min}(v) \cup O$ with $\emptyset \neq O \subseteq x_{\mathrm{out}}(v)$;*
- *for $c_{\mathrm{in}}(v)$, such that $x(v) = x_{\min}(v) \cup I$ with $\emptyset \neq I \subseteq x_{\mathrm{in}}(v)$;*
- *for $c_{\mathrm{in\,out}}(v)$, such that $x(v) = x_{\min}(v) \cup O \cup I$ with O and I as above.*

If v has a child v_c, define $\Delta_\sigma(v_c)$ as the minimum cost of an explanatory hanging subhistory rooted at v such that v is in state σ relative to v_c, i.e.,

- *for $\Delta_{\min}(v_c)$, such that $x(v) = x_{\min}(v)$;*
- *for $\Delta_{\mathrm{out}}(v_c)$, such that $x(v) = x_{\min}(v) \cup O$ with $\emptyset \neq O \subseteq x_{\mathrm{out}}(v_c) - x_{\min}(v)$;*
- *for $\Delta_{\mathrm{in}}(v_c)$, such that $x(v) = x_{\min}(v) \cup I$ with $\emptyset \neq I \subseteq x_{\mathrm{in}}(v_c) \cup x_{\mathrm{lca}}(v_c)$;*
- *for $\Delta_{\mathrm{in\,out}}(v_c)$, such that $x(v) = x_{\min}(v) \cup O \cup I$ with O and I as above.*

Lemma 2 (Partial recurrences including the root edge). *For any edge* $(v, v_c) \in \mathrm{E}(T)$, *assuming the values for* $c_\sigma(v_c)$, $x_{\min}(v)$, *and* $x_{\min}(v_c)$ *are known, the values of* $\Delta_\sigma(v_c)$ *may be computed as follows:*

$$\Delta_{\min}(v_c) = \min \begin{cases} c_{\min}(v_c) + l(v, v_c) \cdot \delta_{\text{loss}} + g(v, v_c) \cdot \delta_{\text{gain}} & \text{(1a)} \\ c_{\text{out}}(v_c) + g(v, v_c) \cdot \delta_{\text{gain}} & \textit{if } l(v, v_c) = 1 \quad \text{(1b)} \\ c_{\text{in}}(v_c) + l(v, v_c) \cdot \delta_{\text{loss}} + \delta_{\text{gain}} & \text{(1c)} \\ c_{\text{in out}}(v_c) + \delta_{\text{gain}} & \textit{if } l(v, v_c) = 1 \quad \text{(1d)} \end{cases}$$

$$\Delta_{\text{out}}(v_c) = \min \begin{cases} c_{\min}(v_c) + \delta_{\text{loss}} + g(v, v_c) \cdot \delta_{\text{gain}} & \text{(2a)} \\ c_{\text{out}}(v_c) + g(v, v_c) \cdot \delta_{\text{gain}} & \text{(2b)} \\ c_{\text{in}}(v_c) + \delta_{\text{loss}} + \delta_{\text{gain}} & \text{(2c)} \\ c_{\text{in out}}(v_c) + \delta_{\text{gain}} & \text{(2d)} \end{cases}$$

$$\Delta_{\text{in}}(v_c) = \min \begin{cases} c_{\min}(v_c) + l(v, v_c) \cdot \delta_{\text{loss}} & \textit{if } g(v, v_c) = 1 \quad \text{(3a)} \\ c_{\text{out}}(v_c) & \textit{if } g(v, v_c) = 1 \textit{ and } l(v, v_c) = 1 \quad \text{(3b)} \\ c_{\text{in}}(v_c) + l(v, v_c) \cdot \delta_{\text{loss}} & \text{(3c)} \\ c_{\text{in out}}(v_c) & \textit{if } l(v, v_c) = 1 \quad \text{(3d)} \end{cases}$$

$$\Delta_{\text{in out}}(v_c) = \min \begin{cases} c_{\min}(v_c) + \delta_{\text{loss}} & \textit{if } g(v, v_c) = 1 \quad \text{(4a)} \\ c_{\text{out}}(v_c) & \textit{if } g(v, v_c) = 1 \quad \text{(4b)} \\ c_{\text{in}}(v_c) + \delta_{\text{loss}} & \text{(4c)} \\ c_{\text{in out}}(v_c), & \text{(4d)} \end{cases}$$

where $l(v, v_c)$ *and* $g(v, v_c)$ *are indicator functions defined as*

$$l(v, v_c) = \begin{cases} 1 & \textit{if } x_{\min}(v) \not\subseteq x_{\min}(v_c) \\ 0 & \textit{otherwise} \end{cases} \qquad g(v, v_c) = \begin{cases} 1 & \textit{if } x_{\min}(v_c) \not\subseteq x_{\min}(v) \\ 0 & \textit{otherwise.} \end{cases}$$

See an illustration for the $\Delta_{\min}(v_c)$ case in Fig. 2, for $\Delta_{\text{out}}(v_c)$ in Fig. 6 (in Appendix) and for $\Delta_{\text{in}}(v_c)$ in Fig. 7.

We are now ready to state the main recurrence equations allowing us to solve the SMALL GAIN-LOSS PARSIMONY PROBLEM.

Theorem 1 (Main recurrences). *For any node* $v \in \mathrm{V}(T)$, *the values of* $c_\sigma(v)$ *may be computed as follows. If* v *is a leaf,* $c_{\min}(v) = 0$ *and* $c_{\text{out}}(v) = c_{\text{in}}(v) = c_{\text{in out}}(v) = \infty$. *Otherwise,*

$$c_{\min}(v) = \begin{cases} \Delta_{\min}(v_\ell) + \Delta_{\min}(v_r) & \text{if } x_{\min}(v) \neq \emptyset \\ \infty & \text{otherwise} \end{cases}$$

$$c_{\text{out}}(v) = \Delta_{\text{out}}(v_\ell) + \Delta_{\text{out}}(v_r)$$

$$c_{\text{in}}(v) = \min \begin{cases} \Delta_{\text{in}}(v_\ell) + \Delta_{\text{out}}(v_r) \\ \Delta_{\text{out}}(v_\ell) + \Delta_{\text{in}}(v_r) \\ \Delta_{\text{in out}}(v_\ell) + \Delta_{\text{in out}}(v_r) \end{cases}$$

$$c_{\text{in out}}(v) = \min \begin{cases} \Delta_{\text{in out}}(v_\ell) + \Delta_{\text{out}}(v_r) \\ \Delta_{\text{out}}(v_\ell) + \Delta_{\text{in out}}(v_r) \\ \Delta_{\text{in out}}(v_\ell) + \Delta_{\text{in out}}(v_r). \end{cases}$$

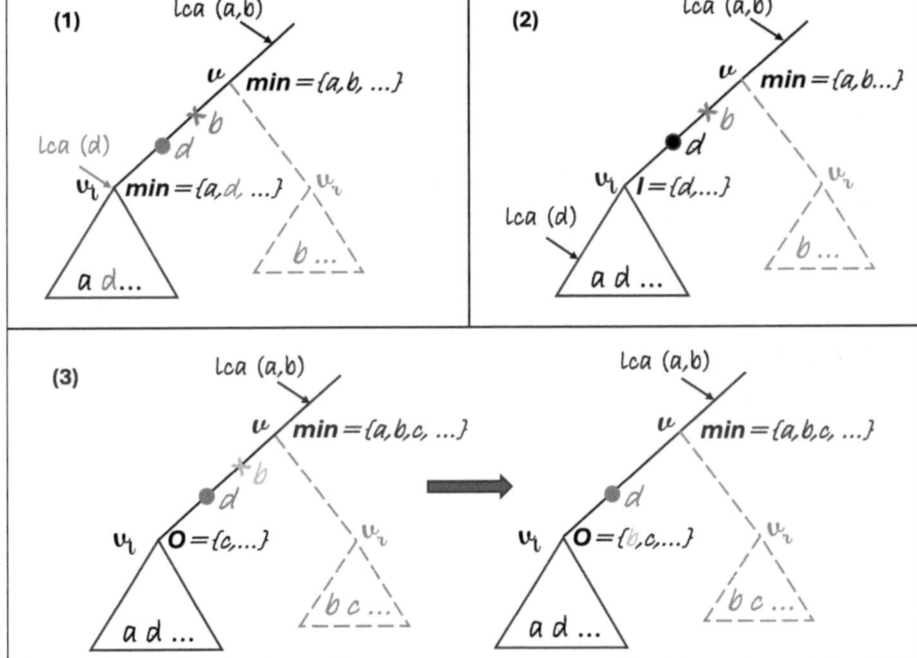

Fig. 2. An illustration of a left scenario (of cost $\Delta_{\min}(v_\ell)$) leading to $c_{\min}(v)$ (a similar scenario can be drawn for the right part, i.e. for $\Delta_{\min}(v_r)$); the characters in a triangle represent the gene families in the leaves of the corresponding subtree; the events and characters in gray can be present or absent, depending on the value of $l(v, v_\ell)$ and $g(v, v_\ell)$. (1): A scenario with $x(v_\ell) = x_{\min}(v_\ell)$ (line 1a). (2): A scenario with $x(v_\ell) = x_{\min}(v_\ell) \cup I$ where $I \subseteq x_{\text{in}}(v_\ell)$, $I \neq \emptyset$ (line 1c). There must be a gain on the edge (v, v_ℓ) because $I \subseteq x(v_\ell) - x(v)$ and therefore $x(v_\ell) \not\subseteq x(v)$. (3): Two scenarios with $x(v_\ell) = x_{\min}(v_\ell) \cup O$ where $O \subseteq x_{\text{out}}(v_\ell)$, $O \neq \emptyset$ (line 1b). The left scenario is not optimal with a loss on the edge (v, v_ℓ) as the lost gene (here b) can always be postponed to be lost later in the same losses of the syntenies containing the extra gene c.

In order to compute $c_{\min}(v)$, $c_{\text{out}}(v)$, $c_{\text{in}}(v)$ and $c_{\text{in out}}(v)$ using Theorem 1, we need to determine the value of $x_{\min}(v)$, which is the purpose of Algorithm 2. This, in turn, requires computing x_{lca}, which is done using Algorithm 1. The correctness of both algorithms is proven in Lemma 3.

Algorithm 1: LCA-Content($\mathcal{T} = \langle T, \tilde{x} \rangle$)

```
 1 for each v ∈ V(T) in post-order do
 2     if v ∈ L(T) then
 3         A(v) ← x̃(v)
 4     else
 5         A(v) ← A(vℓ) ∪ A(vr)
 6 B(r(T)) ← A(r(T))
 7 for each v ∈ V(T) in pre-order do
 8     if v ∈ L(T) then
 9         xlca(v) ← B(v)
10     else
11         B(vℓ) ← B(v) − A(vr)
12         B(vr) ← B(v) − A(vℓ)
13         xlca(v) ←
               B(v) − B(vℓ) − B(vr)
14 return xlca
```

Algorithm 2: Min-Content($\mathcal{T} = \langle T, \tilde{x} \rangle, x_{\text{lca}}$)

```
 1 for each v ∈ V(T) in post-order do
 2     if v ∈ L(T) then
 3         xmin(v) ← x̃(v)
 4     else
 5         xmin(v) ←
               (xmin(vℓ) − xlca(vℓ)) ∪
               (xmin(vr) − xlca(vr))
 6 return xmin
```

Lemma 3. (Computing x_{lca} and $x_{\min}$). *For all $v \in V(T)$, Algorithm 1 returns $x_{\text{lca}}(v)$ and Algorithm 2 returns $x_{\min}(v)$.*

Using the equations from Theorem 1, Algorithm 3 returns the cost of an optimal history explaining $\mathcal{T}$.

Algorithm 3: InOutParsimonyCost($\mathcal{T} = \langle T, \tilde{x} \rangle, x_{\min}, \delta_{\text{loss}}, \delta_{\text{gain}}$)

```
 1 for each v ∈ V(T) in post-order do
 2     compute cmin(v), cout(v), cin(v), cin out(v) according to Theorem 1
 3 return δgain + min{cmin(r(T)), cin(r(T))}
```

Corollary 1. *Algorithm 3 returns $\min_{\mathcal{H} = \langle \dot{T}, \epsilon, x \rangle \in \mathbb{H}(\mathcal{T})} \{c(\mathcal{H})\}$.*

From the result of Algorithm 3 and through a standard backtracking procedure, we can output ϵ, i.e. the positions of gain and loss events in an optimal history. The remaining information, i.e. the content x of ancestral syntenies, is then returned by Algorithm 4.

Algorithm 4: InOutParsimonyContent($\mathcal{T} = \langle T, \tilde{x}\rangle, \epsilon, x_{\min}$)

---

1 **for** *each* $v \in \mathrm{V}(T)$ *in post-order* **do**
2 $\quad$ $x(v) \leftarrow x_{\min}(v)$
3 $\quad$ **if** $v \notin L(T)$ **then**
4 $\quad\quad$ **if** Gain $\notin \epsilon((v, v_\ell))$ **then**
5 $\quad\quad\quad$ $x(v) \leftarrow x(v) \cup x(v_\ell)$
6 $\quad\quad$ **if** Gain $\notin \epsilon((v, v_r))$ **then**
7 $\quad\quad\quad$ $x(v) \leftarrow x(v) \cup x(v_r)$
8 **for** *each* $v \in \mathrm{V}(T) - \{\mathrm{r}(T)\}$ *in pre-order* **do**
9 $\quad$ **if** Loss $\notin \epsilon((p(v), v))$ **then**
10 $\quad\quad$ $x(v) \leftarrow x(v) \cup x(p(v))$
11 $x(\mathrm{r}(\dot{T})) \leftarrow \emptyset$
12 **return** $\langle \dot{T}, \epsilon, x\rangle$

Theorem 2. (Correctness of Algorithm 4**).** *If ϵ is such that there exists $\langle \dot{T}, \epsilon, x^*\rangle \in \mathrm{argmin}_{\mathcal{H} \in \mathbb{H}(\mathcal{T})}\{c(\mathcal{H})\}$, then Algorithm 4 returns a history $\langle \dot{T}, \epsilon, x\rangle \in \mathrm{argmin}_{\mathcal{H} \in \mathbb{H}(\mathcal{T})}\{c(\mathcal{H})\}$ on input $(\mathcal{T} = \langle T, \tilde{x}\rangle, \epsilon, x_{\min})$.*

As stated in the next theorem, the history returned by Algorithm 4 is unique for a given input. More precisely, given ϵ leading to an optimal history for $\mathcal{T}$, there is a unique mapping x such that $\langle \dot{T}, \epsilon, x\rangle \in \mathrm{argmin}_{\mathcal{H} \in \mathbb{H}(\mathcal{T})}\{c(\mathcal{H})\}$.

Theorem 3. (Uniqueness of x given ϵ). *Let $\langle \dot{T}, \epsilon, x\rangle \in \mathrm{argmin}_{\mathcal{H} \in \mathbb{H}(\mathcal{T})}\{c(\mathcal{H})\}$. There exists no history $\langle \dot{T}, \epsilon, x'\rangle \in \mathbb{H}(\mathcal{T})$ such that $x' \neq x$.*

We are now ready to state our main InOutParsimony algorithm (Algorithm 5 below) solving the SMALL GAIN-LOSS PARSIMONY PROBLEM. Its correctness directly follows from all previous results of this section. See Fig. 3 for an example of the execution of the algorithm.

Algorithm 5: InOutParsimony($\mathcal{T} = \langle T, \tilde{x}\rangle, \delta_{\mathrm{loss}}, \delta_{\mathrm{gain}}$)

---

1 $x_{\mathrm{lca}} \leftarrow$ LCA-Content($\mathcal{T} = \langle T, \tilde{x}\rangle$)
2 $x_{\min} \leftarrow$ Min-Content($\mathcal{T} = \langle T, \tilde{x}\rangle, x_{\mathrm{lca}}$)
3 $\epsilon \leftarrow$ backtracking procedure from the result of
$\quad$ InOutParsimonyCost($\mathcal{T} = \langle T, \tilde{x}\rangle, x_{\min}, \delta_{\mathrm{loss}}, \delta_{\mathrm{gain}}$)
4 **return** InOutParsimonyContent($\mathcal{T} = \langle T, \tilde{x}\rangle, \epsilon, x_{\min}$)

Notice that the standard backtracking procedure mentioned above for Algorithm 3 does not necessarily lead to a unique ϵ. Consider breaking the ties in the minimums (see Theorem 1) by selecting the first (from top to bottom) line leading to the minimum. Using this backtracking, that we call *First-Line-Backtracking*, we obtain an optimal history of minimum total size in terms of gene contents, as shown in the next theorem.

Theorem 4 (A solution minimizing gene contents). *If $x_{min}(v) \neq \emptyset$ for all $v \in V(T)$, then Algorithm InOutParsimony (Algorithm 5) using First-Line-Backtracking returns the solution $\langle \dot{T}, \epsilon, x \rangle$ minimizing $\sum_{v \in V(T)} |x(v)|$ among all possible solutions* $\mathrm{argmin}_{\mathcal{H}=\langle \dot{T}, \epsilon, x \rangle \in \mathbb{H}(T)} \{c(\mathcal{H})\}$.

We finally state the time complexity of Algorithm 5, assuming operations on sets can be done in constant time (by encoding the gene family sets as binary vectors fitting into a constant number of 64-bit words).

Theorem 5 (Time complexity). *Assuming operations on sets can be performed in constant time, the* SMALL GAIN-LOSS PARSIMONY PROBLEM *can be solved in $O(n)$ time where $n = |V(T)|$.*

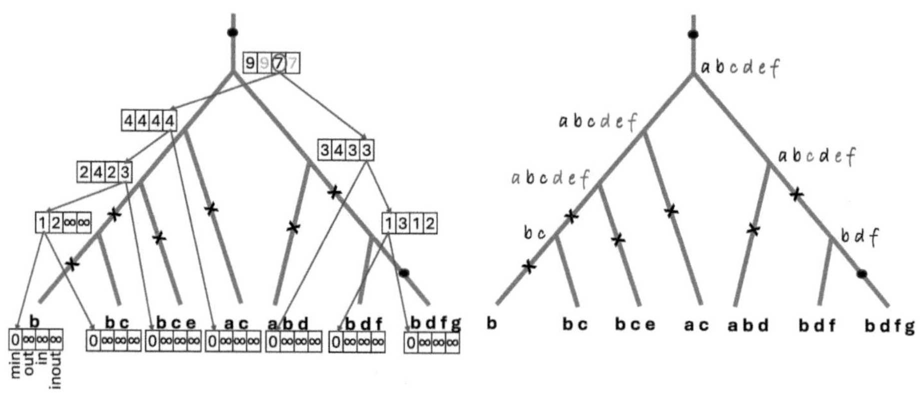

Fig. 3. (Left) the synteny tree $\langle T, \tilde{x} \rangle$ from Figure 1 with each node v labeled with a vector of size four containing the values (respectively from left to right) of $c_{min}(v)$, $c_{out}(v)$, $c_{in}(v)$ and $c_{in\,out}(v)$. Those values are computed by Algorithm 3, considering a unit cost $\delta_{loss} = \delta_{gain} = 1$. Notice that the $c_{out}(v)$ and $c_{in\,out}(v)$ positions in the vector of $r(T)$ (shown in light gray) do not lead to a valid solution. From Line 4 of Algorithm 3, as $\min\{c_{min}(r(T)), c_{in}(r(T))\} = 7$ and $\delta_{gain} = 1$, the cost of a minimum solution returned by the algorithm is 8. The *First-Line-Backtracking* (see Theorem 4) is illustrated by the red pointers. It leads to the optimal labeling ϵ (positions of the 2 gains and 6 losses). (Right) The optimal history output by Algorithm 4 when executed on the tree and events on the left. The gene families in black and red are added to the content of internal nodes during the bottom-up part of Algorithm 4, while those in blue are added during the top-down part of Algorithm 4. This optimal history corresponds to $\mathcal{H}^*$ from Fig. 1.

4   Application

We implemented InOutParsimony in Python. Given a binary tree, a content labeling for the tree leaves, and the costs for gains and losses, the program returns an optimal history as a fully labeled tree in Newick format. The source code can be accessed at https://github.com/UdeM-LBIT/InOutParsimony/tree/recombcg2026.

We applied InOutParsimony to a dataset of mitochondrial-encoded protein families, called MitoCOGs, taken from [17]. This dataset was considered in a previous paper of our group [3] for studying the exchange of genes between the mitochondrial and nuclear eukaryotic genomes. In fact, it is largely established that all extant mitochondria originated from a unique endosymbiotic event integrating an $\alpha-$proteobacterial genome into an eukaryotic cell, creating an ancestral mitochondrial genome. Subsequently, eukaryote evolution was marked by episodes of Endosymbiotic Gene Transfers (EGT), meaning the transfer of genes between the mitochondrial and nuclear genomes of the same species, mainly from the mitochondria to the nucleus, eventually leading to the disappearance of the mitochondria. However, a high variability of gene repertoire distribution indicating an ongoing EGT process in both directions is still observable in some lineages such as in land plants.

We took the same dataset of 11 plants considered in [3] chosen to have MitoCOGs in both their nuclear and mitochondrial genomes. This set includes nine *Viridiplantae*, one *Rhodophyta* (*Cyanidioschyzon merolae*) and one *Glaucophyta* (*Cyanophora paradoxa*). The selected MitoCOGs were those appearing in either the mitochondrial or nuclear genomes of those 11 plants, yielding 81 MitoCOGs.

We ran InOutParsimony with two models: (1) the *unitary-cost model* with $\delta_{\text{gain}} = \delta_{\text{loss}}$ minimizing the number of events, and (2) a *gains-at-LCA model*, based on the previous model of *Synesth* [11], in which the gain point of each gene family is positioned at its LCA and gains are excluded from the optimization function, only minimizing losses. Figure 4 for the unitary-cost and Fig. 5 for the gains-at-LCA model illustrate the predicted mitochondrial and nuclear contents for the seven multisubunits complexes given in Kannan's paper [17, Table S8].

The scenario with the unit-cost model involves two large gains, one at the root and the other at one of its children, while those gains are rather split into smaller ones closer to the leaves with the gains-at-LCA model. The first scenario avoids splitting the genes of a given complex into different gain episodes, which may be considered more biologically relevant. For example, the SDH complex in *Cyanidioschyzon merolae* is gained in two episodes with the unit-cost model, while it is gained in three episodes with the gains-at-LCA model (i.e. with an additional gain on the terminal edge). In addition, the disappearance of this complex in the mitochondrial genome of *Chlamydomonas reinhardtii*, or similarly in the nuclear genome of Cyanidioschyzon merolae, is explained by a single loss.

We wanted to see how well the two scenarios reflect the EGTs predicted on the same dataset with EndoRex [3]. This software takes as input a species tree and a gene tree where each leaf is additionally labeled 0 or 1 depending on whether the corresponding gene appears in the mitochondrial or nuclear genome, and returns a reconciliation involving EGT events. Figure 4 in [3] shows such predicted reconciliations obtained with different cost functions. The most supported events are an EGT of MitoCOG0014 (corresponding to a gene family of the ATP complex) from mitochondria to nucleus in *Chlamydomonas reinhardtii*, and another EGT of MitoCOG0072 (belonging to the RPS complex) from nucleus to mitochondria in the LCA of *Ostreococcus tauri* and *Micromonas sp. RCC299*. These two events are more consistent with the scenario obtained with the gains-at-LCA model as, in each case, the loss in one genome (mitochondria or nucleus) is accompanied with a parallel gain in the other genome, while on the scenario obtained with the unit-cost model the complex it is gained earlier in the tree.

We also compared the output of our algorithm with the output of Count [8] under parsimony and probabilistic models. For the latter, similarly to Kannan [17], because gene gain in mitochondrial genome is rare, a pure-loss model architecture was assumed and the prior distribution at the root was assumed to be Poisson. InOutParsimony's results for the unit-cost model are comparable to Count's results for the probabilistic pure-loss model (see Fig. 4 and Fig. 8 in Appendix), while InOutParsimony's results for the gains-at-LCA model are comparable to Count's results under Dollo parsimony (see Fig. 5 and Fig. 9 in Appendix). For both models, Count's inferred sizes are almost always slightly lower than those inferred with InOutParsimony. This is due to the fact that each gene family is considered independently one from the other by Count, while InOutParsimony is able to consider them together and group individual losses in unique loss episodes placing them lower in the tree, thus keeping more genes at internal nodes.

While grouping losses can more appropriately reflect the unique transfer of a full gene complex, as illustrated above with the SDH complex, it can also lead to artificially large genomic sizes on internal nodes. This argues for an intermediate model, going beyond Count's single loss model (a loss per gene family) but penalizing each segmental loss according to its size.

Finally, prohibiting parallel gains (Dollo's parsimony) can also artificially increase genomic sizes, in addition to losing the possibility of identifying multiple EGT events. For example, gene ATP9 (MitoCOG0014), only found in the nuclear genomes of Arabidopsis and Chlamydomonas, is explained by the EndoRex software [3] with two EGT events copying the gene family from the mitochondria to the nucleus in both species. This gene is also identified as two separate gains with Count under the Wagner parsimony (see Fig. 10 in Appendix, where ATP9 is the last vertical strip in the ATP complex rectangles). However, these two EGT events are rather interpreted as a single gain with InOutParsimony.

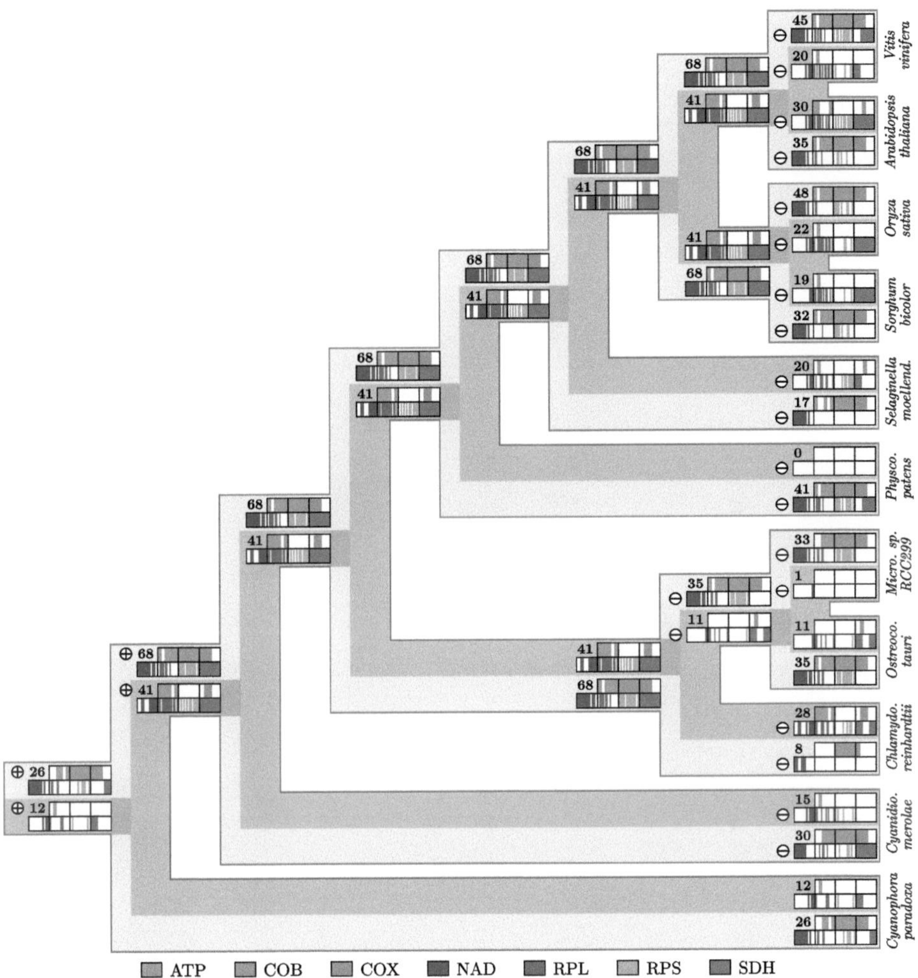

Fig. 4. Results obtained with InOutParsimony for the *unitary-cost* model on the species tree of the considered 11 plants. The topology of the tree is based on [17]. Light gray edges (the outer structure) represent the mitochondrial tree, while dark gray edges (the inner structure) represent the nuclear tree. Each internal node of the mitochondrial/nuclear tree contains seven rectangles representing, in order, the seven multisubunits complexes given by Table S8 in [17] and represented on the bottom of this figure. Each rectangle is divided into strips corresponding to the genes of the complex and each strip is colored when the gene is present at that node. Gains and losses are signaled by ⊕ and ⊖ symbols respectively.

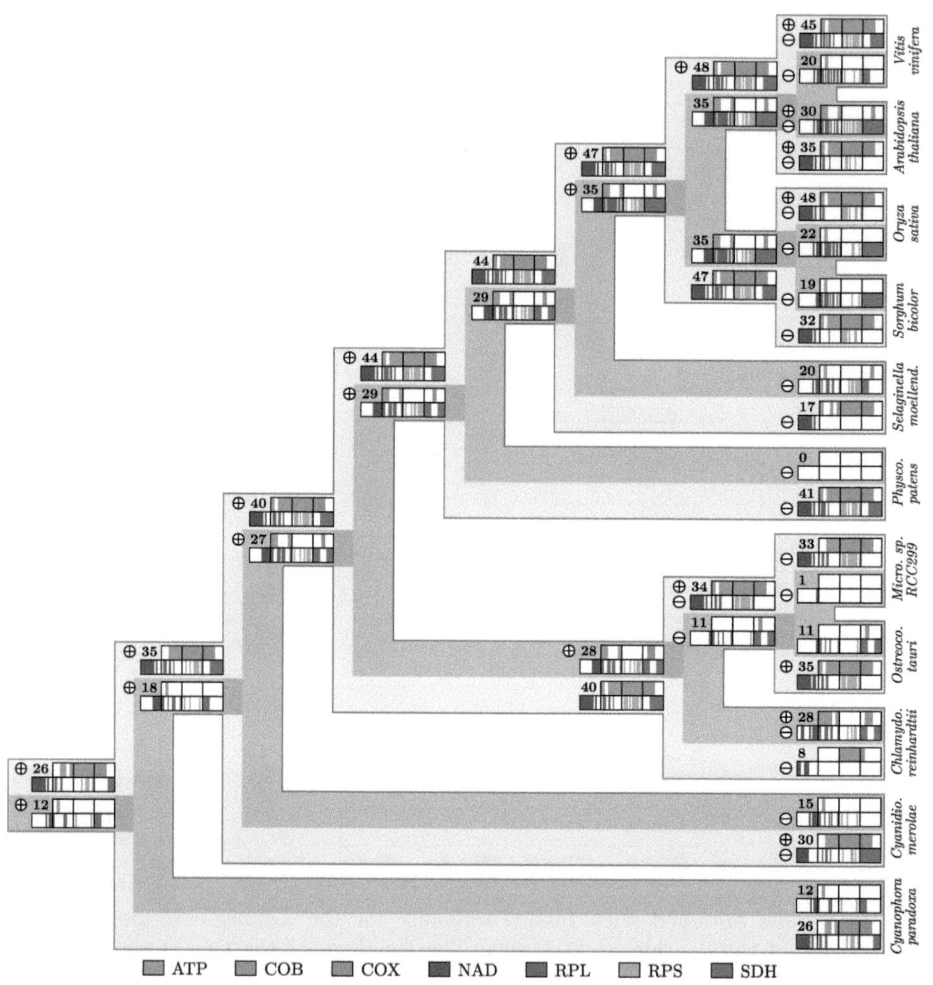

Fig. 5. Results obtained with InOutParsimony for the *gains-at-LCA model* on the species tree of the considered 11 plants. Representation is the same as in Fig. 4.

5   Conclusion

We developed the first linear-time algorithm that predicts a most parsimonious history scenario of segmental gains and losses for a given phylogenetic tree leaf-labeled by sets of genes without duplicates. Linearity is achieved by testing only four cases at internal nodes rather than all possible genetic contents. The algorithm can be used with arbitrary, but constant, costs for gains and losses.

It will be interesting to extend the model to more general weights, such as affine or convex weight function accounting for the size of a gain or a loss. This could help strike a balance between individual loss and gain scenarios, as inferred by Count, and overly broad events leading to redundant segmental losses in the terminal edges of the tree.

It will be also interesting to generalize InOutParsimony to enable exploring the whole solution space through algebraic dynamic programming. For now, our method outputs a single solution among all optimal ones, the one minimizing the total size of gene contents.

Another promising extension of this work would be to allow multiple gain points for a given gene family, extending the model from Dollo parsimony to Wagner parsimony. In our MitoCOGs case study, this would lead to a more precise and robust identification of EGTs.

Finally, the results presented in this paper pave the way to the integration of segmental gain minimization into the Synesth reconciliation framework, enabling the study of gene repertoire's evolution through a comprehensive model involving duplications and horizontal gene transfers.

Acknowledgments. We would like to thank Corélie Godefroid for her important work in clearing up the problem. We also thank the community of the "Novel Mathematical Paradigm for Phylogenomics" workshop (hosted by the Banff International Research Station) for fruitful discussions. This research was funded by the *Fonds de recherche du Québec Nature et technologies* [grant numbers 335135 (MG) and 335893 (MD)] and the Natural Sciences and Engineering Research Council of Canada [grant number RN000743 (NEM)].

Disclosure of Interests. The authors have no competing interests to declare that are relevant to the content of this article.

Appendix

Additional Notation

Let T be a tree. For any two nodes v and v' of T, there exists a unique path from v to v' that we denote $\mathrm{P}_T(v, v') \subseteq \mathrm{E}(T)$.

Let $v \in \mathrm{V}(T)$. The depth of v is defined as $|\mathrm{P}_T(\mathrm{r}(T), v)|$ and the height of v is defined as $\max_{l \in L(T_v)} |\mathrm{P}_T(v, l)|$.

Additional Definitions

The following definition is used in the proofs in Appendix.

Definition 5 (Cost of a subhistory with content X). *For any $X \subseteq \mathcal{F}$, let $c(v, X)$ be the minimum cost of an explanatory subhistory rooted at v such that $x(v) = X$, i.e. $c(v, X) = \min\{c(\mathcal{H}_v) \mid \mathcal{H} = \langle \dot{T}, \epsilon, x \rangle \in \mathbb{H}(\mathcal{T}) \wedge x(v) = X\}$.*

Additional Lemmas

The following lemmas are used in the proofs of the results stated in the paper.

Lemma 4. *Let v be an internal node of T and let $I \subseteq x_{\mathrm{in}}(v)$ be non empty. For any non empty $O \subseteq x_{\mathrm{out}}(v)$:*

- *There exists a subhistory $\mathcal{H}_v$ of a history $\mathcal{H} = \langle \dot{T}, \epsilon, x \rangle$ explaining $\mathcal{T}$ such that $x(v) = x_{\min}(v) \cup O$ and $c(\mathcal{H}_v) = c_{\mathrm{out}}(v)$.*
- *There exists a subhistory $\mathcal{H}_v$ of a history $\mathcal{H} = \langle \dot{T}, \epsilon, x \rangle$ explaining $\mathcal{T}$ such that $x(v) = x_{\min}(v) \cup O \cup I$ and $c(\mathcal{H}_v) = \min_{\substack{O' \subseteq x_{\mathrm{out}}(v) \\ O' \neq \emptyset}} c(v, x_{\min}(v) \cup O' \cup I)$.*

Proof. Let $O \subseteq x_{\mathrm{out}}(v)$ be non empty. First notice that there exists a history $\mathcal{H} = \langle \dot{T}, \epsilon, x \rangle$ explaining $\mathcal{T}$ such that $x(v) \cap x_{\mathrm{out}}(v) = O$. In fact, it is possible to gain all content in $x_{\mathrm{out}}(v)$ at the root of T and then not lose any content from O and lose all content from $x_{\mathrm{out}}(v) - O$ on the path from the root to v.

Now, let $\mathcal{H}'_v$ be a subhistory of cost $c_{\mathrm{out}}(v)$ (reps. $\min_{\substack{O' \subseteq x_{\mathrm{out}}(v) \\ O' \neq \emptyset}} c(v, x_{\min}(v) \cup O' \cup I)$) of a history $\mathcal{H}' = \langle \dot{T}, \epsilon', x' \rangle$ explaining $\mathcal{T}$. Notice that there must be a loss event on every path from v the leaves in $L(\dot{T}_v)$ as otherwise $\mathcal{H}$ would not be explaining $\mathcal{T}$. Therefore, we can add any given subset of $x_{\mathrm{out}}(v)$ to $x'(v)$ and simply lose this new content at those loss events without changing the cost of the subhistory. The result follows. $\qquad\square$

Lemma 5. *Let $\mathcal{H} = \langle \dot{T}, \epsilon, x \rangle \in \mathbb{H}(\mathcal{T})$. If there is an edge $(p(v), v) \in \mathrm{E}(\dot{T})$ such that $\mathrm{Gain} \in \epsilon((p(v), v))$ and $x(p(v)) \cap (x_{\mathrm{in}}(v) \cup x_{\mathrm{lca}}(v)) \neq \emptyset$, then*

$$\mathcal{H} \notin \mathrm{argmin}_{\mathcal{H}' \in \mathbb{H}(\mathcal{T})}\{c(\mathcal{H}')\}$$

and

$$c(\mathcal{H}_{p(v)}) \notin \min_{I \subseteq x_{\mathrm{in}}(v),\ I \neq \emptyset}\{c(p(v), x(p(v)) \cup I)\}.$$

Proof. Let $\mathcal{H} = \langle \dot{T}, \epsilon, x \rangle \in \mathbb{H}(\mathcal{T})$ and let's assume that there exists an edge $(p(v), v) \in \mathrm{E}(\dot{T})$ such that $\mathrm{Gain} \in \epsilon((p(v), v))$ and $x(p(v)) \cap (x_{\mathrm{in}}(v) \cup x_{\mathrm{lca}}(v)) \neq \emptyset$. As $x(p(v)) \cap (x_{\mathrm{in}}(v) \cup x_{\mathrm{lca}}(v)) \neq \emptyset$, there is at least one gene family g in $x(p(v))$ such that $lca(g) \geq v$ and $\mathrm{Gain}_{\mathcal{H}}(g)$ is an ancestor of $p(v)$. As $lca(g) \geq v$, $g \notin \cup_{l \in L(T) - L(T_v)} \tilde{x}(l)$. Therefore, g must be lost at some point between $\mathrm{Gain}_{\mathcal{H}}(g)$

and each leaves in $L(T) - L(T_v)$ descending from $\text{Gain}_{\mathcal{H}}(g)$ because otherwise $\mathcal{H}$ would not be explaining T. We can thus obtain another history $\mathcal{H}^*$ explaining T from $\mathcal{H}$ by removing Gain from $\epsilon((p(v), v))$ and adding the content that was gained in this event in the content of every descendant node of $\text{Gain}_{\mathcal{H}}(g)$ containing the gene family g that is not a descendant of v. Therefore, $c(\mathcal{H}^*) = c(\mathcal{H}) - \delta_{\text{gain}}$ and $\mathcal{H} \notin \text{argmin}_{\mathcal{H}' \in \mathbb{H}(T)}\{c(\mathcal{H}')\}$. Furthermore, $c(\mathcal{H}^*_{p(v)}) = c(\mathcal{H}_{p(v)}) - \delta_{\text{gain}}$ and $x^*(p(v)) = x(p(v)) \cup I$ from some non empty set $I \subseteq x_{\text{in}}(v)$ by construction and therefore

$$c(\mathcal{H}_{p(v)}) \notin \min_{I \subseteq x_{\text{in}}(v),\ I \neq \emptyset}\{c(p(v), x(p(v)) \cup I)\}.$$

$\square$

Lemma 6. *Let* $\mathcal{H} = \langle \dot{T}, \epsilon, x \rangle \in \mathbb{H}(T)$. *If there is an edge* $(p(v), v) \in \text{E}(\dot{T})$ *such that* $\text{Loss} \in \epsilon((p(v), v))$ *and* $x(v) \cap x_{\text{out}}(v) \neq \emptyset$, *then*

$$\mathcal{H} \notin \text{argmin}_{\mathcal{H}' \in \mathbb{H}(T)}\{c(\mathcal{H}')\} \text{ and } c(\mathcal{H}_{p(v)}) \notin c(p(v), x(p(v))).$$

Proof. Let $\mathcal{H} = \langle \dot{T}, \epsilon, x \rangle \in \mathbb{H}(T)$ and let's assume that there exists an edge $(p(v), v) \in \text{E}(\dot{T})$ such that $\text{Loss} \in \epsilon((p(v), v))$ and $x(v) \cap x_{\text{out}}(v) \neq \emptyset$. As $x(v) \cap x_{\text{out}}(v) \neq \emptyset$, there is at least one gene family g in $x(v)$ such that $g \notin \cup_{l \in L(T_v)} \tilde{x}(l)$ and therefore this gene family must be lost at some point between v and each leaves in $\text{L}(T_v)$ because otherwise $\mathcal{H}$ would not be explaining T. We can thus obtain another history $\mathcal{H}^*$ explaining T from $\mathcal{H}$ by removing Loss from $\epsilon((p(v), v))$ and adding the content that was lost in this event in the content of every descendant node of v containing the gene family g. Therefore, $c(\mathcal{H}^*) = c(\mathcal{H}) - \delta_{\text{loss}}$ and $\mathcal{H} \notin \text{argmin}_{\mathcal{H} = \langle \dot{T}, \epsilon, x \rangle \in \mathbb{H}(T)}\{c(\mathcal{H})\}$. Furthermore, $c(\mathcal{H}^*_{p(v)}) = c(\mathcal{H}_{p(v)}) - \delta_{\text{loss}}$ and $x^*(p(v)) = x(p(v))$ by construction and thus $c(\mathcal{H}_{p(v)}) \notin c(p(v), x(p(v)))$. $\square$

Proofs

Proof of Lemma 1

Proof. Let $\mathcal{H} = \langle \dot{T}, \epsilon, x \rangle \in \mathbb{H}(T)$ and let v be a node of T. Let $g \in \mathcal{F}$.

If $lca(g) \leq v$, then $\text{Gain}_{\mathcal{H}}(g) \leq v$, and if $g \in \tilde{x}(l)$ for a given $l \in L(T_v)$, then it cannot be lost on the path $\text{P}_{\dot{T}}(\text{Gain}_{\mathcal{H}}(g), l)$ containing v by condition (3) of Definition 1. Therefore, for any $g \in x_{min}(v)$, $g \in x(v)$.

The result follows from the fact that for any $g \notin x_{min}(v)$, either $lca(g) > v$ in which case $g \in x_{\text{in}}(v)$, or $g \notin \cup_{l \in L(T_v)} \tilde{x}(l)$ in which case $g \in x_{\text{out}}(v)$. $\square$

Proof of Lemma 2

Proof. Let v be an internal node of T and let v_c be a child of v. Then for any history $\mathcal{H} = \langle \dot{T}, \epsilon, x \rangle$ explaining T, $x(v_c) = x_{min}(v_c) \cup O_c \cup I_c$ for some sets $O_c \subseteq x_{\text{out}}(v_c)$ and $I_c \subseteq x_{\text{in}}(v_c)$ by Lemma 1. For each $\Delta_\sigma(v_c)$ ($\sigma \in \{\text{min}, \text{out}, \text{in}, \text{in out}\}$), we will test each possible case of the sets O_c and I_c being empty or not:

- $\Delta_{\min}(v)$: In this case, $x(v) = x_{\min}(v)$. Then,
  - If $O_c = \emptyset$ and $I_c = \emptyset$: In that case, $x(v_c) = x_{min}(v_c)$. The cost of an optimal subhistory rooted at v_c is therefore $c_{\min}(v_c)$ by Definition 4. By Definition 1, Loss $\in (v, v_c)$ if and only if $x_{min}(v) \not\subseteq x_{min}(v_c)$ and Gain $\in (v, v_c)$ if and only if $x_{min}(v_c) \not\subseteq x_{min}(v)$. This leads to a hanging subhistory of cost $c_{\min}(v_c) + l(v, v_c) \cdot \delta_{\text{loss}} + g(v, v_c) \cdot \delta_{\text{gain}}$ (line 1a). See Fig. 2 (1) for an illustration.
  - If $O_c \neq \emptyset$ and $I_c = \emptyset$: This case is valid if and only if $x_{min}(v) \not\subseteq x_{min}(v_c)$ as otherwise there exists no history explaining $\mathcal{T}$ such that $x(v) = x_{min}(v)$ and $x(v_c) = x_{min}(v_c) \cup O_c$ for a non empty set $O_c \subseteq x_{\text{out}}(v_c)$. If this is the case, the cost of an optimal subhistory rooted at v_c is $c_{\text{out}}(v_c)$ by Lemma 4. Loss $\notin (v, v_c)$ by Lemma 6 and Gain $\in (v, v_c)$ if and only if $x_{min}(v_c) \not\subseteq x_{min}(v)$ by Definition 1. This leads to a hanging subhistory of cost $c_{\text{out}}(v_c) + g(v, v_c) \cdot \delta_{\text{gain}}$ (line 1b). See Fig. 2 (3) for an illustration.
  - If $O_c = \emptyset$ and $I_c \neq \emptyset$: In that case, the cost of an optimal subhistory rooted at v_c is $c_{\text{in}}(v_c)$ by Definition 4. Gain $\in (v, v_c)$ because the content in $I_c \subseteq x_{\text{in}}(v_c)$ is not in $x(v) = x_{\min}(v)$ by definition and therefore $x(v_c) \not\subseteq x(v)$. Loss $\in (v, v_c)$ if and only if $x_{min}(v) \not\subseteq x_{min}(v_c)$ by Definition 1. This leads to a hanging subhistory of cost $c_{\text{in}}(v_c) + l(v, v_c) \cdot \delta_{\text{loss}} + \delta_{\text{gain}}$ (line 1c). See Fig. 2 (2) for an illustration.
  - If $O_c \neq \emptyset$ and $I_c \neq \emptyset$: This case is valid if and only if $x_{min}(v) \not\subseteq x_{min}(v_c)$ as otherwise there exists no history explaining $\mathcal{T}$ such that $x(v) = x_{min}(v)$ and $x(v_c) = x_{min}(v_c) \cup O_c \cup I_c$ for some non empty sets $O_c \subseteq x_{\text{out}}(v_c)$ and $I_c \subseteq x_{\text{in}}(v_c)$. If this is the case, the cost of the optimal subhistory rooted at v_c is $c_{\text{in out}}(v_c)$ by Lemma 4. Gain $\in (v, v_c)$ because the content in $I_c \subseteq x_{\text{in}}(v_c)$ is not in $x(v) = x_{\min}(v)$ by definition and therefore $x(v_c) \not\subseteq x(v)$. Loss $\notin (v, v_c)$ by Lemma 6. This leads to a hanging subhistory of cost $c_{\text{in out}}(v_c) + \delta_{\text{gain}}$ (line 1d).
- $\Delta_{\text{out}}(v)$: In this case, $x(v) = x_{\min}(v) \cup O$ with $\emptyset \neq O \subseteq x_{\text{out}}(v_c) - x_{\min}(v)$. This case is similar to the previous case ($\Delta_{\text{in}}(v)$) with the difference that here we are sure that there is some content at v that is not in $x_{\min}(v_c) \cup x_{\text{in}}(v_c)$ (Definition 3). Thus, everywhere in the equation for $\Delta_{\text{in}}(v)$ that we test if there is a need to lose some content between v and v_c, we know this is the case here and therefore the result is obtained by replacing $l(v, v_c)$ by 1 in the equation for $\Delta_{\min}(v)$. See Fig. 6 for an illustration of some of the cases.
- $\Delta_{\text{in}}(v)$: In this case, $x(v) = x_{\min}(v) \cup I$ with $\emptyset \neq I \subseteq x_{\text{in}}(v_c) \cup x_{\text{lca}}(v_c)$. This case is similar to the case $\Delta_{\min}(v)$, with the difference that Gain $\notin (v, v_c)$ by Lemma 5 and that the case $I_c = \emptyset$ is only allowed if $x_{min}(v_c) \not\subseteq x_{min}(v)$ as otherwise it would be impossible that $x(v) = x_{\min}(v) \cup I$ where $I \subseteq x_{\text{in}}(v_c) \cup x_{\text{lca}}(v_c)$ is non empty. See Fig. 7 for an illustration of some of the cases.

- $\Delta_{\text{in out}}(v)$: In this case, $x(v) = x_{\min}(v) \cup O \cup I$ with $\emptyset \neq O \subseteq x_{\text{out}}(v_c) - x_{\min}(v)$ and $\emptyset \neq I \subseteq x_{\text{in}}(v_c) \cup x_{\text{lca}}(v_c)$. This case is similar to the previous case $(\Delta_{\text{in}}(v))$ with the difference that here we are sure that there is some content at v that is not in $x_{\min}(v_c) \cup x_{\text{in}}(v_c)$ (Definition 3). Thus, everywhere in the equation for $\Delta_{\text{in}}(v)$ that we test if there is a need to lose some content between v and v_c, we know this is the case here and therefore the result is obtained by replacing $l(v, v_c)$ by 1 in the equation for $\Delta_{\text{in}}(v)$.

$\square$

Proof of Theorem 1

Proof. Let $v \in V(T)$.

If v is a leaf, then the results follows from Definitions 1 and 4. In fact, there is no history $\mathcal{H} = \langle \dot{T}, \epsilon, x \rangle$ explaining $\mathcal{T}$ such that $x(v) \neq x_{\min}(v)$ and thus $c_{\text{in}}(v) = c_{\text{out}}(v) = c_{\text{in out}}(v) = \infty$.

Let v be an internal node and let show the result for every possible state.

For $c_{\min}(v)$, $x(v) = x_{\min}(v)$ by definition. Then, either $c_{\min}(v) = \infty$ if $x_{\min}(v) = \emptyset$ by Definition 1 or $c_{\min}(v) = \Delta_{\min}(v_\ell) + \Delta_{\min}(v_r)$ by Definition 4.

For $c_{\text{out}}(v)$, $x(v) = x_{\min}(v) \cup O$ with $\emptyset \neq O \subseteq x_{\text{out}}(v)$ by definition. In this case, $O \subseteq x_{\text{out}}(v_c) - x_{\min}(v)$ for $c \in \{\ell, r\}$ by definition and thus $c_{\text{out}}(v) = \Delta_{\text{out}}(v_\ell) + \Delta_{\text{out}}(v_r)$.

For $c_{\text{in}}(v)$, $x(v) = x_{\min}(v) \cup I$ with $\emptyset \neq I \subseteq x_{\text{in}}(v)$ by definition. In this case, either $I \subseteq x_{\text{in}}(v_c) \cup x_{\text{lca}}(v_c)$ and $I \subseteq x_{\text{out}}(v_{c'}) - x_{\min}(v)$ for $c \neq c'$, $c, c' \in \{\ell, r\}$ or $I \cap (x_{\text{in}}(v_c) \cup x_{\text{lca}}(v_c)) \neq \emptyset$ and $I \cap (x_{\text{out}}(v_c) - x_{\min}(v)) \neq \emptyset$ for both $c = \ell$ and $c = r$. Therefore, $c_{\text{in}}(v) = \min\{\Delta_{\text{in}}(v_\ell) + \Delta_{\text{out}}(v_r), \Delta_{\text{out}}(v_\ell) + \Delta_{\text{in}}(v_r), \Delta_{\text{in out}}(v_\ell) + \Delta_{\text{in out}}(v_r)\}$.

For $c_{\text{in out}}(v)$, $x(v) = x_{\min}(v) \cup O \cup I$ with $\emptyset \neq O \subseteq x_{\text{out}}(v)$ and $\emptyset \neq I \subseteq x_{\text{in}}(v)$ by definition. Here, $O \subseteq x_{\text{out}}(v_c) - x_{\min}(v)$ for $c \in \{\ell, r\}$ by definition and either $I \subseteq x_{\text{out}}(v_c) - x_{\min}(v)$ and $I \subseteq x_{\text{in}}(v_{c'}) \cup x_{\text{lca}}(v_{c'})$ for $c \neq c'$, $c, c' \in \{\ell, r\}$ or $I \cap (x_{\text{in}}(v_c) \cup x_{\text{lca}}(v_c)) \neq \emptyset$ for both $c = \ell$ and $c = r$. Therefore, $c_{\text{in}}(v) = \min\{\Delta_{\text{in out}}(v_\ell) + \Delta_{\text{out}}(v_r), \Delta_{\text{out}}(v_\ell) + \Delta_{\text{in out}}(v_r), \Delta_{\text{in out}}(v_\ell) + \Delta_{\text{in out}}(v_r)\}$. $\square$

Proof of Lemma 3

Proof. We first show that Algorithm 1 returns $x_{\text{lca}}(v)$ for all $v \in V(T)$. Notice that after the first **for** loop, $A(v) = \cup_{l \in L(T_v)} \tilde{x}(l)$ for all $v \in L(T)$. We now show that in the second **for** loop, $B(v)$ is set to $\{g \in \mathcal{F} \mid lca(g) \geq v\}$ for all $v \in V(T)$ by induction on the depth of v. If $v = r(T)$, then $\{g \in \mathcal{F} \mid lca(g) \geq v\} = \cup_{l \in L(T_v)} \tilde{x}(l) = A(v)$. Hence, $B(r(T)) = A(r(T))$. Now, we suppose that $B(v) = \{g \in \mathcal{F} \mid lca(g) \geq v\}$ for $v \in V(T) - L(T)$ by induction hypothesis (IH) and we want to show that $B(v_c) = \{g \in \mathcal{F} \mid lca(g) \geq v_c\}$ for $v_c \in \text{ch}(v)$. Without loss of generality, we show the result for v_l. Notice that if $g \in \{g \in \mathcal{F} \mid lca(g) \geq v_r \vee lca(g) = v\}$, then $g \in \cup_{l \in L(T_{v_r})} \tilde{x}(l)$ because if $lca(g)$ is a descendant of v_r or $lca(g) = v$ there must be at least one leaf l in T_{v_r} such that $g \in \tilde{x}(l)$ by definition of $lca(g)$. Therefore,

$$\{g \in \mathcal{F} \mid lca(g) \geq v_r \vee lca(g) = v\} - \cup_{l \in L(T_{v_r})} \tilde{x}(l) = \emptyset \tag{5}$$

Also notice that

$$\{g \in \mathcal{F} \mid lca(g) \geq v_\ell\} \cap \cup_{l \in L(T_{v_r})} \tilde{x}(l) = \emptyset \tag{6}$$

because if $g \in \cup_{l \in L(T_{v_r})} \tilde{x}(l)$, then $lca(g)$ cannot be a descendant of v_l. Then,

$$
\begin{aligned}
B(v) - A(v_r) &= \{g \in \mathcal{F} \mid lca(g) \geq v\} - \cup_{l \in L(T_{v_r})} \tilde{x}(l) && \text{by IH} \\
&= (\{g \in \mathcal{F} \mid lca(g) \geq v_\ell\} \\
&\quad \cup \{g \in \mathcal{F} \mid lca(g) \geq v_r \vee lca(g) = v\}) - \cup_{l \in L(T_{v_r})} \tilde{x}(l) \\
&= \{g \in \mathcal{F} \mid lca(g) \geq v_\ell\} && \text{by (5) and (6)}
\end{aligned}
$$

Thus, $B(v_\ell) = \{g \in \mathcal{F} \mid lca(g) \geq v_\ell\}$. By definition of x_{lca}, $x_{\mathrm{lca}}(v) = B(v)$ if v is a leaf and $x_{\mathrm{lca}}(v) = B(v) - B(v_\ell) - B(v_r)$ otherwise. The result follows.

We now show that Algorithm 2 returns $x_{\min}(v)$ for all $v \in V(T)$. Let $v \in V(T)$. If $v \in L(T)$, then for any $g \in \tilde{x}(v)$, $lca(g) \leq v$ thus $g \in x_{min}(v)$, and conversely for any $g \in x_{min}(v)$, g should appear in the subtree rooted at v which is v itself and thus $g \in \tilde{x}(v)$. Now suppose $v \notin L(T)$. Let $g \in x_{min}(v)$, thus $lca(g) \leq v$ and $g \in \cup_{l \in L(T_{v_\ell})} \tilde{x}(l)$ or $g \in \cup_{l \in L(T_{v_r})} \tilde{x}(l)$. In the first case, $g \in x_{min}(v_\ell)$ but as $lca(g) < v_\ell$, $g \in x_{min}(v_\ell) - x_{\mathrm{lca}}(v_\ell)$. In the second case, $g \in x_{min}(v_r)$ but as $lca(g) < v_r$, $g \in x_{min}(v_r) - x_{\mathrm{lca}}(v_r)$. Conversely, for any $c \in \{\ell, r\}$ and $g \in x_{min}(v_c) - x_{\mathrm{lca}}(v_c)$, we have $lca(g) \leq v$ and $g \in \cup_{l \in L(T_v)} \tilde{x}(l)$, thus $g \in x_{min}(v)$. $\qquad\square$

Proof of Corollary 1

Proof. Notice that for any history $\mathcal{H} = \langle \dot{T}, \epsilon, x \rangle$, $\mathrm{Gain} \in \epsilon((\mathrm{r}(\dot{T}), \mathrm{r}(\dot{T})_c))$ because otherwise point (4) of Definition 1 would not be respected. Also notice that $x_{\mathrm{out}}(\mathrm{r}(T)) = \emptyset$ by definition and therefore the content of the node $\mathrm{r}(T)$ in any history explaining T is $x_{min}(\mathrm{r}(T)) \cup I$ for some set $I \subseteq x_{\mathrm{in}}(\mathrm{r}(T))$ by Lemma 1. Therefore, the cost of the optimal subhistory rooted at $\mathrm{r}(T)$ explaining $\mathcal{T}$ is either $c_{\min}(\mathrm{r}(T))$ or $c_{\mathrm{in}}(\mathrm{r}(T))$ and the result follows. $\qquad\square$

Proof of Theorem 2

Proof. As ϵ is given in input and lead to an optimal cost by hypothesis, it is sufficient to show that $\langle \dot{T}, \epsilon, x \rangle \in \mathbb{H}(\mathcal{T})$. Notice that $x(v) \in \mathcal{P}(\mathcal{F})$ for $v \in V(\dot{T})$ by construction because the only gene families appearing in the history are the genes families appearing at the leaves of T. We now need to show that $\langle \dot{T}, \epsilon, x \rangle$ respects the conditions of Definition 1.

- Condition (1): Notice that for each leaf $l \in L(T)$, $x(l)$ is set to $x_{\min}(l)$ on line 2. By Lemma 3, $x_{\min}(l) = \tilde{x}(l)$ because l is leaf. We now show that $x(l)$ is not modified afterward on line 10. Let us assume, for the sake of contradiction, that there is a leaf $l \in L(T)$ such that $x(l)$ is modified on line 10. In that case, Loss $\notin \epsilon((p(l), l))$ and $x(p(l)) \not\subseteq \tilde{x}(l)$. Thus, there exists a gene family $g \in x(p(l))$ such that $g \notin \tilde{x}(l)$. There are three possible cases as to why g is in $x(p(l))$:

- Case (1): $g \in x_{\min}(p(l))$. In that case, for any history $\mathcal{H} = \langle \dot{T}, \epsilon, x \rangle$ explaining $\mathcal{T}$, $g \in x(p(l))$ by Lemma 1. As Loss $\notin \epsilon((p(l), l))$, $g \in x(l)$ by Definition 1. But $g \notin \tilde{x}(l)$ and we conclude that there exists no history $\mathcal{H} = \langle \dot{T}, \epsilon, x \rangle$ explaining $\mathcal{T}$ which is a contradiction.
- Case (2): g was added to $x(p(l))$ on either line 5 or line 7. In that case, notice that for any history $\mathcal{H} = \langle \dot{T}, \epsilon, x \rangle$ explaining $\mathcal{T}$, g must be in $x(p(l))$ because in that case there is some leaf l' descendant of $p(l)$ such that $g \in x(l')$ and there is no gain event in the path from $p(l)$ to l' allowing to gain g. As Loss $\notin \epsilon((p(l), l))$, $g \in x(l)$. But $g \notin \tilde{x}(l)$ and we conclude that there exists no history $\mathcal{H} = \langle \dot{T}, \epsilon, x \rangle$ explaining $\mathcal{T}$ which is a contradiction.
- Case (3): g was added to $x(p(l))$ on line 10. Here, Loss $\notin \epsilon((p(p(l)), p(l)))$ and we can repeat the same argument as to why g is in $p(p(l))$ (and thus in $p(l)$ and in l) until we get to the root to obtain a contradiction. If $p(p(l)) = r(T)$, then only cases (1) and (2) apply, leading to a contradiction.

Therefore, for each leaf $l \in L(T)$, $x(l)$ equals $x_{\min}(l)$. By Lemma 3, $x_{\min}(l) = \tilde{x}(l)$ because l is a leaf and thus, condition (1) is respected.

- Condition (2): Let $(v, v_c) \in E(\dot{T})$. If Loss $\notin \epsilon((v, v_c))$, then all the content of v is added to v_c on line 10 and afterward the contents of v and v_c are not modified. Therefore, $x(v) \subseteq x(v_c)$. Otherwise, if Loss $\in \epsilon((v, v_c))$, the fact that $x(v) \not\subseteq x(v_c)$ comes from the fact that ϵ leads to an optimal cost by hypothesis. In fact, if $x(v) \subseteq x(v_c)$ and Loss $\in \epsilon((v, v_c))$ in the output of Algorithm 4, we can obtain a new history $\langle \dot{T}, \epsilon', x' \rangle$ explaining $\mathcal{T}$ by simply removing the useless loss. This new history has a lower cost than any history $\langle \dot{T}, \epsilon, x \rangle$ and thus ϵ does not lead to an optimal cost, which is a contradiction. We conclude that condition 2.(a) is respected. If Gain $\notin \epsilon((v, v_c))$, then all the content of v_c is added to v on either line 5 or line 7 and afterward the only possible modification to the content of v is to add some new content on line 10 and then the only possible modification to the content of v_c is to add some new content on line 10, but this content is already in $p(v)$. Thus, $x(v_c) \subseteq x(v)$. Otherwise, if Gain $\in \epsilon((v, v_c))$, the fact that $x(v_c) \not\subseteq x(v)$ comes from the fact that ϵ leads to an optimal cost by hypothesis, similarly to the previous case. Therefore, condition 2.(b) is respected.
- Condition (3): Let $g \in \mathcal{F}$. Let's first show that there is a node $v \in V(\dot{T}) - \{r(\dot{T})\}$ such that $g \in x(v) - x(p(v))$. By construction, the first node v on the path from $lca(g)$ to $r(T)$ such that Gain $\in \epsilon(p(v), v)$ is such that $g \in x(v) - x(p(v))$. If there is no such node, then there exists no history $\langle \dot{T}, \epsilon, x \rangle$ explaining $\mathcal{T}$ because there is then no gain event on the edges of the path from $lca(g)$ to $r(T)$ and thus Gain $\notin \epsilon(r(T), r(T)_c)$ and then it is impossible for Conditions (2) and (4) of Definition 1 to be respected simultaneously. This is a contradiction and we conclude that such a node must exist. We now

show that this node v is unique. Let $v' \in V(\dot{T}) - \{r(\dot{T})\}$ different from v. If v' is a strict ancestor of v, then $g \notin x(v')$ and $g \notin x(p(v'))$ by construction. Otherwise, if v' is on the path from $lca(g)$ (included) to v (excluded), then there is no gain on the edges of this path and thus $g \in x(v')$ and $g \in x(p(v'))$ by construction. Otherwise, if v' is strict descendant of $lca(g)$ such that $g \in \cup_{l \in L(T_{v'})} x(l)$, then $g \in x_{\min}(v')$ and $g \in x_{\min}(p(v'))$ by definition and thus $g \in x(v')$ and $g \in x(p(v'))$ by construction. Finally, if v' is such that $g \notin \cup_{l \in L(T_{v'})} x(l)$, then $g \in x(v')$ if and only if $g \in x(p(v'))$ by construction. In all cases, we conclude that $g \notin x(v') - x(p(v'))$.

- Condition (4): Notice that $x(r(\dot{T}))$ is set to $\emptyset$ on line 11. Let us assume, for the sake of contradiction, that there is a node in $V(\dot{T}) - \{r(\dot{T})\}$ with an empty synteny content. Let v be the highest node in $\dot{T}$ different from the root such that this is the case (i.e. $x(v) = \emptyset$). First notice that $x_{\min}(v) = \emptyset$ as otherwise $x(v)$ would not be empty by construction. Also notice that for $v_c \in \{v_\ell, v_r\}$, either $x(v_c) = \emptyset$ or $\text{Gain} \in \epsilon((v, v_c))$ as otherwise $x(v)$ would not be empty by construction. If $\text{Gain} \in \epsilon((v, v_c))$, then $x(v) \cap (x_{\text{in}}(v_c) \cup x_{\text{lca}}(v_c)) = \emptyset$ by Lemma 5 otherwise ϵ would not lead to an optimal history. If $x(v_c) = \emptyset$, then we can repeat the same argument (i.e. for $v_{c_c} \in \{v_{c_\ell}, v_{c_r}\}$ either $x(v_{c_c}) = \emptyset$ or $\text{Gain} \in \epsilon((v_c, v_{c_c})))$ until we get to a leaf l for which we know that $\text{Gain} \in \epsilon((p(l), l))$ because $x(l) \neq \emptyset$ by construction. In both case, we conclude that $x(v) \cap (x_{\text{in}}(v_c) \cup x_{\text{lca}}(v_c)) = \emptyset$ and we can deduce that $x(v) \cap x_{\text{in}}(v) = \emptyset$ for any history explaining $\mathcal{T}$ with events ϵ (given that ϵ leads to an optimal history). As v is the highest node in $\dot{T}$ different from the root such that $x(v) = \emptyset$, either v is the child of the root or $\text{Loss} \in \epsilon((p(v), v)$ as otherwise $x(v)$ would not be empty by construction. Thus, $x(v) \cap x_{\text{out}}(v) = \emptyset$ by Lemma 6 otherwise ϵ would not lead to an optimal history. But it is impossible that simultaneously $x_{\min}(v) = \emptyset$, $x(v) \cap x_{\text{out}}(v) = \emptyset$ and $x(v) \cap x_{\text{in}}(v) = \emptyset$. Therefore, ϵ does not lead to an optimal history, which is a contradiction. Thus, we conclude that condition (4) is respected.

$\square$

Proof of Theorem 3

Proof. Let $\mathcal{H} = \langle \dot{T}, \epsilon, x \rangle \in \text{argmin}_{\hat{\mathcal{H}} \in \mathbb{H}(\mathcal{T})} \{c(\hat{\mathcal{H}})\}$. Let us assume, for the sake of contradiction, that there exists a history $\mathcal{H}' = \langle \dot{T}, \epsilon, x' \rangle \in \mathbb{H}(\mathcal{T})$ such that $x' \neq x$. Let v be the highest node in $\dot{T}$ such that $x(v) \neq x'(v)$. Notice that $v \neq r(\dot{T})$ because $x(r(\dot{T})) = \emptyset$ and $x'(r(\dot{T})) = \emptyset$ by Definition 1. We will show that a contradiction arises in each possible cases for the set $\epsilon((p(v), v))$:

- $\epsilon((p(v), v)) = \emptyset$: In this case, $x(v) = x(p(v))$ and $x'(v) = x'(p(v))$ by Definition 1. But as $x(p(v)) = x'(p(v))$, this implies that $x(v) = x'(v)$ which is a contradiction.
- $\epsilon((p(v), v)) = \{\text{Loss}\}$: In this case, $x(v) \subsetneq x(p(v))$ and $x'(v) \subsetneq x'(p(v))$ by Definition 1. By lemma 6, $x(v) \cap x_{\text{out}}(v) = \emptyset$ and $x'(v) \cap x_{\text{out}}(v) = \emptyset$. Notice that any gene family $g \in x_{min}(v) \cup x_{\text{in}}(v)$ such that $g \in x(p(v))$ cannot be lost by the event on the edge $(p(v), v)$ because each gene family is gained

only once in a history by Definition 1. As $x(p(v)) = x'(p(v))$ and as for each $g \in \mathcal{F}$, either $g \in x_{min}(v) \cup x_{\mathrm{in}}(v)$ or $g \in x_{\mathrm{out}}(v)$ by definition, $x(v) = x'(v)$ which is a contradiction.

- $\epsilon((p(v), v)) = \{\mathrm{Gain}\}$: As $x(v) \neq x'(v)$, there is at least one gene family in either $x(v)$ or $x'(v)$ that is not in the other set. Without loss of generality, we will assume that there exists a gene family $g \in x(v)$ such that $g \notin x'(v)$. As $x(p(v)) = x'(p(v))$ this implies that $\mathrm{Gain}_{\mathcal{H}}(g) = v$. Thus, $lca(g) \geq v$ because each gene family is gained only once in a history by Definition 1. Therefore, $\mathrm{Gain}_{\mathcal{H}'}(g) > v$ because $g \notin x'(v)$. Let $w = \mathrm{Gain}_{\mathcal{H}'}(g)$. This implies that $lca(g) \geq w$ and thus there is at least one loss event in $\mathcal{H}'$ (and $\mathcal{H}$) in each of the paths from v to a leaf $l \in \mathrm{L}(\dot{T}_v) - \mathrm{L}(\dot{T}_w)$. Therefore, we can construct a new history $\mathcal{H}^*$ explaining $\mathcal{T}$ from $\mathcal{H}$ by removing the gain event on the edge $(p(w), w)$ and adding the content that was gained in this event in the content of every node containing the gene family g that is not a descendant of w. Thus, $c(\mathcal{H}^*) = c(\mathcal{H}) - \delta_{\mathrm{gain}}$ and $\mathcal{H} \notin \mathrm{argmin}_{\hat{\mathcal{H}} \in \mathbb{H}(\mathcal{T})}$ which is a contradiction.

- $\epsilon((p(v), v)) = \{\mathrm{Gain}, \mathrm{Loss}\}$: As shown in the case $\epsilon((p(v), v)) = \{\mathrm{Loss}\}$, any gene family $g \in x_{\mathrm{out}}(v)$ must be lost in the loss event on the edge $(p(v), v)$ and any gene family not in $x_{\mathrm{out}}(v)$ cannot be lost in the loss event on the edge $(p(v), v)$. Therefore, if $x(v) \neq x'(v)$, there must be at least one gene family in either $x(v)$ or $x'(v)$ that is not in the other set and that is gained at v. We can thus use the same argument as in the case $\epsilon((p(v), v)) = \{\mathrm{Gain}\}$.

$\square$

Proof of Theorem 5

Proof. This result follows from Lemma 3, Theorems 1 and 2, Corollary 1 and from the fact that each step of the algorithm consists in a constant number of traversal of T in which each node is processed in $O(1)$ time because we suppose by hypothesis that operations on sets of gene families can be done in $O(1)$ time. $\square$

Additional Figures

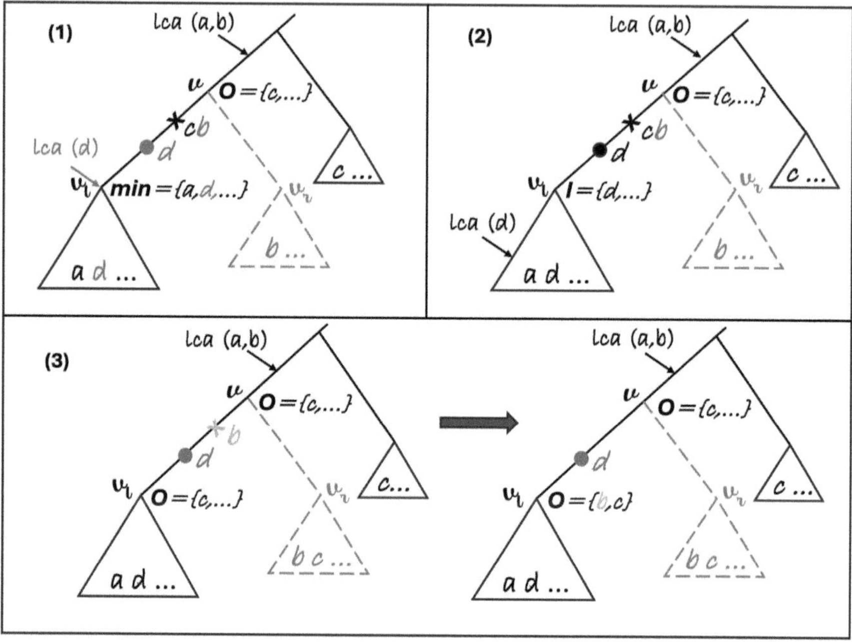

Fig. 6. An illustration of a left scenario (of cost $\Delta_{\mathrm{out}}(v_\ell)$) leading to $c_{\mathrm{out}}(v)$ (a similar scenario can be drawn for the right part, i.e. for $\Delta_{\mathrm{out}}(v_r)$); the characters in a triangle represent the gene families in the leaves of the corresponding subtree; the events and characters in gray can be present or absent, depending on the value of $g(v, v_\ell)$. (1): A scenario with $x(v_\ell) = x_{\min}(v_\ell)$ (line 1a) (2): A scenario with $x(v_\ell) = x_{\min}(v_\ell) \cup I$ where $I \subseteq x_{\mathrm{in}}(v_\ell)$ (line 1c) (3): Two scenarios with $x(v_\ell) = x_{\min}(v_\ell) \cup O$ where $O \subseteq x_{\mathrm{out}}(v_\ell)$, $O \neq \emptyset$ (line 1b). The left scenario is not optimal with a loss on the edge (v, v_ℓ) as the lost gene (here b) can always be postponed to be lost later in the same losses of the syntenies containing the extra gene c.

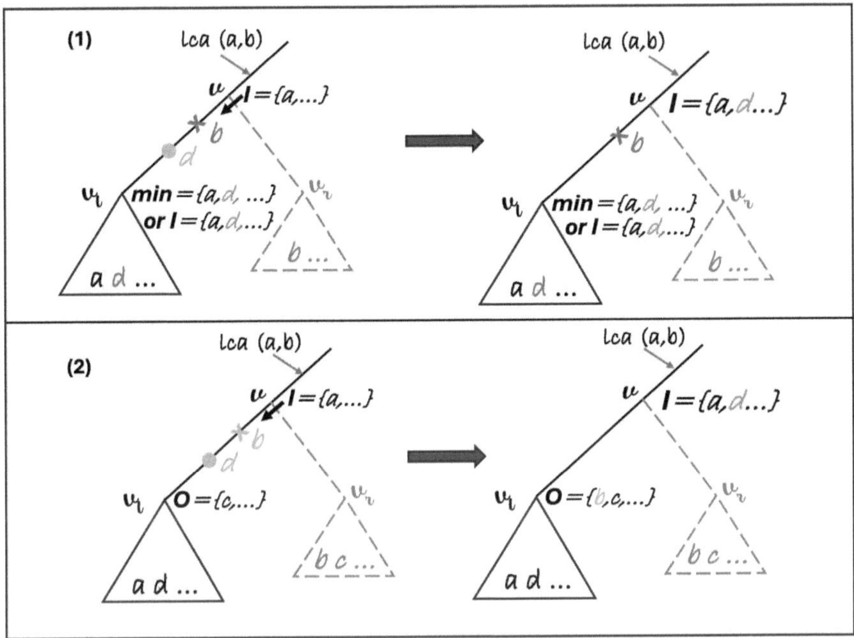

Fig. 7. An illustration of a left scenario (of cost $\Delta_{in}(v_\ell)$) leading to $c_{in}(v)$ in the case where the extra content I comes from the left side (a scenario similar to those of Fig. 6 can be drawn for the right part, i.e. for $\Delta_{out}(v_r)$); the characters in a triangle represent the gene families in the leaves of the corresponding subtree; the events and characters in gray can be present or absent, depending on the value of $l(v, v_\ell)$. (1): Two scenarios with either $x(v_\ell) = x_{min}(v_\ell)$ or $x(v_\ell) = x_{min}(v_\ell) \cup I$ where $I \subseteq x_{in}(v_\ell)$ (line 3a or 3c). The left scenario is not optimal with a gain on the edge (v, v_ℓ) as the gained gene (here d) can always be gained earlier in this history at the same gain point as the gene a. (2): Two scenarios with $x(v_\ell) = x_{min}(v_\ell) \cup O$ where $O \subseteq x_{out}(v_\ell)$, $O \neq \emptyset$ (line 3b). The left scenario is not optimal with a loss and a gain on the edge (v, v_ℓ) as the lost gene (here b) can always be postponed to be lost later in the same losses of the syntenies containing the extra gene c and the gained gene (here d) can always be gained earlier in this history at the same gain point as the gene a.

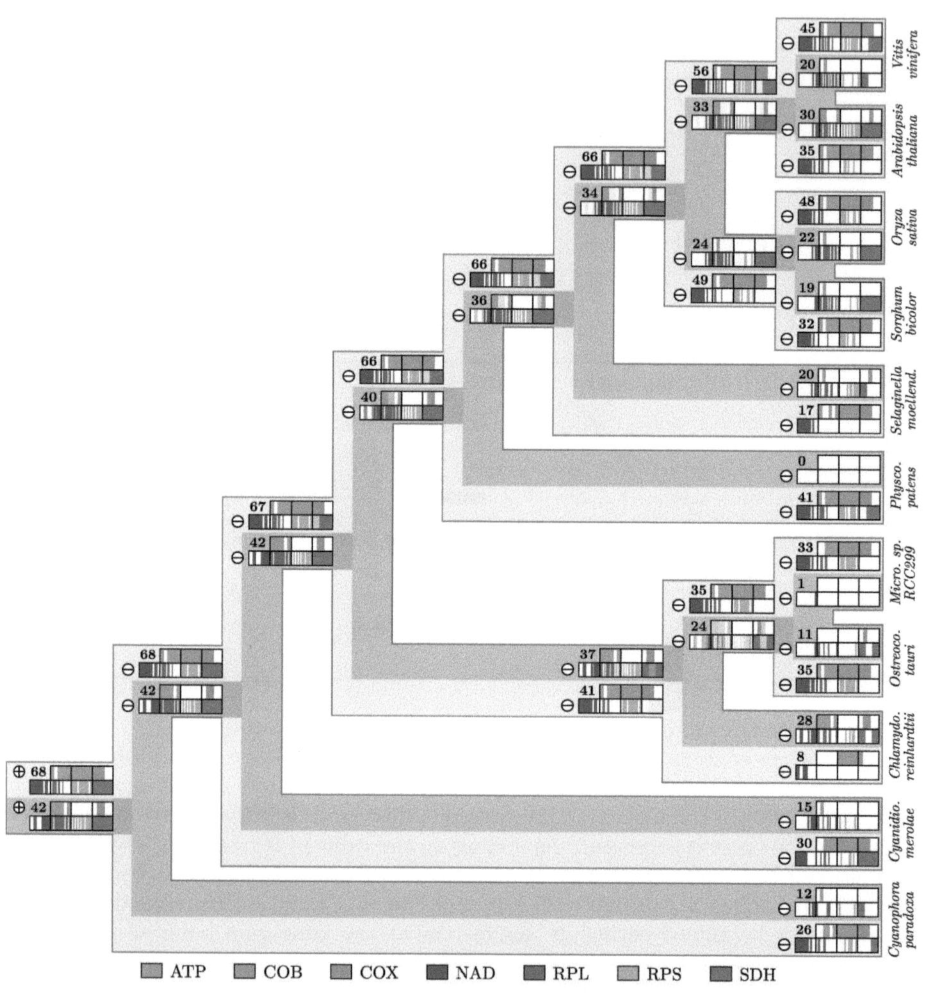

Fig. 8. Results obtained with Count for a probabilistic pure-loss model on the species tree of the considered 11 plants. Representation is the same as in Fig. 4, except that vertical strips are represented with a level of transparency illustrating their inferred probability of presence.

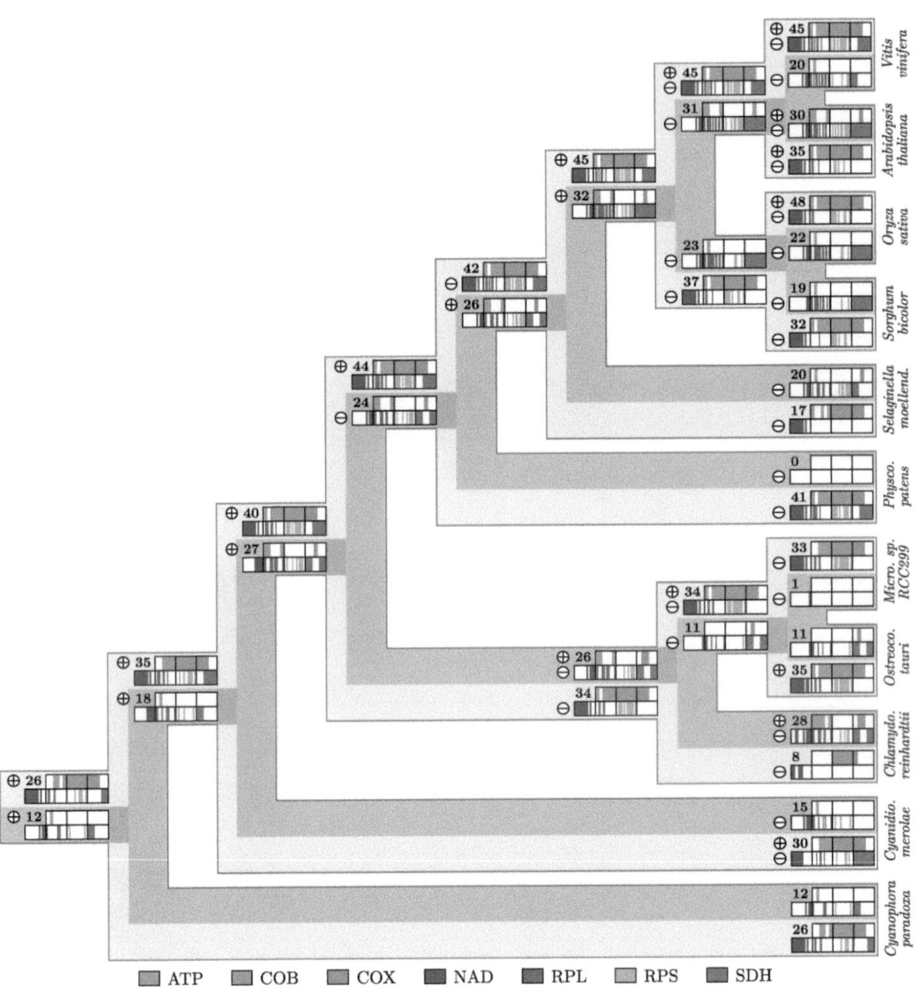

Fig. 9. Results obtained with Count for the Dollo parsimony model on the species tree of the considered 11 plants. Representation is the same as in Fig. 4.

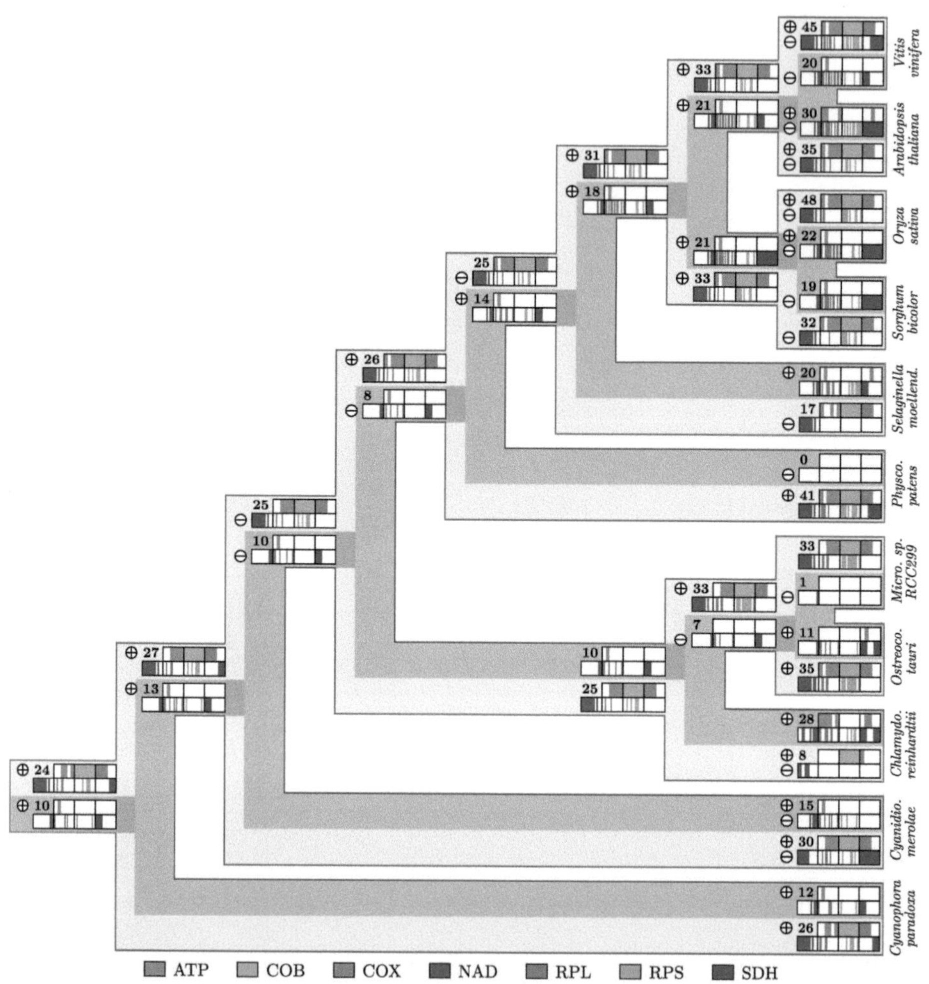

Fig. 10. Results obtained with Count for the Wagner parsimony model with unitary costs on the species tree of the considered 11 plants. Representation is the same as in Fig. 4.

References

1. Angibaud, S., Fertin, G., Rusu, I., Thévenin, A., Vialette, S.: On the approximability of comparing genomes with duplicates. J. Graph Algorithms Appl. **13**(1), 19–53 (2009)
2. Anselmetti, Y., Delabre, M., El-Mabrouk, N.: Reconciliation with segmental duplication, transfer, loss and gain. In: Jin, L., Durand, D. (eds.) Comparative Genomics, pp. 124–145. Cham (2022)
3. Anselmetti, Y., El-Mabrouk, N., Lafond, M., Ouangraoua, A.: Gene tree and species tree reconciliation with endosymbiotic gene transfer. Bioinformatics **37**(SI-1), i120–i132 (2021)
4. Blin, G., Chauve, C., Fertin, G., Rizzi, R., Vialette, S.: Comparing genomes with duplications: a computational complexity point of view. IEEE/ACM Trans. Comput. Biol. Bioinf. **4**(4), 523–534 (2007)
5. Bohnenkamper, L., Stoye, J., Doerr, D.: Reconstructing rearrangement phylogenies of natural genomes. Algorithms Mol. Biol. **20**(10) (2025)
6. Caprara, A.: The reversal median problem. INFORMS J. Comput. **15**(1), 93–113 (2003)
7. Csűrös, M.: Ancestral reconstruction by asymmetric Wagner parsimony over continuous characters and squared parsimony over distributions. In: Proceedings of the Sixth RECOMB Comparative Genomics Satellite Workshop. LNB, vol. 5267, pp. 72–86 (2008)
8. Csűrös, M.: Count: evolutionary analysis of phylogenetic profiles with parsimony and likelihood. Bioinformatics **26**(15), 1910–2 (2010)
9. Csűrös, M.: Gain-loss-duplication models for copy number evolution on a phylogeny: exact algorithms for computing the likelihood and its gradient. Theor. Popul. Biol. **145**, 80–94 (2022)
10. Csűrös, M., Miklós, I.: A probabilistic model for gene content evolution with duplication, loss, and horizontal transfer. In: Proceedings of the Tenth Annual International Conference on Research in Computational Molecular Biology (RECOMB). LeNB, vol. 14899, pp. 206–220 (2006)
11. Delabre, M., El-Mabrouk, N.: Synesth: comprehensive syntenic reconciliation with unsampled lineages. Algorithms **17**(5) (2024)
12. Doerr, D., Chauve, C.: Small parsimony for natural genomes in the DCJ-indel model. J. Bioinf. Comput. Biol. **19**(6) (2021)
13. Feijao, P., Meidanis, J.: SCJ: a breakpoint-like distance that simplifies several rearrangement problems. IEEE/ACM Trans. Comput. Biol. Bioinf. **8**(5), 1318–1329 (2011)
14. Fitch, W.M.: Toward defining the course of evolution: minimum change for a specific tree topology. Syst. Zool. **20**(4), 406–416 (1971)
15. Gascon, M., Delabre, M., El-Mabrouk, N.: FullSynesth: syntenic reconciliation of a set of consistent gene trees. Theory Comput. Syst. **70**(8) (2026)
16. Goodman, M., Czelusniak, J., Moore, G.W., Romero-Herrera, A.E., Matsuda, G.: Fitting the gene lineage into its species lineage, a parsimony strategy illustrated by cladograms constructed from globin sequences. Syst. Zool. **28**, 132–163 (1979)
17. Kannan, S., Rogozin, I., Koonin, E.: MitoCOGs: clusters of orthologous genes from mitochondria and implications for the evolution of eukaryotes. BMC Evol. Biol. **14**(11), 1–16 (2014)
18. Kovác, J.: On the complexity of rearrangement problems under the breakpoint distance. J. Comput. Biol. **21**(1), 1–15 (2014)

19. Luhmann, N., Lafond, M., Thévenin, A., Ouangraoua, A., Wittler, R., Chauve, C.: The SCJ small parsimony problem for weighted gene adjacencies. IEEE/ACM Trans. Comput. Biol. Bioinf. **16**(4), 1364–1373 (2019)
20. Pe'er, I., Shamir, R.: The median problems for breakpoints are NP-complete. In: Proceedings of the Electronic Colloquium on Computational Complexity. ECCC 1998, vol. 5 (1998)
21. Sankoff, D.: Minimal parsimony is not enough. Syst. Zool. **24**(2), 158–164 (1971)
22. Tannier, E., E., C.Z., Sankoff, D.: Multichromosomal median and halving problems under different genomic distances. BMC Bioinformatics **10**, 120 (2009)

Phylogenetics II: Comparing and Evaluating Phylogenies

Similarities, Differences and Biases in Cophylogenetic Models for Host-Symbiont Coevolution

Gabriele Di Palma[1,2] (ID), Catherine Matias[3,4,5] (ID), and Blerina Sinaimeri[2(✉)] (ID)

[1] Università Campus Bio-Medico di Roma, Rome, Italy
dipalmag@luiss.it
[2] LUISS University, Rome, Italy
bsinaimeri@luiss.it
[3] Sorbonne Université, Paris, France
[4] Université Paris Cité, Paris, France
[5] CNRS, Laboratoire de Probabilités, Statistique et Modélisation, LPSM, 75005 Paris, France

Abstract. Synthetic data are becoming increasingly important for computational studies of cophylogeny, including machine learning, benchmarking, and method testing. Several generators have been proposed to produce such data, but each relies on different assumptions about host-symbiont coevolution. These assumptions are often implicit and rarely examined, even though results can depend strongly on the synthetic model being used. In this article, we present a systematic structural analysis of representative cophylogeny generators under controlled scenarios. The goal is to make their assumptions explicit and to understand how these choices shape the synthetic data they produce as well as the conclusions that may be drawn from them.

Keywords: cophylogeny · synthetic data · host–symbiont evolution

1 Introduction

Reconstructing the shared evolutionary history of symbionts and their hosts is central in several application domains, including the identification and tracking of emerging infectious diseases [12,14,22]. The growing availability of public sequence data has made these analyses increasingly feasible at scale. A standard way to formalize host–symbiont coevolution is through *cophylogeny* models, which aim to explain the histories of hosts and symbionts using their evolutionary trees (typically inferred from DNA sequences). In this framework, coevolution is often expressed as the problem of mapping the symbiont phylogeny onto the host phylogeny (see, e.g., [6,9,16,21,28,30]). This mapping, called a *reconciliation*, describes the relationship between the two trees via biologically meaningful events such as cospeciation, host-switching, duplication, and loss. Because real

M. Lafond (Ed.): RECOMB-CG 2026, LNBI 16569, pp. 213–233, 2026.
https://doi.org/10.1007/978-3-032-26891-4_11

host–symbiont datasets are limited in number, size, and diversity, synthetic data are an important resource for developing and validating computational methods in cophylogeny. Several synthetic generators for host–symbiont systems have been proposed, based on different modeling assumptions and simulation strategies. These generators are used to evaluate reconciliation methods and statistical tests of congruence, and they are expected to become even more important for data-driven pipelines, especially machine learning, where large, controlled datasets are needed for training and benchmarking. Despite this growing role, synthetic generators are often treated as interchangeable, and their outputs are rarely analyzed in detail beyond the specific task they are used for. In practice, different generators may encode different assumptions about evolutionary processes and may constrain the space of host-symbiont structures they can produce. As a result, datasets generated under comparable evolutionary scenarios may still differ substantially in their trees structures, detectable coevolutionary signal, and association network topology. Such differences can directly affect the conclusions drawn from synthetic-based computational studies.

In this paper we study the structural properties of synthetic host-symbiont datasets produced by commonly used cophylogeny generators. Thus, we adopt a regime-based evaluation framework, considering cospeciation-dominated, host-switch-dominated, and mixed coevolutionary regimes, and analyze each simulated dataset at three complementary levels. **First**, we examine coevolution-based summaries of the generated data, focusing on the relative frequencies of key events (in particular cospeciation and host-switching) to verify how closely each generator follows the intended regime. **Second**, we characterize the host and symbiont phylogenies through size and shape descriptors, including the number of leaves and standard tree-balance indices such as Cherry, Sackin, and Colless. **Third**, we study the association structure by analysing the bipartite network defined by the host-symbiont nowadays associations. We measure classical network parameters such as the connectivity and heterogeneity (e.g., density, assortativity, and hosts hotspot concentration). Finally, we compare these structural characteristics of the synthetic data to those observed in real host-symbiont datasets, which we use more as a reference region rather than as ground truth. This study is guided by two questions:

(Q1) When we fix the same high-level coevolutionary regime, do different generators produce datasets with comparable structure, or do they systematically induce different tree shapes, congruence signals, and association-network patterns?
(Q2) How much do the structural patterns produced by each generator overlap with those observed in real host–symbiont datasets, and do the generators fully cover this structural space or leave significant regions occupied by real data unexplored?

Our goal is not to identify the "best" generator but to explore how different modeling choices determine the structure of synthetic cophylogenetic data. By making these differences clear, we aim to help researchers choose synthetic data in a careful way for cophylogeny studies, and to point out the biases that can come from using one generator instead of another.

2 Methods

Here we focus on the host–symbiont setting, where the generators simulate a host phylogeny, a symbiont phylogeny, and the associations between them. Existing generators in the literature follow different modeling choices (for example in the set of macro-evolutionary events they include, or whether they use time and branch lengths). We therefore compare the following host–symbiont generators, namely `Coala` [4], `treeducken` [10] and the model described in [1] that has no name and that we call `cophylo` (from the name of the main function provided by the authors). We also include `AsymmeTree` [26], originally developed in the gene–species setting. Although its gene-family simulation builds on ideas from tools such as SaGePhy [13], `AsymmeTree` provides a simpler "tree evolving on a tree" framework that produces reconciliation-style histories without introducing additional layers such as domain evolution or population processes. This makes it structurally closer to host–symbiont generators and allows a more direct comparison in our setting. We do not consider gene–species simulators more broadly, since many of them include processes specific to gene evolution, such as horizontal gene transfer with replacement or incomplete lineage sorting, which do not have clear counterparts in host–symbiont macro-evolutionary models. Conversely, host–symbiont systems may involve *multiple associations*, where a symbiont is linked to more than one host at a time, a situation that does not arise in the gene–species setting.

2.1 Description of the Evaluated Generators

We now describe the 4 generators considered in this study. Notice that most of these generators are by-products of more global coevolution or cophylogenetics methods that we do not describe here. We rather only focus on the parts of these tools that enable generating a pair of host and symbiont trees together with coevolution history and extant leaves associations. Some of these tools only generate a symbiont tree *conditional* on a given input host tree (and additional parameters), while other *jointly* generate the host and symbiont tree (in general, from a birth and death process). In order to get a full generating process of host and symbiont pair, we combine the former tools with a simple birth-death process with parameters (λ_H, μ_H) (speciation and extinction, respectively) on the host tree when it should be given as input.

Coala. The `Java` software `Coala` [4,5] contains a generator (`TGLGenerator.jar`) for simulating symbiont coevolution along a given host tree and allowing 4 different types of events: *cospeciation, duplication, host-switch* and *loss*. The input host tree does not need to be dated (if branch lengths exist, they will be discarded) so we rely on the function `TreeSim` from the `DendroPy python` library [17,18], using parameters (λ_H, μ_H) and conditioned on the number of extant leaves to produce the input host tree. The output symbiont tree will similarly be undated. The generator uses 3 parameters resulting in a vector $\langle p_c, p_d, p_s, p_l \rangle$ of 4 probabilities constrained to sum to 1. The model is based on the Duplication–Transfer–Loss (DTL) model [3,29]. Conditional on a given host tree H, the generation of the symbiont tree proceeds by recursively considering its unmapped nodes that are (temporarily) "positioned" on branches and for which one of the 4 different events occurs. The process starts by positioning the symbiont root on the branch before the root of H (by default, although the user may choose any host branch), where this initial branch is fictitious and serves only to initialise the simulation. For every unmapped symbiont node that is temporarily positioned on a host branch, the 3 events cospeciation, duplication and switch (so except in case of a loss) will induce its mapping on the host node below the branch it was positioned on, as well as its speciation. Now, in case of a cospeciation (occurring with probability p_c), each descendant symbiont lineage is positioned on one of the 2 host branches below the associated host. Conversely, when duplication occurs (with probability p_d), each descendant symbiont lineage stays positioned on the same host branch its parent was on. A host-switch (occurring with probability p_s) induces one descendant lineage to be positioned on the same host branch its parent was on, while the other descendant is positioned on a host branch chosen randomly, under the constraint that this does not violate the time feasibility of the reconstruction so far [28]. If such a branch does not exist, the switch does not happen and a new event i is drawn among the 3 others (namely $i \in \{c, d, l\}$) with rescaled probability $p_i/(p_c + p_d + p_l)$. Finally when a loss occurs (with probability $p_l = 1 - p_c - p_d - p_s$), the symbiont node is not mapped and rather re-positioned on one of the 2 host branches (chosen randomly) below the branch it was. Note that a loss event can represent three distinct yet indistinguishable scenarios: (a) speciation of the host species independently of the symbiont, which then remains associated with only one of the new host species; (b) cospeciation of the host and symbiont, "immediately" followed by the extinction of one of the newly formed symbiont species; or (c) the same as (b), but with a failure to detect the symbiont in one of the two new host species. The process continues until no unmapped symbiont nodes remain; the evolution of a lineage stops once it is mapped to a host leaf. For a fixed host tree H and a given set of event probabilities, the simulator generates multiple symbiont trees (100 by default). A representative tree is then selected as the median tree, i.e., the tree that minimises the distance to the others under a chosen tree-distance measure. Note that the set of events included in `Coala` result in host-symbiont coevolved trees with no multiple associations between their leaves (so that a symbiont has at most one host). The method `AmoCoala` [27] later developed to overcome this limitation relies on an initial pair of host-symbiont trees and their leaves associations to set additional local probabilities of new events (called *spreads*) that produce these multiple associations and are defined at every internal node of the host tree. This very-fine tuning of the model (allowing heterogeneous probabilities of spreads along the tree) becomes a drawback in our study, as the sheer number of additional choices would complicate our analysis. For this reason, we choose not to include this generator in our comparison.

treeducken. The `sim_cophyBD` function from the R package `treeducken` [11] implements a cophylogenetic birth-death process that simulates, in forward time, a pair of phylogenies (host tree H and symbiont tree S) together with their extant ecological interactions [10]. Note that it's the only tool out of the four tested here that simulates the two trees jointly. The model allows hosts and symbionts to undergo speciation and extinction independently, and it also includes coupled events such as *cospeciation* (with explicit parameter) and *coextinction* (implicit).

Table 1. Correspondence between biological processes, event names and parameters across models. AS stands for associated symbionts; MA for multiple associations.

Coevolution event	TreeSim +Coala	treeducken	species_tree_n_age +cophylo	AsymmeTree
(Independent) Host speciation	host speciation λ_H; induces a loss for AS	host speciation λ_H; induces MA for AS	host speciation λ_H or no cospeciation $1 - c$; induces either a loss (if specialist) or MA (if generalist) for AS	species tree speciation λ_H
(Independent) Host extinction	host extinction μ_H	host extinction μ_H	host extinction μ_H	species tree extinction μ_H
Independent symbiont speciation	duplication p_d	symbiont speciation λ_S	symbiont speciation λ_S	duplication λ_D
Independent symbiont extinction	–	symbiont extinction μ_S	symbiont extinction μ_S	loss λ_L
Induced symbiont extinction	–	host extinction (μ_H) of the unique host (depends on past history) of that symbiont	–	–
Cospeciation (joint host–symbiont speciation)	cospeciation p_c	cospeciation λ_c	cospeciation c	(gene) speciation; prob. 1
Symbiont speciates and one descendant moves to another host lineage	switch p_s	switch χ	–	horizontal transfer λ_{HGT}
Association gain at host split (one symbiont associated to both descendant hosts)	–	spread χ	switch s or no cospeciation $(1 - c)$ of a generalist symbiont $(1 - f_d(1) \simeq 0.54$ by default)	–
Association loss at host split	loss p_l	–	no cospeciation $(1 - c)$ of a specialist symbiont $(f_d(1) \simeq 0.46$ by default)	–

The generator is based on a continuous-time Markov process and is conditioned on a simulation time horizon. Both trees will start at the same date and evolve during that time horizon, creating trees with the same stem age. The generator contains a total of six competing events, parameterized by their rates $(\lambda_H, \mu_H, \lambda_S, \mu_S, \chi, \lambda_c)$. Symbiont evolution is governed by the speciation rate λ_S and the extinction rate μ_S. A symbiont speciation event is similar to a duplication-like event from `Coala`, in which both descendant lineages inherit the ancestral host associations, while a symbiont extinction corresponds to a (particular case of a) loss-like event in that setting. The model includes a *host-expansion* event that takes 2 different forms (*switch* and *spread*), both being (additional) symbiont speciation events occurring with rate χ. The switch event creates one symbiont speciation with one descendant that has a new host association (among hosts not associated to its parent) while the other descendant retains the ancestral host repertoire. In spread event, one randomly chosen descendant gains a novel host association in addition to the ancestral host repertoire (thus creating a multiple association). The tool has three options (accessible through `hs_mode` $\in \{$`"switch"`,`"spread"`,`"both"`$\}$), where only switches, spreads or both events (with half probability for each) may occur, respectively. The user may also enforce a *host limit*, i.e., a maximum number of host associations per symbiont lineage at any given time. Cospeciation occurs at rate λ_c when a host lineage bifurcates and induces a simultaneous symbiont speciation event, triggered on one randomly selected symbiont lineage currently associated with that host. After cospeciation, each descendant host lineage is associated with one of the two new symbiont descendants, and remaining ancestral associations are redistributed at random among descendants. Hosts can also speciate independently of symbionts at rate λ_H and any associated symbiont will then remain associated to each descendant (thus creating a multiple association). This is analogous to a failure-to-diverge event as described in [7] and implemented in [9,15]. Finally, host extinction occurs at rate μ_H and may implicitly induce coextinction of symbiont lineages that are left without any associated hosts.

cophylo. The article [1] comes with a code as Supplementary Information [2]. It includes a C program (`main_cophylo.c`) generator for simulating a symbiont tree S, conditional on the input host H, using both a continuous-time Markov process with three competing events: *speciation, extinction* and *host shift*; as well as an additional process of *cospeciation* that may occur at each host speciation. The input host should be a dated tree and we simulate it using the auxiliary function `species_tree_n_age` from the tool `AsymmeTree` (described below), that generates a birth-death process, conditional on the number of extant leaves but also on a time horizon T. Then the conditional simulation of the symbiont tree is obtained by successively running the former Markov process on each time interval defined by two successive host speciations. The generator has 5 evolutionary parameters: λ_S, μ_S and s are the symbiont speciation, extinction and switch rates, respectively. Additionally, c is a cospeciation probability and T_{MRCA} a time to the most recent common ancestor (TMRCA). It enables to start simulating the symbiont tree at a different age than the host tree.

The first symbiont is linked to a random number d (drawn from parameter distribution f_d on the number of hosts per symbiont at any time) of contemporary hosts, or all available hosts if d exceeds their number. From the initial time to the next host speciation event, or between any 2 successive host speciation events, a continuous-time Markov process with the 3 above mentioned competing events is run. With rate λ (resp. μ) times the number of existing symbionts, a speciation (resp. extinction) event occurs to one symbiont randomly chosen.

Table 2. Comparison of simulation design across cophylogeny generators.

Model	Host phylo.	Symb.phylo.	Associations	Time	Symb. root
Coala	fixed input (indep.)	cond.	cond.	no	mapped to a host node
treeducken	joint	joint	joint	yes	same origin as host
cophylo	fixed input (indep.)	cond.	cond. with constraint on nb of hosts per symbiont at any time (input possible)	yes	independent origin time
AsymmeTree	birth–death process	cond.	cond.	yes	same origin as host

indep. = simulated independently; cond. = conditionally simulated; joint = jointly simulated

In case of a speciation, a first descendant inherits all associations of its parent while the other inherits a random number d (drawn with f_d) among a subsample of these. In case of an extinction, that symbiont is simply removed. With rate s times the total number of current associations, a switch event occurs, randomly chosen among symbionts weighted by their number of current hosts. This symbiont gains a new host among those not yet associated to it. Note that in that switch event, contrarily to what happens in Coala or treeducken, the symbiont does not undergo a speciation. So that this event creates a multiple association. It rather corresponds to the spread event from treeducken. Additionally, at each host speciation time, their associated symbionts may cospeciate with probability c, in which case two symbiont descendants are created and each one follows one of the two descendant host. If there is no cospeciation, then the symbiont may either be a *specialist* (with probability $f_d(1)$ which is user-chosen or ≈ 0.46 by default) or a *generalist*. A specialist will follow only one of the two descendant hosts, while a generalist will follow both and thus create a new multiple association. Note that in the latter case, the symbiont did not speciate and remains the same in the two descendant hosts, differentiating this case from the cospeciation one.

AsymmeTree. The article [26] contains a general model for gene family history simulation along a species tree with *duplication, loss, horizontal transfer and conversion*. Putting aside the conversion event that is specific to the gene-species context, the model is similar to DTL mentioned above and may be used in the host-symbiont context. In the corresponding python package [25], we focus here on two functions, namely `species_tree_n_age` that uses a birth-death model conditioned on time and number of extant leaves for simulating a species tree (host tree in our context), and `dated_gene_tree` that simulates a gene tree (symbiont tree in our context), conditional on a dated species tree. Note that in the former, the branches leading to extinct species are pruned from the output. We use the latter with only three events: duplication with rate λ_D, loss with rate λ_L and horizontal (gene) transfer with rate λ_{HGT}. The root of the gene (symbiont) tree is placed at the root of the species (host) tree. The host-speciation events are ordered and processed sequentially. The simulation alternates between two cases: a host-speciation event and random events occurring in the symbiont tree during two successive host speciation times. When host-speciation occurs, the gene necessarily "cospeciates" at the same time; in our host-symbiont case, that means the symbiont cospeciates with its host with probability 1. Then, and similarly to `cophylo`, between two successive host-speciations, a continuous-time Markov process is run with the three competing events: duplication, loss and horizontal transfer. Duplication is the same event as in `Coala`, corresponding to symbiont speciation in `treeducken` or `cophylo`. A loss event, is the same as a loss in `Coala` or as a symbiont extinction in `treeducken` or `cophylo`. Finally, horizontal transfer corresponds to a switch in `Coala` or `treeducken` (and does not have an equivalent event in `cophylo`, since it implies a symbiont speciation). Note that this generator never produces multiple associations (that do not exist in the gene-species context).

Summary of the Generators Characteristics. Table 1 summarizes the characteristics of the tools with a focus on the different coevolutionary events, while Table 2 contrasts these characteristics by emphasizing joint or conditional simulation of the trees as well as the role of time. Note that in the former, the event "Induced symbiont extinction" only exists for `treeducken` because the host and symbiont trees are jointly simulated, while for the other tools, the host tree is generated and pruned for extinct branches before being input in the conditional generator for the symbiont tree.

2.2   Measures Used to Compare the Generators

In this section, we describe 2 different type of characteristics that we have measured on each pair of host and symbiont trees, across the different generators. While the coevolution-based measures where used in a calibration process, to tune the parameters so as to obtain comparable situations across generators, the tree-based and association-based measures are used to understand the characteristics of the data they produce.

Coevolution-Based Measures. For each simulated host–symbiont pair, we record the number of cospeciation and host-switch events appearing in the coevolution process between the two trees. We focus on these two events because they are present, with comparable interpretation, across all generators. This persistence and stability also enables us to rigorously analyze simulation regimes defined by their relative frequencies (see Sect. 3.1).

Tree-Based Measures. For each simulated pair of host and symbiont trees, we first record basic size and scale characteristics. In particular, for each tree (host or symbiont) we measure the number of leaves and the tree height (i.e. the distance from the root to its deepest leaf). For each simulated pair, we also compute the difference between the number of leaves in the host and symbiont tree. We then compute standard shape and balance indices, through normalized versions of the *Cherry, Colless* [8] and *Sackin* [24] indices (see Appendix A for definitions).

All distances and heights reported in those measures are defined in the topological sense (number of edges), rather than using branch lengths. This is because not all generators produce comparable temporal information, and some models enforce a fixed time horizon while others allow simulated times to extend beyond it. Thus, using topological distances ensures that the measures are comparable across generators.

Association-Based Measures. For each simulated dataset, we consider the bipartite host–symbiont interaction network defined on the leaves of the two trees. On this interaction network, we measure the *density*, the *degree-assortativity* and the frequency of *host hotspots* (see Appendix A for definitions). The density is the average number of connections in the ecological interaction network of hosts and symbionts. The degree-assortativity [19] measures the tendency of generalist (resp. specialist) species to be associated to other generalist (resp. specialists). It corresponds to a correlation so that it can exhibit positive or negative values. The frequency of hosts hotspots captures the presence of hosts that are highly connected [20,23].

2.3   Experimental Setup

We explored a high-cospeciation (regime R1), a high host-switch (regime R2), and a mixed situation (regime R3). The exact parameter choices are given in Table 4 in Appendix B. These parameters were tuned so that each generator reproduces as closely as possible the intended coevolutionary regime, defined in terms of the relative prevalence of cospeciation and host-switch events. Comparability across generators was therefore monitored using these coevolution-based measures (see Sect. 3.1). Moreover, to make the tools as comparable as possible, we choose to set up the time T_{MRCA} in `cophylo` equal to the horizon time used to generate the symbiont tree. As a consequence, the host and the symbiont trees have the same age (as in the two other tools where time is considered). We also need to set up the distribution f_d on the number of hosts per symbiont at any time for that tool. Following the suggestion of the authors on their own dataset example, we rely on a power law with exponent parameter $\alpha = 1.6$ and support in $\{1, \ldots, n_H\}$ where n_H is the number of extant hosts (*i.e.* the probability to have k hosts is proportional to $k^{-\alpha}$).

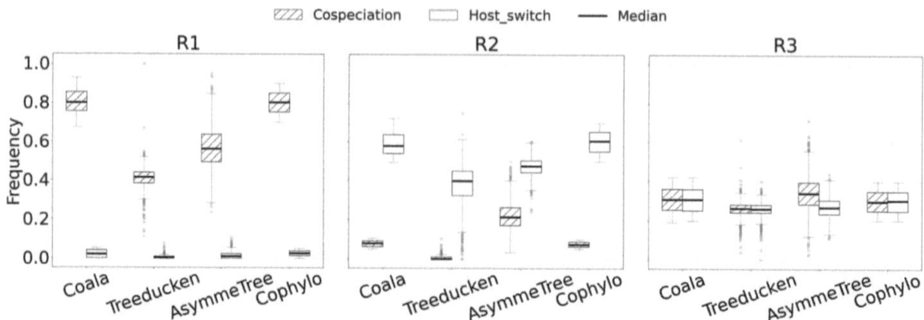

Fig. 1. Boxplots of cospeciation and host-switch frequencies across generators in regimes R1–R3, shown after parameter tuning. The figure indicates how closely each generator attains the intended regime.

3   Results and Discussions

3.1   Coevolution-Based Analysis

Figure 1 shows how the observed frequencies of cospeciation and host-switch events vary across generators and regimes. We stress that these measures were first used to tune the parameters of the generators (described in Table 4) in order to best reproduce each regime. Hence, the results shown correspond to the best match obtained for each generator. Because cospeciation and host-switch frequencies were used during parameter tuning, Fig. 1 should be interpreted mainly as showing how closely each generator matches the intended regime after calibration. By contrast, the analyses in the following subsections rely on measures that were not used for tuning and therefore more directly reflect structural differences between generators.

In R1, where cospeciation is expected to dominate, `Coala` and `cophylo` closely follow the intended regime, producing high cospeciation frequencies with very low host-switch values and relatively concentrated distributions. In contrast, `treeducken` generates substantially lower cospeciation frequencies and almost no host-switches, indicating that a significant portion of events is allocated to other processes not shown in the figure. `AsymmeTree` produces intermediate cospeciation values with a broader spread and more outliers, suggesting less tightly controlled outcomes. In R2, where host-switches are expected to dominate, `Coala` and `cophylo` again reproduce the intended behaviour with high host-switch frequencies and low cospeciation. `treeducken` still shows a noticeable deviation, with host-switch frequencies that are lower and more dispersed than in the other models and cospeciation values close to zero, while `AsymmeTree` displays intermediate values with moderate dispersion. In R3, corresponding to a mixed regime, all generators move toward more balanced frequencies between cospeciation and host-switches. Here the differences between generators are smaller: `treeducken` produces relatively concentrated distributions, while `AsymmeTree` exhibits the widest spreads and more outliers, and `Coala` and `cophylo` remain comparatively stable.

3.2   Tree-Based Analysis

Basic tree parameters. Figures 2, and 3 show how strongly the regime parameters overall constrain tree growth across the different regimes and generators. In R1, high cospeciation and low host-switch should keep symbiont and host tree sizes similar, and this is what we observe for `Coala`, `cophylo` and `treeducken`. However `treeducken` still produces much larger trees, indicating that additional events drive tree growth even in scenarios where cospeciation dominates. In R2, frequent switches allow symbionts to diversify independently, and this may lead to a dramatic increase in symbiont size for `AsymmeTree`, while `Coala` and `cophylo` remain constrained. In R3, differences are smaller but `treducken` continues to generate the largest trees, showing that its growth is less tightly controlled by the regime. Finally, especially in the R2 and R3 regimes, the simulated symbiont trees are often larger than the corresponding host trees, indicating that under these parameter settings, symbiont lineages diversify more rapidly than host lineages. In R1, this effect is less pronounced overall, but it is still very visible for `AsymmeTree`, where symbiont trees can grow substantially even under high cospeciation. This behaviour is expected in that model, which was originally designed for gene-species evolution, where gene lineages can diversify more independently from the species tree.

Shape and (Im)Balance-Measures. Figure 4 shows that, across the three regimes, the balance indices reflect how strongly each generator constrains the symbiont tree to follow the host tree. Generators that simulate host and symbiont jointly (such as `treducken`) tend to produce more similar host and symbiont shapes, since both trees are built under the same process and remain tightly coupled. In contrast, generators where the host is simulated independently and the symbiont is added conditionally (such as `Coala` and `cophylo`), or where additional independent events are allowed (such as `AsymmeTree`), show larger differences between the two trees. In the high–cospeciation regime (R1), host and symbiont shapes are generally close for `Coala` and `treducken`, indicating that strong cospeciation induces strong topology similarities, while `cophylo` produces more imbalanced symbiont trees. In the switch-dominated regime (R2), the coupling weakens: frequent switches allow the symbiont tree to diverge more from the host, and several generators produce noticeably more imbalanced symbiont topologies. The mixed regime (R3) lies between these behaviours, with moderate similarity but persistent model-dependent differences. We can conclude that matching cospeciation and switch settings does not guarantee similar tree shapes: the resulting topology depends both on how the trees are generated (jointly or independently) and on how strongly symbiont evolution is constrained by the host.

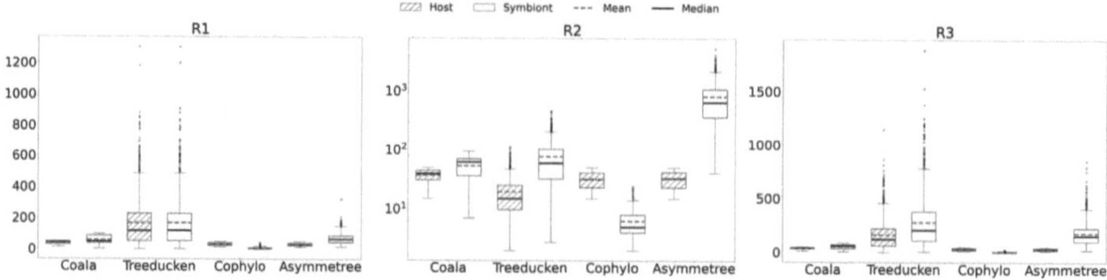

Fig. 2. Boxplots of the number of leaves for each regime R1–R3.

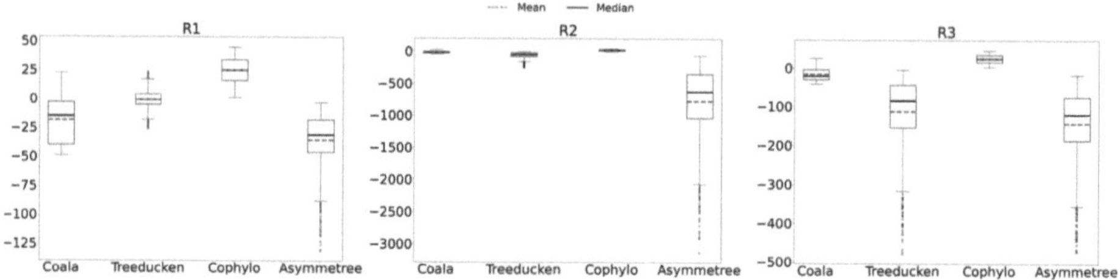

Fig. 3. Boxplots of the difference in the number of leaves between host and symbiont trees for each simulated dataset, across regimes R1–R3 (without 1% extremes; see Figure 8 for complete image).

3.3   Association-Based Measures

In Table 3 and Figs. 5–6, the network-level measures show that each generator produces markedly different host–symbiont association structures across regimes, reflecting how associations are constructed in the simulation. Notice that `cophylo` is excluded from this analysis. This is because we observed a memory-related bug in `cophylo` when reporting host–symbiont associations: in some cases, the output includes hosts or symbionts that do not appear in the corresponding phylogenetic trees. We reported this issue to the authors, but a fix is not yet available. Because this problem affects only the association output, it may bias network-based metrics. For this reason, we use `cophylo` results only for the previously defined tree-based measures, and exclude it from the association-based comparisons. In `Coala`, each symbiont is associated with exactly one host, which keeps the association network sparse across regimes. Thus, the degree variation comes mainly from hosts accumulating symbionts, so the proportion of host hotspots remains similar in R1–R3. Assortativity is consistently negative, indicating that edges tend to link nodes with different degrees. Because each symbiont can attach to only one host, this dis-assortative structure is expected and remains stable across regimes. In `treeducken`, density stays low in all regimes, but it is much more stable in the R2 than in the other two. Indeed, when switching dominates, events mainly change which host a lineage is linked to, so repeated runs tend to give similar density values, whereas the high-cospeciation and mixed regimes lead to larger run-to-run differences in how associations are distributed.

Assortativity is close to zero in the high-cospeciation and mixed regimes, but becomes clearly negative in the high-switch regime, because frequent switching produces a stronger generalist–specialist structure where edges tend to connect nodes with different degrees. Host hotspots are similar in the high-cospeciation and high-switch regimes but drop in the mixed regime, because R1 and R2 produce more uneven host degrees, while R3 makes degrees more even, leaving fewer hosts above mean+1sd. In `AsymmeTree`, density stays low in all regimes, but it is lowest in the R2 regime. A simple way to read this is that, under high switching, the generator tends to increase the number of possible host–symbiont pairs faster than it increases the number of realised links, so the network becomes sparser. Assortativity is negative in all regimes, and it is most negative in R2. This matches the regime meaning: when switching dominates, links are more likely to connect nodes with very different degrees, which pushes assortativity down; when cospeciation dominates, this effect is weaker and assortativity moves closer to zero. Host hotspots follow the same general idea: they are lowest in R1, highest in R2, and remain relatively high in R3, which is consistent with switching creating a more uneven host-degree distribution (more "outlier" hosts above the mean+1sd threshold), while cospeciation spreads associations more evenly across hosts.

3.4   Comparison to Real Data

We considered 37 real host–symbiont datasets from the literature. These datasets span a broad range of biological systems, including plant–fungus interactions, insect–parasite and vertebrate–parasite associations, as well as endosymbiotic relationships (e.g., *Wolbachia*), and display substantial variation in tree sizes and association patterns. To compare real and synthetic datasets in one quantitative view, we embed each dataset in a common feature space and project it to two dimensions using PCA. Each dataset is represented by a seven-dimensional vector of tree size/shape summaries (host–symbiont leaf difference and the Cherry, Sackin, and Colless indices). Network features such as assortativity and host hotspots could also be included, but here we use only tree-based features so that `cophylo` can be compared fairly. Features are standardized (zero mean, unit variance), PCA is fit on the union of real and synthetic datasets, and we report the first two principal components. It is important to note that the real datasets do not come with regime labels (e.g., "cospeciation-dominated" vs "switch-dominated"), so we cannot assign each real dataset to a specific simulated setting. A second issue is that we cannot fairly evaluate a generator by sampling its parameters at random: most random combinations produce biologically implausible trees or association patterns, so the task would be computationally hard. That's why we restrict attention to three interpretable regimes that represent common coevolutionary scenarios.

In Fig. 7, the real datasets (black) sit in a fairly compact area (delineated by the dashed ellipse). The synthetic datasets do not all fall there: where they land depends much more on which generator produced them than on whether we used one of the three regimes. `cophylo` (green) overlaps the real region the

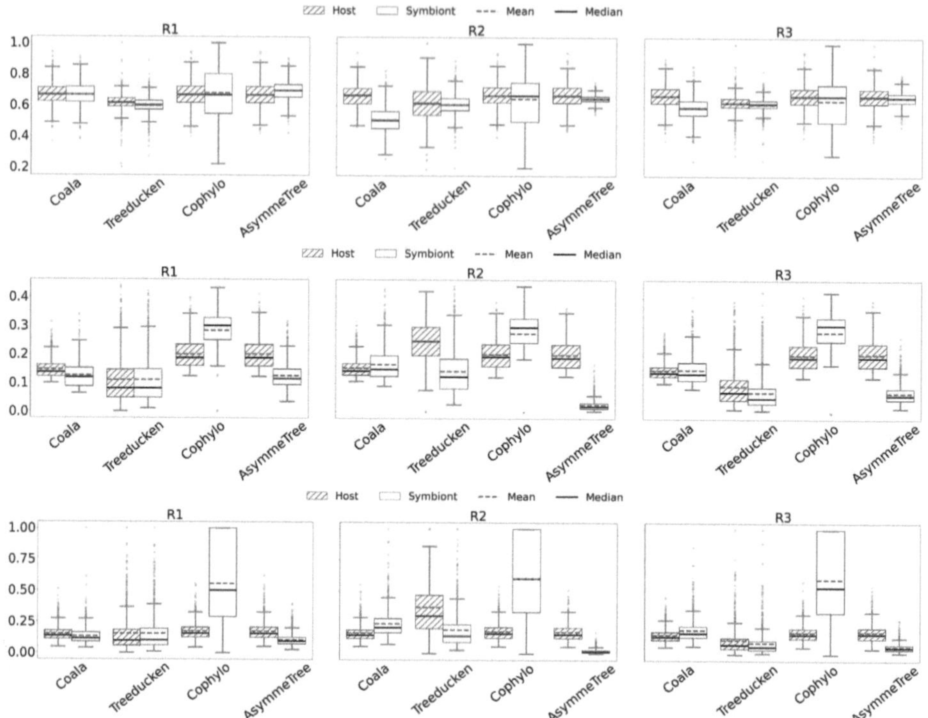

Fig. 4. Normalized Cherry (first row), Sackin (second row) and Colless (third row) indices across regimes R1–R3.

Table 3. Mean, Median, and Standard Deviation of association network density across regimes R1–R3.

Method	Mean/Median (sd)	Mean/Median (sd)	Mean/Median (sd)
	R1	R2	R3
Coala	0.032/0.028 (0.014)	0.041/0.032 (0.018)	0.041/0.033 (0.019)
treeducken	0.039/0.016 (0.067)	0.030/0.025 (0.019)	0.033/0.018 (0.054)
AsymmeTree	0.031/0.027 (0.011)	0.022/0.019 (0.008)	0.026/0.023 (0.01)

most, but also shows a clear downward extension (lower PC2) outside the ellipse. treeducken (red) is the most dispersed, reaching the real region but also extending far upward/right with several outliers. Coala (orange) is the most concentrated, clustering in the lower-left/central area and covering only part of the real region. AsymmeTree (blue) is mostly shifted left (negative PC1) and slightly higher on PC2, with only limited overlap near the left edge of the real region.

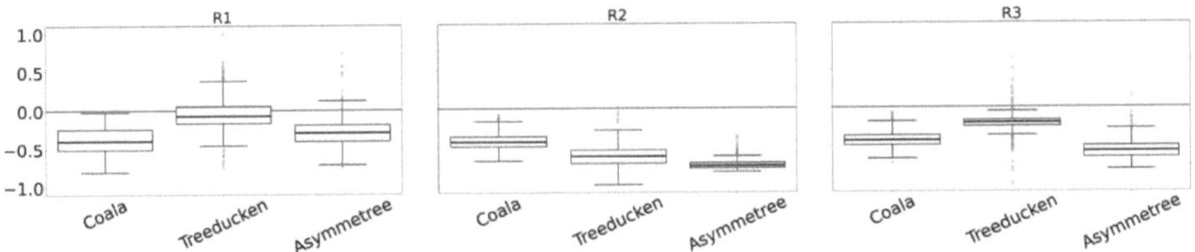

Fig. 5. Degree assortativity across regimes R1–R3.

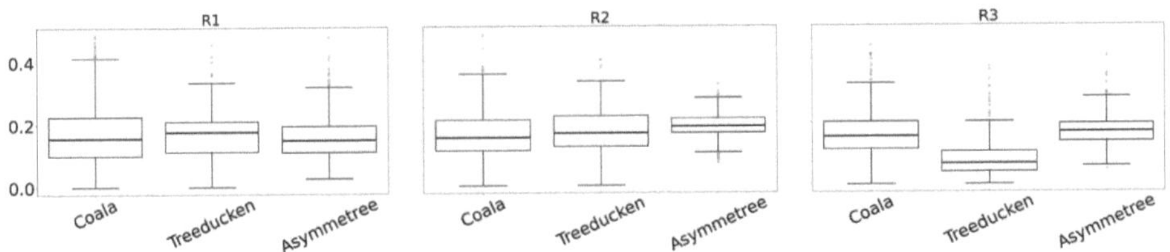

Fig. 6. Host hotspots frequency across regimes R1–R3.

4 Conclusion

This study shows that modeling choices affect the synthetic datasets used in cophylogeny. By comparing four common generators (`Coala`, `treeducken`, `cophylo`, and `AsymmeTree`) under the same three regimes, we observe systematic differences in the resulting feature profiles, including tree-shape summaries and, when available, association structure. This should be taken into account when using synthetic data to evaluate methods, because benchmarks may mix the effect of the intended regime with generator-specific patterns. A clear next step is to make "realism" and "coverage" measurable: define simple feature-based distances to quantify how close simulated datasets are to real ones and how well each generator covers the region occupied by real data, and use these scores to guide parameter calibration and to build comparable benchmark sets. This would make simulation-based comparisons more reliable and easier to interpret across studies. The code for reproducing those experiments as well as the real-datasets used are available at https://github.com/Gabbo240900/synthetic_cophylo

Disclosure of Interests. The authors have no competing interests.

A Formal Definitions of Measures

For any rooted binary tree T, let $L(T)$ and $V(T)$ denote its sets of leaves and vertices, respectively.

A *cherry* is a pair of leaves that are adjacent to a common ancestor node. The *normalized Cherry index* of a tree T is defined as the number of cherries

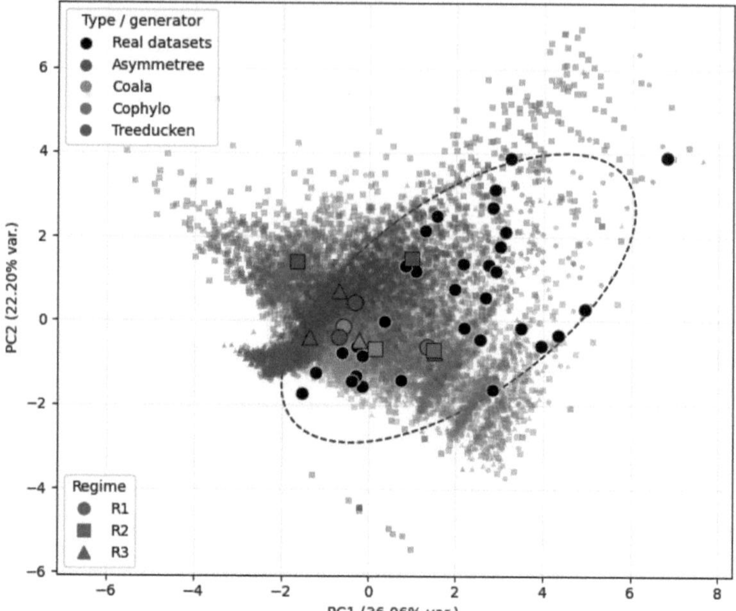

Fig. 7. Projection of real (black) and synthetic datasets (colored by generator, shaped by regime) onto PC1–PC2 from standardized feature vectors. Large markers show generator–regime centroids; the dashed ellipse summarizes real-data dispersion.

divided by $|L(T)|/2$ (corresponding to maximal value). The *normalized Colless index* [8] is defined as

$$C(T) = \frac{2}{|L(T)|(|L(T)| - 1)} \sum_{v \in \mathring{V}(T)} |n_L(v) - n_R(v)|,$$

where $\mathring{V}(T)$ is the set of internal nodes of T, and $n_L(v)$ and $n_R(v)$ denote the number of descendant leaves in the left and right subtrees of node v. The *normalized Sackin index* [24] is defined as

$$S(T) = \frac{1}{|L(T)|^2} \sum_{\ell \in L(T)} d(r, \ell),$$

where $d(r, \ell)$ is the distance from the root r to leaf ℓ. Smaller values of $S(T)$ correspond to more balanced trees.

For a pair of host–symbiont trees (H, S), let $A \subseteq L(H) \times L(S)$ denote the set of associations, where $(h, s) \in A$ if symbiont s is associated with host h. We write $G = (L(H) \cup L(S), A)$ for the resulting bipartite graph, which corresponds to the ecological interaction network of interest.

The *density* of G is defined as

$$\delta(G) = \frac{|A|}{|L(H)| \cdot |L(S)|}.$$

Let $d(v)$ denote the degree of a node v in G. The *degree assortativity* (see Eq. (21) in [19]) of G measures the tendency of nodes to connect to other nodes with similar degree and is defined as the Pearson correlation coefficient between the degrees at the endpoints of edges:

$$r(G) = \frac{\sum_{(h,s)\in A}\big(d(h) - \mu_H\big)\big(d(s) - \mu_S\big)}{\sqrt{\sum_{(h,s)\in A}\big(d(h) - \mu_H\big)^2}\sqrt{\sum_{(h,s)\in A}\big(d(s) - \mu_S\big)^2}},$$

where

$$\mu_H = \frac{1}{|A|}\sum_{(h,s)\in A} d(h), \qquad \mu_S = \frac{1}{|A|}\sum_{(h,s)\in A} d(s).$$

A *host hotspot* is defined as a host leaf $h \in L(H)$ whose degree exceeds the average host degree plus one standard deviation, a standard threshold used to identify highly connected nodes in ecological and interaction networks. Formally, letting

$$\mu = \frac{1}{|L(H)|}\sum_{h\in L(H)} d(h), \qquad \sigma = \sqrt{\frac{1}{|L(H)|}\sum_{h\in L(H)} (d(h) - \mu)^2},$$

we define the set of host hotspots as

$$\text{Hot}(G) = \{h \in L(H) : d(h) > \mu + \sigma\},$$

and record the frequency $|\text{Hot}(G)|/|V(H)|$.

B Parameter Settings

Table 4 shows the parameter choices for the generators across the three regimes.

C Removal of Extreme Values

In order to increase the readability of the plots, Fig. 8 was produced on filtered data, where 1% extreme values were removed from all generators. The most important example is `AsymmeTree` in regime R2 where the range of data removed goes from -5000 to -3000.

D Real Data Charts

In this section, we provide the characteristics measured on the set of 37 real host-symbiont datasets (Fig. 9, 10 and 11).

Table 4. Parameter settings for regimes R1–R3 across generators.

R1: high cospeciation and low switching

Generator	Parameter settings
Coala	$\lambda_H = 0.7$; $\mu_H = 0.63$; Nb extant host leaves $\sim X \in [15, 50]$; $p_c \sim U(0.7, 0.9)$; $p_s \sim U(0, 0.05)$; $(p_d, p_l) \sim (1 - p_c - p_s) *$ Dirichlet$(1, 1)$.
treeducken	$\lambda_H = 0.7$; $\mu_H = 0.63$; $\lambda_S = 0.7$; $\mu_S = 0.645$; $\lambda_c \sim U(2, 2.5)$; $\chi \sim U(0, 0.05)$; $T = 2$; hs-mode="both"
cophylo	$\lambda_H = 0.7$; $\mu_H = 0.63$; Nb extant host leaves $\sim X \in [15, 50]$; $\lambda_S = 0.7$; $\mu_S = 0.645$; $c \sim U(0.7, 0.9)$; $s \sim U(0, 0.05)$; $f_d(k) \propto k^{-1.6}$; $T = T_{MRCA}2$ (host and symbiont).
AsymmeTree	$\lambda_H = 0.7$; $\mu_H = 0.63$; $\lambda_D = 0.7$; $\lambda_L = 0.645$; $\lambda_{HGT} \sim U(0, 0.05)$; $T = 2$ (host and symbiont).

R2: high host-switches

Generator	Parameter settings
Coala	$\lambda_H = 0.7$; $\mu_H = 0.24$; Nb extant host leaves $\sim X \in [15, 50]$; $p_c \sim U(0.05, 0.1)$; $p_s \sim U(0.5, 0.7)$; $(p_d, p_l) \sim (1 - p_c - p_s) *$ Dirichlet$(1, 1)$.
treeducken	$\lambda_H = 0.7$; $\mu_H = 0.24$; $\lambda_S = 0.7$; $\mu_S = 0.66$; $\lambda_c \sim U(0, 0.05)$; $\chi \sim U(2, 2.5)$; $T = 2$; hs-mode="both"
cophylo	$\lambda_H = 0.7$; $\mu_H = 0.24$; Nb extant host leaves $\sim X \in [15, 50]$; $\lambda_S = 0.7$; $\mu_S = 0.66$; $c \sim U(0.05, 0.1)$; $s \sim U(0.5, 0.7)$; $f_d(k) \propto k^{-1.6}$; $T = T_{MRCA} = 2$ (host and symbiont).
AsymmeTree	$\lambda_H = 0.7$; $\mu_H = 0.24$; $\lambda_D = 0.7$; $\lambda_L = 0.66$; $\lambda_{HGT} \sim U(2, 2.5)$; $T = 2$ (host and symbiont).

R3: mixed coevolution

Generator	Parameter settings
Coala	$\lambda_H = 0.7$; $\mu_H = 0.45$; Nb extant host leaves $\sim X \in [15, 50]$; $p_c \sim U(0.2, 0.4)$; $p_s \sim U(0.2, 0.4)$; $(p_d, p_l) \sim (1 - p_c - p_s) *$ Dirichlet$(1, 1)$.
treeducken	$\lambda_H = 0.7$; $\mu_H = 0.45$; $\lambda_S = 0.7$; $\mu_S = 0.6$; $\lambda_c \sim U(1.5, 1.8)$; $\chi \sim U(1.5, 1.8)$; $T = 2$; hs-mode="both"
cophylo	$\lambda_H = 0.7$; $\mu_H = 0.45$; Nb extant host leaves $\sim X \in [15, 50]$; $\lambda_S = 0.7$; $\mu_S = 0.6$; $c \sim U(0.2, 0.4)$; $s \sim U(0.2, 0.4)$; $f_d(k) \propto k^{-1.6}$; $T = T_{MRCA} = 2$ (host and symbiont).
AsymmeTree	$\lambda_H = 0.7$; $\mu_H = 0.45$; $\lambda_D = 0.7$; $\lambda_L = 0.6$; $\lambda_{HGT} \sim U(1.5, 1.8)$; $T = 2$ (host and symbiont).

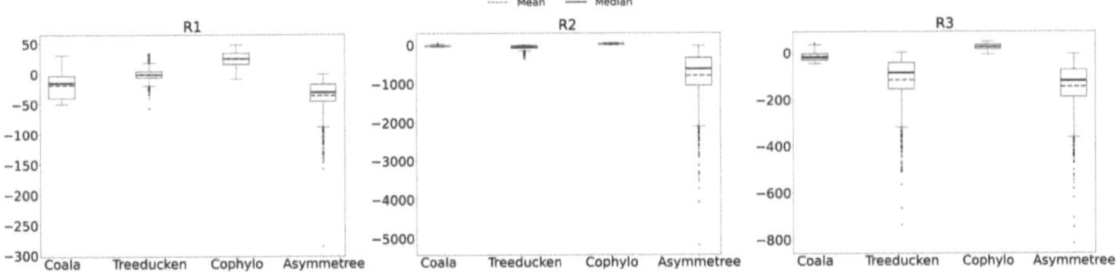

Fig. 8. Boxplots of the difference in the number of leaves between host and symbiont trees for each simulated dataset, across regimes R1–R3.

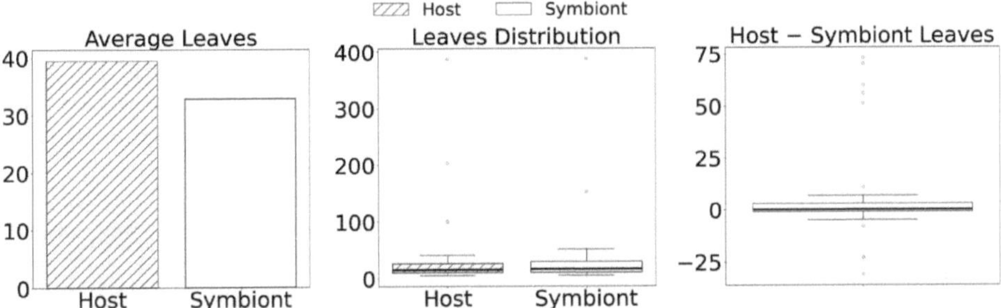

Fig. 9. Distribution of the number of leaves on the 37 real datasets (average, boxplot and difference between Host and Symbiont).

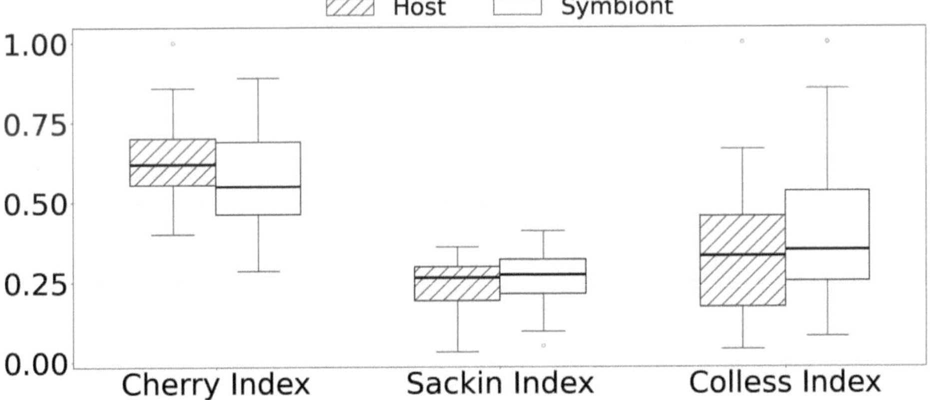

Fig. 10. Boxplots of the (Im)Balance indices measured on the 37 real datasets.

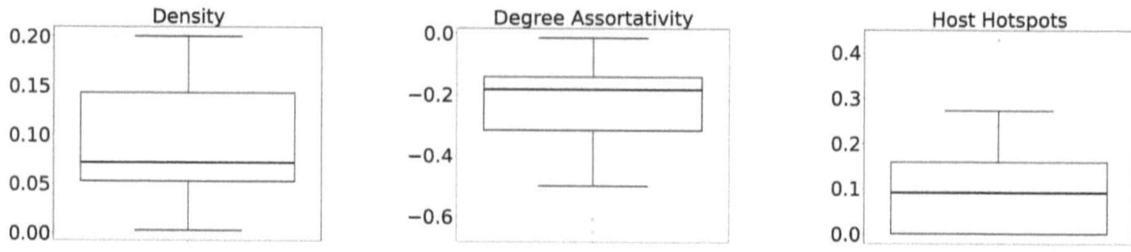

Fig. 11. Boxplots of the association-based measures for the 37 real datasets.

References

1. Alcala, N., Jenkins, T., Christe, P., Vuilleumier, S.: Host shift and cospeciation rate estimation from co-phylogenies. Ecol. Lett. **20**, 1014–1024 (2017)
2. Alcala, N., Jenkins, T., Christe, P., Vuilleumier, S.: Simulation software for host-parasite cophylogenies (2017). https://onlinelibrary.wiley.com/doi/full/10.1111/ele.12799, Supplementary Information
3. Bansal, M.S., Alm, E., Kellis, M.: Efficient algorithms for the reconciliation problem with gene duplication, horizontal transfer and loss. Bioinformatics **28**(12), i283–i291 (2012)
4. Baudet, C., et al.: Cophylogeny reconstruction via an approximate Bayesian computation. Syst. Biol. **64**(3), 416–431 (2014)
5. Baudet, C., et al.: AmoCoala (2023). https://github.com/sinaimeri/AmoCoala, Java software
6. Charleston, M.A.: Recent results in cophylogeny mapping. Adv. Parasitol. **54**, 303–330 (2003)
7. Charleston, M.A.: A new likelihood method for cophylogenetic analysis (2009). http://www.it.usyd.edu.au/research/tr/tr636.pdf, unpublished
8. Colless, D.H.: Review of: Phylogenetics: the theory and practice of phylogenetic systematics. Syst. Zool. **31**(1), 100–104 (1982)
9. Conow, C., Fielder, D., Ovadia, Y., Libeskind-Hadas, R.: Jane: a new tool for the cophylogeny reconstruction problem. Algorithms Mol. Biol. **5**(1) (2010)
10. Dismukes, W., Heath, T.A.: treeducken: an R package for simulating cophylogenetic systems. Methods Ecol. Evol. **12**(8), 1358–1364 (2021)
11. Dismukes, W., Justison, J.: treeducken (2021). https://github.com/wadedismukes/treeducken/tree/main, R package version 1.1.0
12. Etherington, G.J., Ring, S.M., Charleston, M.A., Dicks, J., Rayward-Smith, V.J., Roberts, I.N.: Tracing the origin and co-phylogeny of the caliciviruses. J. Gen. Virol. **87**(5), 1229–1235 (2006)
13. Kundu, S., Bansal, M.S.: SaGePhy: an improved phylogenetic simulation framework for gene and subgene evolution. Bioinformatics **35**(18), 3496–3498 (2019)
14. Lei, B.R., Olival, K.J.: Contrasting patterns in mammal-bacteria coevolution: bartonella and Leptospira in bats and rodents. PLoS Negl. Trop. Dis. **8**(3), 1–11 (2014)
15. Libeskind-Hadas, et al.: Jane 4 (2019). https://www.cs.hmc.edu/~hadas/jane/, Java software
16. Merkle, D., Middendorf, M., Wieseke, N.: A parameter-adaptive dynamic programming approach for inferring cophylogenies. BMC Bioinf. **11**(Suppl 1), S60 (2010)
17. Moreno, M.A., Holder, M.T., Sukumaran, J.: DendroPy 5 (2024). https://jeetsukumaran.github.io/DendroPy/index.html, Python library - version 5.0.8

18. Moreno, M.A., Holder, M.T., Sukumaran, J.: Dendropy 5: a mature python library for phylogenetic computing. J. Open Source Software **9**(101), 6943 (2024)
19. Newman, M.E.J.: Mixing patterns in networks. Phys. Rev. E **67**(2) (2003)
20. Newman, M.: Networks. Oxford University Press (2018)
21. Page, R.D.M.: Parallel phylogenies: reconstructing the history of host-parasite assemblages. Cladistics **10**(2), 155–173 (1994)
22. Pennington, P.M., et al.: The Chagas disease domestic transmission cycle in Guatemala: Parasite-vector switches and lack of mitochondrial co-diversification between Triatoma dimidiata and Trypanosoma cruzi subpopulations suggest non-vectorial parasite dispersal across the Motagua valley. Acta Tropica **151**, 80–87 (2015). ecology and diversity of Trypanosoma cruzi
23. Poulin, R.: Network analysis shining light on parasite ecology and diversity. Trends Parasitol. **26**(10), 492–498 (2010)
24. Sackin, M.J.: "Good" and "bad" phenograms. Syst. Biol. **21**(2), 225–226 (1972)
25. Schaller, D., Hellmuth, M., Stadler, P.F.: AsymmeTree (2022). https://github.com/david-schaller/AsymmeTree, Python package version 2.2
26. Schaller, D., Hellmuth, M., Stadler, P.F.: AsymmeTree: a flexible python package for the simulation of complex gene family histories. Software **1**(3), 276–298 (2022)
27. Sinaimeri, B., Urbini, L., Sagot, M.F., Matias, C.: Cophylogeny reconstruction allowing for multiple associations through approximate Bayesian computation. Syst. Biol. **72**(6), 1370–1386 (2023)
28. Stolzer, M.L., Lai, H., Xu, M., Sathaye, D., Vernot, B., Durand, D.: Inferring duplications, losses, transfers and incomplete lineage sorting with nonbinary species trees. Bioinformatics **28**(18), i409–i415 (2012)
29. Tofigh, A., Hallett, M., Lagergren, J.: Simultaneous identification of duplications and lateral gene transfers. IEEE/ACM Trans. on Comput. Biol. Bioinf. **8**(2), 517–535 (2011)
30. Wang, Y., Mary, A., Sagot, M.F., Sinaimeri, B.: Capybara: equivalence class enumeration of cophylogeny event-based reconciliations. Bioinformatics **36**(14), 4197–4199 (2020)

On the Comparison of LGT Networks and Tree-Based Networks

Bertrand Marchand[1](✉), Nadia Tahiri[2], Olivier Tremblay-Savard[3], and Manuel Lafond[2](✉)

[1] Université du Québec à Montréal, Montreal, Canada
`marchand.bertrand@uqam.ca`
[2] Université de Sherbrooke, Sherbrooke, Canada
[3] University of Manitoba, Winnipeg, Canada

Abstract. Phylogenetic networks are widespread representations of evolutionary histories for taxa that undergo hybridization or Lateral-Gene Transfer (LGT) events. There are now many tools to reconstruct such networks, but no clearly established metric to compare them. Such metrics are needed, for example, to evaluate predictions against a simulated ground truth. Despite years of effort in developing metrics, known dissimilarity measures either do not distinguish all pairs of different networks, or are extremely difficult to compute. Since it appears challenging, if not impossible, to create the ideal metric for all classes of networks, it may be relevant to design them for specialized applications. In this article, we introduce a metric on LGT networks, which consist of trees with additional arcs that represent lateral gene transfer events. Our metric is based on edit operations, namely the addition/removal of transfer arcs, and the contraction/expansion of arcs of the base tree, allowing it to connect the space of all LGT networks. We show that it is linear-time computable if the order of transfers along a branch is unconstrained but NP-hard otherwise, in which case we provide a fixed-parameter tractable (FPT) algorithm in the level. We implemented our algorithms and demonstrate their applicability on three numerical experiments.

Full online version:https://www.biorxiv.org/content/10.1101/2025.11.20.689557

1 Introduction

Phylogenetic networks are often seen as more accurate representations of evolution than trees, especially for species that undergo hybridization or lateral gene transfer, and one can even argue that the "tree of life" hypothesis should be updated to incorporate network-like structures [1]. Several approaches now aim to reconstruct networks, for example NeighborNet [2], PhyloNet [3], SNaQ [4], PhyNest [5], T-Rex [6], and more. Despite the importance of predicting such networks, there is still no clearly established method to evaluate the quality of the reconstructions. For phylogenetic *trees*, the Robinson-Foulds (RF) metric and

M. Lafond (Ed.): RECOMB-CG 2026, LNBI 16569, pp. 234–264, 2026.
https://doi.org/10.1007/978-3-032-26891-4_12

extensions [7,8] are widely used, and despite some shortcomings are accepted as a good measure of similarity. Therefore, when a new tree prediction tool is developed, it can be evaluated against a gold standard or simulated dataset using RF, providing insights on its strengths and weaknesses.

In the case of networks, this is not so easy. Several network measures have been attempted in the last two decades, but all of them have drawbacks. This includes structural measures that compare small components of the networks, such as softwired or hardwired clusters [9–11], rooted triples and trinets [12–14], displayed trees [15], and others [9,16], as well as metrics based on counting paths, such as the μ-distance [17,18], which was recently generalized to semi-directed networks [19]. All of these metrics suffer from the fact that, in all generality, they may say two networks are the same when they are actually different (see e.g. [9, 16]). Another family of measures are operational metrics, which are based on a number of operations needed to transform one network into the other (e.g., SPR [20–22], NNI [23,24], contractions [25], or cherry-picking [26,27] operations). However, these tend to be very hard to compute and, to our knowledge, not yet used frequently in practice. Overall, it appears extremely difficult to design a "one-size-fits-all" metric for networks, and it may be more relevant to design metrics that are tailored for specific applications.

Towards this goal, our aim in this paper is to compare networks in which reticulations represent lateral gene transfer (LGT) events. Specialized approaches can be used to reconstruct such networks, since the problem can be formulated in terms of starting with a species *tree*, and inserting arcs between branches to represent genetic exchanges between co-existing species. This results in what is called an *LGT network*, which consisting of the known original *base tree* and extra *transfer arcs* [28]. There are many LGT network reconstruction approaches: they can be built using tri-nets [29]; using so-called *Duplication-Transfer-Loss (DTL) reconciliation* [30,31]; using orthology and xenology data [32]; or using the presence/absence of character traits to predict clades that exchanged characters [33,34]. Moreover, LGT networks are a specialized form of *tree-based networks*, which also consist of base trees with additional arcs, but in which the base tree is unknown [35]. As tree-based networks form a popular class that generalizes many others [36], there is also a need to develop metrics for them.

Even in the case of LGT networks, there appears to be no ideal comparison metric. As a result, LGT network predictors have mostly been evaluated in an ad hoc manner in previous work. For example in [34], LGT networks were reconstructed to predict bacterial transfers, notably interphylum transfers between Proteobacteria and Actinobacteria, and the results were compared to published studies on this set of species. Unfortunately, this methodology is qualitative and does not easily generalize to other datasets. As already mentioned, another way of reconstructing LGT networks is by computing *DTL reconciliations* between species and gene trees [31,37]. These infer events on gene trees and the predicted gene transfers can be used to add LGT arcs on the species tree. These approaches typically measure the accuracy of the events on the gene trees, but not the accuracy of the underlying species LGT network—which is more relevant in some contexts such as transfer highway identification.

Our Results. In this work, we address the aforementioned gaps by proposing a novel metric to compare LGT networks. We exploit the distinction between the base tree and the transfer arcs and compare these two components in an independent fashion. The "base tree components" of the networks are compared using the established Robinson-Foulds metric, which reduces the problem to comparing only the "transfer components". An intuitive way of comparing the sets of transfer arcs in two networks would be to count the number of "same" transfers, but as we show this can be ambiguous. More specifically, we obtain the following results: We introduce a new dissimilarity measure d_{LGT} and show that it defines a metric space on the set of *all* LGT networks with the same leaf taxa. Our measure can also be extended to tree-based networks.

The metric is based on operations to transform one network into the other, and so it can be used to explore the space of all LGT networks on the same taxa (which is useful for stochastic and hill-climbing reconstruction methods that make local moves on networks, the currently dominant approach [4,5]). We obtain a complexity dichotomy: d_{LGT} can be computed in linear time if the order of transfers along a species branch does not matter, and NP-hard if it does. In the latter case, we show that d_{LGT} is fixed-parameter tractable (FPT) in the *level* (maximum number of reticulations in a biconnected component) of the input networks. We implemented our algorithms and perform three sets of experiments. First, we show on random simulated networks that our implementation is usable in practice for large networks (up to ~ 1800 vertices). We then show its usefulness in comparing LGT network predictions in two proofs-of-concept. In the first one, we compute d_{LGT} values between the outputs of the character-based methods presented in [34] and show quantitatively that transfer highway predictions are highly dependent on the approach and parameters used. In the other application, we look at networks predicted with a reconciliation tool (Ranger-DTL [31]), and show how our metric can be used to tune cost values within the reconciliation.

Due to space constraints, some of the content (proofs, pseudo-code...) has been deferred to the Appendix, or is only present in the full online version (https://www.biorxiv.org/content/10.1101/2025.11.20.689557).

2   LGT Networks and Tree-Based Networks

We first introduce all mathematical notions required to define our metric. The terminology is based on the definition of LGT networks presented in [28,38]. In a directed graph, the in-neighbors of a vertex are called its *parents* and its out-neighbors are *children*. A *phylogenetic network*, or just *network* for short, is a *directed acyclic graph* that has a single vertex of indegree 0, called its *root*. We also require that each vertex of outdegree 0, called a *leaf*, has indegree 1. Note that we allow vertices of indegree 1 and outdegree 1, which we call *subdivision vertices*. Given a phylogenetic network N, we denote its set of vertices by $V(N)$, its set of arcs by $E(N)$, and its set of leaves by $L(N)$. The *contraction* of an arc (u, v) of a network N is an operation that identifies u and v, that is, it: adds the arc (w, u) for each in-neighbor w of v; adds the arc (u, w) for each out-neighbor w of v; removes v and its incident arcs. If v is a vertex with a single parent, contracting v means contracting the only arc entering v.

Two networks N_1, N_2 are *isomorphic*, denoted $N_1 \simeq N_2$, if $L(N_1) = L(N_2)$ and there exists a bijective function $\phi : V(N_1) \to V(N_2)$ such that: (1) for each $\ell \in L(N_1)$, $\phi(\ell) = \ell$; (2) $(u, v) \in E(N_1)$ if and only if $(\phi(u), \phi(v)) \in E(N_2)$. Furthermore, the two networks are *homeomorphic* if they are isomorphic when we ignore subdivision vertices. That is, N_1 and N_2 are homeomorphic if N_1' and N_2' are isomorphic, where N_1' (resp. N_2') is obtained from N_1 (resp. N_2) by contracting each subdivision vertex.

Trees and clusters. A *tree* T is a special type of network in which the underlying undirected graph has no cycle. For a vertex v of T, we write $L_T(v)$ for the set of leaves descending from v, which we call the *cluster* of v in T. Note that two trees are isomorphic if and only if they have the same clusters, and each cluster occurs with the same multiplicity in both trees (recall that we allow subdivision vertices). Two trees are homeomorphic if and only if their sets of clusters are the same, without considering multiplicities.

LGT Networks. An *LGT network* is a pair $\mathcal{N} = (N, (E_p, E_t))$ where $N = (V, E)$ is a network, (E_p, E_t) is a pair of subsets of E such that $E_p \cup E_t = E$ and $E_p \cap E_t = \emptyset$, with E_p the *principal arc set* and E_t the *transfer arc set*; and the subgraph $T_0(\mathcal{N}) = (V, E_p)$ consisting only of principal arcs is a tree with the same set of leaves as N. The tree $T_0(\mathcal{N})$ is called the *base tree* of $\mathcal{N}$ (note that $T_0(\mathcal{N})$ may contain subdivision vertices, see Fig. 1). A vertex with at least two children in $T_0(\mathcal{N})$ is called a *tree vertex* (of either N or $T_0(\mathcal{N})$). Moreover, a vertex of N (or $T_0(\mathcal{N})$) that is incident to a transfer arc in N is called an *attachment point* (named as such since that endpoint that exists to attach a transfer). Since one goal of LGT networks is to distinguish between vertical and horizontal evolution, we assume the following.

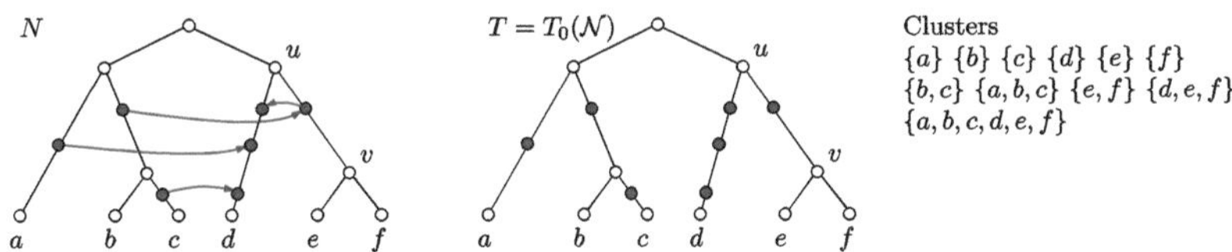

Fig. 1. Left is an LGT network $\mathcal{N} = N|T$. The base tree $T = T_0(\mathcal{N})$ is shown in the middle. In N, black arcs are those of the base tree T (their direction is omitted, they point downwards) and red arcs are transfer arcs. White vertices are tree vertices and filled red vertices are attachment points. The set of clusters of T is shown on the right. Multiplicities are ignored, but for instance $\{d\}$ occurs four times and $\{e, f\}$ two times.

Assumption 1 *In an LGT network $\mathcal{N}$, all attachment points are subdivision vertices in $T_0(\mathcal{N})$.*

The example in Fig. 1 satisfies this assumption. Note, this is not required in the original definition of LGT networks. It always holds in binary LGT networks, but our networks are not necessarily binary.

It will often be more useful to refer to the base tree directly instead of using the set of arcs E_p and the tree T_0. We therefore use the notation $N|T$ to denote the unique LGT network $\mathcal{N}$ whose network is N and whose base tree $T_0(\mathcal{N})$ is T, and stick with this alternate notation for the rest of the paper. We sometimes use the calligraphic $\mathcal{N}$ to denote an LGT network without specifying N and T. Note that the $N|T$ notation lets us deduce the arc partition of $\mathcal{N}$ with $E_p = E(T)$ and $E_t = \mathrm{E}(N) \setminus E(T)$. Using this notation, given an LGT network $N|T$, we write $v \preceq_T u$ to indicate that v is a descendant of u in T.

A *tree pair* in $N|T$ is an ordered pair of vertices from $V(N)$, denoted $[u, v]$, such that u, v are tree vertices, $v \prec_T u$, and the directed path from u to v in T contains only attachment points, except u and v (we use brackets to distinguish tree pairs from arcs of T). Intuitively, a tree pair is just a branch linking tree vertices between which transfers occurred. Any attachment point on the $u - v$ path is said to be *on* $[u, v]$. In Fig. 1, $[u, v]$ forms a tree pair, as well as $[u, d]$. Finally, we need a notion of identical LGT networks, since isomorphims may not preserve principal and transfer arcs. We say that two LGT networks $\mathcal{N}_1 = N_1|T_1, \mathcal{N}_2 = N_2|T_2$ are *LGT-isomorphic*, denoted $\mathcal{N}_1 \simeq_{LGT} \mathcal{N}_2$, if there exists an isomorphism ϕ between N_1 and N_2 such that ϕ is also an isomorphism between T_1 and T_2.

Tree-Based Networks. A network $N = (V, E)$ is a *tree-based network* if there *exists* a pair (E_p, E_s) such that $(N, (E_p, E_s))$ is an LGT network with the same leaves as N. In this case, (V, E_p) is called a *base tree* of N. This is equivalent to asking for a subset of arcs E_p such that (V, E_p) is a tree with the same set of leaves as N [35]. The main difference is that in an LGT network, the base tree is given, whereas a tree-based network may contain many base trees. The set of all base trees of a tree-based network N will be denoted $B(N)$. Thus, the set of all LGT networks contained in a tree-based network N is $\{N|T : \mathrm{T} \in B(N)\}$.

Time Consistent LGT Networks. We say that an LGT network $N|T$ is *time-consistent* if their exists a labeling function $\lambda : V(N) \to \mathbb{N}^+$ such that: (1) for any transfer arc (u, v), $\lambda(u) = \lambda(v)$; and (2) for any other arc (u, v), $\lambda(u) < \lambda(v)$. In other words, λ represents time that increases from past to present. The arcs of the base tree go in strictly increasing times, as they represent vertical evolution, whereas arcs representing horizontal evolution require co-existence and thus have the same time. The LGT network in Fig. 1 is easily seen to be time-consistent.

## 3	A Metric for LGT Networks and Tree-Based Extensions

Given two LGT networks on the same leaves, we want to compare both their base trees and their transfer arcs. For the latter, we delete transfers to eliminate disagreements. As for base trees, they can be compared using established tree metrics, and we here use Robinson-Foulds. To avoid situations where non-attachment points become incident to transfer arcs, we only allow contraction of arcs whose two ends are not attachment points. Given an LGT network $N|T$, we thus define the two following operations (see Fig. 2):

- the *contraction* of an arc (u, v) of $E(N)$, where u and v are both non-leaf tree vertices of $N|T$. Since u and v are not incident to a transfer, (u, v) is also in $E(T)$. This operation results in another LGT network $N'|T'$, where N' (resp. T') is obtained by contracting (u, v) in N (resp. in T);
- the *deletion* of a transfer arc (u, v) of $E(N) \setminus E(T)$. This operation results in another LGT network $N'|T'$ where N' is obtained by removing arc (u, v) from N, and contracting any resulting subdivision vertices of N (note, only u or v can become subdivision vertices); T' is obtained from T by contracting the same subdivision vertices that were contracted in N.

An LGT network $N'|T'$ is an *LGT reduction* of $N|T$ if $N'|T'$ can be obtained from $N|T$ by applying a sequence of contractions and transfer deletions, as defined above. We define $\delta(N|T, N'|T')$ as the number of operations required to transform $N|T$ into $N'|T'$. Given two LGT networks $\mathcal{N}_1 = N_1|T_1, \mathcal{N}_2 = N_2|T_2$, we say that $N|T$ is a *common LGT reduction* of $\mathcal{N}_1, \mathcal{N}_2$ if $N|T$ is LGT-isomorphic to an LGT reduction of both $\mathcal{N}_1$ and $\mathcal{N}_2$. The LGT distance is:

$$d_{LGT}(N_1|T_1, N_2|T_2) = \min_{N|T}(\delta(N_1|T_1, N|T) + \delta(N_2|T_2, N|T))$$

where the minimization is over all common LGT reductions $N|T$. When such an $N|T$ minimizes the above expression, it is called a *maximum common LGT reduction*. See Fig. 2.

Remark on Exploring Spaces of LGT Networks. A common LGT reduction always exists if the leaf sets are the same, since on both networks we can delete every transfer arc, resulting in a tree, and then contract all remaining arcs to obtain a star tree. By defining the reverse of contractions and transfer deletions, one could therefore transform any LGT network into another (as long as leaves are the same). Thus our operations and their reverse can be used to traverse the space of all LGT networks, as claimed in the introduction.

Extending to Tree-Based Networks. Note, d_{LGT} translates naturally to tree-based networks, by asking for the base trees that minimize d_{LGT}. That is, if N_1 and N_2 are tree-based networks, the distance is:

$$d_{TB}(N_1, N_2) = \min_{T_1 \in B(N_1), T_2 \in B(N_2)} d_{LGT}(N_1|T_1, N_2|T_2).$$

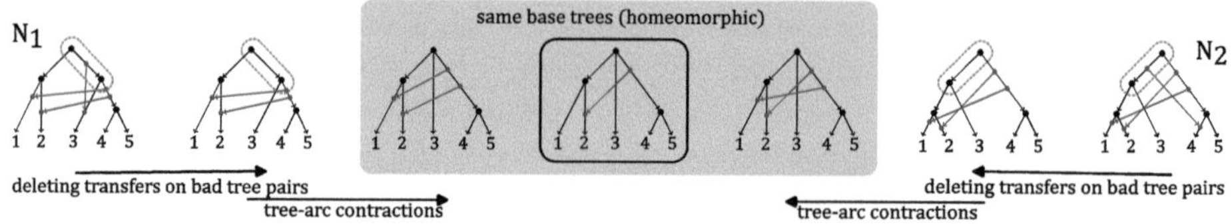

Fig. 2. Illustration of the different steps of an edition pathway between two networks $\mathcal{N}_1, \mathcal{N}_2$ (leftmost and rightmost networks). We want to contract "bad" tree pairs (encircled in $\mathcal{N}_1$ and $\mathcal{N}_2$, which respectively correspond to bad clusters $\{3,4,5\}$ and $\{1,2,3\}$). Because tree-arc contractions forbid attachment points, we must first delete the transfer arcs on those tree-pairs. Once bad tree pairs are removed, the base trees are homeomorphic (grey area), and one must then compute d_{TR}. As a side note, here $\mathcal{N}_1$ is not time-consistent because of the vertical transfer arc.

3.1   A Special Case: The Transfer Reduction Distance

A special case of d_{LGT} occurs when the base trees of the given LGT networks are the same (i.e., homeomorphic), and all disagreements are just between transfer arcs. An LGT network $N'|T'$ is a *transfer reduction* of $N|T$ if it is obtained from $N|T$ by applying a sequence of transfer deletions only. Observe that the number of transfer deletions needed to transform $N|T$ into $N'|T'$ is $|E(N) \setminus E(T)| - |E(N') \setminus E(T')|$. We say that $N|T$ is a *common transfer reduction* of two LGT networks $N_1|T_1$ and $N_2|T_2$ if $N|T$ is LGT-isomorphic to a transfer reduction of both $N_1|T_1$ and $N_2|T_2$ We can define the *transfer reduction* distance as

$$d_{TR}(N_1|T_1, N_2|T_2) = \min_{N|T}(|E(N_1) \setminus E(T_1)| + |E(N_2) \setminus E(T_2)| - 2|E(N) \setminus E(T)|)$$

where the minimization is over all common transfer reductions $N|T$. An $N|T$ that minimizes this expression is called a *maximum common transfer reduction*. Note that a common transfer reduction does not always exist, and in fact it exists if and only if T_1 and T_2 are homeomorphic.

3.2   Useful Properties of d_{LGT}

We show that in order to compute d_{LGT}, it suffices to compare the base tree and the transfer arcs separately. We first define a weighted RF (wRF) distance for the tree part. Let $\mathcal{N}_1 = N_1|T_1$ and $\mathcal{N}_2 = N_2|T_2$ be two LGT networks. A vertex v of N_1 is *bad* (w.r.t. $\mathcal{N}_1$ and $\mathcal{N}_2$) if v is a tree vertex and the cluster $L_{T_1}(v)$ is not a cluster of T_2. Likewise, a vertex v of N_2 is bad if it is a tree vertex and $L_{T_2}(v)$ is not a cluster of T_1. Now let v be a non-root tree vertex of either network, and let u be the unique tree vertex such that $[u, v]$ is a tree-pair (recall, tree-pairs have only attachment points between them). Assign the weight $w(v)$ as the number of transfer arcs that have an endpoint on $[u, v]$, plus one. Then the wRF distance is defined as the sum of weights of bad vertices, i.e.,

$$wRF(N_1|T_1, N_2|T_2) = \sum_{v \in V(N_1): v \text{ is bad}} w(v) + \sum_{v \in V(N_2): v \text{ is bad}} w(v).$$

In Fig. 2, each network has a bad vertex of weight two, so $wRF(\mathcal{N}_1, \mathcal{N}_2) = 4$. For the transfer component, let (u, v) be a transfer arc of $N_1|T_1$. Let $[u_1, u_2]$ be the unique tree pair that u is on, and let $[v_1, v_2]$ be the unique tree pair that v is on. We say that (u, v) is a *bad transfer* if one of u_2 or v_2 is a bad vertex. We also say that (u, v) is a *doubly bad transfer* if both u_2, v_2 are bad. Apply the analogous definitions to the transfers of $N_2|T_2$. In Fig. 2, each network has one bad transfer, but no doubly bad ones.

Theorem 1. *Given two LGT networks* $\mathcal{N}_1 = N_1|T_1, \mathcal{N}_2 = N_2|T_2$ *on the same leafsets,*

$$d_{LGT}(\mathcal{N}_1, \mathcal{N}_2) = wRF(\mathcal{N}_1, \mathcal{N}_2) + d_{TR}(N_1^*|T_1^*, N_2^*|T_2^*) - D(\mathcal{N}_1, \mathcal{N}_2), \ \ where$$

- $N_1^*|T_1^*$ *(resp. $N_2^*|T_2^*$) is the LGT network obtained from $N_1|T_1$ (resp. $N_2|T_2$) by applying a transfer deletion on each of its bad transfers (including doubly bad ones), and then contracting every bad vertex.*
- $D(\mathcal{N}_1, \mathcal{N}_2)$ *is the number of doubly bad transfer arcs that are in $\mathcal{N}_1$ or in $\mathcal{N}_2$.*

In essence, wRF counts the number of bad transfers and tree-arc contractions that must be performed, and d_{TR} adds the transfer deletions needed after (wRF double-counts doubly bad transfers so $D(\mathcal{N}_1, \mathcal{N}_2)$ is subtracted). Day's algorithm can be used to compute wRF in linear time [39]. The core of the problem is therefore to compute d_{TR}. We conclude this section by showing that d_{LGT} is a true mathematical metric. Recall that this requires showing that: (1) $d_{LGT}(\mathcal{N}_1, \mathcal{N}_2) = 0$ if and only if the two networks are LGT-isomorphic; (2) the distance is symmetric; and (3) it satisfies the triangle inequality.

Theorem 2. *The d_{LGT} distance satisfies all conditions of a metric.*

4 The Complexity of Computing d_{TR} (and d_{LGT})

We show that there is a dichotomy in the complexity of computing d_{TR}, depending on whether we care about the order of transfers along a tree pair. That is, suppose we have an LGT network $N|T$ and that there is a tree pair $[u, v]$ with more than one attachment points on it. This assumes an ordering of the transfers that occurred between u and v. However, in many cases, this order is unknown and very difficult to predict accurately, and it may be sufficient to contract all these attachment points into one due to uncertainty. This leads to LGT networks with at most one attachment point per tree pair. This matters, because we can compute d_{TR} in linear time under this assumption, and otherwise computing d_{TR} is NP-hard. This is true even if there are at most 3 attachment points on tree pairs, leaving the case of 2 attachment points open. Unless stated otherwise, we now deal with pairs of LGT networks whose base trees are homeomorphic. To simplify notation, we shall assume that two networks have the same set of tree vertices (instead of relying on an homeomorphism). Thus for two LGT networks $N_1|T_1, N_2|T_2$, we may assume that $[u, v]$ is a tree pair of T_1 if and only if $[u, v]$ is a tree pair of T_2 (though the number of attachment points may vary).

4.1 One Attachment Point per Tree Pair

Let $N|T$ be an LGT network, possibly non-binary. We define the set of tree pairs that share the two ends of a transfer, as follows:

$$tr(N|T) = \{([u_1, u_2], [v_1, v_2]) : \exists \text{ transfer } (u, v) \text{ s.t. } u \text{ is on } [u_1, u_2], v \text{ is on } [v_1, v_2]\}.$$

In words, $([u_1, u_2], [v_1, v_2]) \in tr(N|T)$ means that some transfer has its tail between u_1 and u_2, and its head between v_1 and v_2. Note, $tr(N|T)$ contains ordered pairs (of tree pairs), and $([v_1, v_2], [u_1, u_2])$ is different. Since we assume that tree vertices of $\mathcal{N}_1$ and $\mathcal{N}_2$ are the same, we can compare $tr(\mathcal{N}_1)$ and $tr(\mathcal{N}_2)$ directly. When there is at most one attachment point per tree pair, it suffices to delete transfers whose corresponding pair in the tr set is not in the other. We can thus use Day's algorithm for the wRF part [39], and compute the symmetric difference for the transfer part. Recall that $A \triangle B = (A \setminus B) \cup (B \setminus A)$.

Theorem 3. *Let $\mathcal{N}_1 = N_1|T_1$, $\mathcal{N}_2 = N_2|T_2$ be two LGT networks such that T_1, T_2 are homeomorphic. Suppose that for any tree pair $[u, v]$ of either network, at most one attachment point is on $[u, v]$. Then:*

$$d_{TR}(\mathcal{N}_1, \mathcal{N}_2) = |tr(\mathcal{N}_1) \triangle tr(\mathcal{N}_2)|.$$

Consequently, $d_{LGT}(\mathcal{N}_1, \mathcal{N}_2)$ can be computed in time $O(m_1 + m_2)$, where $m_1 = |E(N_1)|$, $m_2 = |E(N_2)|$.

Beyond one attachment point, the symmetric difference $tr(\mathcal{N}_1) \triangle tr(\mathcal{N}_2)$ may fail because the order may disagree. Figure 3 illustrates this phenomenon, which is exploited in the NP-hardness proof that follows.

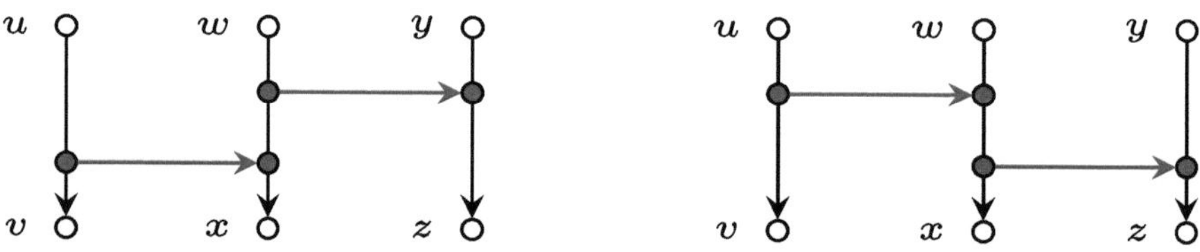

Fig. 3. With more than one attachment point per tree pair, one may have $tr(\mathcal{N}_1) = tr(\mathcal{N}_2)$ even though the networks are different. We have $tr(\mathcal{N}_1) = tr(\mathcal{N}_2) = \{([u, v], [w, x]), ([w, x], [y, z])\}$, and yet the two LGT networks cannot be isomorphic.

4.2 Three Attachment Points, When the Order of Transfers Matters

We show that computing d_{LGT} is NP-hard, even if the input is restricted to time-consistent networks. The hardness arises when many transfers are in reverse order as in Fig. 3. Such pairs of transfers are incompatible and we can keep at most one in each network. When many such incompatibilities arise, we have to find a maximum number of pairwise-compatible transfers, a difficult problem. Inspired by [40], our reduction is from 3-SAT, so that maximally compatible transfers correspond to satisfying a formula.

The Reduction (see Fig. 4). In 3-SAT, we receive a boolean formula ϕ over variables $x_1, \ldots, x_n$, with ϕ a conjunction of clauses $c_1, \ldots, c_m$, each containing three literals (a literal is a variable x or its negation $\bar{x}$). We must decide if there is a variable assignment that makes every clause *true*. From ϕ, we construct time-consistent LGT networks $N_1|T_1, N_2|T_2$ and an integer k, such that ϕ is satisfiable $\Leftrightarrow d_{TR}(N_1|T_1, N_2|T_2) \le k$.

For a variable x, we write $d(x)$ for the number of clauses in which x appears. We denote the clauses that contain x as $c_{x,0}, c_{x,1}, \ldots, c_{x,d(x)-1}$, ordered arbitrarily. To construct $N_1|T_1$ and $N_2|T_2$, the any base tree will do, so we let T be any tree with $\sum_{i=1}^{n} 2d(x_i) + m$ leaves. Denote by E the set of arcs of T whose head is a leaf. To obtain T_1 and T_2, we start from T and add attachment points only to arcs in E (this is for time-consistency). The arcs in E will therefore become tree pairs. Given a transfer arc $t = (u, v)$, we denote $t.\text{tail} = u$ and $t.\text{head} = v$.

Variable Gadget. We describe variable gadgets in $N_1|T_1$. The construction of the base tree starts from T. For each variable x, choose $2d(x)$ arcs in E that we denote $e_x[0], e_x'[0], e_x[1], e_x'[1], \ldots, e_x[d(x) - 1], e_x'[d(x) - 1]$ (note, each variable has its own distinct arcs). Subdivide each of these arcs twice to create two attachment points per arc. For $i \in [0; d(x) - 1]$, we introduce a transfer arc $t_x[i]$ from the highest attachment point on $e_x[i]$ to the highest attachment point on $e_x'[i]$. These represent assigning x to true. Then, for all $i \in [0; d(x) - 1]$, add a transfer arc $f_x[i]$ from the lowest attachment point on $e_x'[i]$ to the lowest attachment point on $e_x[i + 1]$ (we take $i + 1$ modulo $d(x)$). In $N_2|T_2$, we do the same, except that we **reverse** the order of attachment points of $t_x[i]$ and $f_x[i]$. That is, the $f_x[i]$'s take the highest attachment points from $e_x'[i]$ to $e_x[i + 1]$, and the $t_x[i]$'s take the lowest from $e_x[i]$ to $e_x'[i]$.

Clause Gadget. For each clause c of ϕ, we denote the variables that occur in c as $x_{c,0}, x_{c,1}, x_{c,2}$, with the only constraint that in this ordering, all positive literals appear before all negative literals. Let us now get back to modifying $N_1|T_1$. For each clause c, choose an arc e_c of E that is not used for a variable gadget or for another clause gadget. Add three attachment points $v_c[0], v_c[1], v_c[2]$ on e_c, in this order from top to bottom. For $j \in \{0, 1, 2\}$, consider the j-th variable $x := x_{c,j}$. Let i be the rank of c among the clauses in which x appears, i.e., $c = c_{x,i}$. If x appears as a positive literal in c, we add an attachment point $a_{c,x}$ above $t_x[i].\text{head}$ on $e_x'[i]$. If x appears as a negative literal in c, $a_{c,x}$ is added below $f_x[i].\text{tail}$ on $e_x'[i]$. We add a transfer arc from $a_{c,x}$ to $v_c[j]$. In $N_2|T_2$, add attachment points $v_c[2], v_c[1], v_c[0]$ on e_c (so, in reverse order). For $j \in \{0, 1, 2\}$ and variable $x := x_{c,j}$, let i be such that $c = c_{x,i}$. Add attachment point $a_{c,x}$ on $e_x'[i]$ above $t_x[i].\text{head}$ and below $f_x[i].\text{tail}$ (whether x is positive or not). Then add a transfer arc from $a_{c,x}$ to $v_c[j]$.

$$\phi = \underbrace{(a \vee b \vee \bar{c})}_{c_1} \wedge \underbrace{(b \vee c \vee \bar{d})}_{c_2} \wedge \underbrace{(a \vee \bar{b} \vee d)}_{c_3}$$

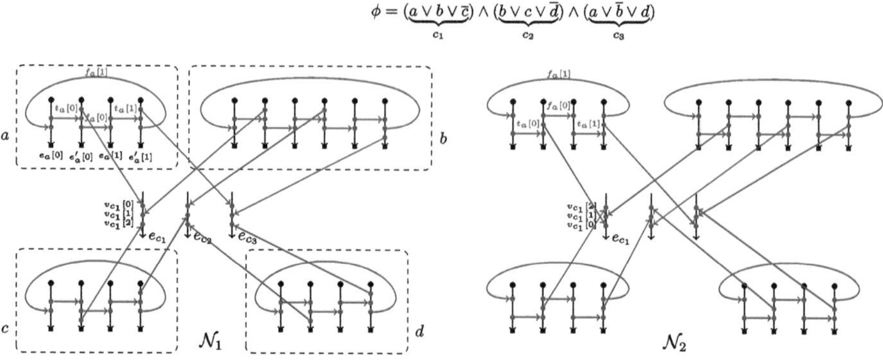

Fig. 4. Illustration of the reduction used in the proof of Theorem 4 (NP-completeness of d_{LGT}).

We denote $N = \sum_{i=1}^{n} d(x_i)$. We show that ϕ is satisfiable if and only if $d_{TR}(N_1|T_1, N_2|T_2) \leq 2N + 4m$.

Theorem 4. *Computing d_{LGT} is NP-complete, even on binary time-consistent networks with at most 3 transfers per attachment point.*

Proof (sketch). Time-consistency can be established by providing an explicit time-labeling, see Appendix. For the equivalence with the satisfiability problem, first assume that ϕ is satisfied by some assignment. If variable x is assigned true, then we delete all $f_x[i]$ arcs and keep only the $t_x[i]$ arcs, in both networks. Do the opposite if x is false. This amounts to $2N$ deletions (N per network). Then for every clause c, take a variable x that satisfies it, and on both networks delete every transfer arc that enters e_c, except the arc from $a_{c,x}$ to $v_c[j]$ (where $j \in \{0, 1, 2\}$). Thus for each clause we perform two transfer deletions, which amounts to $2m$ deletions per network. The total is $2N + 4m$, as desired. One can see in the Appendix (Fig. 8) that doing these deletions on both networks results in the same transfer reduction. In the other direction, assume that $d(N_1|T_1, N_2|T_2) \leq 2N + 4m$. One can argue that any solution must essentially mimic the previous direction. In an x gadget, we cannot keep both a $t_x[i]$ and $f_x[i]$ arc because their orders are reversed in the two networks. The best we can do is keep all $t_x[i]$ or all $f_x[i]$ arcs. The ones we keep correspond to an assignment of x. In a c gadget, we can keep at most one transfer arc going on e_c because the order of attachment points are reversed in the two networks. To perform at most $2N + 4m$ deletions, one such arc must be kept, which corresponds to the variable that can satisfy c.

This reduction can also be adapted to prove the NP-hardness of d_{TB}, the distance on tree-based networks. The idea is to add a gadget to N_1 and N_2 to constraint their base trees, such that transfer arcs are the same.

Theorem 5. *Computing d_{TB} is NP-hard.*

4.3   Parameterized Algorithms and Pre-processing Rules

We first observe that to compute d_{TR} in practice, and thus d_{LGT}, one can design a simple pre-processing rule as follows.

Transfer Cleaning Rule. Given two networks $\mathcal{N}_1 = N_1|T_1$ and $\mathcal{N}_2 = N_2|T_2$, with T_1, T_2 homeomorphic, if $\mathcal{N}_1$ has a transfer from tree pair $[a, b]$ to tree pair $[c, d]$, but $\mathcal{N}_2$ has no such transfer, then we delete this transfer immediately (and remember to count it in the distance).

The transfer cleaning rule is easily seen to be correct by observing that no edge contraction nor transfer deletion may introduce new transfer arcs between tree pairs of $\mathcal{N}_2$. We note that this pre-processing rule is critical in achieving the practical scalability results presented in the experiments section. After cleaning, we can test all ways of deleting transfers.

Proposition 1. *The distance $d_{LGT}(\mathcal{N}_1, \mathcal{N}_2)$ can be computed in time $O(2^t \cdot m)$, where t (resp. m) is the number of transfer arcs (resp. arcs) in $\mathcal{N}_1$ and $\mathcal{N}_2$.*

On binary networks, we can do better by handling matching blobs in the two networks independently (see appendix and [11] for a definition of the level).

Theorem 6. *Given two binary LGT networks $\mathcal{N}_1, \mathcal{N}_2$ both of level at most ℓ, one can compute $d_{TR}(\mathcal{N}_1, \mathcal{N}_2)$ and thus $d_{LGT}(\mathcal{N}_1, \mathcal{N}_2)$ in time $O(4^\ell \cdot m^2)$, where m is the total number of edges in both networks.*

5   Experiments

All of the experiments use an implementation of the algorithms presented in Sect. 4.1 and 4.2, publicly-available at https://github.com/bmarchand/transfer-edition_distance. The algorithms were implemented in Rust and are available in Python with PyO3 (https://github.com/PyO3/pyo3).

We remark that for the first experiment, other measures designed to compare phylogenetic networks (reviewed in the introduction [9–17]) are not applicable, as the task is specifically to compare LGT networks, in which a distinction between tree arcs and transfer arcs is made. To illustrate the problem, consider the fact that a given tree-based network N may give rise to two different LGT networks $N|T$ and $N|T'$ if T, T' are two different base trees. The metrics mentioned above do not distinguish these situations. While this argument does not apply to the second and third experiment, we note that in both of them, the networks predicted by the methods are tree-based, but not necessarily orchard networks [26] (see Fig. 11 in Appendix for more details). To our knowledge, none of the existing implemented metrics [3, 17] are capable of distinguishing all tree-based networks.

5.1   Benchmarks on Random LGT Networks

Overview. We first evaluate the computational scalability of our metric on random LGT networks, using the simulator from [38]. It outputs binary LGT networks, potentially with several transfers along a tree-pair, with their orders specified. Our results indicate that, although d_{LGT} is NP-hard to compute, Algorithm 1 (Appendix) is fast in practice up to very large network sizes. Indeed, Fig. 5 shows that under reasonable parameter choices, two output networks of size $\simeq 1800$ may be compared in $\lesssim 0.1$ seconds.

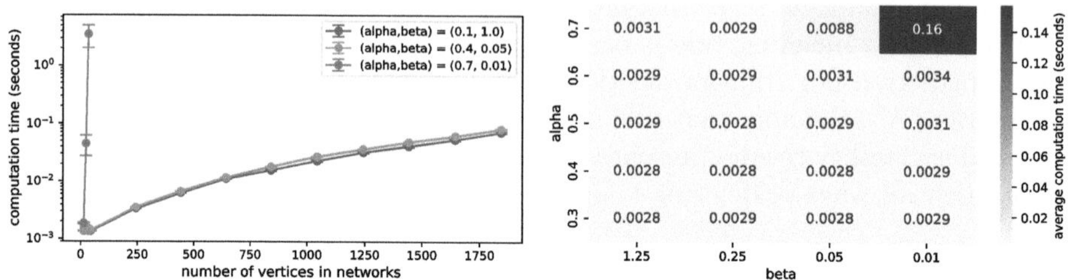

Fig. 5. Run-time benchmark of the computation of our metric on random LGT networks. (left) For several combinations of (α, β), average computation time of the distance between two random networks, as a function of the number of vertices. (right) For a fixed network size (200), table giving the average computation time of the distance between two random networks, for various combinations of (α, β). Note that $\alpha = 0.7$ is higher than values explored in [38] and arguably leads to unrealistic networks.

Simulator. The simulator in [38] takes three parameters: the "number of steps" N (each step adds two vertices to the network, see below), a transfer probability α, and a "level coefficient" β. Starting with a rooted tree with two leaves, at each step a random choice is made between a *speciation* and a *transfer*, with α the probability of a transfer. If "speciation" is picked, a leaf is chosen uniformly at random and a new leaf is attached on its parent arc. If "transfer" is picked, then two leaves l and l' are chosen and a transfer is added on their parent arcs. Parameter β influences the choice of l and l', with a higher β increasing the chances of choosing two leaves in different biconnected components. In essence, a higher α increases the number of transfers, and β controls the *level* (the maximum number of reticulations in a biconnected component).

Results. Figure 5 shows that our algorithm is fast on a wide range of parameters, except when $\alpha = 0.7$ and $\beta = 0.01$. This can be seen both on Fig. 5 (a), where $(\alpha, \beta) = (0.7, 0.01)$ exhibits a stronger exponential scalin, and on Fig. 5 (b) for a fixed network size. We note that $\alpha = 0.7$ is beyond the range of values explored in [38], and may be considered unrealistically high (recall that $\alpha = 0.7$ means that, in expectation, 70% of events are transfers, and 30% speciations). Even then, hard instances are only produced if $\beta \leq 0.01$, i.e., when transfers are 100× more likely between two leaves whose parents are in the same biconnected component, producing "blobs" with exceedingly many transfers.

5.2   Comparison of Character-Based Transfer Reconstructions

Some models such as *perfect phylogenetic networks* [41] and *perfect transfer networks* [33,42] can predict transfers on a species tree by explaining the presence/absence of given characters at the leaves (here, characters could be genes, functions, traits, ...). We revisit the experiments and data used in one of the latest iterations of this line of work [34] to illustrate a use-case of d_{LGT}. The authors of [34] were only able to compare their predictions qualitatively, and here our metric may be used to compare predictions *quantitatively*.

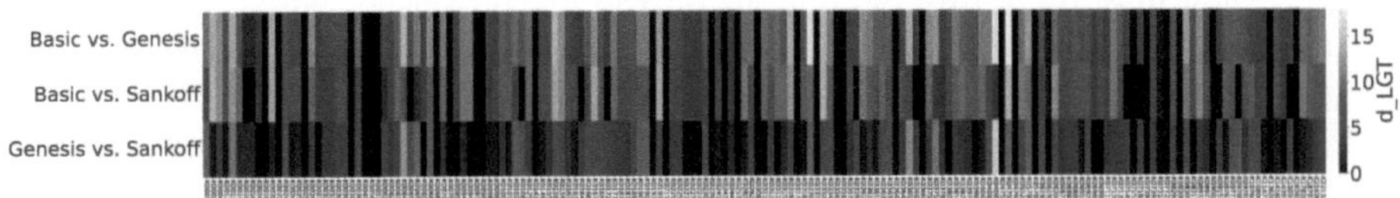

Fig. 6. Matrix of distances between reconstructed phylogenetic networks, using the character-based methods presented in [34], dubbed "Basic", "Sankoff" and "Genesis". Each column of the matrix corresponds to a character. For each character, each method reconstructs a network. We observe that Genesis and Sankoff tend to give network reconstructions that are closer to one another, compared with Basic. In [34], only qualitative comparisons had been reported.

Overview. Given a species tree and characters at the leaves, [34] considers three different approaches for reconstructing the history of individual characters. For each character, they assume a *single origin* and *losses* and *transfers* as evolutionary events. A first approach is called *Basic labeling*, and infers a character at an internal vertex if and only if it is present in all of its descendants. The assumption is then that maximal subtrees containing a character are involved in an HGT. Such a labeling has sometimes been used in hypothesis-testing, where biologists want to check whether a gene could have emerged in a de novo fashion, or was rather acquired through HGT (see [43,44]). The two other approaches, *Genesis* and *Sankoff*, are more sophisticated dynamic programming schemes computing ancestral assignments of the characters minimizing the costs of losses and transfers. Details can be found in [34]. For each prediction method and each character, the set of inferred transfers yields an LGT network with the species tree as base tree (plus subdivisions). Note that the inferred transfers along a tree pair are *unordered*, so we may use Theorem 3.

Data. We use the same data as in [34]. It consists of a set of 180 functional characters from the KEGG database (Kyoto Encyclopedia of Genes and Genomes [45]) within a set of 45 bacterial species. We take the species tree connecting these species from the NCBI Taxonomy Browser [46].

Results. In [34], reconstructed networks were combined into sets of "transfer highways", that were then compared regardless of their weights (i.e. regardless of how many transfers use the highway). Having a metric such as d_{LGT} allows for a finer quantitative comparison. On Fig. 6, we report computations of d_{LGT} values between network predictions reconstructed with each of the three methods. For each of the 180 KEGG characters mentioned above (x-axis on Fig. 6), the reconstructions are compared in a pairwise fashion: Basic vs. Genesis, Basic vs. Sankoff, and Genesis vs. Sankoff (y-axis on Fig. 6). We note that the methods "Genesis" and "Sankoff" yield predictions that are much closer to one another, compared to "Basic". While this analysis does not tell which method is more accurate, it indicates that the choice of predictor significantly alters the results. Hence, care must be taken when making biological conclusions from HGT predictions as they may be highly dependent on the method used. Nevertheless, these results are an example of how a metric may be used to "cluster" predictions into similar groups, so that a consensus may be identified.

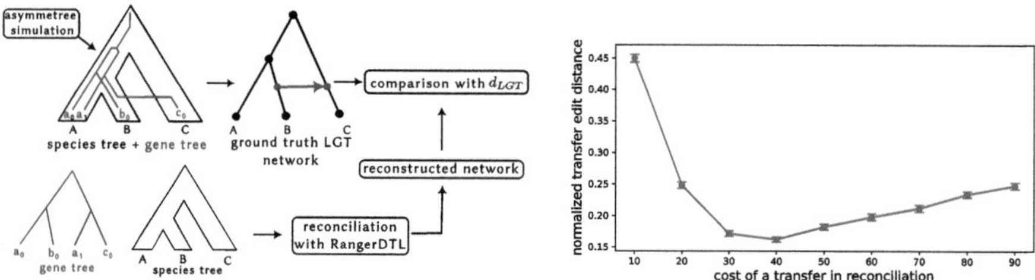

Fig. 7. (left) A set of species tree along with a gene tree evolving within it (with duplications, transfers and losses), are simulated using Asymmetree [47]. Adding the transfers to the species tree yields a ground-truth LGT network. A reconciliation of each gene-tree/species-tree pair is also computed using RangerDTL [31], yielding a reconstructed LGT network. (right) Average normalized d_{LGT} values between the reconstructed and ground truth networks, as a function of the cost associated to transfer events in the reconciliation (the costs of duplications and losses is kept constant). This plot allows the identification of cost values that minimize the average reconstruction distance, offering a way to calibrate reconciliation costs.

5.3   Optimizing Reconciliation Parameters for Transfer Prediction

Overview. Reconciliation addresses the following problem: given a species tree S and a gene tree G, with a map from $L(G)$ to $L(S)$, find an evolutionary history on G that explains their differences, typically with duplication, loss, and transfer events. This is often done parsimoniously by minimizing the weights of inferred events. A crucial problem in this context is to know what cost to associate to each allowed type of event (see e.g. [48]). We show how metrics may help tune the parameters of reconciliation tools, to optimize for transfer inference.

Experimental Protocol. A set of $N = 100$ histories, each consisting of a species tree with $L = 50$ leaves and a gene tree evolving inside it were sim-

ulated with Asymmetree [47], using rates for duplications, losses, and transfers of respectively 0.5, 0.5 and 0.1 (see the documentation of Asymmetree for details). This simulator includes speciations, duplications, losses and transfers events. By adding arcs corresponding to the transfers in the simulation to the species tree, we get a "ground truth" network. We also give each pair of species tree and gene tree to RangerDTL [31], a leading tool for reconciliation. Its output is a mapping of each vertex of the gene tree into the species tree, along with a labeling with one of the events (speciation, duplication, transfer). We run RangerDTL $p = 20$ times per pair of species/gene trees, to get a variety of optimal reconciliations (there may be many). Likewise, by adding the transfers onto the species tree, we get a "reconstructed LGT network", see Fig. 7 (a). Note that both the ground truth and reconstructed networks are unordered, as described in Sect. 4.1, i.e. with at most one attachment point per tree pair. In the (rare) case that two transfer events take place between the same tree pair, we treat it as a single transfer. RangerDTL also supports cost values for losses, duplications and transfers (the default being 1,2 and 3). We fix the costs of duplications and losses in RangerDTL to 10, and use transfer costs ranging from 10 to 100. Figure 7 (b) shows normalized d_{LGT} values between the reconstructed and ground truth networks, as a function of the cost of transfers, averaged over N and p. species as base tree and (2) are unordered. Formally, for the normalization we compute
$$\frac{d_{LGT}(\mathcal{N}_1,\mathcal{N}_2)}{|tr(\mathcal{N}_1)|+|tr(\mathcal{N}_2)|} = \frac{|tr(\mathcal{N}_1)\Delta tr(\mathcal{N}_2)|}{|tr(\mathcal{N}_1)|+|tr(\mathcal{N}_2)|}.$$

Results. Fig. 7 (b) displays the normalized distance as a function of the transfer cost chosen in the reconciliation. We observe the presence of an optimal cost of 40, yielding reconstructions closest to the ground truth. Such a setup may be used to tune the parameters of RangerDTL (or other similar tools), prior to applications in cases where the ground truth is not known.

6   Discussion and Conclusion

Future research directions could include methods for computing the *consensus* of more than two networks, for instance through a *median* under our distance. One could also explore other popular choices of edit operations for the species tree, such as *subtree-prune and regraft* (SPR) or *nearest-neighbor interchange* (NNI). It should also be noted that the complexity of d_{TR} when tree pairs have at most 2 attachment points is open. Another algorithmic question of interest is whether computing our metric is fixed-parameter tractable in d_{LGT}, e.g., with a complexity that is exponential in d_{LGT} only and not the sizes of the networks. To finish, one could also explore generalizations of our metric that would take into account the "distance" between disagreeing transfers, to better distinguish a slightly misplaced transfer from one further away.

Appendix: Complete Proofs

A   A Proofs for Section 3 (A Metric for LGT Networks and Tree-Based Extensions)

Theorem 1. *Given two LGT networks $\mathcal{N}_1 = N_1|T_1, \mathcal{N}_2 = N_2|T_2$ on the same leafsets,*

$$d_{LGT}(\mathcal{N}_1, \mathcal{N}_2) = wRF(\mathcal{N}_1, \mathcal{N}_2) + d_{TR}(N_1^*|T_1^*, N_2^*|T_2^*) - D(\mathcal{N}_1, \mathcal{N}_2), \ \ where$$

- *$N_1^*|T_1^*$ (resp. $N_2^*|T_2^*$) is the LGT network obtained from $N_1|T_1$ (resp. $N_2|T_2$) by applying a transfer deletion on each of its bad transfers (including doubly bad ones), and then contracting every bad vertex.*
- *$D(\mathcal{N}_1, \mathcal{N}_2)$ is the number of doubly bad transfer arcs that are in $\mathcal{N}_1$ or in $\mathcal{N}_2$.*

Proof. First notice that for any LGT network $N|T$, by applying a transfer deletion or base tree arc contraction to create another LGT network $N'|T'$, we cannot create a new clusters in the base tree, i.e., every cluster of T' is also in T. In fact, transfer deletions preserve all clusters, and a base tree arc contraction of (u, v) removes the cluster of v and preserves every other cluster.

Now let v be a bad vertex of $N_1|T_1$, and let u be the unique vertex such that $[u, v]$ is a tree pair. Notice that any LGT reduction of $\mathcal{N}_1$ that contains the cluster $L_{T_1}(v)$ cannot be isomorphic to an LGT reduction of $N_2|T_2$. By our above observations, this means that v needs to be contracted eventually. Before this can be achieved, every transfer arc with an attachment point on $[u, v]$ needs to be deleted. Counting these deletions, plus the contraction of (u, v), amounts to $w(v)$ operations. This is true for every bad vertex, and one can easily see that any minimum sequence of operations can be altered, if necessary, so that it first deletes every bad transfer arc, then contracts every bad vertex with its parent. The same must be done with $N_2|T_2$. The number of operations needed to perform this is

$$\sum_{v \in V(N_1) \cup V(N_2):v \text{ is bad}} w(v) - D(\mathcal{N}_1, \mathcal{N}_2) = wRF(\mathcal{N}_1, \mathcal{N}_2) - D(\mathcal{N}_1, \mathcal{N}_2)$$

where we subtract $D(\mathcal{N}_1, \mathcal{N}_2)$ because in the summation, every doubly bad transfer arc is counted twice (and those that are bad but not doubly bad are counted once). Once this step is done, we obtain LGT reductions $N_1^*|T_1^*$ and $N_2^*|T_2^*$ of $N_1|T_1$ and $N_2|T_2$, respectively, as described in our statement. Note that T_1^* and T_2^* now have the same clusters. They are thus homeomorphic, and all that remains is to apply transfer deletions to make the LGT networks make isomorphic. By definition, this requires $d_{TR}(N_1^*|T_1^*, N_2^*|T_2^*)$ additional operations. This shows that any minimum sequence must perform at least the number of operations in our statement, and that number can easily be seen to also be an upper bound, since it corresponds to deleting bad transfers, contracting bad vertices, and handling remaining transfers.

Theorem 2. *The d_{LGT} distance satisfies all conditions of a metric.*

Proof. The fact that $d_{LGT}(N_1|T_1, N_2|T_2) = 0 \Leftrightarrow N_1|T_1 \simeq_{LGT} N_2|T_2$ holds by definition, since a common LGT reduction can be reached by 0 operations if and only if the two networks are already LGT-isomorphic. Symmetry is also easily verified.

Let us focus on the triangle inequality. We want to show that for any triple of LGT networks $\mathcal{N}_1 = N_1|T_1, \mathcal{N}_2 = N_2|T_2, \mathcal{N}_3 = N_3|T_3$ on the same leaves, we have

$$d_{LGT}(\mathcal{N}_1, \mathcal{N}_3) \leq d_{LGT}(\mathcal{N}_1, \mathcal{N}_2) + d_{LGT}(\mathcal{N}_2, \mathcal{N}_3).$$

Let $\mathcal{N}_{12} = N_{12}|T_{12}$ be a maximum common LGT reduction of $\mathcal{N}_1, \mathcal{N}_2$ and let $\mathcal{N}_{13} = N_{13}|T_{13}$ be a maximum common LGT reduction of $\mathcal{N}_2, \mathcal{N}_3$. Observe that to reach a common LGT reduction of $\mathcal{N}_1$ and $\mathcal{N}_3$, we can first transform $\mathcal{N}_1$ into $\mathcal{N}_{12}$ and $\mathcal{N}_3$ into $\mathcal{N}_{23}$, and then find a common LGT reduction of the two LGT reductions. Therefore we get:

$$
\begin{aligned}
d_{LGT}(\mathcal{N}_1, \mathcal{N}_3) &\leq \delta(\mathcal{N}_1, \mathcal{N}_{12}) + \delta(\mathcal{N}_3, \mathcal{N}_{23}) + d_{LGT}(\mathcal{N}_{12}, \mathcal{N}_{23}) \\
&= d_{LGT}(\mathcal{N}_1, \mathcal{N}_2) - \delta(\mathcal{N}_2, \mathcal{N}_{12}) + d_{LGT}(\mathcal{N}_2, \mathcal{N}_3) - \delta(\mathcal{N}_2, \mathcal{N}_{23}) + d_{LGT}(\mathcal{N}_{12}, \mathcal{N}_{23}) \\
&= d_{LGT}(\mathcal{N}_1, \mathcal{N}_2) + d_{LGT}(\mathcal{N}_2, \mathcal{N}_3) + (d_{LGT}(\mathcal{N}_{12}, \mathcal{N}_{23}) - \delta(\mathcal{N}_2, \mathcal{N}_{12}) - \delta(\mathcal{N}_2, \mathcal{N}_{23})).
\end{aligned}
$$

We next claim that $d_{LGT}(\mathcal{N}_{12}, \mathcal{N}_{23}) \leq \delta(\mathcal{N}_2, \mathcal{N}_{12}) + \delta(\mathcal{N}_2, \mathcal{N}_{23})$, which will prove the triangle inequality. To see this, we view $\mathcal{N}_{12}$ and $\mathcal{N}_{23}$ as obtained from $\mathcal{N}_2$ by applying operations, which lets us assume that $V(N_{12}) \subseteq V(N_2)$ and $V(N_{23}) \subseteq V(N_2)$ (this is without loss of generality, it just lets us avoid introducing an isomorphism in the notation). Also note that we may assume that to go from $\mathcal{N}_2$ to $\mathcal{N}_{12}$, all transfer deletions are done first, followed by all tree arc contractions (this is because if a contraction is followed by a transfer deletion, we can just move the transfer deletion before since contractions do not create nor destroy transfers). The same holds for turning $\mathcal{N}_2$ into $\mathcal{N}_{12}$. Let A_{12} (resp. A_{23}) be the set of transfer arcs deleted from $\mathcal{N}_2$ to $\mathcal{N}_{12}$ (resp. $\mathcal{N}_{23}$). Likewise, let C_{12} (resp. C_{23}) be the set of arcs of the base tree contracted from $\mathcal{N}_2$ to $\mathcal{N}_{12}$ (resp. $\mathcal{N}_{23}$) after all transfer deletions were applied. Then, consider the LGT reduction $\mathcal{N}' = N'|T'$ obtained from $\mathcal{N}_2$ by deleting every transfer arc in $A_{12} \cup A_{23}$, and then contracting every base tree arc in $C_{12} \cup C_{23}$ (note that these arcs have no attachment points since we delete all transfer arcs first).

Notice that $\mathcal{N}'$ is an LGT reduction of $\mathcal{N}_{12}$, since we can start from $\mathcal{N}_{12}$ and delete every arc from A_{23} that it still contains, and then contract the base tree arcs from C_{23} that remain (note, these are well-defined since we assume that $V(N_{12}), V(N_{23}) \subseteq V(N_2)$). Likewise, $\mathcal{N}'$ can be obtained from $\mathcal{N}_{23}$ by deleting the transfer arcs in A_{12} and contracting the tree arcs in C_{12}. Thus, $\mathcal{N}'$ is a common LGT reduction of $\mathcal{N}_{12}$ and $\mathcal{N}_{23}$. The distance $\delta(\mathcal{N}_{12}, \mathcal{N}')$ is at most $\delta(\mathcal{N}_2, \mathcal{N}_{23})$ since we apply at most $|A_{23}| + |C_{23}|$ operations. Likewise, $\delta(\mathcal{N}_{23}, \mathcal{N}')$ is at most $\delta(\mathcal{N}_2, \mathcal{N}_{12})$. We thus get the desired claim, since

$$d(\mathcal{N}_{12}, \mathcal{N}_{23}) \leq \delta(\mathcal{N}_{12}, \mathcal{N}') + \delta(\mathcal{N}_{23}, \mathcal{N}') \leq \delta(\mathcal{N}_2, \mathcal{N}_{12}) + \delta(\mathcal{N}_2, \mathcal{N}_{23}).$$

B   Proofs for Section 4 (The Complexity of Computing d_{TR} (and d_{LGT}))

Theorem 3. *Computing d_{LGT} is NP-complete, even on binary time-consistent networks with at most 3 transfers per attachment point.*

Proof. We prove the correctness of the reduction given above.

Time-Consistency. We first verify that the networks are time-consistent. We build time maps λ_1 for $N_1|T_1$ and λ_2 to $N_2|T_2$. First note that we only add attachment points to arcs in E, which are arcs of the initial tree T whose heads are leaves. Thus, if we take the subgraph induced by the ancestors of the tails of arcs in E (inclusively) in either $N_1|T_1$ or $N_2|T_2$, we have a tree. It is easy to assign times in λ_1 and λ_2 such that all tails of arcs in E have time less than 0 (their ancestors will have negative times, which we allow for simplicity). It therefore suffices to assign a time above 0 to attachment points in a time-consistent manner.

In $N_1|T_1$, assign in λ_1 a time of 4 to the two ends of every $t_x[i]$ transfer arc, and a time of 5 to the two ends of each $f_x[i]$ transfer arc. Now consider a clause c and recall that on arc e_c, we added three attachment points $v_c[0], v_c[1], v_c[2]$. Those correspond to the variables $x_{c,0}, x_{c,1}, x_{c,2}$, which we recall are ordered so that positive literals appear first. Assign a time to the $v_c[j]$ vertices that correspond to positive literals a time in $\{1, 2, 3\}$, and a time to those for negative vertices a time of 6 or more (by our ordering, this can be done in a time-consistent manner). Finally, for each transfer arc whose head is a $v_c[j]$ vertex, assign the tail the same time as $v_c[j]$.

Let us argue that λ_1 labeling is time-consistent. By construction, the ends of each transfer arc have equal times. Then, for $i \in [0; d(x) - 1]$, there are exactly 2 attachment points on $e_x[i]$, namely $t_x[i]$.tail and $f_x[i - 1]$.head, with $t_x[i]$.tail above $f_x[i]$.head. Those have time-labels by λ_1 equal to 4 and 5, respectively, which is time-consistent. On $e_x[i']$ for $i \in [0; d(x) - 1]$, there are exactly 3 attachment points: $t_x[i]$.head followed by $f_x[i]$.tail, again assigned respective times 4 and 5. There is a third attachment point $a_{c,x}$, whose time is either in $\{1, 2, 3\}$ or greater than 5, corresponding respectively to a positive occurrence of x in c (in which case $a_{c,x}$ is above $t_x[i]$.head) and a negative occurrence of x in c (in which case $a_{c,x}$ is below $f_x[i]$). In both cases, the ordering of time labels is respected. Our construction of λ_1 shows that for each tree pair corresponding to an arc of E, the times of the attachment points go in increasing order, and it follows that $N_1|T_1$ is time-consistent.

Now let us build time map λ_2 for $N_2|T_2$. For each $i \in [0, d(x) - 1]$, set the time of both ends of $f_x[i]$ to 1, and the time of both ends of $t_x[i]$ to 5. Then for each clause c and arc e_c, the attachment points are, in order, $v_c[2], v_c[1], v_c[0]$, which we assign respective times $2, 3, 4$. The tail of each transfer ending at one of these $v_c[j]$ vertices is assigned the same time as $v_c[j]$.

To see that this makes $N_2|T_2$ time-consistent, for $e_x[i]$ with $i \in [0; d(x) - 1]$, the attachment points occuring on this arc have times 1 and 5. For $e'_x[i]$, $a_{c,x}$ is below $f_x[i]$.tail and above $t_x[i]$.head, and is assigned one of the times in $\{2, 3, 4\}$. This is between 1 and 5, and so the times along $e'_x[i]$ are in increasing order. This concludes the proof of time-consistency.

We next show that ϕ is satisfiable if and only if a common transfer reduction can be achieved using a total of $2N + 4m$ deletions.

$$\phi = (a \vee b \vee \overline{c}) \wedge (\underbrace{b \vee c \vee \overline{d}}_{c_2}) \wedge (\underbrace{a \vee \overline{b} \vee d}_{c_3})$$

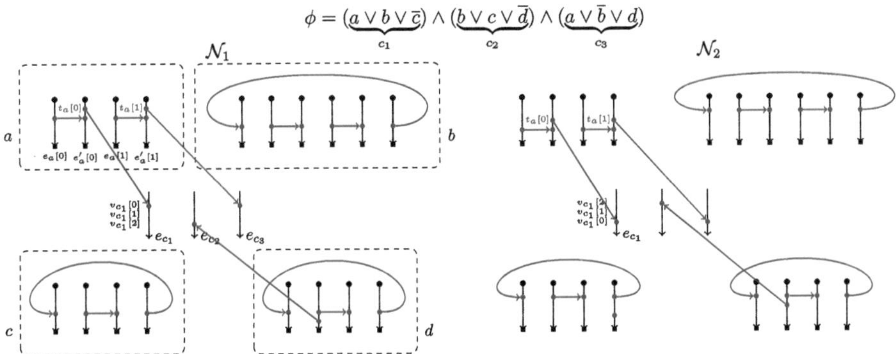

Fig. 8. The solution obtained in the first direction, corresponding the the satisfying assignment that puts $a = true, b = c = d = false$. Here, c_1 and c_3 are satisfied by $a = true$ and c_2 by $d = false$.

ϕ **Satisfiable** $\Rightarrow$ **Existence of a Large Common Reduction.** Consider an assignment of each variable to true or false that satisfies ϕ. We show how to delete transfers from both networks, as illustrated in Fig. 8. If a variable x is set to true, we remove the transfers $f_x[i]$ for all $i \in [0; d(x)-1]$ in both networks. If it is set to false, likewise we remove $t_x[i]$ for $i \in [0; d(x) - 1]$ in both networks. This amounts to $2\sum_x d(x) = 2N$ deletions. Then, for each clause c of ϕ, we pick a variable x whose assignment satisfies ϕ (as true if x appears positively, as false otherwise) and remove in both networks all transfers arriving at e_c, except the one originating at $a_{c,x}$. For each clause, a total of 4 transfer deletion occurs this way in both networks (we delete two incoming transfers per clause per network). We have thus performed $2N + 4m$ deletions.

We now argue that after these deletions on $N_1|T_1$ and $N_2|T_2$, we have achieved the same LGT network. Note that the two networks are identical above the tails of E, so we must only consider the transfer arcs between the tree pairs corresponding to arcs of E. To start with, for each variable x, either the set $\{t_x[i]\}_{i \in [0;d(x)-1]}$ or $\{f_x[i]\}_{i \in [0;d(x)-1]}$ is kept, and the same choice is made in both networks, following the variable assignments. Therefore, between the tree pairs corresponding to the $e_x[i]$ and $e'_x[i]$ arcs, for all x, there is in the reductions of both networks exactly one transfer between $e_x[i]$ and $e'_x[i]$ for $i \in [0; d(x) - 1]$ in the former case, and between $e'_x[i]$ and $e_x[i+1]$ in the latter (addition modulo $d(x)$). Moreover, if x is the variable chosen to satisfy c, and c is the i-th clause in which x appears, there is also $a_{c,x}$ kept on $e'_x[i]$. If x appears positively in

c, then $a_{c,x}$ is above $t_x[i]$.head on $e'_x[i]$ in N_1. This is also the case in N_2, regardless of whether x appears positively or negatively in c. Likewise, if x appears negatively in c, then $a_{c,x}$ is below $f_x[i]$.tail in N_1. This is also the case in N_2, again regardless of whether x appears positively or negatively in c. As for the edges e_c for each clause c, exactly 1 attachment point remains on it in each network, and is the end-point of a transfer originating from the same arc $e_x[i]$, with x the variable that was picked to witness that c is satisfied, and with c the i-th clause in which x appears. Therefore, the reductions applied to N_1 and N_2 have given two networks in which transfers occur between exactly the same set of pairs of edges. In addition, the ordering of the attachment points of these transfers is also the same in both networks. Overall, a common reduction has been reached, with a total of $2N + 4m$ transfer deletions.

Existence of a Large Common Reduction $\Rightarrow \phi$ satisfiable. Conversely, suppose that a common transfer reduction M has been obtained from $N_1|T_1$ and $N_2|T_2$ with less than $2N + 4m$ transfer deletions. We need to show that ϕ is satisfiable. This direction is a bit more complicated and we introduce a useful tool before proceeding.

Transferring Above Other Transfers. Let $N|T$ be an LGT network. We call a pair of attachment points uv a *transfer pair* if either (u, v) or (v, u) is a transfer arc of N. Note, the uv notation emphasizes that the direction is ignored. Given two tree pairs $[x, y]$ and $[z, t]$ of the base tree T, and $e = [p, q]$ another tree pair of T, we say that $[x, y]$ is *has a transfer above* $[z, t]$ *in* e if there are 2 transfer pairs ab, cd such that (1) a is on $[x, y]$ (2) c is on $[z, t]$ (3) b and d are on e with $d \prec_T b$. We denote this relationship by $[x, y] \Rightarrow_e [z, t]$. Note that we may have both $[x, y] \Rightarrow_e [z, t]$ and $[z, t] \Rightarrow_e [x, y]$ in a network with base tree T. This notion is illustrated on Fig. 9.

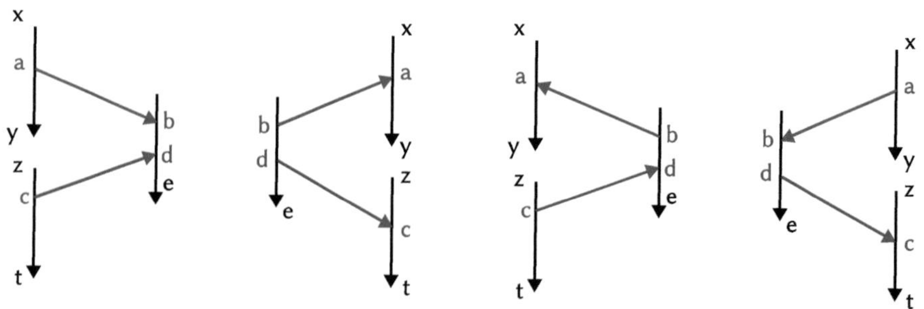

Fig. 9. The different ways the relation $[x, y] \Rightarrow_e [z, t]$ can materialize. $[x, y]$ and $[z, t]$ essentially play the role of labels of the end points of the transfers. With this point of view, the relation $[x, y] \Rightarrow_e [z, t]$ simply states that "label $[x, y]$ is above label $[z, t]$ in e".

When taking reductions of a network, the following property stating that $\Rightarrow_e$ relations cannot be created will be useful:

Property 1. Let $\mathcal{N}' = N'|T'$ be a transfer reduction of $\mathcal{N} = N|T$. If we do not have $[x, y] \Rightarrow_e [z, t]$ in $N|T$, then we also do not have $[x, y] \Rightarrow_e [z, t]$ in $N'|T'$.

Proof. We prove the result if $\mathcal{N}$, $\mathcal{N}'$ differ by 1 transfer deletion, and the result follows by induction. We show the contraposition of our statement, i.e., that given two tree pairs $[x, y]$ and $[z, t]$ such that $[x, y] \Rightarrow_e [z, t]$ for some tree pair e in $\mathcal{N}'$, we also have $[x, y] \Rightarrow_e [z, t]$ in $\mathcal{N}$. Let (u, v) be the transfer that is deleted to go from $\mathcal{N}$ to $\mathcal{N}'$. We also denote $s_1 t_1$ and $s_2 t_2$ transfer pairs in $N'|T$ such that s_1 is on $[x, y]$, t_1 is on e, s_2 is on $[z, t]$, t_2 is on e and $t_2 \preceq_T t_1$ in $\mathcal{N}'$ (such transfer pairs exist since $[x, y] \Rightarrow_e [z, t]$ in $\mathcal{N}'$). To finish, we denote the transfer arc sets of $\mathcal{N}$ and $\mathcal{N}'$ by E_t and E_t', respectively. We get that $E_t' = E_t \setminus \{(u, v)\}$, and $\{(s_1, t_1), (s_2, t_2)\} \subseteq E_t' \subseteq E_t$. We also have $t_2 \preceq_{T'} t_1$ in $\mathcal{N}$. Indeed, recall that to go from $\mathcal{N}'$ to $\mathcal{N}$, two arcs of $E(T')$ are possibly subdivided to create u and v, and then an arc is added between u and v. Whether one of the subdivided arcs is on the path from t_1 to t_2 in T' or not, we have $t_1 \preceq_T t_2$ in $\mathcal{N}$ as well. Overall, $[x, y] \Rightarrow_e [z, t]$ in $\mathcal{N}$ as well. $\qquad\square$

Let us now go back to our main proof. Property 1 will be used to show that if $[x, y] \Rightarrow_e [z, t]$ holds in $\mathcal{N}_1$ but not in $\mathcal{N}_2$, then we must apply a deletion that makes $[x, y] \Rightarrow_e [z, t]$ become false in $\mathcal{N}_1$ (since the property says that the $\Rightarrow_e$ relation can never be created in $\mathcal{N}_2$).

We first argue that any common transfer reduction of $\mathcal{N}_1$ and $\mathcal{N}_2$ requires *at least* $2N + 4m$ deletions. Let us first note that a transfer arc $t_x[i]$ is deleted in $\mathcal{N}_1$ if and only if that same transfer arc $t_x[i]$ is deleted in $\mathcal{N}_2$, since this is the only transfer arc between the tree pairs of $e_x[i]$ and $e_x'[i]$. The same holds for the $f_x[i]$ arcs.

Next, note that for any variable x and for $i \in [0; d(x)-1]$, $e_x[i] \Rightarrow_{e_x'[i]} e_x[i+1]$ in $\mathcal{N}_1$ (as $t_x[i]$.head is above $f_x[i]$.tail on $e_x'[i]$ in $\mathcal{N}_1$), and not the reverse, whereas $e_x[i+1] \Rightarrow_{e_x'[i]} e_x[i]$ in $\mathcal{N}_2$ ($t_x[i]$.head is below $f_x[i]$.tail on $e_x'[i]$ in $\mathcal{N}_2$), and not the reverse (as usual, additions are to be taken modulo $d(x)$). In both cases, the reverse relation is not true, so if both $t_x[i]$ and $f_x[i]$ are kept in a common transfer reduction $\mathcal{N} = N|T$ of $\mathcal{N}_1$ and $\mathcal{N}_2$, we would have both $e_x[i] \Rightarrow_{e_x'[i]} e_x[i+1]$ and $e_x[i+1] \Rightarrow_{e_x'[i]} e_x[i]$ in $\mathcal{N}$, contradicting Property 1. Likewise, for $i \in [0; d(x)-1]$, if both $f_x[i]$ and $t_x[i+1]$ are kept in a common transfer reduction $\mathcal{N}$, then we have both $e_x'[i] \Rightarrow_{e_x[i+1]} e_x'[i+1]$ and $e_x'[i+1] \Rightarrow_{e_x[i+1]} e_x'[i]$ in $\mathcal{N}$, contradicting Property 1.

In other words, any two transfers that have their attachment point on the same tree pair are incompatible: one of the two must be deleted. This yields at least $\sum_{i=1}^{n} d(x_i) = N$ deletions in the variable gadgets in each network. In addition, we show that if exactly N deletions occur in a variable gadget associated to x, then either $\{t_x[i]\}_{i \in [0;d(x)-1]}$ or $\{f_x[i]\}_{i \in [0;d(x)-1]}$ is entirely kept. If exactly N deletions occur, then for each x and for each $i \in [0; d(x) - 1]$, either $t_x[i]$ or $f_x[i]$ is removed but not both. Likewise, either $f_x[i]$ or $t_x[i + 1]$ must be deleted, but not both. Suppose $t_x[0]$ is kept, then $f_x[d(x) - 1]$ and $f_x[0]$ are removed. This implies $t_x[1]$ is kept. Iterating this argument shows that exactly $\{t_x[i]\}_{i \in [0;d(x)-1]}$ is kept. We can show similarly that if only N deletions occur and $f_x[0]$ is kept in the clause gadget of x, then $\{f_x[i]\}_{i \in [0;d(x)-1]}$ is kept.

As for the clause gadget, consider a clause c, and two variables $x = x_{c,j}$ and $x' = x_{c,j'}$ appearing in c. We suppose w.l.o.g. that x is before x', that is $j < j'$. We denote i and i' the integers such that c is the i-th clause in which x appears, and the i'-th clause in which x' appears (so, $c = c_{x,i} = c_{x',i'}$). Note that we have $e'_x[i] \Rightarrow_{e_c} e'_{x'}[i']$ in $\mathcal{N}_1$ and not the reverse, while the opposite is true in $\mathcal{N}_2$, i.e., $e'_{x'}[i'] \Rightarrow_{e_c} e'_x[i]$ in $\mathcal{N}_2$ because we reversed the order of attachment points on e_c. As this is true for any two variables x, x' that appear in c, it implies through Property 1 that at most 1 transfer entering e_c may be kept in each network. It implies at least $2m$ deletions in each network. In addition, if only $2m$ deletions occur, exactly one transfer arriving in each e_c must be kept.

Finally, let x be a variable in which we kept all $t_x[i]$ arcs. One can see that if, in the gadget for x, we kept an arc of $\mathcal{N}_1$ from some $a_{c,x}$ vertex to some $v_c[j]$ vertex, then x must occur positively in c. This is because if the occurrence was negative, $a_{c,x}$ would be below $t_x[i]$.head in $\mathcal{N}_1$ but above $t_x[i]$.head in $\mathcal{N}_2$ and we would again violate Property 1. For the same reasons, if all $f_x[i]$ arcs were kept instead, then all remaining arcs starting from $a_{c,x}$ vertices correspond to negative occurrences. Thus in a common transfer reduction, all clause arcs e_c whose incoming transfer arcs are from the same x gadget have the same type of occurrence, all positive or all negative. This means that we can assign $x = true$ if the $t_x[i]$'s were kept and $x = false$ if the $f_x[i]$'s were kept. Since every e_c arc has an incoming transfer from some x gadget that corresponds to satisfying c, this assignment shows that ϕ is satisfiable.

Proof That Computing d_{TB} is NP-Hard.

Construction of the Proof of Hardness for d_{TB}. We now modify the reduction described in Sect. 4.2 to show that d_{TB} is NP-hard. Letting ϕ be an instance of 3-SAT, we let $\mathcal{N}_1 = N_1|T_1$ and $\mathcal{N}_2 = N_2|T_2$ be the LGT networks constructed from ϕ as described previously. Recall that both have a base tree that originated from a tree T that contains a set of arcs E, all having their head vertex being a leaf, to which we added attachment points.

Let us obtain a tree-based network N'_1 from N_1. Let (u, v) be an arc in E, and note that $[u, v]$ is a tree-pair of $N_1|T_1$. Now make the transformation to N_1 that is illustrated in Fig. 10. That is, denoting by p_u the parent of u in N_1 (and T_1), the arc (p_u, u) is replaced with a subgraph with 5 vertices where u is now a reticulation (in the figure, the possible u vertices are the larger black circles—p_u is not shown, but instead of having u as a child it would have the root of the inserted subnetwork in blue as a child). Denote by N'_1 the network obtained from N_1 after applying this transformation to every (u, v) in E. Likewise, denote by N'_2 the network obtained from N_2 by applying the same transformations. It is not difficult to show that N'_1 and N'_2 are tree-based networks. Moreover, the point of this transformation is that the set of transfer arcs of our original reduction are enforced.

Lemma 1. *Let (x, y) be a transfer arc of $N_1|T_1$. Then for any base tree T'_1 in $B(N'_1)$, the arc (x, y) is a transfer arc of $N'_1|T'_1$. Likewise, if (x, y) is a transfer*

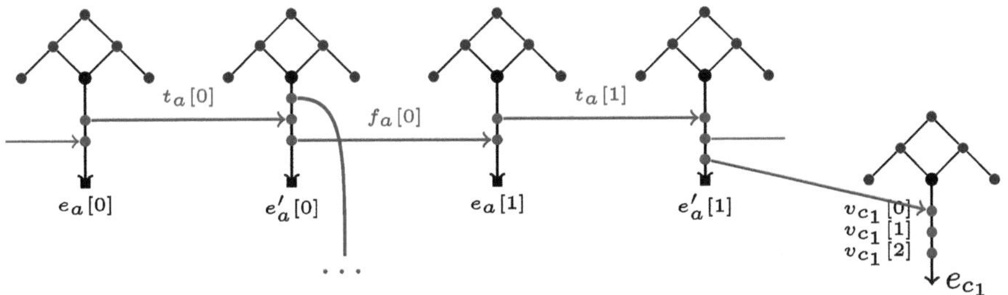

Fig. 10. Modification of the gadgets of $N_1|T_1$ for a variable a and a clause c_1. For each tail u of an arc in E with parent p_u (p_u is not shown): we remove the arc (p_u, u), we add the subgraph consisting of the five blue vertices as shown, then add the root of that subgraph as a child of p_u. In each such subgraph, the two dangling vertices are newly introduced leaves.

arc of $N_2|T_2$, then for any base tree T_2' in $B(N_2')$, the arc (x, y) is a transfer arc of $N_2'|T_2'$.

Proof. We prove the result for $N_1|T_1$ first. Let T_1' be a base tree of N_1'. Let $e_c \in E$ be an arc of the initial tree corresponding to a clause c, and let $e_c = (u, v)$. Note that $[u, v]$ is a tree pair of $N_1|T_1$ on which we added attachment points $v_j[0], v_j[1], v_j[2]$. Because in N_1', u is a reticulation, the arc $(u, v_j[0])$ must be in the base tree T_1' of N_1' (recall that by definition, the base tree must have the same leaves as N_1', and so every non-leaf of N_1' needs to have at least one child in the base tree). Thus, the other arc entering $v_c[0]$, namely some $(a_{c,x}, v_c[0])$ arc, cannot also be in the base tree and thus it must be a transfer of $N_1'|T_1'$. Following the same logic, $(v_c[0], v_c[1])$ must be in T_1' and the other arc entering $v_c[1]$ is a transfer arc, and the same holds for $v_c[2]$. Thus, the transfer arcs with an end on e_c are the same in $N_1|T_1$ and $N_1'|T_1'$.

We next argue that each $t_x[i]$ arc cannot be in the base tree. Consider the arc $e_x'[i] = (u, v)$ and note that $t_x[i]$.head is on the tree pair corresponding to $e_x'[i]$. By construction, in N_1 either $t_x[i]$.head has u as a parent, or has an attachment point $a_{c,x}$ for a clause c as a parent. In the former case, $(u, t_x[i]$.head$)$ must be in the base tree because u has one child, implying that $t_x[i]$ cannot also be in the base tree. In the latter case, the arc from $a_{c,x}$ to e_c was argued to be a transfer, so $(a_{c,x}, t_x[i]$.head$)$ must be in the base tree since $a_{c,x}$ needs a child, implying in turn that $t_x[i]$ cannot be in the base tree. In all cases, $t_x[i]$ must be a transfer arc. Now consider a $f_x[i]$ arc. In N_1, $f_x[i]$.head has $t_x[i+1]$.tail as a parent. Since we now know that $t_x[i]$ is not in the base tree, then $(t_x[i]$.tail, $f_x[i]$.head$)$ is in the base tree and $f_x[i]$ must be a transfer arc. Thus, all transfer arcs of $N_1|T_1$ are also transfers of $N_1'|T_1'$.

For $N_2|T_2$, we use similar ideas to argue that any base tree of N_2 results in the same transfer arcs. The transfer arcs entering e_c cannot be in any base tree of N_2 for the same reasons as before. Each $f_x[i]$ cannot be in the base tree because each reticulation u on the upper end of $e_x[i+1]$ must have $f_x[i]$.head as its child. Finally, each $t_x[i]$ cannot be in the base tree, because in N_2' the parent of $t_x[i]$.head is some $a_{c,x}$ attachment point, which has its arc going into e_c a transfer and which therefore needs $t_x[i]$.head as its child, preventing $t_x[i]$ from being in the base tree.

Theorem 4. *Computing d_{TB} is NP-hard.*

Proof. We propose a reduction from 3-SAT. Given an instance ϕ of 3-SAT, first construct the networks $\mathcal{N}_1 = N_1|T_1$ and $\mathcal{N}_2 = N_2|T_2$ from Sect. 4.2, and then apply the modifications described above to obtain tree-based networks N_1' and N_2', resulting in the d_{TB} instance. We claim that $d_{LGT}(N_1|T_1, N_2|T_2) = d_{TB}(N_1', N_2')$, which is sufficient to prove NP-hardness since it implies ϕ is satisfiable if and only if $d_{TB}(N_1', N_2') \leq 2N + 4m$ as before.

Let us first show that $d_{TB}(N_1', N_2') \geq d_{LGT}(N_1|T_1, N_2|T_2)$. Let T_1', T_2' be base trees of N_1, N_2 such that $d_{LGT}(N_1'|T_1', N_2'|T_2')$ is minimum. Let D_1 be the set of transfer arcs deleted from $N_1'|T_1'$ and let D_2 be those deleted from $N_2'|T_2'$ to obtain LGT networks $\mathcal{N}_1''$ and $\mathcal{N}_2''$ that form a maximum common LGT reduction (note, contractions could have been performed, but we don't even count them). Recall that by Lemma 1, all transfer arcs in $N_1|T_1$ and $N_2|T_2$ are also in $N_1'|T_1'$ and $N_2'|T_2'$, respectively. Consider the LGT networks obtained from $N_1|T_1$ by deleting all of its transfers that are also in D_1, and the LGT network obtained from $N_2|T_2$ by deleting all those that are also in D_2, resulting in new LGT networks $\mathcal{N}_1^*$ and $\mathcal{N}_2^*$. As a result, the subgraph induced by the descendants of tails of elements from E is identical in $\mathcal{N}_1^*$ and in $\mathcal{N}_1''$, and identical in $\mathcal{N}_2^*$ and in $\mathcal{N}_2''$ (recall that N_1', N_2' are obtained by replacing the arcs incoming into the tails of elements from E by a subgraph, so everything "below" tails of E was unaltered, and thus we get the same subgraphs in these parts after making the same deletions). Moreover, since $\mathcal{N}_1''$ and $\mathcal{N}_2''$ are LGT-isomorphic, the subgraphs induced by the descendants of tails of elements from E are identical in both LGT networks, and thus the same holds for $\mathcal{N}_1^*$ and $\mathcal{N}_2^*$. It follows that $\mathcal{N}_1^*$ and $\mathcal{N}_2^*$ are themselves LGT-isomorphic. Since we made at most $|D_1| + |D_2| \leq d_{TB}(N_1', N_2')$ transfer deletions to reach $\mathcal{N}_1^*$ and $\mathcal{N}_2^*$, we get $d_{TB}(N_1', N_2') \geq d_{LGT}(N_1|T_1, N_2|T_2)$.

We show the converse inequality $d_{TB}(N_1', N_2') \leq d_{LGT}(N_1|T_1, N_2|T_2)$. In N_1', we take the base tree T_1' with the same transfer arcs as $\mathcal{N}_1$, and for each reticulation above E we choose the transfer arc arbitrarily. For N_2', we take the base tree T_2' with the same transfer arcs as $\mathcal{N}_2$, and for the reticulations above E we make the same arbitrary choice as in T_1'. It then becomes easy to see that a sequence of transfer deletions that makes $\mathcal{N}_1$ and $\mathcal{N}_2$ LGT-isomorphic can also be used to make $N_1'|T_1'$ and $N_2'|T_2'$ LGT-isomorphic. This yields $d_{TB}(N_1', N_2') \leq d_{LGT}(N_1|T_1, N_2|T_2)$.

Proof that d_{LGT} is FPT in the Number of Transfers.

We recall our proposition of interest here.

Proposition 1. *The distance $d_{LGT}(\mathcal{N}_1, \mathcal{N}_2)$ can be computed in time $O(2^t \cdot m)$, where t (resp. m) is the number of transfer arcs (resp. arcs) in $\mathcal{N}_1$ and $\mathcal{N}_2$.*

Proof. Using the ideas in the proof of Theorem 3, one may calculate wRF and $D(\mathcal{N}_1, \mathcal{N}_2)$ in linear time, and it remains to state how to calculate $d_{TR}(\mathcal{N}_1, \mathcal{N}_2)$. Suppose that $\mathcal{N}_1$ has t_1 transfer arcs and $\mathcal{N}_2$ has t_2. One may iterate over all 2^{t_1} subsets of transfers of $\mathcal{N}_1$ and all 2^{t_2} sets of transfers of $\mathcal{N}_2$ in time $2^{t_1} \times 2^{t_2} = 2^{t_1+t_2} = 2^t$. For each subset, we can obtain transfer reductions $\mathcal{N}_1', \mathcal{N}_2'$ in linear time (technically, we must copy each network before doing so, which just adds another linear time term to the complexity). Then, one needs to check whether the two reductions are LGT-isomorphic. This may also be done in linear time as follows. We already know at this point that the two networks are homeomorphic, so we only need to check that the transfer arcs are identical. For that, one may simply store, for each transfer arc (u, v): (1) the tree pairs $[u_1, u_2]$ and $[v_1, v_2]$ that u and v are on; (2) the rank of u on $[u_1, u_2]$ and the rank of v on $[v_1, v_2]$ (i.e., the number of attachment points to traverse on the tree pair to reach the vertex). We then ensure LGT-isomorphic by checking, for each transfer arc (u, v) in either network, that the corresponding transfer arc with the same extra information exists in the other network. To do this, assuming that vertices are labeled 1 to $n_1 + n_2$, with $n_i = |V(N_i)|$ as in the proof of Theorem 3, we can store each transfer arc in the form $([i, j], [k, l], r, r')$, where r, r' are the aforementioned ranks. This is a sextuple with integers bounded by $n_1 + n_2$. These can be sorted in linear time using radix sort, and then we can check that the lists of transfers are identical.

Algorithms for Sect. 4

Here we describe the algorithm from Proposition 1. The same ideas can be used for Theorem 6. The algorithm computes d_{LGT} as the sum of the weight of bad vertices (wRF), the number of extra transfer deletions to make (D_{tr}), and the number of doubly bad transfers (D). Bad vertices are found using Day algorithm. Algorithm 1 integrates these components and returns the combined contribution of tree disagreement and transfer disagreement.

When transfer arcs are treated as unordered, the distance reduces to counting the arcs that appear in exactly one of the two networks. In this setting, no additional structural constraints need to be verified once the relative complements are computed. Algorithm 2 therefore returns the transfer distance directly as the total number of non-shared arcs. Note, the relative complement of A in B is the set difference $B \setminus A$. We can obtain the symmetric difference $A \triangle B$ by computing both relative complements $A \setminus B$ and $B \setminus A$. The algorithm for GetRelativeComplements is shown afterward.

When transfers are ordered, our FPT algorithm simply explores all subsets of the remaining transfer arcs and finds the smallest pair of deletion sets that yields equivalent LGT structures, as shown in Algorithm 3.

Algorithm 4 performs this computation of the relative complements, so that the returned pair of sets can be used to get the symmetric difference, or to delete the transfers in both networks. This is done by sorting both arc sets and scanning them in parallel to isolate the non-shared arcs. The transfers in

---

Algorithm 1: getD_LGT($N_1|T_1, N_2|T_2$)

Data: LGT networks $N_1|T_1$ and $N_2|T_2$
Result: $d_{LGT}(N_1|T_1, N_2|T_2)$

1 $badVertices \leftarrow \emptyset$; $wRF \leftarrow 0$; $D\_tr \leftarrow 0$; $D \leftarrow 0$;
2 $bad^1_{tail} \leftarrow \emptyset$; $bad^1_{head} \leftarrow \emptyset$; $bad^2_{tail} \leftarrow \emptyset$; $bad^2_{head} \leftarrow \emptyset$;
3 Compute $badVertices$ and wRF using Day's algorithm;
4 **for** $i \in \{1,2\}$ **do**
5     $\quad bad^i_{tail} \leftarrow$ transfers of $N_i|T_i$ whose tail is on $[u,v]$ with v in $badVertices$;
6     $\quad bad^i_{head} \leftarrow$ transfers of $N_i|T_i$ whose head is on $[u,v]$ with v in $badVertices$;
7     $\quad D \leftarrow D + |bad^i_{tail} \cap bad^i_{head}|$;   // counting the doubly bad transfers
8     $\quad$ Delete from $N_i|T_i$ all transfer arcs in $bad^i_{tail} \cup bad^i_{head}$;
9 **for** $i \in \{1,2\}$ **do**
10    $\quad$ **foreach** $v \in V(T_i)$ *in pre-order* **do**
11       $\quad\quad$ **if** $v \in badVertices$ **then**
12          $\quad\quad\quad$ Contract v in T_i;
13 **if** $N_1|T_1, N_2|T_2$ *have unordered transfers* **then**
14    $\quad D\_tr \leftarrow getD\_trUnordered(N_1|T_1, N_2|T_2)$;
15 **else**
16    $\quad D\_tr \leftarrow getD\_trOrdered(N_1|T_1, N_2|T_2)$;
17 **return** $wRF + D\_tr - D$

Algorithm 2: getD_trUnordered($N_1|T_1, N_2|T_2$)

Data: LGT networks $N_1|T_1$ and $N_2|T_2$ with unordered transfers where T_1 and T_2 are homeomorphic
Result: $d_{TR}(N_1|T_1, N_2|T_2)$

1 $(tr_{rc}(N_1|T_1), tr_{rc}(N_2|T_2)) \leftarrow$ GetRelativeComplements($N_1|T_1, N_2|T_2$);
2 **return** $|tr_{rc}(N_1|T_1)| + |tr_{rc}(N_2|T_2)|$

Algorithm 3: getD_trOrdered($N_1|T_1, N_2|T_2$)

Data: LGT networks $N_1|T_1$ and $N_2|T_2$ with ordered transfers where T_1 and T_2 are homeomorphic
Result: $d_{TR}(N_1|T_1, N_2|T_2)$

1 $min_d \leftarrow \infty$;
2 //Pre-processing: remove transfers that have no chance of being kept;
3 $(tr_{rc}(N_1|T_1), tr_{rc}(N_2|T_2)) \leftarrow$ GetRelativeComplements($N_1|T_1, N_2|T_2$);
4 Delete transfer arcs from $N_1|T_1$ corresponding to the set $tr_{rc}(N_1|T_1)$;
5 Delete transfer arcs from $N_2|T_2$ corresponding to the set $tr_{rc}(N_2|T_2)$;
6 //Brute-force the subsets of remaining transfers to find the optimal deletion sets **foreach** $R_1 \subseteq E(N_1) \setminus E(T_1)$ **do**
7     $\quad$ **foreach** $R_2 \subseteq E(N_2) \setminus E(T_2)$ **do**
8        $\quad\quad N_1'|T_1' \leftarrow$ copy of $N_1|T_1$ with transfers from R_1 deleted;
9        $\quad\quad N_2'|T_2' \leftarrow$ copy of $N_2|T_2$ with transfers from R_2 deleted;
10       $\quad\quad$ **if** $N_1'|T_1' \simeq_{LGT} N_2'|T_2'$ **then**
11          $\quad\quad\quad min_d \leftarrow min(min_d, |R_1| + |R_2|)$;
12 **return** min_d

$tr(N_i|T_i)$ have the form $([u_1, u_2], [v_1, v_2])$, which can be seen as a quadruple of integers between 0 and the sizes of the networks, in which case radix sort takes linear time.

Algorithm 4: getRelativeComplements($N_1|T_1$, $N_2|T_2$)

Data: LGT networks $N_1|T_1$ and $N_2|T_2$ with T_1, T_2 homeomorphic

Result: A pair $(tr_{rc}(N_1|T_1), tr_{rc}(N_2|T_2))$ corresponding to the relative complements between $N_1|T_1$ and $N_2|T_2$.

```
 1  tr_ord(N1|T1) ← radixSort(tr(N1|T1)); tr_ord(N2|T2) ← radixSort(tr(N2|T2));
 2  i ← 0; j ← 0;
 3  tr_rc(N1|T1) ← ∅; tr_rc(N2|T2) ← ∅ ;      // To store the relative complements
 4  while i < |tr_ord(N1|T1)| and j < |tr_ord(N2|T2)| do
 5      if tr_ord(N1|T1)[i] == tr_ord(N2|T2)[j] then      // tr_ord(N1|T1)[i]: ith pair
 6          i ← i + 1; j ← j + 1;
 7      else if tr_ord(N1|T1)[i] < tr_ord(N2|T2)[j] then
 8          tr_rc(N1|T1) ← tr_rc(N1|T1) + tr_ord(N1|T1)[i];
 9          i ← i + 1;
10      else
11          tr_rc(N2|T2) ← tr_rc(N2|T2) + tr_ord(N2|T2)[j];
12          j ← j + 1;
13  while i ≤ |tr_ord(N1|T1)| do
14      tr_rc(N1|T1) ← tr_rc(N1|T1) + tr_ord(N1|T1)[i];
15      i ← i + 1;
16  while j ≤ |tr_ord(N2|T2)| do
17      tr_rc(N2|T2) ← tr_rc(N2|T2) + tr_ord(N2|T2)[j];
18      j ← j + 1;
19  return (tr_rc(N1|T1), tr_rc(N2|T2))
```

C   Proofs for Section 5 (Experiments)

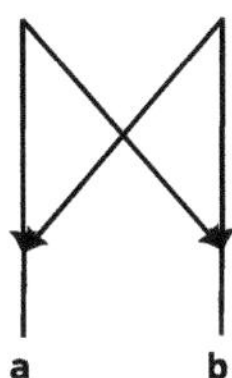

Fig. 11. The pattern shown the left is not allowed for the sub-class of orchard networks (see for instance [26] for a definition). Although it can be considered non-realistic (it is not time-consistent). Yet, the network prediction methods used in Sects. 5.2 and 5.3(namely [31,34]) could in principle return networks exhibiting this pattern, requiring comparison methods that are capable of comparing ideally all tree-based networks [36], and not just orchard networks.

References

1. Doolittle, W.F., Bapteste, E.: Pattern pluralism and the tree of life hypothesis. Proc. Nat. Acad. Sci. **104**(7), 2043–2049 (2007)
2. Bryant, D., Moulton, V.: Neighbor-Net: an agglomerative method for the construction of phylogenetic networks. Mol. Biol. Evol. **21**(2), 255–265 (2004)
3. Wen, D., Yun, Yu., Zhu, J., Nakhleh, L.: Inferring phylogenetic networks using PhyloNet. Syst. Biol. **67**(4), 735–740 (2018)
4. Solís-Lemus, C., Bastide, P., Ané, C.: Phylonetworks: a package for phylogenetic networks. Mol. Biol. Evol. **34**(12), 3292–3298 (2017)
5. Kong, S., Swofford, D.L., Kubatko, L.S.: Inference of phylogenetic networks from sequence data using composite likelihood. Syst. Biol. **74**(1), 53–69 (2025)
6. Boc, A., Philippe, H., Makarenkov, V.: Inferring and validating horizontal gene transfer events using bipartition dissimilarity. Syst. Biol. **59**(2), 195–211 (2010)
7. Robinson, D.F., Foulds, L.R.: Comparison of phylogenetic trees. Math. Biosci. **53**(1-2), 131–147 (1981)
8. Lin, Y., Rajan, V., Moret, B.M.: A metric for phylogenetic trees based on matching. IEEE/ACM Trans. Comput. Biol. Bioinf. **9**(4), 1014–1022 (2011)
9. Cardona, G., Llabrés, M., Rosselló, F., Valiente, G.: Metrics for phylogenetic networks I: generalizations of the Robinson-Foulds metric. IEEE/ACM Trans. Comput. Biol. Bioinf. **6**(1), 46–61 (2008)
10. Huson, D.H., Rupp, R., Berry, V., Gambette, P., Paul, C.: Computing galled networks from real data. Bioinformatics **25**(12), i85–i93 (2009)
11. Huson, D.H., Rupp, R., Scornavacca, C.: Phylogenetic Networks: Concepts, Algorithms and Applications. Cambridge University Press (2010)
12. Jansson, J., Nguyen, N.B., Sung, W.K.: Algorithms for combining rooted triplets into a galled phylogenetic network. SIAM J. Comput. **35**(5), 1098–1121 (2006)
13. Jansson, J., Sung, W.-K.: Inferring a level-1 phylogenetic network from a dense set of rooted triplets. Theoret. Comput. Sci. **363**(1), 60–68 (2006)
14. Huber, K.T., Moulton, V.: Encoding and constructing 1-nested phylogenetic networks with trinets. Algorithmica **66**, 714–738 (2013)
15. Huson, D.H., Scornavacca, C.: A survey of combinatorial methods for phylogenetic networks. Genome Biol. Evol. **3**, 23–35 (2011)
16. Cardona, G., Llabrés, M., Rosselló, F., Valiente, G.: Metrics for phylogenetic networks ii: Nodal and triplets metrics. IEEE/ACM Trans. Comput. Biol. Bioinf. **6**(3), 454–469 (2008)
17. Cardona, G., Pons, J.C., Ribas, G., Coronado, T.M.: Comparison of orchard networks using their extended μ-representation. IEEE/ACM Trans. Comput. Biol. Bioinf. **21**(3), 501–507 (2024)
18. Bai, A., Erdős, P.L., Semple, C., Steel, M.: Defining phylogenetic networks using ancestral profiles. Math. Biosci. **332**, 108537 (2021)
19. Maxfield, M., Xu, J., Ané, C.: A dissimilarity measure for semidirected networks. IEEE Trans. Comput. Biol. Bioinf. (2025)
20. Allen, B.L., Steel, M.: Subtree transfer operations and their induced metrics on evolutionary trees. Ann. Combinatorics **5**(1), 1–15 (2001)
21. Bordewich, M., Semple, C.: On the computational complexity of the rooted subtree prune and regraft distance. Ann. Comb. **8**(4), 409–423 (2005)
22. Bordewich, M., Linz, S., Semple, C.: Lost in space? Generalising subtree prune and regraft to spaces of phylogenetic networks. J. Theor. Biol. **423**, 1–12 (2017)

23. Huber, K.T., Linz, S., Moulton, V., Wu, T.:0 Spaces of phylogenetic networks from generalized nearest-neighbor interchange operations. J. Math. Biol. **72**(3), 699–725 (2016)
24. Gambette, P., Van Iersel, L., Jones, M., Lafond, M., Pardi, F., Scornavacca, C.: Rearrangement moves on rooted phylogenetic networks. PLoS Comput. Biol. **13**(8), e1005611 (2017)
25. Marchand, B., Tahiri, N., Fard, S.G., et al.: Finding maximum common contractions between phylogenetic networks. Algorithms Mol. Biol. **20**(1), 18 (2025). https://doi.org/10.1186/s13015-025-00283-9
26. Landry, K., Teodocio, A., Lafond, M., Tremblay-Savard, O.: Defining phylogenetic network distances using cherry operations. IEEE/ACM Trans. Comput. Biol. Bioinf. **20**(3), 1654–1666 (2022)
27. Landry, K., Tremblay-Savard, O.: Fast calculation of cherry distance on level-1 orchard networks: Optimization, heuristic and implementation. In: RECOMB International Workshop on Comparative Genomics, pp. 157–177. Springer (2025). https://doi.org/10.1007/978-3-031-94928-9_12
28. Cardona, G., Pons, J.C., Rosselló, F.: A reconstruction problem for a class of phylogenetic networks with lateral gene transfers. Algorithms Mol. Biol. **10**(1), 28 (2015)
29. Gabriel Cardona and Joan Carles Pons: Reconstruction of LGT networks from tri-LGT-nets. J. Math. Biol. **75**(6), 1669–1692 (2017)
30. Jacox, E., Chauve, C., Szöllősi, G.J., Ponty, Y., Scornavacca, C.: ECCETERA: comprehensive gene tree-species tree reconciliation using parsimony. Bioinformatics **32**(13), 2056–2058 (2016)
31. Bansal, M.S., Kellis, M., Kordi, M., Kundu, S.: Ranger-DTL 2.0: rigorous reconstruction of gene-family evolution by duplication, transfer and loss. Bioinformatics **34**(18), 3214–3216 (2018)
32. Jones, M., Lafond, M., Scornavacca, C.: Consistency of orthology and paralogy constraints in the presence of gene transfers. Peer Community Math. Comput. Biol. (2022)
33. Sánchez, A., Lafond, M.: Predicting horizontal gene transfers with perfect transfer networks. Algorithms Mol. Biol. **19**(1), 6 (2024)
34. Sánchez, A., Scholz, G.E., Stadler, P.F., Lafond, M.: A SANKOFF-ROUSSEAU-like algorithm for minimizing lateral gene transfers and losses on single origin characters. In: RECOMB International Workshop on Comparative Genomics, pp. 69–86. Springer (2025). https://doi.org/10.1007/978-3-031-94928-9_5
35. Francis, A.R., Steel, M.: Which phylogenetic networks are merely trees with additional arcs? Syst. Biol. **64**(5), 768–777 (2015)
36. Kong, S., Pons, J.C., Kubatko, L., Wicke, K.: Classes of explicit phylogenetic networks and their biological and mathematical significance. J. Math. Biol. **84**(6), 47 (2022)
37. Doyon, JP., Scornavacca, C., Gorbunov, K.Y., Szöllősi, G.J., Ranwez, V., Berry, V.: An efficient algorithm for gene/species trees parsimonious reconciliation with losses, duplications and transfers. In: RECOMB International Workshop on Comparative Genomics, pp. 93–108. Springer (2010). https://doi.org/10.1007/978-3-642-16181-0_9
38. Pons, J.C., Scornavacca, C., Cardona, G.: Generation of level-k k lgt networks. IEEE/ACM Trans. Comput. Biol. Bioinf. **17**(1), 158–164 (2019)
39. Day, W.H.E.: Optimal algorithms for comparing trees with labeled leaves. J. Classif. **2**(1), 7–28 (1985)

40. Sipser, M.: Introduction to the theory of computation. ACM SIGACT News **27**(1), 27–29 (1996)
41. Nakhleh, L., Ringe, D.A., Warnow, T.: Perfect phylogenetic networks: a new methodology for reconstructing the evolutionary history of natural languages. Language **81**(2), 382–420 (2005)
42. Sánchez, A., Lafond, M.: Galled perfect transfer networks. In: RECOMB International Workshop on Comparative Genomics, pp. 24–43. Springer (2024). https://doi.org/10.1007/978-3-031-58072-7_2
43. Wu, L., Lambert, J.D.: Clade-specific genes and the evolutionary origin of novelty; new tools in the toolkit. In: Seminars in Cell & Developmental Biology, vol. 145, pp. 52–59 (2023)
44. Wang, Y.W., Hess, J., Slot, J.C., Pringle, A.: De novo gene birth, horizontal gene transfer, and gene duplication as sources of new gene families associated with the origin of symbiosis in amanita. Genome Biol. Evol. **12**(11), 2168–2182 (2020)
45. Kanehisa, M.: The KEGG database. In: In silico'simulation of biological processes: Novartis Foundation Symposium 247, vol. 247, pp. 91–103. Wiley Online Library (2002)
46. Schoch, C.L., et al.: NCBI taxonomy: a comprehensive update on curation, resources and tools. Database, 2020:baaa062 (2020)
47. David Schaller, Marc Hellmuth, and Peter F Stadler. Asymmetree: A flexible python package for the simulation of complex gene family histories. *Software*, 1(3):276–298, 2022
48. Bulteau, L., Weller, M.: Parameterized algorithms in bioinformatics: an overview. Algorithms **12**(12), 256 (2019)

k-Nearest Common Leaves Algorithm for Phylogenetic Tree Completion

Aleksandr Koshkarov⬤ and Nadia Tahiri$^{(\boxtimes)}$⬤

Department of Computer Science, University of Sherbrooke, Sherbrooke, QC, Canada
{Aleksandr.Koshkarov,Nadia.Tahiri}@USherbrooke.ca

Abstract. Phylogenetic trees represent the evolutionary histories of taxa and support tasks such as clustering and Tree of Life reconstruction. Many established comparison methods, including the Robinson-Foulds (RF) distance, assume identical taxon sets. A methodological gap remains for trees with distinct but overlapping taxa. Existing approaches either prune non-common leaves, which can discard information, or complete both trees such that they share the same taxa. Completion is more comprehensive, but current methods typically ignore branch lengths, which are essential for identifying evolutionary patterns. This paper introduces k-Nearest Common Leaves (k-NCL), an algorithm for completing rooted phylogenetic trees defined on different but overlapping taxa. The method uses branch lengths and topological characteristics and does not rely on a specific distance measure. The k-NCL algorithm is designed to preserve evolutionary relationships in the trees under comparison. The running time is $O(n^2)$, where n is the size of the union of the two leaf sets. Additional properties include preservation of original distances and topology, symmetry, and uniqueness of the completion. Implemented in Python, k-NCL is evaluated on biological datasets of amphibians, birds, mammals, and sharks. Experimental results show that RF combined with k-NCL improves phylogenetic tree clustering performance compared to the RF(+) tree completion approach.

Keywords: Algorithm · Phylogenetic tree comparison · Tree completion · Tree distance · Supertree

1 Introduction

Phylogenetic trees are widely used to study the diversity of life, providing simplified representations of complex evolutionary relationships. They are utilized in fields such as comparative genomics and evolutionary biology. A common task in these areas is measuring distances between phylogenetic trees, often for purposes such as clustering trees and evaluating phylogenetic inference methods. Such distance measurements are especially informative when the trees share the same set

Supplementary Information The online version contains supplementary material available at https://doi.org/10.1007/978-3-032-26891-4_13.

of species. However, in many practical cases, the trees are defined on different but overlapping sets of taxa. In order to compare such trees, researchers often use methods such as pruning non-overlapping leaves or completing both trees to make them defined on the same set of taxa [1,35]. Pruning can result in the loss of valuable evolutionary information by removing unique taxa, whereas tree completion retains all taxa and allows for a more comprehensive comparison of evolutionary relationships. However, topology-only tree completion approaches may overlook important information, such as branch lengths, potentially limiting the accuracy of comparisons. This highlights the need for more advanced methods that use both topological and branch length information to provide a more accurate representation of evolutionary signals.

Practical tasks such as supertree construction [2,29,34], phylogenetic tree clustering [29,30], the assembly of the Tree of Life [11], and searches within phylogenetic databases [5,33] require the computation of distances between trees with different numbers of overlapping leaves. Approaches that can handle trees defined on different but overlapping sets of taxa include the Robinson-Foulds(-) method, which prunes both trees to a shared taxa set before computing the classical Robinson-Foulds (RF) distance [7], the generalized RF distance [17], and the vectorial tree distance [21]. The tree completion-based methods involve the addition of distinct leaves from one tree to another, thereby defining both trees on the same set of taxa, which is the union of the leaf sets of the original trees. These methods include the RF(+) approach [1,35], geodesic in the extended Billera-Holmes-Vogtmann (BHV) tree space [10,24], and several others [14,16].

The RF-based tree completion approach, detailed and expanded in various studies [1,6–8,15,35], uses a completion method aiming to minimize the RF distance. The RF(+) method described in [35] implements a polynomial-time dynamic programming algorithm for computing the RF(+) distance on completed trees. The RF(+) method only accounts for the topology of the trees being compared. The exact algorithm for the RF(+) problem has a complexity of $O(n \cdot k^2)$, where n is the size of the union of the leaf sets of both trees under comparison, and k is the number of maximal subtrees unique to one input tree.

The geodesic distance in the BHV tree space [4] considers both the topology and branch lengths of the trees. Extensions of this metric to process trees with different but overlapping sets of taxa have been proposed in [10,24], introducing techniques to compute distances in an extended BHV tree space. The introduction of additional structures within the BHV tree space, such as a connection cluster, a connection space, a connection graph, and the incorporation of new leaves, which requires a transition between lower and higher dimensional orthants in the BHV space, results in an increase in computational complexity to $O(n^{\ell+2})$, where n is the size of the union of the leaf sets of the trees being compared and ℓ is the number of new leaves to be added to the tree. Although these methods allow for comprehensive distance calculations, the computational intensity required for large trees with numerous non-common leaves is a significant limitation. It should be noted that the extended BHV geodesic defines a pairwise distance via extension spaces rather than yielding a unique completed tree for each input, and therefore is not a tree completion method.

Several researchers have developed different strategies for imputing missing taxa to phylogenetic trees [18,19,22,36,37]. The approach introduced by Yasui et al. [36] involves an optimization-based method to handle missing data in gene trees using a mixed integer non-linear programming model. The method is designed to impute missing pairwise distances between leaves in gene trees. It uses a two-stage optimization process, where the first stage involves assigning individuals to hypothetical groups or clades based on the available genetic distances, and the second stage focuses on estimating the missing distances.

Yoshida [37] addresses missing data in phylogenetic trees using tropical geometry. The approach constructs a tropical polytope from known gene trees that are structurally close to the incomplete tree, then projects the incomplete tree onto this polytope in a tropical metric space to estimate the missing parts. The method is restricted to equidistant trees, where all leaves are equidistant from the root. Other approaches address missing data in gene trees using quartet-based optimization or imputation techniques based on tree topology [18,19,22], but do not focus on pairwise tree completion for comparison.

Our Contributions. In this work, we introduce a novel algorithm named *k-Nearest Common Leaves* (*k*-NCL), designed to complete phylogenetic trees that are defined over distinct but overlapping sets of taxa. Our contributions can be summarized as follows. (1) We incorporated branch lengths into the tree completion process, resulting in completed trees that represent both structural and evolutionary relationships, in contrast to topology-only approaches. (2) We introduced a scaling-based strategy to account for differences in evolutionary rates between the two trees. (3) We designed the *k*-NCL algorithm to be independent of any specific distance metric, i.e., it does not assume or optimize for a predefined tree distance, such as RF or geodesic distance in BHV space. (4) We provide a Python implementation of the *k*-NCL algorithm, with open-source code available on GitHub. (5) We evaluated the proposed method on multiple biological datasets (amphibians, birds, mammals, and sharks), where the trees have partial taxon overlap.

2   Preliminaries and Notation

Let T be a rooted phylogenetic tree. Let $V(T)$ be its set of nodes (both internal and terminal). The leaves (or terminal nodes) of T, each denoted by l, represent the individual species or taxa under study, and their collection is denoted by $L(T) \subset V(T)$. The branches (or edges), denoted by $E(T)$, connect the nodes and represent evolutionary relationships. We assume that each branch $e = (u, w) \in E(T)$ has an associated strictly positive length between nodes u and w, indicating the amount of evolutionary change or the time that has passed along that lineage.

Definition 1 (Distance Between Nodes). *Let u and w be two nodes (internal or terminal) in T. The distance between nodes u and w, denoted by*

$d^{(T)}(u, w)$, *is defined as the cumulative sum of branch lengths along the unique path connecting u to w.*

Definition 2 (Common and Distinct Leaves). *In two phylogenetic trees T_1 and T_2 with different but overlapping sets of taxa, the term common leaves refers to the leaves that are present in both trees. The set of common leaves shared by trees T_1 and T_2 is denoted as $CL(T_1, T_2) = L(T_1) \cap L(T_2)$. Conversely, leaves present in one tree but absent in the other are termed distinct leaves, with the set of such leaves for tree T_1 related to tree T_2 denoted as $DL(T_1|T_2) = L(T_1) \backslash L(T_2)$.*

Throughout the paper, we assume that $|CL(T_1, T_2)| \geq 2$ in order for pairwise distances over common leaves to be defined. Let $\mathrm{idx}(\cdot)$ denote the rank of an item in a predetermined linear order, defined separately on the common leaves and the original branches of each input tree. (i) Let $CL(T_1, T_2)$ be ordered by ascending lexicographic order of taxon labels. For any $l \in CL(T_1, T_2)$, let $\mathrm{idx}(l)$ denote its position in this order. This rank is used only for deterministic tie-breaking and remains fixed throughout the algorithm. (ii) For each input tree T_i, a fixed rank $\mathrm{idx}(e)$ is assigned to every original branch $e \in E(T_i)$ via a single depth-first enumeration established once at the start. If an original branch is split during insertion, all resulting sub-branches inherit the same rank $\mathrm{idx}(e)$.

Definition 3 (Maximal Distinct-Leaf Subtree). *A subtree S of T_1 is called a maximal distinct-leaf subtree if all leaves of S belong to the set $DL(T_1|T_2)$, and S is not strictly contained in any larger subtree of T_1 whose leaves are also entirely included in $DL(T_1|T_2)$.*

Conceptually, a maximal distinct-leaf subtree S is the largest possible subtree in T_1 that contains only leaves not shared with T_2, and cannot be extended without including shared leaves. S may consist of one or more distinct leaves, i.e., $|L(S)| \geq 1$. The *root* of S in T_1 is defined as the lowest common ancestor (LCA) of all its leaves, denoted by $lca_{T_1}(L(S))$. In the special case where S contains only a single leaf, its root is that leaf itself. The *root branch* of S is the branch that connects the root of S to its most immediate ancestor node in the entire tree T_1. We refer to this parent node as the *attachment node* of S, denoted $att_{T_1}(S)$. For any given tree T, the collection of all its maximal distinct-leaf subtrees is denoted as $\mathcal{S}_T$.

Definition 4 (Nearest Common Leaves). *For a distinct-leaf subtree $S \in \mathcal{S}_{T_1}$, the set of k nearest common leaves, denoted $\mathcal{N}_k(S, T_1)$, consists of the k leaves in T_1 that are also present in T_2 and are closest to the root of S, ordered by increasing distance from it. The parameter k must satisfy $2 \leq k \leq |CL(T_1, T_2)|$.*

Let $T \in \{T_1, T_2\}$ and let $S \in \mathcal{S}_T$. If several common leaves have the same distance as the k-th closest leaf to the root of S, the ties are resolved using the fixed order of $CL(T_1, T_2)$. Specifically, the common leaves $l \in CL(T_1, T_2)$ are ordered by the pair $\left(d^{(T)}(l, lca_T(L(S))), \mathrm{idx}(l)\right)$ in lexicographic (increasing) order, and the first k leaves are taken. Therefore, $|\mathcal{N}_k(S, T)| = k$, and the selection is deterministic.

Problem 1 (k-NCL Tree Completion). Let T_1 and T_2 be rooted phylogenetic trees with strictly positive branch lengths, leaf sets $L(T_1)$ and $L(T_2)$, and common leaves $CL(T_1, T_2) = L(T_1) \cap L(T_2)$ with $|CL(T_1, T_2)| \geq 2$. Let $k \in [2, |CL(T_1, T_2)|]$. The objective is to construct completions $T_1^{\uplus}$ and $T_2^{\uplus}$ on the unified leaf set $L(T_1) \cup L(T_2)$ such that: (i) the restriction of $T_i^{\uplus}$ to the leaf set $L(T_i)$ is identical to T_i in both topology and branch lengths, for $i \in \{1, 2\}$, and (ii) for each $i \in \{1, 2\}$ and each maximal distinct-leaf subtree S present in T_{3-i} but absent from T_i, S is attached to T_i at a location determined from the k nearest common leaves of S in T_{3-i}. Among all possible attachment locations on the original branches of T_i (including existing internal nodes), the selected location minimizes the discrepancy between distances measured in the target tree and the corresponding position distances induced from T_{3-i}. The completion is obtained by attaching each such subtree in this manner, preserving the original topology and branch lengths on the original leaves.

3   Methods

3.1   *k*-Nearest Common Leaves Algorithm

The k-NCL algorithm addresses the tree completion problem by constructing completed trees $T_1^{\uplus}$ and $T_2^{\uplus}$ on the unified leaf set $L(T_1) \cup L(T_2)$. The method inserts maximal distinct-leaf subtrees from each input tree to the other on the basis of common leaves and adjustment rates, which calibrate branch lengths during subtree insertion.

A high-level description of the k-NCL algorithm is given in Appendix D, and Fig. 1 presents its main components. The procedure begins with the identification of common leaves, distinct leaves (Definition 2), and maximal distinct-leaf subtrees (Definition 3) in both trees.

Remark 1. For $i \in \{1, 2\}$, the set $\mathcal{S}_{T_i}$ relative to $DL(T_i \mid T_{3-i})$ can be computed by a post-order traversal of T_i. A node is marked if all of its descendant leaves belong to $DL(T_i \mid T_{3-i})$. The roots of maximal distinct-leaf subtrees are precisely the marked nodes whose parent, if any, is unmarked. Each such subtree is then recorded together with its root branch.

Remark 2. Subsequent steps of the k-NCL algorithm involve a large number of node-to-node distance computations, specifically on the order of $O(n^2)$ in total, where n is the total number of taxa across both trees. These include the distances between common leaves, distances from common leaves to subtree roots, and evaluations of candidate insertion points across all branches. To reduce the cost of these operations, we associate each tree with a distance oracle [31], which is rebuilt after every maximal distinct-leaf subtree insertion. The distance oracle is constructed using a depth-first traversal that generates an Euler tour of the tree, along with the depth of each visit and the first occurrence index of every node. A sparse table is built over the depth array to efficiently answer range minimum queries (RMQ), which support constant-time LCA queries using the method of Bender and Farach-Colton [3]. Based on these LCA results, the distance between any two nodes can be computed in $O(1)$.

The tree completion procedure continues by adjusting the branch lengths of the maximal distinct-leaf subtrees, including their root branches, using the corresponding branch adjustment rate (Eq. 1).

Let $CL(T_1, T_2)$ be ordered and indexed. We introduce a global adjustment rate $r(T_1|T_2)$, which is defined as the ratio between the total pairwise distances over all common leaves in T_1 and the corresponding total pairwise distances in T_2 (Eq. 1). This rate represents the relative scaling of branch lengths between the two trees based on their shared taxa.

$$r(T_1|T_2) = \frac{\displaystyle\sum_{\substack{l_i, l_j \in CL(T_1, T_2) \\ i < j}} d^{(T_1)}(l_i, l_j)}{\displaystyle\sum_{\substack{l_i, l_j \in CL(T_1, T_2) \\ i < j}} d^{(T_2)}(l_i, l_j)}. \tag{1}$$

Next, all branch lengths (including the root branch) in the maximal distinct-leaf subtree $S \in \mathcal{S}_{T_2}$, which is to be inserted into T_1, are scaled using the adjustment rate $r(T_1|T_2)$. For each branch $(u, w) \in E(S)$ with $E(S) \subseteq E(T_2)$, the adjusted branch length is given by Eq. 2.

$$d^{(S)}(u, w) \leftarrow d^{(S)}(u, w) \cdot r(T_1|T_2). \tag{2}$$

This adjustment is crucial for preserving the phylogenetic relationships and distances within the subtree as it is integrated into the new tree structure. However, the original branch lengths in the initial trees remain unaltered.

Subsequently, the k nearest common leaves, $\mathcal{N}_k(S, T_2)$ (Definition 4) are identified for each subtree $S \in \mathcal{S}_{T_2}$ (see Fig. 1). Given a selected common leaf, $l_c \in \mathcal{N}_k(S, T_2)$, the leaf-based adjustment rate, $r^{(l_c)}(T_1|T_2)$, is computed as the ratio of cumulative distances from l_c to all other common leaves in T_1 relative to the corresponding sum in T_2 (Eq. 3).

$$r^{(l_c)}(T_1|T_2) = \frac{\displaystyle\sum_{l_i \in CL(T_1, T_2)} d^{(T_1)}(l_c, l_i)}{\displaystyle\sum_{l_i \in CL(T_1, T_2)} d^{(T_2)}(l_c, l_i)}. \tag{3}$$

The insertion process is initiated with T_1, and T_1' is created as the insertion target for maximal distinct-leaf subtrees from T_2. Subtrees are iteratively inserted into T_1' until all subtrees from $\mathcal{S}_{T_2}$ have been added, at which point T_1' becomes the completed tree $T_1^{\uplus}$. The notation T_1' emphasizes that the original tree T_1 remains unchanged throughout the subtree insertion process.

At a high level, the attachment point is chosen by first using the k nearest common leaves of the subtree in the source tree to estimate its position relative to the common leaves in the working tree, and then selecting the location on the original branches of the target tree whose distances to these leaves best match the estimated position.

As part of identifying the optimal insertion point for a subtree $S \in \mathcal{S}_{T_2}$, a *position distance* is computed relative to each selected common leaf l_c. This distance is calculated as follows (Eq. 4).

$$d^{(T_1')}(l_c, att_{T_1'}(S)) = d^{(T_2)}(l_c, att_{T_2}(S)) \cdot r^{(l_c)}(T_1 | T_2), \tag{4}$$

where $att_{T_2}(S)$ denotes the original attachment node of S in T_2, and $att_{T_1'}(S)$ denotes the target attachment location for S in T_1'.

In order to integrate S into T_1', an insertion point v^* is selected by minimizing the discrepancy between the observed distances in T_1' and the computed position distances for the selected common leaves. This procedure is performed as follows.

Every candidate insertion point can lie on an existing node or on a branch connecting two nodes, u (the parent node) and w (the child node). Fix a candidate branch $e = (u, w)$ of T_1'. Let $v_e(x)$ be a point on branch $(u, w) \in E(T_1')$, located at a distance $x \cdot d^{(T_1')}(u, w)$ from the parent node u along the branch, where $x \in [0, 1)$ (see Fig. 1 (a)). In this parameterization setting, the boundary case $x = 0$ corresponds to the node u (an existing node), while x approaching 1 corresponds to a point arbitrarily close to the child node w. The interval $[0, 1)$ is used to avoid double counting existing nodes.

Then, the *observed distance* from a selected common leaf $l_c \in \mathcal{N}_k(S, T_2)$ to the candidate point $v_e(x)$ is given by Eq. 5.

$$d^{(T_1')}(l_c, v_e(x)) = d^{(T_1')}(l_c, u) + \sigma_{c,e} \cdot x \cdot d^{(T_1')}(u, w), \tag{5}$$

where $\sigma_{c,e} = \begin{cases} -1, & \text{if } l_c \text{ is a descendant of } w, \\ +1, & \text{if } l_c \text{ is not a descendant of } w. \end{cases}$ The discrepancy between the observed distance and the position distance is measured using the following objective function (Eq. 6).

$$f_e(x) = \sum_{l_c \in \mathcal{N}_k(S, T_2)} \left(d^{(T_1')}(l_c, v_e(x)) - d^{(T_1')}(l_c, att_{T_1'}(S)) \right)^2. \tag{6}$$

In Eq. 6, the sum is taken over $l_c \in \mathcal{N}_k(S, T_2)$ (the k closest common leaves to S in T_2), thus k sets the number of terms contributing to $f_e(x)$ and controls how local versus broader neighborhood information influences the placement. Since $f_e(x)$ aggregates k leaf-wise discrepancies, increasing k reduces sensitivity to any single leaf while increasing computation proportionally to k.

The function $f_e(x)$ is quadratic in x, and its minimum can be found by taking the derivative with respect to x, setting the derivative to zero, and solving for the optimal x_e^*. The resulting value x_e^* determines the candidate insertion point on the branch as $v_e(x_e^*)$.

For each original branch e in T_1, as represented in T_1', its unique per-branch minimizer x_e^* and minimum value $f_e(x_e^*)$ are computed. Let $f_{\min}$ be the smallest $f_e(x_e^*)$ among all original branches. Among all candidates that attain $f_{\min}$, the insertion point is chosen by minimizing $d^{(T_1')}(lca_{T_1'}(L(T_1'), v_e(x_e^*)))$. If a tie remains, the candidate on the original branch with the smallest fixed rank idx(e) is chosen. This yields a unique insertion point. Then S with adjusted branch lengths is placed at this insertion point (see Fig. 1).

This process is repeated for each maximal distinct-leaf subtree $S \in \mathcal{S}_{T_2}$ until all have been inserted into T_1', resulting in the final completed tree $T_1^{\uplus}$. The same

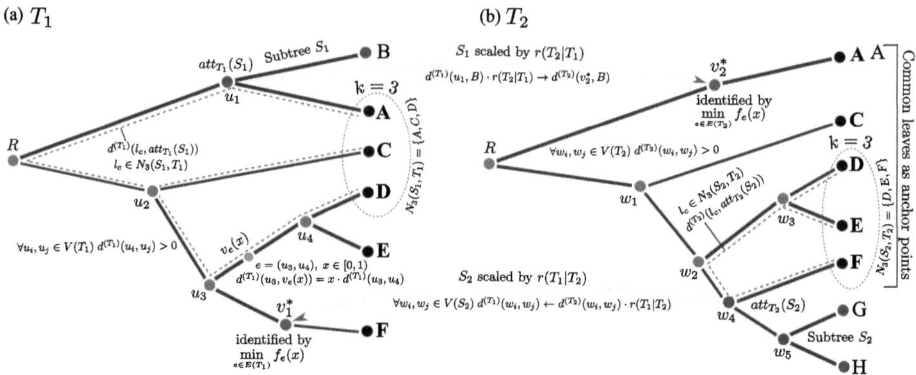

Fig. 1. Illustration of the main parts of the k-NCL algorithm. The common leaves of trees T_1 and T_2 are marked in bold black. The distinct leaves of each tree are highlighted in blue. Tree T_1 has one one-leaf maximal distinct-leaf subtree {B} (Color figure online). Tree T_2 includes one two-leaf maximal distinct-leaf subtree {G, H}. The value $k = 3$ is chosen. The dashed blue lines represent the position distances from the attachment point of a maximal distinct-leaf subtree to the selected nearest common leaves. The red nodes v_1^* and v_2^* are the insertion points for the corresponding subtrees. The gray candidate node $v_e(x)$ in T_1 represents an example of the parameterization setting used to find an optimal insertion point along the original branches. Arrows show the addition of maximal distinct-leaf subtrees, with adjusted branch lengths, from one tree to another.

procedure is then applied to each subtree from $\mathcal{S}_{T_1}$ for insertion into a copy of T_2 to construct the completed tree $T_2^{\uplus}$. The algorithm outputs two completed phylogenetic trees, $T_1^{\uplus}$ and $T_2^{\uplus}$, each defined on the union set of taxa of the initial trees $L(T_1^{\uplus}) = L(T_2^{\uplus}) = L(T_1) \cup L(T_2)$. A practical illustration of the algorithm is provided in Section B of the Appendix.

3.2 Properties

The following theorem, lemma, and propositions represent several properties of the k-NCL tree completion algorithm. Due to the page constraints, the detailed proofs are provided in Section A of the Appendix.

Theorem 1 (Time Complexity for Fixed k). *Let T_1 and T_2 be phylogenetic trees defined on different but overlapping sets of taxa with $L(T_1)$ and $L(T_2)$, respectively. For a fixed $k \in [2, |CL(T_1, T_2)|]$, the k-NCL algorithm completes both trees on the combined taxa set $L(T_1) \cup L(T_2)$ in $O(n^2)$, where $n = |L(T_1) \cup L(T_2)|$.*

Lemma 1 (Time Complexity for Arbitrary k). *For any arbitrary value $k \in [2, |CL(T_1, T_2)|]$ (i.e., k is not fixed), the k-NCL algorithm completes both trees on $L(T_1) \cup L(T_2)$ in $O(n^2 k)$ time.*

Proposition 1 (Preservation of Original Distances and Topology). *Let T_1 and T_2 be rooted phylogenetic trees with branch lengths, defined on overlapping leaf sets $L(T_1)$ and $L(T_2)$, respectively. Let $T_i^{\uplus}$ denote the completed tree produced by the k-NCL algorithm for $i \in \{1, 2\}$. Then, for each i, the completed tree $T_i^{\uplus}$ preserves both the pairwise distances and the topology of the original tree T_i. That is, for every pair of original leaves $l_a, l_b \in L(T_i)$, $d^{(T_i^{\uplus})}(l_a, l_b) = d^{(T_i)}(l_a, l_b)$. Furthermore, the completion process preserves the structure of the input tree in the sense that removing the added leaves from $T_i^{\uplus}$ recovers a tree topologically isomorphic to T_i.*

Proposition 2 (Symmetry in Tree Completion). *Let T_1 and T_2 be two phylogenetic trees defined on different but overlapping sets of taxa, and let $k \in [2, |CL(T_1, T_2)|]$ be fixed. The k-NCL tree completion process is symmetric with respect to the input trees. In particular, the algorithm inserts maximal distinct-leaf subtrees from T_2 into T_1 and, symmetrically, maximal distinct-leaf subtrees from T_1 into T_2. Therefore, applying the algorithm to either input order, (T_1, T_2) or (T_2, T_1), results in the same completed trees $T_1^{\uplus}$ and $T_2^{\uplus}$, appearing in the corresponding output order as $(T_1^{\uplus}, T_2^{\uplus})$ and $(T_2^{\uplus}, T_1^{\uplus})$.*

Proposition 3 (Uniqueness of Tree Completion in the k-NCL Algorithm). *Let T_1 and T_2 be input phylogenetic trees with strictly positive branch lengths, and let $k \in [2, |CL(T_1, T_2)|]$ be fixed. Then the completed trees $T_1^{\uplus}$ and $T_2^{\uplus}$ produced by the k-NCL algorithm are uniquely determined by T_1, T_2, and k, regardless of the order in which maximal distinct-leaf subtrees are inserted.*

Proposition 4 (Applicability to Binary and Non-binary Trees). *Let T_1 and T_2 be rooted phylogenetic trees with strictly positive branch lengths, defined on overlapping leaf sets $L(T_1)$ and $L(T_2)$, respectively. Trees T_1 and T_2 may be binary or non-binary (multifurcating). Let $k \in [2, |CL(T_1, T_2)|]$. Then, the k-NCL tree completion algorithm is well defined for (T_1, T_2, k) without requiring the trees to be binary. All results in Sect. 3.2 (polynomial-time complexity, preservation of original distances and topology, symmetry, and uniqueness) remain valid for both binary and non-binary input trees.*

4   Results and Discussion

4.1   Methodology for Biological Data Simulation

The methodology for obtaining biological data of phylogenetic trees with different but overlapping taxa involved several main steps. The biological data used in this evaluation part were sourced from the VertLife website [32], which offers a method for acquiring tree distributions with specific subsets of taxa. The tool prunes a comprehensive dataset to a smaller subset and samples trees from it.

Four different taxonomic groups were selected for analysis, including Amphibians, Birds, Mammals, and Sharks. The number of species in the entire datasets varied across groups, with 7239 species of Amphibians [12], 9993 species of Birds [13], 5911 species of Mammals [32], and 1192 species of Sharks [28]. Each group was first reduced to one representative species per genus. A random sample of n species was then drawn from this genus-filtered pool. From these selected species, ten overlapping subsets were generated, with overlap levels ranging from 10% to 90%. The corresponding pruned tree sets were then retrieved from VertLife and combined into the final overlapping datasets.

To introduce diversity into the datasets, different numbers of unique species were selected for each group. In this study, two versions of each biological dataset were prepared (see also Table 2 in the Appendix). Datasets A include 100 phylogenetic trees per group and are used for evaluating the influence of the parameter k in the k-NCL tree completion algorithm. Datasets B contain a larger number of trees (170 for Amphibians, 100 for Birds, 140 for Mammals, and 100 for Sharks) and are used for the comparative analysis of distance metrics between completed trees. These datasets, along with a more detailed description and the script used to prepare the data, can be found in the project repository on GitHub.

4.2   Evaluation Setup

In this part of the study, four distance metrics are applied, involving topology-only and topology and branch length comparisons, along with pruning and completion-based approaches. The Branch Score Distance (BSD) [14] is used in two variants in this section. BSD(k-NCL) measures the distance between trees completed using the k-NCL algorithm, while BSD(-) computes the distance after pruning both trees to their common taxa. Two versions of the RF distance [25] are utilized, each based on a different tree completion strategy. The first, RF($+$), is computed on trees completed using the method proposed in [35]. This metric only considers topological differences and ignores branch lengths, making it suitable for evaluating tree structure without regard to evolutionary distances. The second version, RF(k-NCL), applies the classical RF distance to trees completed using the k-NCL method. While the RF metric itself remains purely topological, this variant reflects how the k-NCL algorithm affects tree topology.

For this analysis, tree pairs are categorized by their levels of overlap, which are quantified using the Jaccard coefficient [23]. For the results in this section, overlap levels are binned into intervals of width 0.1, with each level (e.g., $\left\{\frac{i}{10} \mid i \in \mathbb{N}, 1 \leq i \leq 9\right\}$) representing the center of a bin. For instance, the level 0.1 corresponds to the interval $[0.05, 0.15)$, 0.2 to $[0.15, 0.25)$, continuing up to 0.9, which corresponds to $[0.85, 0.95)$.

4.3   Effect of the Parameter k

Let N_{cl} denote the number of common leaves, i.e., $N_{cl} = |CL(T_1, T_2)|$. For each tree pair, the BSD(k-NCL) distance is computed for a range of k values. These include edge cases such as $k \in [2, 3, N_{cl} - 1, N_{cl}]$, along with intermediate values, including $k = \lfloor \sqrt{N_{cl}} \rfloor$ and $k = \lfloor \frac{N_{cl}+2}{2} \rfloor$. Tree pairs are grouped into intervals based on overlap level, and for each group, the average BSD(k-NCL) is plotted as a function of k. This reveals how the distance changes with varying k values. These results are illustrated in Fig. 2.

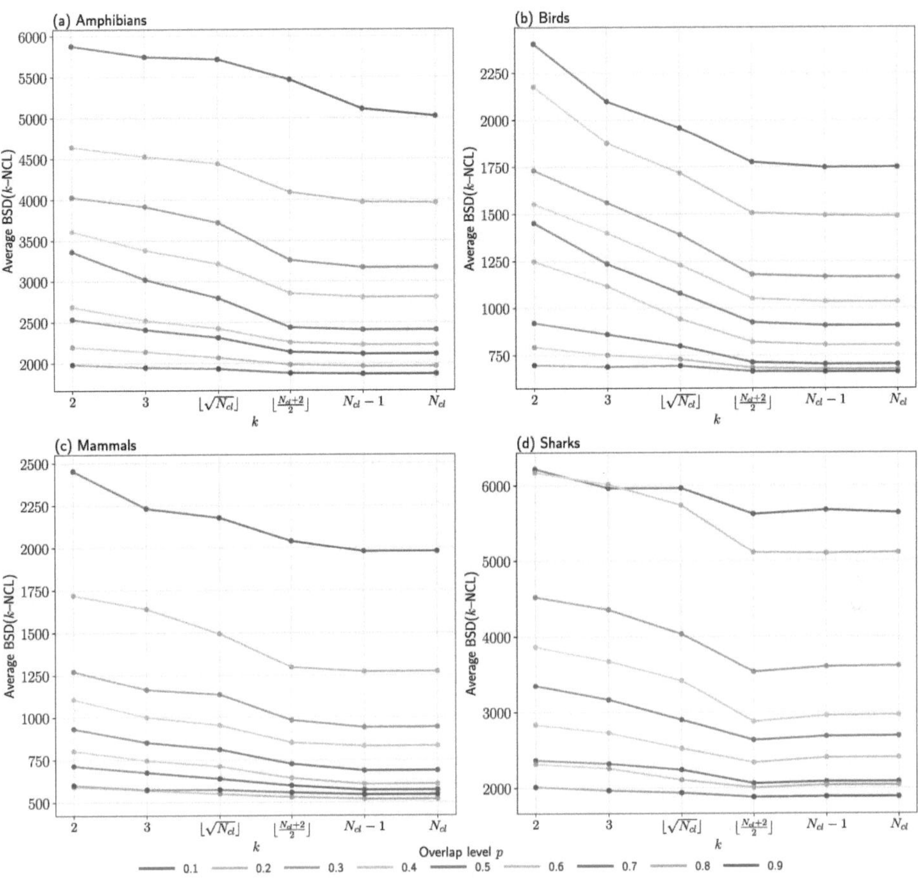

Fig. 2. Average BSD(k-NCL) distance versus k values for various levels of overlap (Datasets A). Edge cases ($k = [2, 3, N_{cl} - 1, N_{cl}]$) and intermediate cases ($k = \lfloor \sqrt{N_{cl}} \rfloor$ and $k = \lfloor \frac{N_{cl}+2}{2} \rfloor$) are used to complete tree pairs using the k-NCL algorithm.

The results show a consistent and decreasing trend in general. Across all four datasets, the average BSD(k-NCL) distance decreases as k increases. The decrease is gradual at first and then largely flattens as k approaches $\lfloor \frac{N_{cl}+2}{2} \rfloor$ and N_{cl}. This pattern holds across different overlap levels, with one exception for Sharks, where the lowest average BSD(k-NCL) occurs at $k = \lfloor \frac{N_{cl}+2}{2} \rfloor$ rather than at the largest k.

Analysis of the optimal k for individual tree pairs supports this trend. Empirically, $k = \lfloor \frac{N_{cl}+2}{2} \rfloor$ can be used as a suitable default. It achieves nearly minimal distances, and larger k provides only minor additional improvements, on average. Accordingly, $k = \lfloor \frac{N_{cl}+2}{2} \rfloor$ is used as the default value in subsequent evaluations in this study.

4.4   Comparison Between Completion and Pruning

In order to evaluate the agreement between BSD(k-NCL) and BSD(-), three scenarios are defined based on how each distance interprets the similarity of two trees, T_1 and T_2, relative to a reference tree T^*. A supertree with branch lengths was constructed as a reference tree for this evaluation part. The overall approach to obtaining a reference tree for each dataset involved first inferring the supertree topology, followed by assigning branch lengths by averaging values across consistent splits from the source trees. The spectral cluster supertree algorithm [20] was applied to build the supertree from a collection of input trees, each defined on different but overlapping subsets of taxa and containing branch length information. To assign branch lengths to the supertree, a post-processing step was performed. Each internal branch (split) of the inferred supertree was matched against the corresponding splits in the original input trees based on their bipartition structure. For every matching split identified, the associated branch lengths were recorded. The final branch length for each supertree branch was calculated as the arithmetic mean of all matching branch lengths across the source trees.

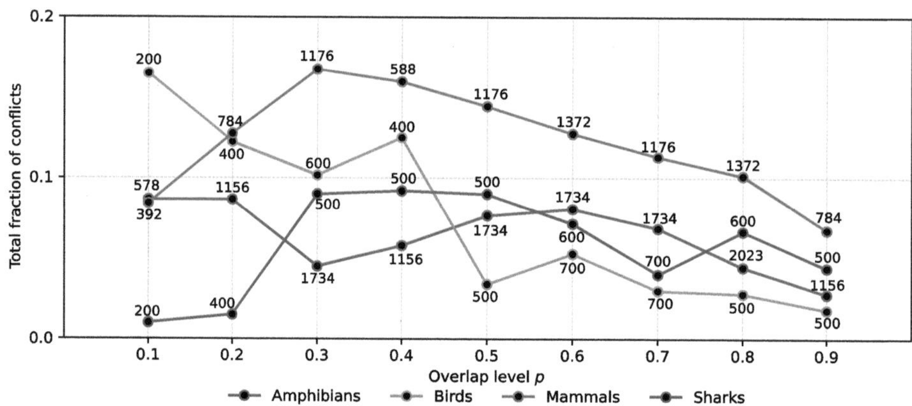

Fig. 3. Fraction of conflicting tree pairs for BSD(k-NCL) versus BSD(-) for all scenarios (Datasets B). $k = \lfloor \frac{N_{cl}+2}{2} \rfloor$ is used for BSD(k-NCL). Numbers above points show the number of tree pairs in each bin.

The following three scenarios are similar to the evaluation procedures described in [35]. Scenario 1 includes cases where the two metrics provide opposite conclusions. One metric indicates that T_1 is more similar to the reference tree T^*, while the other indicates that T_2 is more similar to T^*. Scenario 2 identifies cases where BSD(-) gives equal distances for both trees, but BSD(k-NCL) assigns different distances. This indicates that BSD(k-NCL) detects a difference between the trees that BSD(-) does not. Scenario 3 captures the opposite of Scenario 2, where BSD(k-NCL) gives equal distances, but BSD(-) differentiates the two trees. These scenarios are used to identify conflicting pairs of trees under each comparison, depending on their level of leaf overlap. Results comparing BSD(k-NCL) and BSD(-) are shown in Fig. 3.

Across all four datasets, 8.02% of tree pair comparisons revealed conflicts between BSD(k-NCL) (completion) and BSD(-) (pruning). Among those conflicts, 98.9% are Scenario 1, while Scenarios 2 and 3 contribute only 0.69% and 0.41% of conflicts, respectively. The conflict rates by dataset are as follows: 12.4% for Mammals, 6.2% for Amphibians, 6.5% for Birds, and 6.2% for Sharks. Scenario 1 dominates in each dataset (99.5% for Amphibians, 97.3% for Birds, 98.6% for Mammals, and 100% for Sharks).

Conflict levels vary with leaf overlap p. Aggregating across datasets, the conflict fraction peaks around $p = 0.4$ (9.7%), and falls to 3.9% by $p = 0.9$ from 8.6% at $p = 0.1$. Dataset specific patterns align with Fig. 3. Birds decline steadily from 16.5% at $p = 0.1$ to 1.8% at $p = 0.9$. Mammals show a peak of 16.8% at $p = 0.3$. Amphibians are generally lower, decreasing from 8.7% at $p = 0.1$ to 2.8% at $p = 0.9$. Sharks show a moderate peak at $p = 0.4$ (9.2%).

The median total conflict fraction is highest at $p = 0.4$ (the median is 10.8%) and decreases to 3.6% at $p = 0.9$, indicating both a reduction in conflict and convergence among datasets as overlap increases. It can be concluded that conflicts between completion-based BSD(k-NCL) and pruning-based BSD(-) are infrequent overall ($\approx 8\%$) and occur predominantly as Scenario 1. The risk of conflict is highest at low and medium levels of overlap ($p \leq 0.4$), notably for Mammals, and decreases with increasing overlap (the median across datasets is approximately $3\% - 4\%$ at $p = 0.9$). In practice, pruning can discard information when the overlap is limited. Therefore, BSD(k-NCL) may provide a more informative comparison at low and medium levels of overlap, whereas at high overlap ($p \geq 0.8$) the two metrics behave similarly.

4.5 Tree Completion Methods Comparison

In order to evaluate the proposed k-NCL tree completion method in comparison to the RF(+) tree completion approach, a clustering-based analysis was conducted. RF(+) was selected as the main external baseline because, to the best of our knowledge, it is the only previously published method outside our earlier method in [14] that directly addresses the tree completion problem, namely the construction of completed trees on the union of taxa from two overlapping input trees. The k-NCL method extends the approach introduced in [14]. Other related approaches either define distances that require identical taxon sets, or operate via pruning or implicit embeddings without explicitly returning completed trees. For this reason, RF(+) is the only direct external algorithmic baseline for tree completion considered here. It is important to note that both RF(+) and k-NCL attach maximal distinct-leaf subtrees, and the primary difference is the placement criterion. RF(+) is topology-driven, whereas k-NCL selects attachment points by minimizing a branch-length-based least-squares discrepancy. This comparison is designed to highlight the differences between topology-based and branch length-based tree completion outputs. The evaluation aims to quantify the ability of three tree distance metrics (RF(+), RF(k-NCL), and BSD(k-NCL)) to identify intended clusters from sets of partially overlapping phylogenetic trees of Amphibians, Birds, Mammals, and Sharks.

For each species group, five empirical base trees were selected, each defining one reference cluster. The base trees were selected such that the leaf set overlap across the resulting clusters ranged from 10% to 90%. From each base tree, four perturbed variants were simulated with AsymmeTree [27], resulting in five related trees per cluster and 25 trees per species group in total. This setup was designed as a controlled benchmark with known cluster membership, rather than as a simulation of fully independent empirical trees. Full simulation details and parameter settings are provided in the project repository.

For every unique tree pair within each cluster dataset, RF(+) and k-NCL completion methods are applied. RF distances are calculated for the trees completed with RF(+), and both RF(k-NCL) and BSD(k-NCL) distances are computed for trees completed with k-NCL ($k = \lfloor \frac{N_{cl}+2}{2} \rfloor$). The resulting pairwise distance matrices for each of the three metrics (RF(+), RF(k-NCL), and BSD(k-NCL)) are visualized in Fig. 4. Heatmaps and boxplots illustrate how well each tree completion method reveals the expected clustering structure within the datasets.

In the RF(+) row panels (a) to (d) of Fig. 4, the blocks are present, but they are less distinct. Panel (c), which represents Mammals, shows fuzzy boundaries with visible overlap between clusters. The RF(+) boxplots (e) also reveal overlap. In the RF(k-NCL) row panels (f) to (i), the block diagonals are clearly defined, with very light within-cluster cells and darker between-cluster bands. The corresponding boxplots (j) demonstrate a distinct gap. In the BSD(k-NCL) panels (k) to (n), the blocks are visible, however the edges are less distinct than in the RF (k-NCL) panels. The boxplots in panel (o) show a small but noticeable overlap.

Cluster separation was quantified from the 25×25 pairwise distance matrices for each method and species group, using the known clustering structure of five clusters, each consisting of five trees. The distributions of within-cluster and between-cluster distances were extracted for each method and species group. The silhouette method [26] and the Dunn index [9] were utilized to evaluate cluster separation. The silhouette coefficient for a single tree is computed from a distance matrix by defining a as the mean distance of that tree to all other trees within its own cluster, and b as the mean distance to all trees in the nearest other cluster. The coefficient is then given by $(b - a)/\max(a, b)$. The overall silhouette score is the average of these coefficients across all trees. The Dunn index equals the minimum inter-cluster distance divided by the maximum intra-cluster diameter. Larger values indicate better cluster separation for all metrics. These values are presented in Table 3 in the Appendix.

RF(k-NCL) shows perfect ordering of distances in all species groups and the largest silhouette scores and Dunn indices (Amphibians silhouette 0.765 and Dunn 1.588, Birds silhouette 0.826 and Dunn 2.654, Mammals silhouette 0.668 and Dunn 1.200, Sharks silhouette 0.734 and Dunn 1.281). The boxplots (Fig. 4)

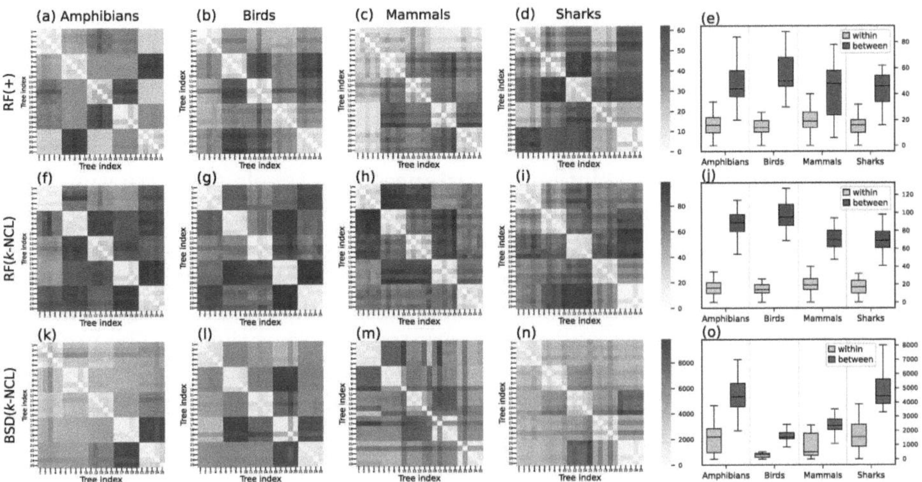

Fig. 4. Pairwise distance heatmaps and summary boxplots for four species groups anthree tree distance metrics. Each row corresponds to a distance metric, including RF(+), RF(k-NCL), and BSD(k-NCL). The first four columns correspond to a species group of Amphibians, Birds, Mammals, and Sharks. There are 25 trees per group, arranged into five clusters of five trees. A value of $k = \lfloor \frac{N_{cl}+2}{2} \rfloor$ was used for tree completion in RF(k-NCL) and BSD(k-NCL).

display a gap between the light within-cluster boxes and the dark between-cluster boxes under RF(k-NCL) with no overlap. BSD(k-NCL) provides clear separation that is close to RF(k-NCL) in Birds and Sharks, but with moderate overlap and lower silhouettes and Dunn indices (Birds silhouette 0.724 versus 0.826 and Dunn 0.866 versus 2.654, Sharks silhouette 0.593 versus 0.734 and Dunn 0.847 versus 1.281), while still outperforming RF(+) in Mammals (silhouette 0.521 versus −0.076 and Dunn 0.456 versus 0.150). RF(+) shows good results for Birds, but demonstrates noticeable overlap in other groups and weaker separation of Mammal clusters as indicated by a negative silhouette and the smallest Dunn value (Birds silhouette 0.645 and Dunn 1.154, Mammals silhouette −0.076 and Dunn 0.150). The most robust method for cluster identification and separation in these data is RF(k-NCL) overall and within Amphibians, Birds, Mammals, and Sharks (overall silhouette 0.748 and Dunn 1.681). BSD(k-NCL) is a second choice, and RF(+) ranks third (overall silhouette 0.599 and Dunn 0.675 for BSD(k-NCL) and overall silhouette 0.344 and Dunn 0.598 for RF(+)).

These results suggest that the k-NCL method, by integrating both topology and branch lengths, can enhance cluster separation for overlapping trees. Furthermore, both topology-based (RF) and branch length-based (BSD) distances benefit from the tree completion results provided by the k-NCL method.

5 Conclusion

This study presents the k-Nearest Common Leaves (k-NCL) algorithm, designed for completing phylogenetic tree pairs that are defined on different but overlapping taxa. The method uses branch lengths and topology to insert non-common leaves while preserving evolutionary distances and topological structure. The resulting completions extend both trees to the same taxon set while retaining their original structure. The algorithm runs in $O(n^2)$ time in the size of the union of leaf sets (Theorem 1) and satisfies the preservation of original distances and topology (Proposition 1), symmetry (Proposition 2), and uniqueness (Proposition 3). It applies to binary and non-binary trees (Proposition 4). The proposed algorithm is evaluated on datasets of amphibians, birds, mammals, and sharks (Sect. 4.1), and improves phylogenetic tree clustering compared to RF(+) completions (Sect. 4.5).

Future work will include additional analysis of how the choice of the parameter k affects the quality of tree completion by relating it to a purely topological distance metric and studying its impact on the robustness and stability of the tree completions. On the algorithmic side, further optimizations of the algorithm to improve scalability, as well as extensions to completing collections of more than two trees, are promising directions. In particular, the approach can be extended to a set of partially overlapping phylogenetic trees in order to obtain a completed set, where each tree is defined on a combined set of taxa. Finally, additional evaluation experiments on larger and more diverse datasets will help to better characterize the behavior of the k-NCL method across a wide range of evolutionary scenarios.

References

1. Bansal, M.S.: Linear-time algorithms for phylogenetic tree completion under robinson-foulds distance. Algorithms Mol. Biol. **15**, 1–15 (2020)
2. Bansal, M.S., Burleigh, J.G., Eulenstein, O., Fernández-Baca, D.: Robinson-foulds supertrees. Algorithms. Mol. Biol. **5**(1), 1–12 (2010)
3. Bender, M.A., Farach-Colton, M.: The LCA problem revisited. In: Latin American Symposium, Punta del Este, Uruguay, 10-14 April 2000 Proceedings 4, pp. 88–94. Springer (2000)
4. Billera, L.J., Holmes, S.P., Vogtmann, K.: Geometry of the space of phylogenetic trees. Adv. Appl. Math. **27**(4), 733–767 (2001)
5. Chen, D., Burleigh, J.G., Bansal, M.S., Fernández-Baca, D.: Phylofinder: an intelligent search engine for phylogenetic tree databases. BMC Evol. Biol. **8**, 1–11 (2008)
6. Christensen, S., Molloy, E.K., Vachaspati, P., Warnow, T.: Octal: optimal completion of gene trees in polynomial time. Algorithms Mol. Biol. **13**(1), 1–18 (2018)
7. Cotton, J.A., Wilkinson, M.: Majority-rule supertrees. Syst. Biol. **56**(3), 445–452 (2007)
8. Dong, J., Fernández-Baca, D.: Properties of majority-rule supertrees. Syst. Biol. **58**(3), 360–367 (2009)
9. Dunn, J.C.: Well-separated clusters and optimal fuzzy partitions. J. Cybern. **4**(1), 95–104 (1974)

10. Grindstaff, G., Owen, M.: Geometric comparison of phylogenetic trees with different leaf sets. arXiv preprint arXiv:1807.04235 (2018)
11. Hinchliff, C.E., et al.: Synthesis of phylogeny and taxonomy into a comprehensive tree of life. PNAS **112**(41), 12764–12769 (2015)
12. Jetz, W., Pyron, R.A.: The interplay of past diversification and evolutionary isolation with present imperilment across the amphibian tree of life. Nat. Ecol. Evol. **2**(5), 850–858 (2018)
13. Jetz, W., Thomas, G.H., Joy, J.B., Hartmann, K., Mooers, A.O.: The global diversity of birds in space and time. Nature **491**(7424), 444–448 (2012)
14. Koshkarov, A., Tahiri, N.: Novel algorithm for comparing phylogenetic trees with different but overlapping taxa. Symmetry **16**(7), 790 (2024)
15. Kupczok, A.: Split-based computation of majority-rule supertrees. BMC Evol. Biol. **11**(1), 1–13 (2011)
16. Li, W., Koshkarov, A., Tahiri, N.: Comparison of phylogenetic trees defined on different but mutually overlapping sets of taxa: a review. Ecol. Evol. **14**(8), e70054 (2024)
17. Llabrés, M., Rosselló, F., Valiente, G.: The generalized robinson-foulds distance for phylogenetic trees. J. Comput. Biol. **28**(12), 1181–1195 (2021)
18. Mahbub, S., Sawmya, S., Saha, A., Reaz, R., Rahman, M.S., Bayzid, M.S.: Quartet based gene tree imputation using deep learning improves phylogenomic analyses despite missing data. J. Comput. Biol. **29**(11), 1156–1172 (2022)
19. Mai, U., Mirarab, S.: Completing gene trees without species trees in sub-quadratic time. Bioinformatics **38**(6), 1532–1541 (2022)
20. McArthur, R.N., Zehmakan, A.N., Charleston, M.A., Lin, Y., Huttley, G.: Spectral cluster supertree: fast and statistically robust merging of rooted phylogenetic trees. Front. Mol. Biosci. **11**, 1432495 (2024)
21. Priel, A., Tamir, B.: A vectorial tree distance measure. Sci. Rep. **12**(1), 5256 (2022)
22. Rabiee, M., Mirarab, S.: Instral: discordance-aware phylogenetic placement using quartet scores. Syst. Biol. **69**(2), 384–391 (2020)
23. Real, R., Vargas, J.M.: The probabilistic basis of jaccard's index of similarity. Syst. Biol. **45**(3), 380–385 (1996)
24. Ren, Y., et al.: A combinatorial method for connecting BHV spaces representing different numbers of taxa. arXiv preprint arXiv:1708.02626 (2017)
25. Robinson, D.F., Foulds, L.R.: Comparison of phylogenetic trees. Math. Biosci. **53**, 131–147 (1981)
26. Rousseeuw, P.J.: Silhouettes: a graphical aid to the interpretation and validation of cluster analysis. J. Comput. Appl. Math. **20**, 53–65 (1987)
27. Schaller, D., Hellmuth, M., Stadler, P.F.: Asymmetree: a flexible python package for the simulation of complex gene family histories. Software **1**(3), 276–298 (2022)
28. Stein, R.W., et al.: Global priorities for conserving the evolutionary history of sharks, rays and chimaeras. Nat. Ecol. Evol. **2**(2), 288–298 (2018)
29. Tahiri, N., Fichet, B., Makarenkov, V.: Building alternative consensus trees and supertrees using k-means and robinson and foulds distance. Bioinformatics **38**(13), 3367–3376 (2022)
30. Tahiri, N., Willems, M., Makarenkov, V.: A new fast method for inferring multiple consensus trees using k-medoids. BMC Evol. Biol. **18**, 1–12 (2018)
31. Thorup, M., Zwick, U.: Approximate distance oracles. J. ACM **52**(1), 1–24 (2005)
32. Upham, N.S., Esselstyn, J.A., Jetz, W.: Inferring the mammal tree: species-level sets of phylogenies for questions in ecology, evolution, and conservation. PLoS Biol. **17**(12), e3000494 (2019)

33. Wang, J.T., Shan, H., Shasha, D., Piel, W.H.: Fast structural search in phylogenetic databases. Evol. Bioinform. **1**, 117693430500100020 (2005)
34. Whidden, C., Zeh, N., Beiko, R.G.: Supertrees based on the subtree prune-and-regraft distance. Syst. Biol. **63**(4), 566–581 (2014)
35. Yao, K., Bansal, M.S.: Optimal completion and comparison of incomplete phylogenetic trees under robinson-foulds distance. In: 32nd Annual Symposium on Combinatorial Pattern Matching (CPM 2021). Schloss Dagstuhl-Leibniz-Zentrum für Informatik (2021)
36. Yasui, N., Vogiatzis, C., Yoshida, R., Fukumizu, K.: Imphy: imputing phylogenetic trees with missing information using mathematical programming. IEEE/ACM Trans. Comput. Biol. Bioinf. **17**(4), 1222–1230 (2018)
37. Yoshida, R.: Imputing phylogenetic trees using tropical polytopes over the space of phylogenetic trees. Mathematics **11**(15), 3419 (2023)

Cherry-Picking Distance Between Binary Orchards Parameterized by Level

Manuel Lafond[1], Kaari Landry[2(✉)], Olivier Tremblay-Savard[3],
Hamza Wahed[4], Christopher Whidden[4], and Norbert Zeh[4]

[1] Université de Sherbrooke, Sherbrooke, Canada
`manuel.lafond@USherbrooke.ca`
[2] Université de Montréal, Montréal, Canada
`kaari.landry@umontreal.ca`
[3] University of Manitoba, Winnipeg, Canada
`olivier.tremblay-savard@umanitoba.ca`
[4] Dalhousie University, Halifax, Canada
`{hm682264,cwhidden,nzeh}@dal.ca`

Abstract. Phylogenetic networks capture complex evolutionary relationships, including hybridization and gene flow, that are not easily represented by tree-based models. Orchards are a class of phylogenetic networks that have proven valuable for reconstructing evolution from ancestral profiles, and in which reticulation arcs model lateral gene transfers. Several methods exist to infer orchards, and distance measures between networks can be used to evaluate them against simulations or gold-standard datasets. One such measure is the maximum agreement cherry-reduced subnetwork (MACRS) of two orchards. Recently, Landry et al. proved that finding such a MACRS of two binary level-1 orchards is fixed-parameter tractable (FPT) with respect to the number of reticulations in both networks. In this paper, we show that this problem can be solved on arbitrary binary orchards, in FPT time with respect to the maximum level of the two networks. In particular, a MACRS of two binary level-1 orchards can be found in polynomial time.

Keywords: Orchards · Level-k networks · Fixed-parameter tractability · Cherry covers · Network distances

1 Introduction

Phylogenetic trees are the traditional model for representing the evolutionary history of a set of taxa by depicting speciation events. However, there are other events that can also drive evolution, such as hybridization, lateral gene transfer, and recombination. For instance, hybridization can produce new species like Triticale, a wheat-rye hybrid [9], while recombination can generate new viral strains through genetic reassortment, as seen with influenza viruses mixing RNA segments in co-infected cells [2]. To represent such non-tree-like events in the evolution of a set of taxa, phylogenetic networks have been introduced, which can include vertices with more than one parent, called *reticulations*.

© The Author(s), under exclusive license to Springer Nature Switzerland AG 2026
M. Lafond (Ed.): RECOMB-CG 2026, LNBI 16569, pp. 283–314, 2026.
https://doi.org/10.1007/978-3-032-26891-4_14

A cherry in a network is a pair of leaves that either have a common parent (a *proper cherry*) or whose parents are connected by a reticulation arc (a *reticulated cherry*). A *cherry-picking operation* removes one cherry from the network, perhaps exposing others. A phylogenetic network reducible to a single leaf by a sequence of such operations is called an orchard. This network class is mathematically attractive, for example, because it takes linear time to decide whether a network is an orchard [6] or whether two orchards are isomorphic [14].

Orchards are also biologically relevant, as they were initially introduced because they can be uniquely reconstructed from their ancestral profiles (see [6]). They can also model horizontal gene transfer, and have in fact been characterized as "trees with additional horizontal arcs" [10]. There are thus several algorithms and software that construct orchards. For instance, methods that add time-consistent transfers to a species tree based on reconciled gene trees [1,13], or based on the presence/absence of character traits [21,22], result in an orchard.

Evaluating the accuracy of such methods typically consists of simulating networks and comparing them against the constructed networks, but there are no established metrics to compare networks, unlike for trees, which have the RF metric [20,23]. There has thus been research on designing (dis)similarity measures between orchards specifically, one example being the extended μ-distance from [5]. In this paper, we focus on another measure, the *maximum agreement cherry-reduced subnetwork* (MACRS), which is the largest common subnetwork that can be obtained from both networks through cherry-picking operations [17]. The distance itself is the quantity of such operations, of which there is a correspondence with the size of the MACRS (Theorem 4 of [17]).

Landry et al. [17] proved that finding the MACRS of two orchards is NP-hard, which raises the question whether it is fixed-parameter tractable (FPT) with respect to common parameters, i.e., whether it can be solved in a running time that depends exponentially only on that parameter. A popular network parameter is the *level*, defined as the maximum number of reticulations in any 2-edge-connected component (2ECC) of the network. The level can be expected to be small in practice [7], and many NP-hard problems are FPT in the level (see for instance [16,25]). Other parameters of interest include the treewidth [12,24], scanwidth [3,4,15], or just the total number of reticulations.

In this paper, we show that the MACRS problem on binary orchards is FPT with respect to level. This solves an open problem by Landry et al. [19], who proved that finding a MACRS of two binary level-1 orchards is FPT with respect to the number of reticulations in the networks. This algorithm was shown to be empirically efficient [18], but practical use is currently limited to level-1 networks.

In comparison, our algorithm applies to *arbitrary* binary orchards and is exponential only in the level ℓ, which is upper-bounded by r and is often much smaller. In particular, our result shows that a MACRS of two level-1 orchards can be found in polynomial time. The algorithm by Landry et al. guessed the reticulated cherries globally, for the entire network, which resulted in a running time of $O(3^r \cdot \text{poly}(n))$. Our contributions can be summarized as follows:

- We show that the reticulated cherries to pick can be guessed one 2ECC at a time, which combined with dynamic programming leads to 3^ℓ guesses per 2ECC instead of 3^r for the whole network.
- For a pair of guesses on two corresponding 2ECCs in the two networks, we must also guess an isomorphism between them to know how vertices are "matched" in an agreement subnetwork. A naive analysis would result in $O(\ell!)$ such isomorphisms, but we show that there are at most $2^{3\ell/2}$ isomorphisms between two 2ECCs of arbitrary networks. This result may be of independent interest for the computation of other (dis)similarity measures.
- To prove the correctness of our algorithm, we generalize the notion of a cherry cover of an orchard [11] to define *partial cherry covers*, which allow only part of the network to be covered with cherry paths. We prove important properties of these partial cherry covers that may be of independent interest.

By combining these techniques, we obtain an $O(14.5^\ell \cdot n^3)$-time algorithm, as developed in Theorem 6 and improved in Appendix A.

2 Preliminaries

In this section, we introduce the necessary terminology and notation regarding networks, cherry picking sequences, and cherry partitions, including preliminary results. Due to space limitations, all proofs can be found in the appendix.

Phylogenetic Networks. A *phylogenetic network* $\mathcal{N}$ (*network* for short) is a directed acyclic graph with a single vertex of in-degree 0, called the *root*, whose out-degree-0 vertices, called *leaves*, have in-degree 1, and whose non-root, non-leaf vertices either have in-degree 1 and out-degree at least 2 (*tree vertices*) or out-degree 1 and in-degree at least 2 (*reticulations*). In the literature, the leaves of a network are labelled bijectively with the elements of some label set X, generally viewed to represent extant taxa. For reasons that will become clear shortly, leaves are labelled with *disjoint non-empty subsets* of X in this paper, as was already done in [17,19]. For a leaf v, we use $X(v)$ to denote its label set.

A network is *binary* if all vertices have in-degree at most 2 and out-degree at most 2. This paper is concerned only with binary networks.

We denote the sets of vertices, arcs, and leaves of a network $\mathcal{N}$ by $V(\mathcal{N})$, $E(\mathcal{N})$, and $\mathcal{L}(\mathcal{N})$, respectively, and extend this notation in the obvious way to subgraphs, sets of arcs, paths, and sets of paths in a network. A vertex v is a *child, parent, descendant* or *ancestor* of another vertex u in a network $\mathcal{N}$ if (u, v) is an arc of $\mathcal{N}$, (v, u) is an arc of $\mathcal{N}$, there exists a directed path from u to v in $\mathcal{N}$ or there exists a directed path from v to u in $\mathcal{N}$, respectively. When v is a non-root leaf or tree vertex, then its parent is unique, and we refer to it as $p(v)$. We extend the definition of $X(v)$ to whole networks and internal vertices of networks by defining $X(v)$ to be the union of the label sets of all descendant leaves of v, and $X(\mathcal{N}) = X(\rho)$, where ρ is the root of $\mathcal{N}$. We call an arc $(u, v) \in E(\mathcal{N})$ a *tree arc* if v is a tree vertex or leaf, and a *reticulation arc* if v is a reticulation.

An undirected graph is *2-edge-connected* if it cannot be disconnected by removing a single edge. A *2-edge connected component* (2ECC) of an undirected graph G is a maximal 2-edge-connected subgraph. An edge of G is a *bridge* if its endpoints belong to different 2ECCs of G. The 2ECCs of a network $\mathcal{N}$ are the 2ECCs of its underlying undirected graph, which is the graph obtained from $\mathcal{N}$ by forgetting the arc directions. We call a 2ECC *trivial* if it consists of a single vertex, and *non-trivial* otherwise. Note that every 2ECC in a network has a single root [8]. For any such root v of a 2ECC, let $\mathcal{N}(v)$ be the subnetwork of N induced by all descendants of v. (If v is not a root of a 2ECC, then the subgraph induced by v's descendants is not a network, as it contains vertices of in-degree 1 and out-degree 1.) The *level* of a network $\mathcal{N}$ is the maximum number of reticulations in its 2ECCs. The top-left network in Fig. 1 is level-2.

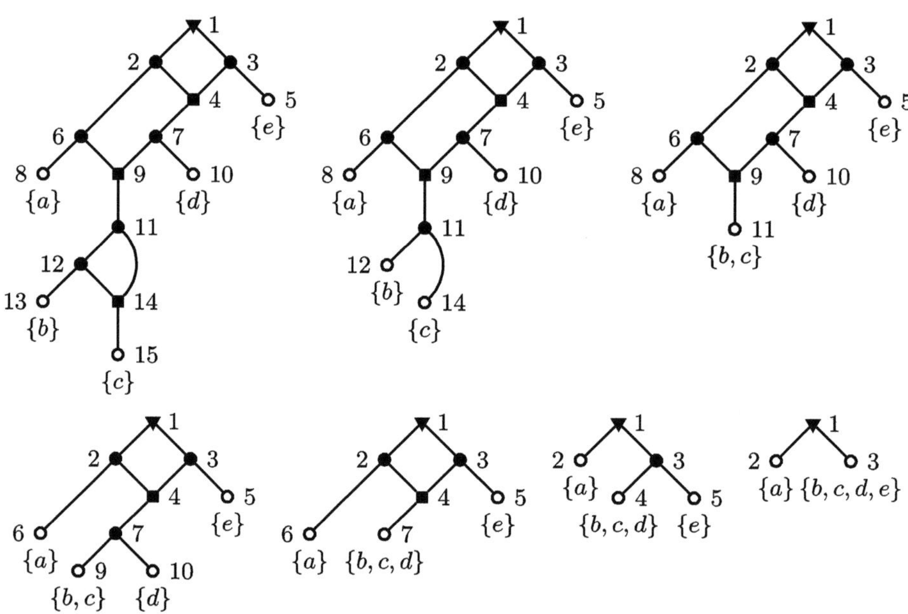

Fig. 1. An illustration of a CPS operating on a level-2 orchard. Leaves are shown as open circles; non-root tree vertices, as filled circles; reticulations, as filled squares; and the root, as a filled, inverted triangle. The sequence of networks produced from the top-left level-2 network by the CPS $\langle (15, 13), (12, 14), (11, 8), (9, 10), (7, 6), (4, 5) \rangle$. Picking the final cherry $(2, 3)$ will result in a network consisting of a single vertex 1 labelled by $\{a, b, c, d, e\}$. Thus, the top-left network is orchard.

In this paper, a *blob* refers to a subgraph of a network $\mathcal{N}$ consisting of a 2ECC of $\mathcal{N}$ and all its pendant bridges (excluding the bridge formed by the root of the 2ECC and its parent). So, blobs are edge disjoint but not necessarily vertex disjoint. A blob can be viewed as a network in which each endpoint v of a pendant bridge is labelled by $X(v)$. A blob is trivial or non-trivial if the 2ECC from which it is constructed is trivial or non-trivial. Note that a leaf of $\mathcal{N}$ forms a trivial 2ECC, which has no pendant bridges. Each such 2ECC still defines a blob, which is identical to its 2ECC. The blobs of $\mathcal{N}$ form a partition of the arcs of $\mathcal{N}$. Every non-leaf vertex of a blob has all its child arcs in the blob, and every non-root vertex has all its parent arcs in the blob.

An isomorphism $\phi = (\phi_V, \phi_E)$ between two directed graphs G_1 and G_2 consists of two set isomorphisms $\phi_V : V(G_1) \rightarrow V(G_2)$ and $\phi_E : E(G_1) \rightarrow E(G_2)$ such that $\phi_E((u,v)) = (\phi_V(u), \phi_V(v))$ for every arc $(u,v) \in E(G_1)$. In the remainder of this paper, we refer to both ϕ_V and ϕ_E simply as ϕ if no confusion can arise. An isomorphism between two networks $\mathcal{N}_1$ and $\mathcal{N}_2$ is a graph isomorphism that additionally respects leaf labels. Since leaves of networks are labelled with subsets of X in this paper, we require a network isomorphism to satisfy the weaker condition that $X(v) \cap X(\phi(v)) \neq \emptyset$ for every leaf of $\mathcal{N}_1$. Landry et al. [19] called this a *weak isomorphism* and showed that using it in the definition of a MACRS, to be recalled shortly, captures exactly the idea that a MACRS can be obtained from two networks whose elements are labelled with individual elements of X by applying sequences of cherry picking operations to them.

Cherry-Picking Sequences, Orchards, and (Agreement) Cherry-Reduced Subnetworks. A *proper cherry* in a network $\mathcal{N}$ is a pair of leaves (a,b) with a common parent. A *reticulated cherry* is a pair of leaves (a,b) such that $(p(b), p(a))$ is a reticulation arc in $\mathcal{N}$. We say *cherry* to refer to a proper or reticulated cherry. Picking a proper cherry (a,b) of $\mathcal{N}$ consists of removing the leaves a and b along with their parent arcs. This makes their common parent a leaf, which is assigned the label set $X(a) \cup X(b)$. Picking a reticulated cherry (a,b) of $\mathcal{N}$ consists of removing the leaves a and b along with their parent arcs and the arc $(p(b), p(a))$. This makes $p(a)$ and $p(b)$ leaves which we assign the label sets $X(a)$ and $X(b)$, respectively.

We call an ordered sequence of cherries $S = \langle (a_1, b_1), \ldots, (a_n, b_n) \rangle$ a *cherry-picking sequence* (CPS) of a network $\mathcal{N}$ if there exists a sequence of networks $\mathcal{N}[0], \mathcal{N}[1], \ldots, \mathcal{N}[n]$ with the following properties:

- $\mathcal{N}[0] = \mathcal{N}$
- (a_i, b_i) is a cherry of $\mathcal{N}[i-1]$, for all $i \in [n]$.
- Picking (a_i, b_i) in $\mathcal{N}[i-1]$ produces $\mathcal{N}[i]$, for all $i \in [n]$.

In this case, we refer to the final network $\mathcal{N}_n$ in this sequence as $\mathcal{N} \wedge S$. We call a CPS *complete* if $\mathcal{N}_n$ has a single leaf, and *partial* otherwise. A network is an *orchard* if it has a complete CPS, an example of which is show in Fig. 1.

A network $\mathcal{N}'$ is a *cherry-reduced subnetwork* (CRS) of another network $\mathcal{N}$ if there exists a CPS S of $\mathcal{N}$ such that there is a network isomorphism between $\mathcal{N}'$ and $\mathcal{N} \wedge S$. A network is an *agreement cherry reduced subnetwork* (ACRS) of two networks $\mathcal{N}_1$ and $\mathcal{N}_2$ if it is a CRS of both $\mathcal{N}_1$ and $\mathcal{N}_2$. It is a *maximum agreement cherry-reduced subnetwork* (MACRS) of $\mathcal{N}_1$ and $\mathcal{N}_2$ if it has the maximum number of arcs among all ACRSs of $(\mathcal{N}_1, \mathcal{N}_2)$.

MAXIMUM AGREEMENT CHERRY-REDUCED SUBNETWORK PROBLEM

Input: Two networks $\mathcal{N}_1$ and $\mathcal{N}_2$

Find: A MACRS of $\mathcal{N}_1$ and $\mathcal{N}_2$

Cherry Partitions. For a network $\mathcal{N}$, we define two collections of *cherry paths* $C(\mathcal{N})$ and $R(\mathcal{N})$. A *proper cherry path* $\langle u, v, w \rangle$ consists of two arcs connecting two vertices u and w to their common parent v. In particular v is a tree vertex. A *reticulated cherry path* $\langle u, v, w, z \rangle$ consists of two arcs connecting two vertices v and w to their children u and z, respectively, and an arc from w to v. In particular, v is a reticulation, w is a tree vertex, and (w, v) is a reticulation arc. $C(\mathcal{N})$ is the collection of all cherry paths in $\mathcal{N}$, and $R(\mathcal{N}) \subseteq C(\mathcal{N})$ is the collection of all reticulated cherry paths in $\mathcal{N}$. The definition of $C(\mathcal{N})$ is illustrated in Fig. 2.

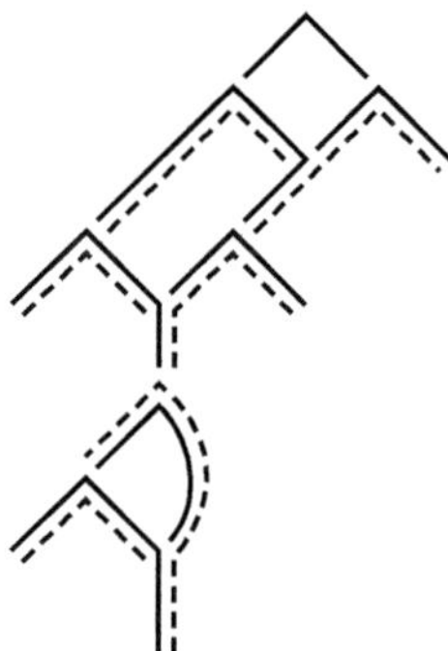

Fig. 2. The complete collection of cherry paths in the top-left network in Fig. 1 represented by both the solid and dashed paths. The solid paths form an acyclic cherry partition corresponding to the complete CPS described in Fig. 1.

A complete CPS $\langle (a_1, b_1), \ldots, (a_n, b_n) \rangle$ of a binary orchard $\mathcal{N}$ defines a partition $\mathcal{P} = \{P_1, \ldots, P_n\} \subseteq C(\mathcal{N})$ of the arcs of $\mathcal{N}$ into arc-disjoint paths, where each path P_i consists of the arcs removed when picking the cherry (a_i, b_i). As illustrated in Fig. 2, $P_1, \ldots, P_n$ are cherry paths. If (a_i, b_i) is a proper cherry of $\mathcal{N}_{i-1}$, then P_i is a proper cherry path; if (a_i, b_i) is a reticulated cherry of $\mathcal{N}_{i-1}$, then P_i is a reticulated cherry path. We call $\mathcal{P}$ a *complete cherry partition* of $\mathcal{N}$. We can define a graph $G_\mathcal{P}$ whose vertices are the paths in $\mathcal{P}$ and with an arc (P_i, P_j) if one of the endpoints of P_i is an internal vertex of P_j. Van Iersel et al. [11] proved that a network is an orchard if and only if $G_\mathcal{P}$ is acyclic. In this case, we call $\mathcal{P}$ an *acyclic complete cherry partition*. In particular, Van Iersel et al. proved that a complete CPS of $\mathcal{N}$ can be built from an acyclic complete cherry partition by picking cherries according to a topological ordering of $G_\mathcal{P}$. They also proved that orchards can be recognized in linear time by picking cherries greedily, i.e., the order in which cherries are picked is unimportant.

This notion extends naturally to partial cherry picking sequences: Given a partial cherry picking sequence $\langle (a_1, b_1), \ldots, (a_n, b_n) \rangle$ of a network $\mathcal{N}$, we can collect the associated cherry paths as for a complete cherry picking sequence to obtain a *partial cherry partition* $\mathcal{P} = \{P_1, \ldots, P_n\}$. This partition consists of arc-disjoint cherry paths such that each endpoint of a path P_i in $\mathcal{P}$ is either a leaf of $\mathcal{N}$ or is an internal vertex of another path $P_j \in \mathcal{P}$. Just as for complete cherry partitions, we can define the graph $G_{\mathcal{P}}$ and say that $\mathcal{P}$ is acyclic if and only if $G_{\mathcal{P}}$ is acyclic. For a network $\mathcal{N}$ and a partial cherry partition $\mathcal{P}$ of $\mathcal{N}$, let $\mathcal{N} - \mathcal{P}$ be the network obtained from $\mathcal{N}$ by deleting all arcs that belong to the paths in $\mathcal{P}$, as well as all isolated vertices this creates. For every leaf v of $\mathcal{N} - \mathcal{P}$, we define $X(v)$ as the union of the label sets of all leaves of $\mathcal{N}$ reachable from v via directed paths that do not contain any of the internal arcs of reticulated cherry paths in $\mathcal{P}$. By the next proposition, finding cherry reduced subnetworks of $\mathcal{N}$ amounts to finding appropriate partial cherry partitions of $\mathcal{N}$.

Proposition 1. *Given two binary networks $\mathcal{N}$ and $\mathcal{N}'$, there exists a cherry picking sequence S with $\mathcal{N} \wedge S = \mathcal{N}'$ if and only if there exists an acyclic partial cherry partition $\mathcal{P}$ of $\mathcal{N}$ such that $\mathcal{N} - \mathcal{P} = \mathcal{N}'$.*

The following lemma shows that the blobs of every orchard $\mathcal{N}$ are orchards and that, therefore, every acyclic cherry partition of $\mathcal{N}$ is the union of acyclic cherry partitions of its blobs.

Lemma 2. *Let $\mathcal{N}$ be a network with blobs $B_1, \ldots, B_t$. Then*

(i) $C(\mathcal{N}) = C(B_1) \cup \cdots \cup C(B_t)$ and $C(B_1), \ldots, C(B_t)$ are pairwise disjoint.
(ii) For any subset $\mathcal{P} \subseteq C(\mathcal{N})$, let $\mathcal{P}_i = \mathcal{P} \cap C(B_i)$, for all $i \in [t]$. Then $\mathcal{P}$ is an acyclic cherry partition of $\mathcal{N}$ if and only if each $\mathcal{P}_i$ is an acyclic cherry partition of B_i and, for each blob B_i whose root is part of some cherry path in $\mathcal{P}$, $\mathcal{P}_i$ is a complete cherry partition of B_i.
(iii) $\mathcal{N}$ is an orchard if and only if $B_1, \cdots, B_t$ are orchards.
(iv) If $\mathcal{P}$ is an acyclic cherry partition of $\mathcal{N}$, $\mathcal{N}' := \mathcal{N} - \mathcal{P}$, x is the root of a blob of $\mathcal{N}$, and $\mathcal{P}' := \mathcal{P} \cap C(\mathcal{N}(x))$, then $\mathcal{P}'$ is an acyclic cherry partition of $\mathcal{N}(x)$. If $x \in \mathcal{N}'$, then $\mathcal{N}'(x) = \mathcal{N}(x) - \mathcal{P}'$; if $x \notin \mathcal{N}'$, then $\mathcal{P}'$ is a complete cherry partition of $\mathcal{N}(x)$, so $\mathcal{N}(x) - \mathcal{P}'$ has x as its only vertex.

3 MACRS Algorithm on Level-k Orchards

In this section, we present our MACRS algorithm for level-k orchards. First, we introduce some necessary background and intuition. This is followed by the algorithm and a detailed description. Finally, we analyze the running time of the algorithm. The correctness proof is deferred to Sect. 4, which develops the recurrence relations used by the algorithm.

Background. A *reticulation selector* is a function $\sigma : R \to \{0, 1, \perp\}$, for some $R \supseteq R(\mathcal{N})$. We also use $\perp$ to denote the reticulation selector that maps every cherry path in R to $\perp$. A reticulation selector σ defines a family $\mathcal{R}_\sigma(\mathcal{N})$ of all subsets $R' \subseteq R(\mathcal{N})$ such that $\sigma^{-1}(1) \cap R(\mathcal{N}) \subseteq R'$ and $\sigma^{-1}(0) \cap R' = \emptyset$. We say that a cherry partition $\mathcal{P}$ is σ-*consistent* if $\mathcal{P} \cap R(\mathcal{N}) \in \mathcal{R}_\sigma(\mathcal{N})$. In words, a reticulation selector σ specifies reticulated cherry paths that must $(\sigma(P) = 1)$ or must not $(\sigma(P) = 0)$ be contained in any σ-consistent cherry partition, and may leave the choice for other paths unspecified $(\sigma(P) = \perp)$. A σ-*CRS* of $\mathcal{N}$ is a CRS $\mathcal{N} - \mathcal{P}$, where $\mathcal{P}$ is σ-consistent. Given two networks $\mathcal{N}_1$ and $\mathcal{N}_2$ and two reticulation selectors σ_1 and σ_2 in $\mathcal{N}_1$ and $\mathcal{N}_2$, respectively, a (σ_1, σ_2)-ACRS of $(\mathcal{N}_1, \mathcal{N}_2)$ is an ACRS $\mathcal{N}^*$ of $(\mathcal{N}_1, \mathcal{N}_2)$ that is a σ_1-CRS of $\mathcal{N}_1$ and a σ_2-CRS of $\mathcal{N}_2$. It is possible that no such ACRS exists. If it does, we define a (σ_1, σ_2)-MACRS to be a (σ_1, σ_2)-ACRS of maximum size (i.e., maximum number of vertices).

Clearly, every partial cherry partition of $\mathcal{N}$ is $\perp$-consistent, so every CRS of $\mathcal{N}$ is a $\perp$-CRS of $\mathcal{N}$, and every ACRS of a pair of networks $(\mathcal{N}_1, \mathcal{N}_2)$ is also a $(\perp, \perp)$-ACRS. Thus, our algorithm needs to search for a $(\perp, \perp)$-MACRS of $(\mathcal{N}_1, \mathcal{N}_2)$. As the search progresses, it will use reticulation selectors that fix more and more cherry paths to guide the search for such a $(\perp, \perp)$-MACRS of $(\mathcal{N}_1, \mathcal{N}_2)$.

We say that a reticulation selector σ *fixes* a path $P \in R(\mathcal{N})$ if $\sigma(P) \in \{0, 1\}$. We say that σ is *blob-respecting* if, for every blob B of $\mathcal{N}$, σ fixes either all or none of the paths in $R(B)$. We use $\mathcal{N}^\sigma$ to denote the σ-CRS such that *every σ-CRS of $\mathcal{N}$ is also a σ-CRS of $\mathcal{N}^\sigma$*. Intuitively, every σ-CRS of $\mathcal{N}$ can be obtained by first constructing $\mathcal{N}^\sigma$ and then picking additional cherries. We show that if σ is blob-respecting and any σ-CRS exists at all, then $\mathcal{N}^\sigma$ exists and is unique.

Lemma 3. *Let σ be a blob-respecting reticulation selector of a network $\mathcal{N}$. If there exists a σ-CRS of $\mathcal{N}$, then there exists a σ-CRS $\mathcal{N}^\sigma$ such that every σ-CRS of $\mathcal{N}$ is also a σ-CRS of $\mathcal{N}^\sigma$. This immediately implies that $\mathcal{N}^\sigma$ is unique. If all reticulated cherry paths fixed by σ belong to the same blob, then one can decide in linear time whether $\mathcal{N}^\sigma$ exists and, if so, construct it.*

We need notation to refer to the blob that contains each vertex v of a network $\mathcal{N}$. We use B_v to denote the unique blob of $\mathcal{N}$ such that v belongs to the underlying 2ECC of B_v. Analogously, for a reticulation selector σ such that $\mathcal{N}^\sigma$ exists and $v \in \mathcal{N}^\sigma$, we use B_v^σ to denote the unique blob of $\mathcal{N}^\sigma$ such that v belongs to the underlying 2ECC of B_v^σ. Note that $B_v^\sigma \subseteq B^v$ in this case. For a blob B of $\mathcal{N}$, we call a reticulation selector σ a B-*reticulation selector* if σ fixes no cherry path in $R(B')$, for any proper descendant blob B' of B, and it fixes all or none of the cherry paths in $R(B)$.

Algorithm. Our algorithm for finding a MACRS of a pair of networks $(\mathcal{N}_1, \mathcal{N}_2)$ is shown in Algorithm 1. Given two vertices $u \in N_1$ and $v \in N_2$, a B_u-reticulation selector σ_1, and a B_v-reticulation selector σ_2 such that u and v are roots of blobs in $\mathcal{N}_1^{\sigma_1}$ and $\mathcal{N}_2^{\sigma_2}$, respectively, the algorithm returns the size of a (σ_1, σ_2)-MACRS of $(\mathcal{N}_1^{\sigma_1}(u), \mathcal{N}_2^{\sigma_2}(v))$. Thus, when called with $u = \rho_1$, $v = \rho_2$, and

---

Algorithm 1. σ-MACRS

Data: $u \in \mathcal{N}_1$, $v \in \mathcal{N}_2$, B_u- and B_v-reticulation selectors σ_1, σ_2
Results: (σ_1, σ_2)-MACRS

1: **if** $D[u, v, \sigma_1, \sigma_2]$ is defined **then return** $D[u, v, \sigma_1, \sigma_2]$
2: **if** $\sigma_1 = \bot$ or $\sigma_2 = \bot$ **then**
3:      $M \leftarrow -\infty$
4:      **for all** σ_1' in $\mathtt{fix}(\sigma_1, B_u)$ **do**
5:          **for all** σ_2' in $\mathtt{fix}(\sigma_2, B_v)$ **do**
6:              $M \leftarrow \max(M,\ \sigma\text{-MACRS}(u, v, \sigma_1', \sigma_2'))$
7:      $D[u, v, \sigma_1, \sigma_2] \leftarrow M$
8: **else**
9:      $\mathcal{N}_1^{\sigma_1} \leftarrow \mathtt{reduce}(\mathcal{N}_1, \sigma_1)$; $\mathcal{N}_2^{\sigma_2} \leftarrow \mathtt{reduce}(\mathcal{N}_2, \sigma_2)$
10:      **if** $\mathcal{N}_1^{\sigma_1}$ is **null** or $\mathcal{N}_2^{\sigma_2}$ is **null then**
11:          $D[u, v, \sigma_1, \sigma_2] = -\infty$; **return** $D[u, v, \sigma_1, \sigma_2]$
12:      Let $B_u^{\sigma_1}$, $B_v^{\sigma_2}$ be the blobs containing u and v in $\mathcal{N}_1^{\sigma_1}$ and $\mathcal{N}_2^{\sigma_2}$
13:      **if subnets-are-reducible-and-overlap**$(u, v, \sigma_1, \sigma_2)$ **then**
14:          $D_1[u, v, \sigma_1, \sigma_2] \leftarrow 1$
15:      **else**
16:          $D_1[u, v, \sigma_1, \sigma_2] \leftarrow -\infty$
17:      $\Phi \leftarrow$ **generate-isomorphisms**$(B_u^{\sigma_1}, B_v^{\sigma_2})$
18:      $M \leftarrow -\infty$
19:      **for all** $\phi \in \Phi$ **do**
20:          $M' \leftarrow |B_u^{\sigma_1}|$
21:          **for all** $x \in \mathcal{L}(B_u^{\sigma_1})$ **do**
22:              $\sigma_1' \leftarrow \sigma_1$; $\sigma_2' \leftarrow \sigma_2$
23:              **if** $x \in \mathcal{L}(B_u)$ **then** $\sigma_1' \leftarrow \bot$
24:              **if** $\sigma(x) \in \mathcal{L}(B_v)$ **then** $\sigma_2' \leftarrow \bot$
25:              $M' \leftarrow M' + \sigma\text{-MACRS}(x, \phi(x), \sigma_1', \sigma_2')$
26:          $M \leftarrow \max(M, M')$
27:      $D_2[u, v, \sigma_1, \sigma_2] \leftarrow M$
28:      $D[u, v, \sigma_1, \sigma_2] \leftarrow \max\left(D_1[u, v, \sigma_1, \sigma_2], D_2[u, v, \sigma_1, \sigma_2]\right)$
29: **return** $D[u, v, \sigma_1, \sigma_2]$

$\sigma_1 = \sigma_2 = \bot$, it returns the size of a MACRS of $(\mathcal{N}_1, \mathcal{N}_2)$. An actual MACRS can be recovered using a standard backtracking approach. We defer a formal correctness proof of the algorithm to Sect. 4. The following gives its intuition.

The algorithm uses memoization by storing the result of each invocation in a table entry $D[u, v, \sigma_1, \sigma_2]$. If it is called with the same argument again, then line 1 of the algorithm simply returns the result computed before.

If σ_1 fixes none of the cherry paths in $R(B_u)$ or σ_2 fixes none of the cherry paths in $R(B_v)$, then it can be shown that a (σ_1, σ_2)-MACRS of $(\mathcal{N}_1^{\sigma_1}(u), \mathcal{N}_2^{\sigma_2}(v))$ is a (σ_1', σ_2')-MACRS of $(\mathcal{N}_1^{\sigma_1}(u), \mathcal{N}_2^{\sigma_2}(v))$, for two reticulation selectors σ_1' and σ_2' obtained from σ_1 and σ_2 by fixing all those paths in $R(B_u)$ and $R(B_v)$ appropriately that are not already fixed by σ_1 and σ_2. Thus, lines 2–7 consider all such "extensions" σ_1' and σ_2' of σ_1 and σ_2 and compute

$D[u, v, \sigma_1, \sigma_2]$ as the maximum of all corresponding table entries $D[u, v, \sigma_1', \sigma_2']$ (returned by the recursive calls σ-MACRS$(u, v, \sigma_1', \sigma_2')$). The $\texttt{fix}(\sigma, B)$ subroutine (not shown) returns all reticulation selectors that can be obtained from σ by fixing the reticulation paths in B not already fixed by σ.

If σ_1 and σ_2 fix all cherry paths in $R(B_u)$ and $R(B_v)$, then a (σ_1, σ_2)-MACRS $\mathcal{N}^*$ of $(\mathcal{N}_1^{\sigma_1}(u), \mathcal{N}_2^{\sigma_2}(v))$ exists only if $\mathcal{N}_1^{\sigma_1}$ and $\mathcal{N}_2^{\sigma_2}$ exist. $\texttt{reduce}(\mathcal{N}, \sigma)$ returns the network $\mathcal{N}^\sigma$ if it exists, or $\texttt{null}$ otherwise. Lines 9–11 use this procedure to construct $\mathcal{N}_1^{\sigma_1}$ and $\mathcal{N}_2^{\sigma_2}$, and return $-\infty$ if one of these networks does not exists.

If $\mathcal{N}^*$ exists, then it either has a single vertex or its root blob is isomorphic to the root blobs of $\mathcal{N}_1^{\sigma_1}(u)$ and $\mathcal{N}_2^{\sigma_2}(v)$. The former is possible only if $\mathcal{N}_1^{\sigma_1}$ and $\mathcal{N}_2^{\sigma_2}$ have σ_1- and σ_2-consistent complete acyclic cherry partitions and their label sets are non-disjoint. $\texttt{subnets-are-reducible-and-overlap}(u, v, \sigma_1, \sigma_2)$ tests this, and we set $D_1[u, v, \sigma_1, \sigma_2]$ to 1 or $-\infty$ accordingly in line 14 or 16.

If $\mathcal{N}^*$ has the same root blob as $\mathcal{N}_1^{\sigma_1}(u)$ and $\mathcal{N}_2^{\sigma_2}(v)$, then there must exist an isomorphism ϕ between the root blobs $B_u^{\sigma_1}$ and $B_v^{\sigma_2}$ of $\mathcal{N}_1^{\sigma_1}(u)$ and $\mathcal{N}_2^{\sigma_2}(v)$, and the pendant network attached to each leaf x of $B_u^{\sigma_1}$ in $\mathcal{N}^*$ must be a (σ_1, σ_2)-MACRS of $(\mathcal{N}_1^{\sigma_1}(x), \mathcal{N}_2^{\sigma_2}(\phi(x)))$. We call $\texttt{generate-isomorphisms}(B_u^{\sigma_1}, B_v^{\sigma_2})$ in line 17 to enumerate the set of all isomorphisms between $B_u^{\sigma_1}$ and $B_v^{\sigma_2}$. For each isomorphism ϕ, lines 19–26 find the size of a corresponding ACRS by calling σ-MACRS$(x, \phi(x), \sigma_1, \sigma_2)$ recursively for each leaf x of $B_u^{\sigma_1}$ and adding the returned sizes of (σ_1, σ_2)-MACRS of $(\mathcal{N}_1^{\sigma_1}(x), \mathcal{N}_2^{\sigma_2}(\phi(x))$ to $|B_u^{\sigma_1}|$. Thus, $D_2[u, v, \sigma_1, \sigma_2]$, defined in line 27, is the size of the biggest (σ_1, σ_2)-ACRS of $(\mathcal{N}_1^{\sigma_1}(u), \mathcal{N}_2^{\sigma_2}(v))$ that has a root blob isomorphic to $B_u^{\sigma_1}$ and $B_v^{\sigma_2}$. Line 28 then defines $D[u, v, \sigma_1, \sigma_2] = \max(D_1[u, v, \sigma_1, \sigma_2], D_2[u, v, \sigma_1, \sigma_2])$ to record the size of the largest of all (σ_1, σ_2)-ACRS of $(\mathcal{N}_1^{\sigma_1}(u), \mathcal{N}_2^{\sigma_2}(v))$.

The attentive reader will have noticed that the recursive call in line 25 does not use σ_1 and σ_2 as arguments but reticulation selectors σ_1' and σ_2' that are reset to $\bot$ whenever $B_x \neq B_u$ or $B_{\phi(x)} \neq B_v$. As we show in Sect. 4, we have $D[u, v, \sigma_1, \sigma_2] = D[u, v, \sigma_1', \sigma_2']$, so this does not affect the result computed by the algorithm, but it ensures that the reticulation selectors passed to each invocation fix cherry paths in at most one blob each. This limits the number of entries of D that need to be computed and is crucial for achieving a running time that is FPT in the level of the networks.

Complexity. Let n be the total size of the two input networks, counting vertices and arcs. The following two lemmas state the costs of key steps in the algorithm.

Lemma 4. *Let $\mathcal{N}$ be a network, let σ be a blob-respecting reticulation selector of $\mathcal{N}$, and let u be the root of a blob of $\mathcal{N}^\sigma$. Then there exists a complete acyclic cherry partition of $\mathcal{N}^\sigma(u)$ that is σ-consistent if and only if there is no path $P \in R(\mathcal{N}^\sigma(u))$ such that $\sigma(P) = 0$. Consequently, each entry $D_1[u, v, \sigma_1, \sigma_2]$ can be computed in linear time.*

Lemma 5. *There are at most $\max(2, 2^{3r/2})$ isomorphisms between two binary blobs B_1 and B_2 with r reticulations. They can be enumerated in $O(2^{3r/2} \cdot |B_1|)$ time. This bound holds even if these blobs are not orchards.*

Theorem 6. *Let $\mathcal{N}_1$ and $\mathcal{N}_2$ be two networks of level at most ℓ. Then the* MACRS *problem can be solved in $O(14.5^\ell \cdot n^3)$ time.*

Section 4 proves the correctness of the algorithm, and the detailed complexity analysis can be found in Appendix A. Here, we sketch a simpler but weaker analysis that provides the intuition. Consider all possible inputs u, v, σ_1, σ_2 on which the algorithm can be called. There are $O(n^2)$ combinations of u and v. Then, σ_1 is either $\bot$, or it fixes all reticulated cherry paths in B_u and in no other blob (since line 23 resets σ_1 to $\bot$ whenever we exit a blob). In a blob with ℓ reticulations, there are 2ℓ reticulated cherry paths, two per reticulation, and so, naively, there are 4^ℓ ways to fix these paths in a blob. However, for a reticulation v, there is no point in fixing both its reticulated cherry paths to 1, since no partial cherry partition can contain both. Hence, the `fix` routine considers, for each reticulation, at most three ways of fixing its reticulated cherry paths. This gives $O(3^\ell)$ possible values for σ_1 on input u. Likewise, σ_2 can take $O(3^\ell)$ values on input v, and so the total number of possible inputs is $O(9^\ell \cdot n^2)$.

The running time is dominated by the calls in which $\sigma_1, \sigma_2 \neq \bot$. In this case, computing the entry $D_1[u, v, \sigma_1, \sigma_2]$ takes linear time, by Lemma 4. To compute $D_2[u, v, \sigma_1, \sigma_2]$, we need to enumerate all the isomorphisms, which takes time $O(2^{3\ell/2} \cdot |B_u^{\sigma_1}|)$ time, by Lemma 5. We then spend $O(|B_u^{\sigma_1}|) = O(n)$ time per isomorphism to make the recursive calls on the leaves of the blobs. The number of possible inputs multiplied by this complexity results in a running time of $O(9^\ell \cdot n^2 \cdot 2^{3\ell/2} \cdot n) = O(25.46^\ell \cdot n^3)$. Appendix A improves the first term to 14.5^ℓ.

4   Correctness of the Algorithm

In this section, we develop the recurrence relations for the D, D_1, and D_2 tables used by Algorithm 1 to prove its correctness. Similar to the algorithm of [19] for level-1 networks, the idea is to guess the reticulated cherries to be included in a partial cherry cover that produces a MACRS and then to complete the partial cherry cover by including the necessary proper cherries. The correctness of this approach is easiest to establish when guessing the reticulated cherries globally, for the whole network, which is what the algorithm of [19] does. Therefore, we start by developing recurrence relations for tables D^*, D_1^*, and D_2^* that are analogous to the tables D, D_1, and D_2 but are indexed by reticulation selectors for the whole networks $\mathcal{N}_1$ and $\mathcal{N}_2$. While this helps with proving the correctness of these recurrence relations, the resulting tables are too big to achieve a running time that is FPT with respect to the level of the input networks. We obtain the tables D, D_1, and D_2 used by the algorithm as compressed versions of these bigger tables based on the observation that the entries $D^*[u, v, \sigma_1, \sigma_2]$ that we care about depend only on the choices made by σ_1 and σ_2 in B_u and B_v.

4.1   Recurrence Relations for Network-Wide Reticulation Selectors

We say a reticulation selector σ_i *exposes* v if

294   M. Lafond et al.

- σ_i either fixes every cherry path in $R(B_v)$ or none of them,
- There exists at least one partial cherry partition that is σ_i-consistent, and
- v is the root of a blob in $\mathcal{N}_i^{\sigma_i}$

In particular, by the last condition, if σ_i fixes none of the cherry paths in $R(B_v)$, then v must be the root of B_v. By the second condition, the network $\mathcal{N}_i^{\sigma_i}$ exists. We can now define our D^* table.

Definition 7. For two vertices u and v of $\mathcal{N}_1$ and $\mathcal{N}_2$, respectively, and two reticulation selectors σ_1 and σ_2 that expose u and v, respectively, we define $D^*[u, v, \sigma_1, \sigma_2]$ to be the size of a (σ_1, σ_2)-MACRS of $\mathcal{N}_1^{\sigma_1}(u)$ and $\mathcal{N}_2^{\sigma_2}(v)$.

If ρ_1 and ρ_2 are the roots of $\mathcal{N}_1$ and $\mathcal{N}_2$, respectively, then $\mathcal{N}_1^{\sigma_1}(\rho_1) = \mathcal{N}_1$ and $\mathcal{N}_2^{\sigma_2}(\rho_2) = \mathcal{N}_2$. Thus, $D^*[\rho_1, \rho_2, \bot, \bot]$ is the size of a MACRS of $\mathcal{N}_1$ and $\mathcal{N}_2$.

Now we call a reticulation selector σ_i' in $\mathcal{N}_i$ a B-*extension*, for a blob B of $\mathcal{N}_i$ if $\sigma_i^{-1}(0) \subseteq (\sigma_i')^{-1}(0)$, $\sigma_i^{-1}(1) \subseteq (\sigma_i')^{-1}(1)$, and $\sigma_i'(P) \in \{0, 1\}$, for all $P \in R(B)$. In words, σ_i' and σ_i agree on the selection of cherry paths fixed by σ_i, and σ_i' fixes all cherry paths in $R(B)$. Let $\text{Ext}(\sigma_i, B)$ be the set of all B-extensions of σ_i.

Observation 8. *For a reticulation selector σ_i of $\mathcal{N}_i$ and a blob B of $\mathcal{N}_i$, a partial cherry partition of $\mathcal{N}_i$ is σ_i-consistent if and only if it is σ_i'-consistent for some $\sigma_i' \in \text{Ext}(\sigma_i, B)$.*

Thus, for any two vertices $u \in \mathcal{N}_1$ and $v \in \mathcal{N}_2$, and reticulation selectors σ_1 and σ_2 that expose u and v, respectively, we have

$$D^*[u, v, \sigma_1, \sigma_2] = \max_{\substack{\sigma_1' \in \text{Ext}(\sigma_1, B_u) \\ \sigma_2' \in \text{Ext}(\sigma_2, B_v)}} D^*[u, v, \sigma_1', \sigma_2']. \tag{1}$$

If σ_1 and σ_2 fix all reticulated cherry paths in $R(B_u)$ and $R(B_v)$, respectively, then this equation holds trivially because $\text{Ext}(\sigma_1, B_u) = \{\sigma_1\}$ and $\text{Ext}(\sigma_2, B_v) = \{\sigma_2\}$, but it defines $D^*[u, v, \sigma_1, \sigma_2]$ in terms of itself and results in an infinite recursion. In this case, we use a different recurrence relation for $D^*[u, v, \sigma_1, \sigma_2]$.

So assume that σ_1 fixes all reticulated cherry paths in $R(B_u)$, and σ_2 fixes all reticulated cherry paths in $R(B_v)$. Assume further that there exists a (σ_1, σ_2)-ACRS $\mathcal{N}^*$ of $\mathcal{N}_1^{\sigma_1}(u)$ and $\mathcal{N}_2^{\sigma_2}(v)$. Then we can view $\mathcal{N}^*$ as a σ_1-CRS of $\mathcal{N}_1^{\sigma_1}(u)$ together with an isomorphism ϕ between $\mathcal{N}^*$ and a σ_2-CRS $\phi(\mathcal{N}^*)$ of $\mathcal{N}_2^{\sigma_2}(v)$. The following lemma helps to characterize what such a network $\mathcal{N}^*$ looks like:

Lemma 9. *If $v \in \mathcal{N}$ and σ is a reticulation selector in $\mathcal{N}$ that exposes v and fixes all reticulated cherry paths in B_v, then for any σ-consistent acyclic cherry partition $\mathcal{P}$ of $\mathcal{N}^\sigma(v)$, the following holds:*

- *If B_v^σ is non-trivial, then it is also the root blob of $\mathcal{N}^\sigma(v) - \mathcal{P}$.*
- *If B_v^σ is trivial, then either it is the root blob of $\mathcal{N}^\sigma(v) - \mathcal{P}$ or $\mathcal{N}^\sigma(v) - \mathcal{P}$ has v as its only vertex.*

By Lemma 9, either the two blobs $B_u^{\sigma_1}$ and $B_v^{\sigma_2}$ are both trivial and u is the only vertex of $\mathcal{N}^*$, or the two blobs are preserved and $\phi|_{B_u^{\sigma_1}}$ is an isomorphism between $B_u^{\sigma_1}$ and $B_v^{\sigma_2}$ (in the latter case, the blobs could be trivial or not). We capture these two cases in two auxiliary tables D_1^* and D_2^*:

- $D_1^*[u, v, \sigma_1, \sigma_2] = 1$ if there exists a (σ_1, σ_2)-ACRS of $\mathcal{N}_1^{\sigma_1}(u)$ and $\mathcal{N}_2^{\sigma_2}(v)$ that has u as its only vertex, and $D_1^*[u, v, \sigma_1, \sigma_2] = -\infty$ otherwise.
- $D_2^*[u, v, \sigma_1, \sigma_2]$ is the size of a (σ_1, σ_2)-MACRS of $\mathcal{N}_1^{\sigma_1}(u)$ and $\mathcal{N}_2^{\sigma_2}(v)$ that has more than one vertex, and $D_2^*[u, v, \sigma_1, \sigma_2] = -\infty$ if no such ACRS exists.

Then, by Lemma 9,

$$D^*[u, v, \sigma_1, \sigma_2] = \max(D_1^*[u, v, \sigma_1, \sigma_2], D_2^*[u, v, \sigma_1, \sigma_2]). \tag{2}$$

Note that, if u or v is a leaf, then $D_2^*[u, v, \sigma_1, \sigma_2] = -\infty$. Thus, $D^*[u, v, \sigma_1, \sigma_2] = D_1^*[u, v, \sigma_1, \sigma_2]$ in this case. This forms the base case of the recurrence.

The entries in the table D_1^* are defined without depending on any other table entries. We do not worry about its computation, as we use D^*, D_1^*, and D_2^* only as vehicles to arrive at the definitions of the tables D, D_1, and D_2. Next, we develop a recurrence relation for the entries in D_2^*:

If $\mathcal{N}^*$ is a (σ_1, σ_2)-ACRS with more than one vertex, then we know by Lemma 9 that $B_u^{\sigma_1}$ must be the root blob of $\mathcal{N}^*$ and must be isomorphic to $B_v^{\sigma_2}$. In particular, the leaves of $B_u^{\sigma_1}$ are mapped onto the leaves of $B_v^{\sigma_2}$. Since these leaves are also vertices of $\mathcal{N}_1^{\sigma_1}(u)$ and $\mathcal{N}_2^{\sigma_2}(v)$, respectively, and their parent arcs are also bridges of $\mathcal{N}_1^{\sigma_1}(u)$ and $\mathcal{N}_2^{\sigma_2}(v)$, this implies that, for every such leaf x of $B_u^{\sigma_1}$, $\mathcal{N}^*(x)$ is a subnetwork of $\mathcal{N}_1^{\sigma_1}(x)$ and $\mathcal{N}_2^{\sigma_2}(\phi(x))$.

The following observation follows immediately from Lemma 2 after observing that, for any pendant subnetwork $\mathcal{N}(x)$ of $\mathcal{N}$ and any σ-consistent partial cherry partition $\mathcal{P}$ of $\mathcal{N}$, $\mathcal{P} \cap C(\mathcal{N}(x))$ is also σ-consistent.

Observation 10. *Let $\mathcal{N}'$ be a σ-CRS of $\mathcal{N}$, and let x be the bottom endpoint of a bridge in $\mathcal{N}$. Then $\mathcal{N}'(x)$ is a σ-CRS of $\mathcal{N}(x)$.*

Observation 10 implies that $\mathcal{N}^*(x)$ is not only a subnetwork of $\mathcal{N}_1^{\sigma_1}(x)$ and $\mathcal{N}_2^{\sigma_2}(\phi(x))$ but a (σ_1, σ_2)-ACRS of these two networks, for every leaf x of $B_u^{\sigma_1}$. This immediately implies that

$$D_2^*[u, v, \sigma_1, \sigma_2] \leq |B_u^{\sigma_1}| + \max_{\phi} \left(\sum_{x \in \mathcal{L}(B_u^{\sigma_1})} D^*[x, \phi(x), \sigma_1, \sigma_2] \right), \tag{3}$$

where the maximum is taken over all isomorphisms ϕ between $B_u^{\sigma_1}$ and $B_v^{\sigma_2}$. If $B_u^{\sigma_1}$ and $B_v^{\sigma_2}$ are non-isomorphic, then $D_2^*[u, v, \sigma_1, \sigma_2] = -\infty$.

Conversely, for every isomorphism ϕ between $B_u^{\sigma_1}$ and $B_v^{\sigma_2}$ and every collection of (σ_1, σ_2)-ACRSs $\mathcal{N}^*(x)$ of $\mathcal{N}_1^{\sigma_1}(x)$ and $\mathcal{N}_2^{\sigma_2}(\phi(x))$, for all $x \in \mathcal{L}(B_u^{\sigma_1})$, $\mathcal{N}^* := B_u^{\sigma_1} \cup \bigcup_{x \in \mathcal{L}(B_u^{\sigma_1})} \mathcal{N}^*(x)$ is a common subnetwork of $\mathcal{N}_1^{\sigma_1}(u)$ and $\mathcal{N}_2^{\sigma_2}(v)$. The next lemma shows that it is in fact a (σ_1, σ_2)-ACRS of these two networks.

Lemma 11. *If ϕ is an isomorphism between $B_u^{\sigma_1}$ and $B_v^{\sigma_2}$ and $\mathcal{N}^*(x)$ is a (σ_1, σ_2)-ACRS of $\mathcal{N}_1^{\sigma_1}(x)$ and $\mathcal{N}_2^{\sigma_2}(\phi(x))$, for all $x \in \mathcal{L}(B_u^{\sigma_1})$, then $\mathcal{N}^* := B_u^{\sigma_1} \cup \bigcup_{x \in \mathcal{L}(B_u^{\sigma_1})} \mathcal{N}^*(x)$ is a (σ_1, σ_2)-ACRS of $\mathcal{N}_1^{\sigma_1}(u)$ and $\mathcal{N}_2^{\sigma_2}(v)$.*

This immediately implies the converse of Eq. (3):

$$D_2^*[u, v, \sigma_1, \sigma_2] \geq |B_u^{\sigma_1}| + \max_\phi \left(\sum_{x \in \mathcal{L}(B_u^{\sigma_1})} D^*[x, \phi(x), \sigma_1, \sigma_2] \right). \tag{4}$$

Together, Eqs. (4) and (3) imply that

$$D_2^*[u, v, \sigma_1, \sigma_2] = |B_u^{\sigma_1}| + \max_\phi \left(\sum_{x \in \mathcal{L}(B_u^{\sigma_1})} D^*[x, \phi(x), \sigma_1, \sigma_2] \right), \tag{5}$$

unless $B_u^{\sigma_1}$ and $B_v^{\sigma_2}$ are non-isomorphic, in which case $D_2^*[u, v, \sigma_1, \sigma_2] = -\infty$.

The table D^* can be filled in using Eqs. (1), (2) and (5), and it is not hard to show that the cost of doing so is FPT with respect to the total number r of reticulations in $\mathcal{N}_1$ and $\mathcal{N}_2$: There are at most 3^r pairs of reticulation selectors (σ_1, σ_2) that have consistent partial cherry partitions. Thus, there are $O(3^r \cdot n^2)$ table entries in D^*, and each can be shown to take $O(f(\ell))$ time to compute, where ℓ is the maximum level of $\mathcal{N}_1$ and $\mathcal{N}_2$.

4.2   Smaller Tables via Reticulation Selectors Restricted to Blobs

To obtain a running time that depends only on the size and maximum level of $\mathcal{N}_1$ and $\mathcal{N}_2$, and not on the total number of reticulations, we need to reduce the sizes of D^*, D_1^*, and D_2^*. Intuitively, for an entry $D^*[u, v, \sigma_1, \sigma_2]$ where u is a blob root of $\mathcal{N}_1$, we can prune away from σ_1 any selection $\sigma_1(P) \in \{0, 1\}$ where $P \notin R(\mathcal{N}_1(u))$, and use $\bot$ instead. The same holds for σ_2, which helps reduce the table sizes significantly. To formalize this, we need two lemmas:

Lemma 12. *Let $u \in \mathcal{N}_1$, $v \in \mathcal{N}_2$, σ_1 and σ_1' reticulation selectors in $\mathcal{N}_1$ that expose u, σ_2 and σ_2' reticulation selectors in $\mathcal{N}_2$ that expose v, and r_u and r_v the roots of B_u and B_v. If $\sigma_1(P) = \sigma_1'(P)$ for every reticulated cherry path $P \in R(\mathcal{N}_1(r_u))$, and $\sigma_2(P) = \sigma_2'(P)$ for every reticulated cherry path $P \in R(\mathcal{N}_2(r_v))$, then $D^*[u, v, \sigma_1, \sigma_2] = D^*[u, v, \sigma_1', \sigma_2']$.*

Consider the dependency graph between all entries in D^*, with an arrow from an entry $D^*[u, v, \sigma_1, \sigma_2]$ to an entry $D^*[x, y, \sigma_1', \sigma_2']$ if the computation of $D^*[u, v, \sigma_1, \sigma_2]$ using Eq. (1) or Eqs. (2) and (5) depends on $D^*[x, y, \sigma_1', \sigma_2']$. We say that the computation of $D^*[\rho_1, \rho_2, \bot, \bot]$ *requires* an entry $D^*[u, v, \sigma_1, \sigma_2]$ if there exists a path from $D^*[\rho_1, \rho_2, \bot, \bot]$ to $D^*[u, v, \sigma_1, \sigma_2]$ in this graph. Clearly, only entries required by $D^*[\rho_1, \rho_2, \bot, \bot]$ need to be computed.

Lemma 13. *For every entry $D^*[u, v, \sigma_1, \sigma_2]$ required by $D^*[\rho_1, \rho_2, \bot, \bot]$, σ_1 fixes all or none of the reticulated cherry paths in $R(B_u)$ and fixes none of the reticulated cherry paths in proper descendant blobs of B_u, and σ_2 fixes all or none of the reticulated cherry paths in $R(B_v)$ and fixes none of the reticulated cherry paths in proper descendant blobs of B_v.*

By Lemma 13, we only need to know $D^*[u, v, \sigma_1, \sigma_2]$ for tuples $(u, v, \sigma_1, \sigma_2)$ such that σ_1 and σ_2 do not fix any cherry paths in proper descendant blobs of B_u and B_v. Thus, for two such entries $D^*[u, v, \sigma_1, \sigma_2]$ and $D^*[u, v, \sigma_1', \sigma_2']$, σ_1 and σ_1' agree on all reticulated cherry paths reachable from the root of B_u if and only if they agree on the reticulated cherry paths in $R(B_u)$, and σ_2 and σ_2' agree on all reticulated cherry paths reachable from the root of B_v if and only if they agree on the reticulated cherry paths in $R(B_v)$. By Lemma 12, $D^*[u, v, \sigma_1, \sigma_2] = D^*[u, v, \sigma_1', \sigma_2']$ in this case. This allows us to replace the entries in Eqs. (1) (2) and (5) with entries in much smaller tables D, D_1, and D_2 indexed by tuples $(u, v, \sigma_1, \sigma_2)$, where σ_1 and σ_2 are reticulation selectors defined only on the paths in $R(B_u)$ and $R(B_v)$ respectively.

We define $\bar{\sigma}_1$ to be the completion of σ_1 to a reticulation selector on all reticulated cherry paths in $R(\mathcal{N}_1)$ by setting

$$\bar{\sigma}_1(P) = \begin{cases} \sigma_1(P) & \text{for all } P \in R(B_u) \\ \bot & \text{for all } P \in R(\mathcal{N}_1) \setminus R(B_u). \end{cases}$$

We define $\bar{\sigma}_2$ analogously. Then the entries of the tables tables D, D_1, and D_2 are defined as

$$D[u, v, \sigma_1, \sigma_2] = D^*[u, v, \bar{\sigma}_1, \bar{\sigma}_2],$$
$$D_1[u, v, \sigma_1, \sigma_2] = D_1^*[u, v, \bar{\sigma}_1, \bar{\sigma}_2],$$
$$D_2[u, v, \sigma_1, \sigma_2] = D_2^*[u, v, \bar{\sigma}_1, \bar{\sigma}_2].$$

By Lemma 12 and 13, if σ_1 does not fix the reticulated cherry paths in B_u or σ_2 does not fix the reticulated cherry paths in B_v, then we have

$$\begin{aligned} D[u, v, \sigma_1, \sigma_2] &= D^*[u, v, \bar{\sigma}_1, \bar{\sigma}_2] \\ &= \max_{\substack{\sigma_1' \in \text{Ext}(\bar{\sigma}_1, B_u) \\ \sigma_2' \in \text{Ext}(\bar{\sigma}_2, B_v)}} D^*[u, v, \sigma_1', \sigma_2'] \\ &= \max_{\substack{\sigma_1' \in \text{Ext}(\sigma_1, B_u) \\ \sigma_2' \in \text{Ext}(\sigma_2, B_v)}} D[u, v, \sigma_1', \sigma_2']. \end{aligned} \tag{6}$$

If σ_1 and σ_2 fix the reticulated cherry paths in B_u and B_v (which includes cases where B_u, B_v or both are trivial because there are no cherry paths to fix then), then

$$\begin{aligned} D[u, v, \sigma_1, \sigma_2] &= D^*[u, v, \bar{\sigma}_1, \bar{\sigma}_2] \\ &= \max(D_1^*[u, v, \bar{\sigma}_1, \bar{\sigma}_2], D_2^*[u, v, \bar{\sigma}_1, \bar{\sigma}_2]) \\ &= \max(D_1[u, v, \sigma_1, \sigma_2], D_2[u, v, \sigma_1, \sigma_2]). \end{aligned} \tag{7}$$

The computation of the D_1 entries was already discussed in Lemma 4.

As for D_2, we have $D_2[u, v, \sigma_1, \sigma_2] = -\infty$ if $B_u^{\sigma_1}$ and $B_v^{\sigma_2}$ are non-isomorphic. Otherwise, we have

$$D_2[u, v, \sigma_1, \sigma_2] = D_2^*[u, v, \bar{\sigma}_1, \bar{\sigma}_2]$$

$$= |B_u^{\sigma_1}| + \max_{\phi} \left(\sum_{x \in \mathcal{L}(B_u^{\sigma_1})} D^*[x, \phi(x), \bar{\sigma}_1, \bar{\sigma}_2] \right)$$

$$= |B_u^{\sigma_1}| + \max_{\phi} \left(\sum_{x \in \mathcal{L}(B_u^{\sigma_1})} D[x, \phi(x), \sigma_1, \sigma_2] \right), \tag{8}$$

where the maximum is taken over all isomorphisms $\phi : B_u^{\sigma_1} \to B_v^{\sigma_2}$. It follows that the value of $D[\rho_1, \rho_2, \bot, \bot]$ obtained from the above recurrences is equal to the desired value of $D^*[\rho_1, \rho_2, \bot, \bot]$. Moreover, one easily verifies that the above recurrences for D are equivalent to those that are computed by the algorithm, and the latter is therefore correct.

5 Discussion

In this paper, we have answered the open problem of [19] by showing that the MACRS problem is FPT with respect to level. This required new ideas on efficient 2ECC isomorphism enumeration and partial cherry partitions that could be useful for other problems. This enumeration makes use of a generator, a graph that represents the internal topology of a 2ECC. In the future, we will aim to implement our algorithm and evaluate it empirically. The algorithm needs to store $O(9^\ell \cdot n^2)$ entries, which could be a practical limitation due to its space requirements. On the other hand, it appears reasonable to expect most entries to have few isomorphisms to enumerate between 2ECCs, with some entries having no isomorphism at all. In this manner, the number of entries that the algorithm needs to compute on practical inputs could be many fewer than our worst-case upper bound, although this remains to be established.

We also believe that the complexity analysis of our algorithm can be improved. In particular, it should be possible to improve the n^3 factor to n^2 with appropriate data structures. It will also be interesting to verify whether our dissimilarity measure can be applied to networks that are not necessarily orchards. Although an ACRS may not exist in this case, our techniques could be generalized to detect whether two non-orchards can be made identical using cherry-picking operations, while doing so in a minimal way.

A Running Time Analysis

The simple analysis given in the main text of the paper only shows a running time of $O((18\sqrt{2})^\ell \cdot \mathrm{poly}(n)) \approx O(25.5^\ell \cdot \mathrm{poly}(n))$ for our algorithm. In this appendix, we prove the $O((6(1+\sqrt{2}))^\ell \cdot n^3) \approx O(14.5^\ell \cdot n^3)$ running time claimed in Theorem 6. To prove it, let $\alpha = 6(1+\sqrt{2})$. Then we assign $O(\alpha^\ell \cdot n^2)$ credits to every vertex in $\mathcal{N}_1$ and show that these credits can pay for the cost of the entire algorithm. As there are $O(n)$ vertices in $\mathcal{N}_1$, this shows that the cost of the algorithm is $O(\alpha^\ell \cdot n^3)$.

Algorithm 1 is a recursive algorithm that computes only table entries that the computation of $D[\rho_1, \rho_2, \bot, \bot]$ relies on. It is easily verified that this means that the only entries that are computed are those entries $D[u, v, \sigma_1, \sigma_2]$ for which u is the root of $B_u^{\sigma_1}$ and v is the root of $B_v^{\sigma_2}$. Consider the cost of computing a single such entry $D[u, v, \sigma_1, \sigma_2]$. If σ_1 does not fix all reticulated cherry paths in B_u and σ_2 does not fix all reticulated cherry paths in B_v, then calculating this entry using Eq. (6) takes $O(9^\ell)$ time, as there are at most 3^ℓ B_u-extensions of σ_1 in $\mathrm{Ext}(\sigma_1, B_u)$ and at most 3^ℓ B_v-extensions of σ_2 in $\mathrm{Ext}(\sigma_2, B_v)$. Summed over all such entries for a fixed choice of u, the cost is $O(9^\ell \cdot n)$ because there are n choices for v and only one choice for each of σ_1 and σ_2.

If σ_1 fixes all reticulated cherry paths in B_u but σ_2 does not fix all reticulated cherry paths in B_v, then calculating $D[u, v, \sigma_1, \sigma_2]$ using Eq. (6) takes $O(3^\ell)$ time, as there are at most 3^ℓ B_v-extensions of σ_2 in $\mathrm{Ext}(\sigma_2, B_v)$. Summed over all such entries for a fixed choice of u, the cost is $O(9^\ell \cdot n)$ because there are n choices for v, at most 3^ℓ choices for σ_1, and only one choice for σ_2. By an analogous argument, computing all entries $D[u, v, \sigma_1, \sigma_2]$ such that σ_1 does not fix all reticulated cherry paths in B_u but σ_2 fixes all reticulated cherry paths in B_v takes $O(9^\ell \cdot n)$ time. Thus, the total cost of computing all entries $D[u, v, \sigma_1, \sigma_2]$ such that σ_1 does not fix all reticulated cherry paths in B_u or σ_2 does not fix all reticulated cherry paths in B_v is $O(9^\ell \cdot n)$, which we pay from the credits assigned to u.

Now assume that σ_1 fixes all reticulated cherry paths in B_u and σ_2 fixes all reticulated cherry paths in B_v. Then for each such entry, computing the value of $D_1[u, v, \sigma_1, \sigma_2]$ takes time $O(n)$ by Lemma 4. We charge this time to u, for each possible v, σ_1, and σ_2, adding a total charge of $O(9^\ell \cdot n^2)$ to u. Then, the rest of the computation for this entry using Eqs. (7) and (8) takes $O(2^{3\ell_1/2} \cdot |B_u^{\sigma_1}|)$ time, where ℓ_1 is the number of reticulations in $B_u^{\sigma_1}$: By Lemma 5, there are at most $2^{3\ell_1/2}$ isomorphisms between $B_u^{\sigma_1}$ and $B_v^{\sigma_2}$, and they can be found in $O(2^{3\ell_1/2} \cdot |B_u^{\sigma_1}|)$ time. For each isomorphism, we need to evaluate the sum in Eq. (8), which takes $O(|B_u^{\sigma_1}|)$ time. We pay for this cost by charging an $O(2^{3\ell_1/2})$ portion of it to each vertex in the 2ECC of $\mathcal{N}^{\sigma_1}$ contained in $B_u^{\sigma_1}$. This is indeed sufficient, as it is easily verified that at least a constant fraction of the vertices in $B_u^{\sigma_1}$ belong to this 2ECC. We need to prove that no vertex is charged for a total cost greater than $O(\alpha^\ell \cdot n)$ in this fashion.

Fix such a vertex x in some blob B of $\mathcal{N}_1$, and consider all tuples $(u, v, \sigma_1, \sigma_2)$ such that we charge x for a portion of the cost of computing $D[u, v, \sigma_1, \sigma_2]$. Then x belongs to the 2ECC of B^{σ_1} contained in $B_u^{\sigma_1}$. Let $r(\sigma_1)$ be the number of paths $P \in R(B)$ such that $\sigma_1(P) = 1$. Then $\ell_1 \le \ell - r(\sigma_1)$, that is, we charge x for an $O(2^{3(\ell-r(\sigma_1))/2})$ portion of the cost of computing $D[u, v, \sigma_1, \sigma_2]$. For a fixed choice of σ_1, there exists at most one vertex $u \in B$ such that x is charged for part of the cost of computing entries $D[u, v, \sigma_1, \sigma_2]$ because x is part of the 2ECC of B^{σ_1} contained in $B_u^{\sigma_1}$ and the 2ECCs of B^{σ_1} are vertex-disjoint. Thus, the total cost for which we charge x is

$$O\left(3^\ell \cdot n \cdot \sum_{r(\sigma_1)=0}^{\ell} \binom{\ell}{r(\sigma_1)} 2^{r(\sigma_1)} 2^{3(\ell-r(\sigma_1))/2}\right). \tag{9}$$

Indeed, there are 3^ℓ choices for σ_2 and n choices for v and, as just argued, only one choice for u. Each choice of σ_1 charges x for a cost of at most $O(2^{3(\ell-r(\sigma_1))/2})$. For each value of $r(\sigma_1)$, there are at most $\binom{\ell}{r(\sigma_1)} 2^{r(\sigma_1)}$ choices of σ_1: First, we choose $r(\sigma_1)$ of the at most ℓ reticulations in B as the reticulations that are part of reticulated cherry paths P with $\sigma_1(P) = 1$. Then, for each such reticulation z, we choose one of its parent arcs as the interior arc of such a cherry path P. There are 2 choices of this arc for every reticulation, $2^{r(\sigma_1)}$ choices for all chosen reticulations. Equation (9) simplifies to

$$O(3^\ell \cdot n \cdot (2 + 2^{3/2})^\ell) = O((6(1 + \sqrt{2}))^\ell n) = O(\alpha^\ell n).$$

This proves that no vertex in $\mathcal{N}_1$ is charged for more than an $O(9^\ell \cdot n^2 + \alpha^\ell \cdot n)$ cost of filling in the table D, which is $O(\alpha^\ell \cdot n^2)$. Since these charges pay for the cost of the algorithm and there are $O(n)$ vertices in $\mathcal{N}_1$, the cost of the algorithm is $O(\alpha^\ell \cdot n^3)$, as claimed.

B   Proof of Section 2 (Preliminaries)

Proof of Proposition 1. First, assume that $\mathcal{N}' = \mathcal{N} \wedge S$, for some cherry picking sequence $S = \langle (a_1, b_1), \ldots, (a_n, b_n) \rangle$. We use induction on n to prove that there exists an acyclic cherry partition $\mathcal{P}$ of $\mathcal{N}$ such that $\mathcal{N}' = \mathcal{N} - \mathcal{P}$.

If $n = 0$, then $\mathcal{N} = \mathcal{N}' = \mathcal{N} - \emptyset$, and $\mathcal{P} = \emptyset$ is clearly acyclic. So assume that $n > 0$, let P_1 be the cherry path in $\mathcal{N}$ corresponding to (a_1, b_1), let $\mathcal{N}'' = \mathcal{N} \wedge (a_1, b_1)$, and let $S' = \langle (a_2, b_2), \ldots, (a_n, b_n) \rangle$. Then $\mathcal{N}'' = \mathcal{N} - \{P_1\}$ and $\mathcal{N}' = \mathcal{N}'' \wedge S'$. Thus, by the induction hypothesis, there exists an acyclic cherry partition $\mathcal{P}'$ of $\mathcal{N}''$ such that $\mathcal{N}' = \mathcal{N}'' - \mathcal{P}'$. Thus, for $\mathcal{P} = \mathcal{P}' \cup \{P_1\}$, we have $\mathcal{N}' = \mathcal{N} - \mathcal{P}$. Since $\mathcal{P}'$ is a cherry partition of $\mathcal{N}''$ and every leaf of $\mathcal{N}''$ is either a leaf of $\mathcal{N}$ or an internal vertex of P_1, $\mathcal{P}$ is a cherry partition of $\mathcal{N}$. Since $\mathcal{P}'$ is acyclic, any cycle in $G_\mathcal{P}$ would have to involve P_1. Since both endpoints of P_1 are leaves of $\mathcal{N}$, no such cycle can exist. Thus, $\mathcal{P}$ is acyclic.

Now, assume that $\mathcal{N}' = \mathcal{N} - \mathcal{P}$, for some acyclic cherry partition $\mathcal{P} = \{P_1, \ldots, P_n\}$ of $\mathcal{N}$. Assume that the cherry paths in $\mathcal{P}$ are indexed according to a topological ordering of $G_\mathcal{P}$. We use induction on n to prove that there exists a partial cherry picking sequence S such that $\mathcal{N}' = \mathcal{N} \wedge S$.

If $n = 0$, then $\mathcal{N}' = \mathcal{N} = \mathcal{N} \wedge \langle \rangle$. So assume that $n > 0$. Then, since $\mathcal{P}$ is a cherry partition of $\mathcal{N}$, each endpoint of P_n is either a leaf of $\mathcal{N}$ or an internal vertex of some other cherry path $P_j \in \mathcal{P}$. In the latter case, $G_\mathcal{P}$ would

contain the arc (P_n, P_j), but this is impossible because the cherry paths in $\mathcal{P}$ are indexed according to a topological ordering of $G_{\mathcal{P}}$. Thus, the two endpoints a_1 and b_1 of P_n are leaves of $\mathcal{N}$ and form a proper or reticulated cherry (a_1, b_1) of $\mathcal{N}$ such that $\mathcal{N}'' := \mathcal{N} \wedge \langle (a_1, b_1) \rangle = \mathcal{N} - P_n$. Now, $\mathcal{N}' = \mathcal{N}'' - \{P_1, \ldots, P_{n-1}\}$. Thus, by the induction hypothesis, there exists a cherry picking sequence S' such that $\mathcal{N}' = \mathcal{N}'' \wedge S'$. Therefore, since $\mathcal{N}'' = \mathcal{N} \wedge \langle (a_1, b_1) \rangle$, $\mathcal{N}' = \mathcal{N} \wedge (\langle (a_1, b_1) \rangle \circ S')$.
□

Proof of Lemma 2 (i) The claim that $C(B_1), \ldots, C(B_t)$ are pairwise disjoint is obvious. To prove that $C(\mathcal{N}) = C(B_1) \cup \cdots \cup C(B_t)$, we need to prove that every cherry path $P \in C(\mathcal{N})$ belongs to $C(B_i)$, for some blob B_i of $\mathcal{N}$. So consider such a cherry path P, and let (u, v) be the arc in P such that u is a tree vertex and v is an endpoint of P. Such an arc exists whether P is a proper or reticulated cherry path. Since the blobs of $\mathcal{N}$ partition the arcs of $\mathcal{N}$, there must exist a blob B_i that contains this arc. We need to prove that B_i also contains the other arcs of P. Since $(u, v) \in B_i$, u is not a leaf of B_i. Since it is a tree vertex, B_i contains both child arcs of u. Let (u, w) be the other child arc of u. If P is a proper cherry path, then $P = \langle v, u, w \rangle$, that is, both arcs of P belong to B_i. If P is a reticulated cherry path, then w is a reticulation whose child is some vertex z, and $P = \langle z, w, u, v \rangle$. Both parent arcs of a reticulation belong to the same 2ECC of $\mathcal{N}$. Thus, neither of these two parent arcs is a bridge, which implies that w is not a leaf of B_i. Thus, its child z belongs to B_i, as does the arc (w, z). Since (u, v) and (u, w) also belong to B_i, this shows that all arcs of P belong to B_i also when P is a reticulated cherry path.

(ii) Consider any subset $\mathcal{P} \subseteq C(\mathcal{N})$ and subsets $\mathcal{P}_1, \ldots, \mathcal{P}_t \subseteq \mathcal{P}$ as defined in the lemma. If $\mathcal{P}$ is an acyclic cherry partition of $\mathcal{N}$, then every subset of $\mathcal{P}$ is acyclic and the paths in every such subset are arc-disjoint. Thus, each set $\mathcal{P}_i$ is an acyclic cherry partition of B_i if every endpoint v of a path in $\mathcal{P}_i$ is a leaf of B_i or an internal vertex of another path in $\mathcal{P}_i$. If v is a leaf of $\mathcal{N}$, then it is also a leaf of B_i. Otherwise, since $\mathcal{P}$ is an acyclic cherry partition of $\mathcal{N}$, there exists a path $P \in \mathcal{P}$ that has v as an internal vertex. If $P \in \mathcal{P}_i$, then v is an internal vertex of another path in $\mathcal{P}_i$. If $P \notin \mathcal{P}_i$, then P must belong to another blob of $\mathcal{N}$, which makes the parent arc of v a bridge in $\mathcal{N}$. Thus, v is a leaf of B_i. This shows that each set $\mathcal{P}_i$ is a partial cherry partition of B_i. If the root of some blob B_i is part of cherry path $P \in \mathcal{P}$, then all arcs in B_i are reachable from P and, therefore, must be covered by cherry paths in $\mathcal{P}$. Since these cherry paths are all part of B_i, by (i), they belong to $\mathcal{P}_i$, that is, $\mathcal{P}_i$ is a complete cherry partition of B_i, as claimed. Conversely, assume that $\mathcal{P}_1, \ldots, \mathcal{P}_t$ are acyclic cherry partitions of $B_1, \ldots, B_t$ and that $\mathcal{P}_i$ is a complete cherry partition of B_i if the root of B_i is contained in some cherry path in $\mathcal{P}$. Then the paths in $\mathcal{P}$ are arc-disjoint because the paths in each of the sets $\mathcal{P}_1, \ldots, \mathcal{P}_t$ are arc-disjoint and the paths in different sets belong to different blobs of $\mathcal{N}$. For any endpoint v of some path $P \in \mathcal{P}_i$, either v is a leaf of B_i or it is an internal vertex of another path $P' \in \mathcal{P}_i \subseteq \mathcal{P}$. If v is a leaf of B_i, then v is either a leaf of $\mathcal{N}$ or the root of another blob B_j. In the latter case, $\mathcal{P}_j$ is a complete cherry partition of B_j, that is, it contains a path P' that has v as an internal vertex. This shows that the endpoint of every path in $\mathcal{P}$ is either a leaf of $\mathcal{N}$ or an internal vertex of another path in $\mathcal{P}$. Since the paths in $\mathcal{P}$ are arc-disjoint, this shows that $\mathcal{P}$ is a cherry partition of $\mathcal{N}$.

It remains to show that $\mathcal{P}$ is acyclic. Assume the contrary. Then there exists a cycle $C = \langle P_0, \ldots, P_k = P_0 \rangle$ in $G_{\mathcal{P}}$. Since each set $\mathcal{P}_i$ is acyclic, this cycle must contain paths from at least two blobs of $\mathcal{N}$. For each path P_j, let B_j be the blob that contains P_j, let $S = \langle B_0, \ldots, B_k \rangle$ be the sequence of blobs corresponding to C, and let $S' = \langle B_{j_0}, \ldots, B_{j_r} = B_{j_0} \rangle$ be the subsequence of S obtained from S by replacing every maximal subsequence of consecutive occurrences of the same blob in S with a single occurrence of this blob. Then, for all, $h \in [r]$, the root of B_{j_h} is a leaf of $B_{j_{h-1}}$. In particular, the root of $B_{j_{h-1}}$ can reach the root of B_{j_h} in $\mathcal{N}$. Since $B_{j_r} = B_{j_0}$, this implies that $\mathcal{N}$ contains a cycle, which is the desired contradiction. Thus, $\mathcal{P}$ is an acyclic cherry partition of $\mathcal{N}$.

(iii) If $\mathcal{N}$ is orchard, then it has a complete acyclic cherry partition $\mathcal{P}$. By (ii) the restrictions $\mathcal{P}_1, \ldots, \mathcal{P}_t$ to the blobs of $\mathcal{N}$ are acyclic cherry partitions of these blobs, and they must be complete by (i) and because $\mathcal{P}$ is complete. Thus, each blob of $\mathcal{N}$ is orchard. Conversely, if the blobs of $\mathcal{N}$ are orchard, then they have complete acyclic cherry partitions. By (ii), their union $\mathcal{P}$ is an acyclic cherry partition of $\mathcal{N}$. By (i), $\mathcal{P}$ is complete, so $\mathcal{N}$ is orchard.

(iv) If x is the root of a blob, then $\mathcal{N}(x)$ is a union of blobs $B_{i_1}, \ldots, B_{i_k}$. If $\mathcal{P}$ is an acyclic cherry partition of $\mathcal{N}$, then the restrictions $\mathcal{P}_{i_1}, \ldots, \mathcal{P}_{i_k}$ of $\mathcal{P}$ to these blobs are acyclic cherry partitions of these blobs, by (ii). Since $\mathcal{N}(x)$ is the union of these blobs, we conclude that $\mathcal{P}' := \mathcal{P} \cap C(\mathcal{N}(x)) = \mathcal{P}_{i_1} \cup \cdots \cup \mathcal{P}_{i_k}$ is an acyclic cherry partition of $\mathcal{N}(x)$, again by (ii). If $x \notin \mathcal{N}'$, then $\mathcal{P}$ must cover all descendant arcs of x in $\mathcal{N}$. But these are exactly the arcs in $\mathcal{N}(x)$. Thus, $\mathcal{P}'$, being the restriction of $\mathcal{P}$ to $\mathcal{N}(x)$, must be a complete acyclic cherry partition of $\mathcal{N}(x)$, and $\mathcal{N}(x) - \mathcal{P}'$ has x as its only vertex. If $x \in \mathcal{N}'$, then $\mathcal{N}'(x)$ is the subnetwork of $\mathcal{N}(x)$ obtained by removing exactly those paths in $\mathcal{P}$ from $\mathcal{N}(x)$ that belong to $\mathcal{N}(x)$. But those are exactly the paths in $\mathcal{P}'$, that is, $\mathcal{N}'(x) = \mathcal{N}(x) - \mathcal{P}'$. $\qquad\square$

C   Proofs for Section 3 (MACRS Algorithm on Level-K Orchards)

Proof of Lemma 3. We call a σ-consistent acyclic cherry partition $\mathcal{P}$ *minimal* if, for every σ-consistent acyclic cherry partition $\mathcal{P}'$, $E(\mathcal{P}) \subseteq E(\mathcal{P}')$. Proposition 14 below proves that, if there exists a σ-consistent acyclic cherry partition, then there exists a minimal σ-consistent acyclic cherry partition $\mathcal{P}$, and such a partition can be found in linear time. By definition, the set $E(\mathcal{P})$ is unique.

Next, we consider two arbitrary minimal σ-consistent partial cherry partitions $\mathcal{P}$ and $\mathcal{P}'$. Our goal is to prove that $N - \mathcal{P} = \mathcal{N} - \mathcal{P}'$. Let $\langle B_1, \ldots, B_t \rangle$ be the sequence of blobs of $\mathcal{N}$, arranged bottom-up in $\mathcal{N}$. By Lemma 2, we can partition $\mathcal{P}$ into acyclic cherry partitions $\mathcal{P}_1, \ldots, \mathcal{P}_t$ of the blobs $B_1, \ldots, B_t$. We arrange these acyclic cherry partitions in the sequence $\langle \mathcal{P}_1, \ldots, \mathcal{P}_t \rangle$ matching the chosen bottom-up ordering of the corresponding blobs. We obtain a similar sequence $\langle \mathcal{P}'_1, \ldots, \mathcal{P}'_t \rangle$ of acyclic cherry partitions of $B_1, \ldots, B_t$ whose union in $\mathcal{P}'$. Next we define two network sequences $\mathcal{N}_0, \ldots, \mathcal{N}_t$ and $\mathcal{N}'_0, \ldots, \mathcal{N}'_t$ defined as $\mathcal{N}_0 = \mathcal{N}'_0 = \mathcal{N}$, and $\mathcal{N}_i = \mathcal{N}_{i-1} - \mathcal{P}_i$ and $\mathcal{N}'_i = \mathcal{N}'_{i-1} - \mathcal{P}'_i$, for all $i \in [t]$. Then $\mathcal{N}_t = \mathcal{N} - \mathcal{P}$ and $\mathcal{N}'_t = \mathcal{N} - \mathcal{P}'$. Thus, our goal is to prove that $\mathcal{N}_t = \mathcal{N}'_t$. We use induction on i to prove that $\mathcal{N}_i = \mathcal{N}'_i$, for all $i \in [t]_0$.

For $i = 0$, the claim is trivial. So assume that $i \in [t]$ and that $\mathcal{N}_{i-1} = \mathcal{N}'_{i-1}$. We distinguish two cases: (i) If $\mathcal{P}_i$ and $\mathcal{P}'_i$ are *complete* cherry partitions of B_i, then $\mathcal{N}_i$ and $\mathcal{N}'_i$ are obtained from $\mathcal{N}_{i-1} = \mathcal{N}'_{i-1}$ by removing all arcs in B_i. In particular, $\mathcal{N}_i$ and $\mathcal{N}'_i$ have the same vertices and arcs. Thus, to prove that $\mathcal{N}_i = \mathcal{N}'_i$, it suffices to prove that every leaf has the same label in $\mathcal{N}_i$ as in $\mathcal{N}'_i$. If r_i is the root of B_i, then $\mathcal{L}(\mathcal{N}_i) = \mathcal{L}(\mathcal{N}'_i) = \mathcal{L}(\mathcal{N}_{i-1}) \setminus \mathcal{L}(B_i) \cup \{r_i\}$. Each leaf in $\mathcal{L}(\mathcal{N}_{i-1}) \setminus \mathcal{L}(B_i)$ has the same label in $\mathcal{N}_{i-1}$, $\mathcal{N}_i$, and $\mathcal{N}'_i$ because picking cherries does not relabel leaves. By Proposition 15 below applied to B_i, r_i also has the same label in $\mathcal{N}_i$ and $\mathcal{N}'_i$, so $\mathcal{N}_i = \mathcal{N}'_i$. (ii) If $\mathcal{P}_i$ and $\mathcal{P}'_i$ are *partial* cherry partitions of B_i, then σ must fix all cherry paths in $R(B_i)$. Indeed, as the proof of Proposition 14 shows, an arc e belongs to $E(\mathcal{P})$ if and only if it is reachable from a cherry path $P' \in R(\mathcal{N})$ such that $\sigma(P') = 1$. For an arc $e \in B_i$, this cherry path P' belongs to $R(B_i)$ or to $R(B)$, for some proper ancestor blob B of B_i in $\mathcal{N}$. In the latter case, all arcs in B_i are reachable from P', that is, $\mathcal{P}$ and, therefore, $\mathcal{P}_i$ would cover all arcs in B_i, that is, $\mathcal{P}_i$ would be a *complete* cherry partition of B_i. Thus, $P' \in R(B_i)$, that is, σ fixes a cherry path in $R(B_i)$. Since σ is blob-respecting, this implies that it fixes all cherry paths in $R(B_i)$. Therefore, Proposition 16 below shows that we have $\mathcal{P}_i = \mathcal{P}'_i$. Thus, $\mathcal{N}_i = \mathcal{N}_{i-1} - \mathcal{P}_i = \mathcal{N}'_{i-1} - \mathcal{P}'_i = \mathcal{N}'_i$ also in this case. This proves the first half of the lemma about the uniqueness of $\mathcal{N}^\sigma$.

Now we prove that we can decide in linear time whether $\mathcal{N}^\sigma$ exists and, if so, construct it. Assume that σ is fixing only the reticulated cherry paths in a given blob B of $\mathcal{N}$. By Proposition 16, we can find a σ-consistent cherry partition $\mathcal{P}$ for B in linear time, if it exists. Next, we find complete cherry partitions for the pendant subnetworks of B whose root is a leaf of a path in $\mathcal{P}$. This is done, in linear time, by picking cherries in any order they come available, and defining the cherry paths as they are picked, producing an acyclic cherry partition (as described by [11, Theorem 4.3]). Partitions produced in this manner are always σ-consistent since σ is unfixed for all pendant subnetworks, and every arbitrary partition is σ-consistent for $\bot$. Also note that cherry paths do not cross blobs (see Lemma 2), i.e. we can take the union of an arbitrary cherry partition of blobs to form a partition for $\mathcal{N}^\sigma$. This means that we can find $\mathcal{N}^\sigma$ in linear time if it exists, and this completes the proof for the second part of the lemma. $\square$

Proposition 14. *If there exists a σ-consistent acyclic cherry partition of $\mathcal{N}$, then there exists a minimal such partition.*

Proof. Let $F = \{P \in R(\mathcal{N}) \mid \sigma(P) = 1\}$, and let D be the set of arcs of $\mathcal{N}$ reachable from paths in F, where we call an arc e reachable from a path $P \in F$ if P contains an arc f such that there exists a directed path in $\mathcal{N}$ that starts with

f and ends with e. We prove that, if there exists a σ-consistent acyclic cherry partition at all, then there exists such a partition $\mathcal{P}$ with $E(\mathcal{P}) = D$. We also prove that every σ-consistent acyclic cherry partition $\mathcal{P}'$ satisfies $E(\mathcal{P}') \supseteq D$. Thus, $\mathcal{P}$ is minimal.

So assume that $\mathcal{P}'$ is a σ-consistent acyclic cherry partition of $\mathcal{N}$. Then $\mathcal{P}' \supseteq F$. Since every endpoint of a path in $\mathcal{P}'$ is either a leaf of $\mathcal{N}$ or an internal vertex of another path in $\mathcal{P}'$, an inductive argument shows that $E(\mathcal{P}') \supseteq D$, as claimed. Now, let $\mathcal{P}'$ be an arbitrary σ-consistent acyclic cherry partition of $\mathcal{N}$, let G' be the maximal subgraph of $G_{\mathcal{P}'}$ such that every vertex in G' is reachable from a vertex in F, and let $\mathcal{P} \subseteq \mathcal{P}'$ be the vertex set of G', that is, $G' = G_{\mathcal{P}}$. Then $F \subseteq \mathcal{P}$ and $G_{\mathcal{P}}$ is acyclic because $G_{\mathcal{P}'}$ is and $G_{\mathcal{P}} \subseteq G_{\mathcal{P}'}$. Moreover, $\mathcal{P}$ is a cherry partition of $\mathcal{N}$. Indeed, since $\mathcal{P}'$ is a cherry partition of $\mathcal{N}$ and $\mathcal{P} \subseteq \mathcal{P}'$, all paths in $\mathcal{P}$ are arc-disjoint. Thus, if $\mathcal{P}$ is not a cherry partition of $\mathcal{N}$, then there exists a path $P_1 \in \mathcal{P}$ with an endpoint v that is not a leaf of $\mathcal{N}$ nor an internal vertex of another cherry path in $\mathcal{P}$. Since $\mathcal{P}'$ is a cherry partition of $\mathcal{N}$ and $\mathcal{P}' \supseteq \mathcal{P}$, this implies that v is an internal vertex of a path $P_2 \in \mathcal{P}' \setminus \mathcal{P}$. This, however, implies that P_2 is reachable from some vertex in F because $P_1 \in \mathcal{P}$ implies that P_1 is. This is a contradiction. Therefore, $\mathcal{P}$ is a cherry partition of $\mathcal{N}$. Since $F \subseteq \mathcal{P} \subseteq \mathcal{P}'$ and $\mathcal{P}'$ is σ-consistent, so is $\mathcal{P}$. As already shown, this implies that $E(\mathcal{P}) \supseteq D$. Conversely, $E(\mathcal{P}) \subseteq D$ because, by the definition of $\mathcal{P}$, every path $P_2 \in \mathcal{P}$ is reachable from a path $P_1 \in F$, which implies that $E(P_2) \subseteq D$. Since this is true fore every path in $\mathcal{P}$, this implies that $E(\mathcal{P}) \subseteq D$.
$\square$

Proposition 15. *For a complete cherry partition $\mathcal{P}$ of a network $\mathcal{N}$, $\mathcal{N} - \mathcal{P}$ is a network that consists of a single leaf with label set $X(\mathcal{N})$.*

Proof. As shown in [10], $\mathcal{P}$ corresponds to a representation of $\mathcal{N}$ as a base tree T plus exactly those reticulation arcs that are internal to reticulated cherry paths in $\mathcal{P}$. For every proper cherry path $\langle u, v, w \rangle$ in $\mathcal{P}$, u and w are the children of v in T. This implies that when we pick a cherry that makes some vertex v a leaf, then v is labelled with the union of the labels of its children in T. Therefore, by a simple inductive argument, the label of every vertex is the union of the labels of its descendant leaves in T. In particular, the label of the root of T, which is the one vertex of $\mathcal{N} - \mathcal{P}$, is $\bigcup_{v \in \mathcal{L}(\mathcal{N})} X(v) = X(\mathcal{N})$.
$\square$

Proposition 16. *Assume that σ is a reticulation selector that fixes all reticulated cherries in $\mathcal{N}$. Then every σ-consistent acyclic cherry partition $\mathcal{P}$ of $\mathcal{N}$ is uniquely determined by $E(\mathcal{P})$. In particular, the minimal σ-consistent acyclic cherry partition of $\mathcal{N}$ is unique if it exists. It takes linear time to decide whether this partition exists and, if so, find it.*

Proof. First assume that $\mathcal{P}$ is a σ-consistent acyclic cherry partition of $\mathcal{N}$. Then the reticulated cherry paths in $\mathcal{P}$ are exactly those paths $P \in R(\mathcal{N})$ that satisfy $\sigma(P) = 1$, as there is no path $P \in R(\mathcal{N})$ with $\sigma(P) = \perp$ and $\mathcal{P}$ is σ-consistent. Thus, $\mathcal{P} \setminus R(\mathcal{N})$ is a set of proper cherries. Since proper cherries do not contain reticulations as internal vertices and no two cherry paths in a cherry partition share an endpoint, this implies that the arcs in $E(\mathcal{P} \setminus R(\mathcal{N}))$ form a binary forest in $\mathcal{N}$ in which every vertex is either a leaf (of the forest) or has two children in the forest. It is easily verified, via an inductive argument that picks cherries bottom-up in this forest, that any binary forest has a unique cherry partition. Thus, $\mathcal{P} = (\mathcal{P} \cap R(\mathcal{N})) \cup (\mathcal{P} \setminus R(\mathcal{N}))$ is also unique, as claimed.

To test whether there exists a minimal σ-consistent acyclic cherry partition $\mathcal{P}$ of $\mathcal{N}$, we start with the set $F = \{P \in R(\mathcal{N}) \mid \sigma(P) = 1\}$ already defined in the proof of Proposition 14. This is the set of reticulated cherries in $\mathcal{P}$. If there are two paths in F that share an arc, which is easy to test in linear time, then there is no cherry partition that is a superset of F, that is, there is no σ-consistent cherry partition of $\mathcal{N}$. Otherwise, the set D of arcs reachable from the paths in F also already defined in the proof of Proposition 14 is easily identified in linear time using a simple graph traversal. By Proposition 14, D is exactly the set of arcs in $E(\mathcal{P})$. Given D and $E(F)$, we can in linear time compute their difference $D' = D \setminus E(F)$. Given D', we can check whether every vertex of $\mathcal{N}$ either has no child arcs in D' or it is a tree vertex with both its child arcs in D'. D' induces a binary forest in $\mathcal{N}$ in which every vertex is either a leaf or has two children if and only if every vertex of $\mathcal{N}$ passes this test. If D' induces such a forest, we can compute the unique partition $\mathcal{P}'$ of the arcs in D' into proper cherry paths by picking cherries bottom-up in this forest. The union $\mathcal{P} = F \cup \mathcal{P}'$ is a σ-consistent cherry partition of $\mathcal{N}$. It remains to check whether this partition is acyclic, which we do by constructing $G_{\mathcal{P}}$ and using depth-first search to try to find a cycle. Since all of the steps in this procedure can be implemented in linear time, this proves the second part of the proposition. $\square$

Proof of Lemma 4. Suppose that $\sigma(P) \in \{1, \perp\}$ for every $P \in R(\mathcal{N}^\sigma(u))$. Then note that, in fact, $\sigma(P) = 1$ is not possible, because $\mathcal{N}^\sigma = \mathcal{N} - \mathcal{P}$, for some σ-consistent acyclic cherry partition $\mathcal{P}$ of $\mathcal{N}$, and $\sigma(P) = 1$ implies that $P \in \mathcal{P}$, so $P \notin R(\mathcal{N}^\sigma) \supseteq R(\mathcal{N}^\sigma(u))$. Thus, $\sigma(P) = \perp$, for all $P \in R(\mathcal{N}^\sigma(u))$, that is, every acyclic cherry partition of $\mathcal{N}^\sigma(u)$ is σ-consistent. Since $\mathcal{N}_1^{\sigma_1}(u)$ is an orchard, it has a complete acyclic cherry partition and, as just argued, this partition is σ-consistent. Conversely, assume that $\sigma(P) = 0$ for some $P \in R(\mathcal{N}^\sigma(u))$. Then P contains a reticulation v. Let P' be the other reticulated cherry path in $R(\mathcal{N}^\sigma(u))$ that contains v. Then $\sigma(P') \neq \perp$ because σ is blob-respecting, and $\sigma(P') \neq 1$ because, as argued above, $\mathcal{P}$ contains all reticulated cherry paths $P \in R(\mathcal{N})$ that satisfy $\sigma(P) = 1$. Thus, $\sigma(P') = 0$. Thus, there is no σ-consistent cherry partition of $\mathcal{N}^\sigma(u)$ that covers v or its child arc, that is, no σ-consistent cherry partition of $\mathcal{N}^\sigma(u)$ is complete.

As for the cost of computing $D_1[u, v, \sigma_1, \sigma_2]$, we have $D_1[u, v, \sigma_1, \sigma_2] = 1$ if and only if $X(\mathcal{N}_1^{\sigma_1}(u)) \cap X(\mathcal{N}_2^{\sigma_2}(v)) \neq \emptyset$ and $\mathcal{N}_1^{\sigma_1}(u)$ and $\mathcal{N}_2^{\sigma_2}(v)$ have complete cherry partitions that are σ_1-consistent and σ_2-consistent, respectively. The intersection property is checked in linear time and, by the previous arguments, we can check for complete cherry covers of $\mathcal{N}_1^{\sigma_1}(u)$ and $\mathcal{N}_2^{\sigma_2}(v)$ in linear time . $\square$

Proof of Lemma 5. If B_1 is a trivial blob, then it can be isomorphic to B_2 only if B_2 is also trivial. In this case, the lemma holds because there are exactly 2 isomorphisms between B_1 and B_2 corresponding to the two possible mappings of the leaves of B_1 onto the leaves of B_2. These two isomorphisms can be found in constant time. For the rest of this proof, therefore, assume that B_1 and B_2 are both non-trivial. For $i \in \{1, 2\}$, let B_i' be the 2ECC obtained by deleting the leaves of B_i and their incident arcs, and let the *generator* G_i of B_i be the graph obtained from B_i' by suppressing all vertices with in-degree 1 and out-degree 1, i.e. we delete such a vertex and all its incident edges, adding a new edge between its former neighbours. Note that each arc e of G_i corresponds to a path $P_i(e)$ in B_i' whose internal vertices have one parent and one child, and G_i may have parallel arcs. Moreover, B_i and G_i have the same number of reticulation.

Restricting any isomorphism $\phi : B_1 \to B_2$ to the vertices of G_1 uniquely defines a corresponding isomorphism $\psi : V(G_1) \to V(G_2)$. Similarly, for every generator arc $e = (u, v) \in E(G_1)$, ϕ must map the path $P_1(e)$ onto $P_2(e')$ in B_2, where $e' = (u', v')$ is an arc of G_2 with $u' = \phi(u)$ and $v' = \phi(v)$. Thus, by setting $\psi(e) = e'$ for every arc of G_1, we obtain a unique isomorphism $\psi : G_1 \to G_2$. Conversely, for every isomorphism $\psi : G_1 \to G_2$, there exists at most one corresponding isomorphism $\phi : B_1' \to B_2'$ defined by setting $\phi(v) = \psi(v)$, for every vertex $v \in V(G)$ and $\phi(P_1(e)) = P_2(\psi(e))$, for every arc $e \in E(G)$. This is an isomorphism exactly if $P_1(e)$ and $P_2(\psi(e))$ have the same length, for every arc $e \in E(G)$. This isomorphism ϕ has a natural extension to a unique isomorphism $B_1 \to B_2$, as no vertex in a non-trivial blob B_1 has more than one child that is a leaf and, therefore, the mapping of each leaf of B_1 is uniquely determined by the mapping of its parent in B_1'. This shows that there exists a 1-to-1 correspondence between isomorphisms $B_1 \to B_2$ and isomorphisms between their generators. Moreover, the construction of an isomorphism $\phi : B_1 \to B_2$ from an isomorphism $\psi : G_1 \to G_2$ just described is easily carried out in linear time, as is the construction of G_1 and G_2 from B_1 and B_2. Thus, it suffices to argue that there are at most $2^{3r/2}$ isomorphisms $\psi : G_1 \to G_2$ and that they can be enumerated in $O(2^{3r/2} \cdot |G_1|)$ time.

Consider a rooted spanning tree T_1 of G_1 and a topological ordering $\sigma = \langle v_1, \ldots, v_n \rangle$ of its vertices. In other words, we have $i < j$ for every arc (v_i, v_j) in T_1. For every isomorphism $\psi : G_1 \to G_2$, the restriction of $\psi|_{T_1}$ is an isomorphism between T_1 and a rooted spanning tree T_2 of G_2. Since all vertices of G_1 are vertices of T_1, $\psi|_{T_1}$ fully defines the action of ψ on the vertices of G_1. In fact, $\psi|_{T_1}$ also fully defines the action of ψ on the arcs of G_1. Clearly, it defines the action of ψ on the arcs of T_1. For every arc $e = (u, v) \in E(G_1) \setminus E(T_1)$, $\psi(e)$ must be an arc with endpoints $\psi(u)$ and $\psi(v)$. If there is only one such arc in G_2, then this arc is fully defined by these two endpoints. If there are two such arcs (because a generator can have parallel arcs, but a generator of a binary network cannot have more than two parallel arcs between two vertices), then, since ψ is an isomorphism, there are also two arcs with endpoints u and v in G_1. This makes u the only parent of v in G_1 (because G_1 is binary), which implies that one of the arcs with endpoints u and v is in T_1. Its image is in T_2. The image of the arc $(u, v) \in E(G_1) \setminus E(T_1)$ is thus the unique arc $(\phi(u), \phi(v)) \in E(G_2) \setminus E(T_2)$. Thus, the image of (u, v) is once again uniquely defined by $\psi|_{T_1}$.

By the argument in the previous paragraph, we can use the following strategy for enumerating all isomorphisms from G_1 to G_2: Conceptually, we enumerate all isomorphisms between T_1 and spanning trees of G_2 and then discard those that do not extend to isomorphisms between G_1 and G_2. However, we will construct these isomorphisms between T_1 and spanning trees of G_2 incrementally, using a recursive search. This makes it more efficient to abort a branch in the search as soon as the choices made so far guarantee that no isomorphism between T_1 and a spanning tree of G_2 that includes these choices can be extended to an isomorphism between G_1 and G_2. In particular, we inspect the vertices of G_1 in the order σ. For all $i \in [n]$, let $V_i = \{v_1, \ldots, v_i\}$. Then, at every point in the recursive search, we will have constructed an isomorphism $\psi|_{V_i}$ between the subtree $T_1|_{V_i}$ and some subtree of G_2 whose root is the root of G_2. We call such an isomorphism *valid* if

- Every vertex $v_j \in V_i$ has the same type in G_1 as $\psi(v_j)$ has in G_2, where the type of a vertex is "tree vertex" or "reticulation" and
- For every pair of vertices $v_h, v_j \in V_i$, there exists an arc $(v_h, v_j) \in E(G_1) \setminus E(T_1)$ if and only if the arc $(\psi(v_h), \psi(v_j))$ exists in $E(G_2) \setminus E(\psi(T_1|_{V_i}))$.

We make recursive calls only on valid isomorphisms $\psi|_{V_i}$ because an invalid isomorphism clearly cannot be extended to an isomorphism ψ between G_1 and G_2. Conversely, by the argument in the 3rd paragraph of this proof, if $\psi|_{V_n}$ is valid, then it has a unique extension to an isomorphism of G_1 and G_2, and it is not hard to verify that this extension can be found in linear time.

There is exactly one valid isomorphism $\psi|_{V_1}$, which maps v_1 to the root of G_2. This is the input to the top-level invocation of the search. Given a valid isomorphism $\psi|_{V_i}$, if $i = n$, then we complete $\psi|_{V_n}$ to an isomorphism between G_1 and G_2 as discussed in the fifth paragraph of this proof. If $i < n$, then we identify all valid isomorphisms $\psi|_{V_{i+1}}$ such $\psi|_{V_i}$ is the restriction of $\psi|_{V_{i+1}}$ to $T_1|_{V_i}$ and make one recursive call on each of them. In particular, let v_j be the parent of v_{i+1} in T_1, and let e be the arc connecting them. Then $\psi(v_{i+1})$ must be a child of $\psi(v_j)$ not in $\psi(T_1|_{V_i})$. We inspect each child arc $e' = (\psi(v_j), z)$ of $\psi(v_j)$, and set $\psi(v_{i+1}) = z$ and $\psi(e) = e'$ for each such arc that satisfies $z \notin \psi(T_1|_{V_i})$. This defines a (not necessarily valid) isomorphism between $T_1|_{V_{i+1}}$ and a subtree of G_2. Next we check that v_{i+1} and $\psi(v_{i+1})$ have the same type and that, if v_{i+1} has an incident arc (v_h, v_{i+1}) or (v_{i+1}, v_h) in $E(G_1) \setminus E(T_1|_{V_{i+1}})$ with $v_h \in V_i$, then there exists the corresponding arc $(\psi(v_h), \psi(v_{i+1}))$ or $(\psi(v_{i+1}), \psi(v_h))$ in $E(G_2) \setminus E(\psi(T_1|_{V_{i+1}}))$. If $\psi|_{V_{i+1}}$ passes both of these tests, then it is valid, so we recurse on it. Otherwise, it is invalid, and we discard it.

The discussion so far shows that this procedure enumerates all isomorphisms between G_1 and G_2. Now we bound its running time and the number of isomorphisms it finds. It is easily seen that the cost per recursive call is constant, as it only needs to find the parent v_j of v_{i+1}, and then it needs to inspect the neighbours of $v_j, v_{j+1}, \psi(v_j)$, and $\psi(v_{i+1})$, of which there are only a constant number. The depth of the recursive search is $O(|G_1|)$. Thus, to bound both the number of isomorphisms found by $2^{3r/2}$ and the time to find them by $O(2^{3r/2} \cdot |G_1|)$, it suffices to prove that there are at most $2^{3r/2}$ leaves in this search tree. To this end, we call a valid isomorphism $\psi|_{V_i}$ a *choice point* if the parent v_j of v_{i+1} in T_1 is a tree vertex in G_1 and neither of its children belongs to V_i. If $\psi|_{V_i}$ is not a choice point, then it is easily seen to have at most one valid extension $\phi|_{V_{i+1}}$. If $\psi|_{V_i}$ is a choice point, then it has at most two valid extensions corresponding to the two possible mappings of v_{i+1} to one of the children of $\psi(v_j)$. Thus, to prove that there are at most $2^{3r/2}$ leaves in the recursion tree, it suffices to prove that any root-to-leaf path in this recursion tree contains at most $3r/2$ choice points.

As an isomorphism $\psi|_{V_i}$ can be a choice point only if v_j is a tree vertex in G_1, the number of tree vertices t in G_1 is an upper bound on the number of choice points on any root-to-leaf path of the recursion tree. Let r_0 be the number of reticulations without children, let r_1 be the number of reticulations with one child and with two distinct parents, and let r_2 be the number of reticulations that are the bottom endpoints of parallel arcs. Since B_1 is a blob, G_1 is 2-edge-connected, which implies that the bottom endpoint of every pair of parallel arcs must be a reticulation that has a child in G_1. Let t_1 be the number of tree vertices that are not top endpoints of parallel arcs, and let t_2 be the number of tree vertices that are top endpoints of parallel arcs. Then $r = r_0 + r_1 + r_2$, $t = t_1 + t_2$, $r_2 = t_2$, and $2(t_1 + t_2) + r_1 + r_2 = t_1 + t_2 - 1 + 2(r_0 + r_1 + r_2)$ because the total number of in-arcs of all vertices must equal the total number of their out-arcs. Thus, since $r_2 = t_2$, we have $t_1 + 1 = 2r_0 + r_1$ and, therefore, $t + 1 = t_1 + t_2 + 1 = 2r_0 + r_1 + r_2 = r + r_0 \leq 2r$. This shows that there are at most $2r$ choice points on any root-to-leaf path in the recursion tree, i.e., the number of isomorphisms $< 4^r$ and they can be enumerated in $O(4^r \cdot |G_1|)$ time.

To obtain a better bound, we need to choose the spanning tree T_1 and the topological ordering σ more carefully. Consider an arbitrary reticulation v_i in G_1 with parents v_j and v_h. Then one of the arcs (v_j, v_i) and (v_h, v_i) belongs to T_1, the other does not. Assume that $(v_j, v_i) \notin T_1$ and that v_j has a child v_k in T_1. If $j > i$, then $k > i$ because $j < k$. Thus, $v_i \in V_{k-1}$ and $\psi|_{V_{k-1}}$ is not a choice point. Our goal now is to construct T_1 and σ so that there are at least $(r_0 - r_2)/2$ tree vertices v_j in G_1 that have an incident arc $(v_j, v_i) \notin T_1$ that satisfies $v_i < v_j$. These tree vertices cannot be choice points, so the number of choice points on any path in the recursion tree is at most $t - (r_0 - r_2)/2 < r + (r_0 + r_2)/2 < 3r/2$.

To construct T_1 and $\sigma = \langle v_1, \ldots, v_n \rangle$, we start by adding the root ρ of G_1 to T_1 and setting $\sigma = \langle \rho \rangle$. Since G_1 is a DAG, there always exists a vertex v that has all its parents in T_1. If v has a single parent (which includes the case when v is the bottom endpoint of two parallel arcs) or at least one of the parents of v is a reticulation, then we choose one of the parents of v—call it u—and make v a child of u in T_1. We add v to the end of σ to ensure that σ remains a topological ordering of the tree constructed so far. If v is a reticulation whose parents u_1 and u_2 in G_1 are both tree vertices, then recall that $u_1, u_2 \in T_1$ at the time we add v to T_1. Assume that u_1 precedes u_2 in σ. Then we make v a child of u_1 and insert v into σ immediately after u_1. This ensures both that σ remains a topological ordering of T_1, as v succeeds its parent, and that the non-tree arc

(u_2, v) has the property we desire: that v precedes u_2 in σ. Thus, whether u_2 ends up having a child in T_1 or not, it does not correspond to a choice point.

Every reticulation with two tree vertices as parents eliminates one choice point, but if there is a tree vertex u whose children v_1 and v_2 are both reticulations, then v_1 and v_2 may both eliminate u as a tree vertex corresponding to a choice point. However, every tree vertex can be eliminated only by its children, that is, every tree vertex is eliminated at most twice. Therefore, the number of choice points eliminated by our construction is at least half the number of reticulations in G_1 that have two distinct tree vertices as parents. The number of parent arcs of reticulations with two distinct parents is $2(r_0 + r_1)$. At most $r_1 + r_2$ of them have reticulations as top endpoints. This leaves at least $2r_0 + r_1 - r_2$ arcs from tree vertices to reticulations with distinct parents. Since there are only $r_0 + r_1$ such reticulations, this implies that there must be at least $r_0 - r_2$ reticulations with two distinct tree vertices as parents. Thus, the chosen tree T_1 and topological ordering ensures that there are at most $t - (r_0 - r_2)/2 < 3r/2$ choice points along any path in the recursion tree. The number of isomorphisms from G_1 to G_2 is thus at most $2^{3r/2}$, and they can be found in $O(2^{3r/2} \cdot |G_1|)$ time. $\square$

D   Proofs for Section 4 (Correctness of the Algorithm)

Proof of Lemma 9. Let $\mathcal{P}$ be any σ-consistent acyclic cherry partition of $\mathcal{N}^\sigma(v)$. Suppose for now that B_v^σ is trivial. If B_v^σ has a single leaf, then that leaf is v, $\mathcal{P} = \emptyset$, and so $\mathcal{N}^\sigma(v) - \mathcal{P} = \mathcal{N}^\sigma(v)$ and has v as its only vertex. Otherwise, the root v of B_v^σ has two children u_1 and u_2. If neither (v, u_1) nor (v, u_2) is in a path in $\mathcal{P}$, then $\mathcal{N}^\sigma(v) - \mathcal{P}$ contains both arcs and, thus, its root blob is B_v^σ, as claimed. If some path $P \in \mathcal{P}$ contains one of the two arcs (v, u_1) and (v, u_2), it must also contain the other, as v is a tree vertex. Since every arc in $\mathcal{N}^\sigma(v)$ is reachable from v, any cherry partition $\mathcal{P}$ of $\mathcal{N}^\sigma(v)$ that contains P must cover all arcs of $\mathcal{N}^\sigma(v)$, that is $\mathcal{N}^\sigma(v) - \mathcal{P}$ has v as its only vertex.

So assume from now on that B_v^σ is non-trivial. Then note that $R(B_v^\sigma) \subseteq R(B_v)$ because removing arcs from $\mathcal{N}$ does not merge 2ECCs, that is, $B_v^\sigma \subseteq B_v$. Thus, since σ fixes all reticulated cherry paths in B_v, we have $\sigma(P) \in \{0, 1\}$, for all $P \in R(B_v^\sigma)$. Since $\mathcal{N}^\sigma$ is a σ-CRS of $\mathcal{N}$, there exists a σ-consistent acyclic cherry partition $\mathcal{P}'$ of $\mathcal{N}$ such that $\mathcal{N}^\sigma = \mathcal{N} - \mathcal{P}'$. In particular, $\mathcal{P}'$ contains all paths $P' \in R(\mathcal{N})$ with $\sigma(P') = 1$, so $R(B_v^\sigma) \subseteq R(\mathcal{N} - \mathcal{P}')$ cannot contain any such path. This shows that $\sigma(P) = 0$ for every reticulated cherry path in $R(B_v^\sigma)$.

Now consider $\mathcal{P}$ again. If no path in $\mathcal{P}$ contains an arc of B_v^σ, then B_v^σ is the root blob of $\mathcal{N}^\sigma(v) - \mathcal{P}$, as claimed. So assume for the sake of contradiction that some path $P \in \mathcal{P}$ contains an arc of B_v^σ. Note that P cannot be a reticulated cherry path, as $\sigma(P') = 0$ for all $P' \in R(B_v^\sigma)$, and $\mathcal{P}$ is σ-consistent. Thus, P is a proper cherry path $\langle x, y,\ z \rangle$. This implies that y and, w.l.o.g., x are non-leaf vertices of B_v^σ because no vertex in a non-trivial binary blob can have two leaves of the blob as children. Because removing the leaves of B_v^σ results in a 2ECC, there must be an undirected cycle $C = \langle v_0, \ldots, v_t \rangle$ in B_v^σ with $v_0 = v_n = y$ and $v_1 = x$. As B_v^σ is acyclic, there exists an index i such that $\langle v_0, \ldots, v_i \rangle$ is a directed path from y to v_i and v_i is a reticulation with v_{i-1} and v_{i+1} as its parents. Since P contains (y, x), $P \in \mathcal{P}$, and $\mathcal{P}$ is an acyclic cherry partition, every arc along this path must be covered by paths in $\mathcal{P}$, including (v_{i-1}, v_i). This implies that some path $P' \in \mathcal{P}$ must cover the child arc of v_i. Since v_i is a reticulation, this path must be reticulated cherry path. Since $\sigma(P') = 0$ for any such path, this contradicts the σ-consistency of $\mathcal{P}$. $\qquad\square$

Proof of Lemma 11. Let $\mathcal{L}(B_u^{\sigma_1}) = \{x_1, \ldots, x_t\}$ and, for all $i \in [i]$, let $\mathcal{P}_i$ be a σ_1-consistent acyclic cherry partition of $\mathcal{N}^{\sigma_1}(x_i)$ such that $\mathcal{N}^{\sigma_1}(x) = \mathcal{N}^*(x)$. These cherry partitions exist because $\mathcal{N}^*(x_i)$ is a σ_1-CRS of $\mathcal{N}^{\sigma_1}(x_i)$, for all $i \in [t]$. As each of the subnetworks $\mathcal{N}_1^{\sigma_1}(x_1), \ldots, \mathcal{N}_1^{\sigma_1}(x_t)$ is rooted in a distinct leaf of $B_u^{\sigma_1}$, no two of these subnetworks have a vertex in common. Thus, $\mathcal{P} := \mathcal{P}_1 \cup \cdots \cup \mathcal{P}_t$ is an acyclic cherry partition of $\mathcal{N}_1^{\sigma_1}(u)$. Our definition of $\mathcal{N}^*$ implies that $\mathcal{N}^* = \mathcal{N}^{\sigma_1}(u) - \mathcal{P}$, so $\mathcal{N}^*$ is a CRS of $\mathcal{N}^{\sigma_1}(u)$. We also need to prove that $\mathcal{P}$ is σ_1-consistent to show that $\mathcal{N}^*$ is a σ_1-CRS of $\mathcal{N}^{\sigma_1}$. As in the proof of Lemma 9, we know that $\sigma_1(P) \in \{0, \bot\}$, for every $P \in R(\mathcal{N}_1^{\sigma_1}(u))$. Thus, it is sufficient to argue that $\mathcal{P}$ contains no path $P \in R(\mathcal{N}_1^{\sigma_1}(u))$ with $\sigma(P) = 0$. This, however, follows from the σ_1-consistency of $\mathcal{P}_1, \ldots, \mathcal{P}_t$, as $\mathcal{P} = \mathcal{P}_1 \cup \cdots \cup \mathcal{P}_t$.

Since $\mathcal{N}^*(x_i)$ is a σ_2-CRS of $\mathcal{N}_2^{\sigma_2}(\phi(x_i))$, for all $i \in [t]$, there exist a network $\mathcal{N}^{**}(\phi(x_i))$, an isomorphism $\psi_i : \mathcal{N}^*(x_i) \to \mathcal{N}^{**}(\phi(x_i))$, and a σ_2-consistent acyclic cherry partition $\mathcal{P}_i'$ of $\mathcal{N}_2^{\sigma_2}(\phi(x_i))$ such that $\mathcal{N}_2^{\sigma_2}(\phi(x_i)) - \mathcal{P}_i' = \mathcal{N}^{**}(\phi(x_i))$. By the same argument as above, $\mathcal{P}' := \mathcal{P}_1' \cup \cdots \cup \mathcal{P}_t'$ is a σ_2-consistent acyclic cherry partition of $\mathcal{N}_2^{\sigma_2}(v)$, and $\mathcal{N}^{**} := B_v^{\sigma_2} \cup \bigcup_{x \in \mathcal{L}(B_u^{\sigma_1})} \mathcal{N}^{**}(\phi(x_i))$ is a σ_2-CRS of $\mathcal{N}_2^{\sigma_2}(v)$, as $\mathcal{N}^{**} = \mathcal{N}_2^{\sigma_2}(v) - \mathcal{P}'$. Since $\phi, \psi_1, \ldots, \psi_t$ are isomorphisms that agree on the mappings of the vertices in $\mathcal{L}(B_u^{\sigma_1}), B_u^{\sigma_1}, \mathcal{N}^{\sigma_1}(x_1), \ldots, \mathcal{N}^{\sigma_1}(x_t)$ share no other vertices nor any arcs, and the same is true for $B_v^{\sigma_2}, \mathcal{N}_2^{\sigma_2}(\phi(x_1)), \ldots, \mathcal{N}_2^{\sigma_2}(\phi(x_t))$, the mapping $\psi : \mathcal{N}^* \to \mathcal{N}^{**}$ that maps vertices and arcs according to $\phi, \psi_1, \ldots, \psi_t$ is an isomorphism. Thus, since $\mathcal{N}^*$ is a σ_1-CRS of $\mathcal{N}_1^{\sigma_1}(u)$ and $\mathcal{N}^{**}$ is a σ_2-CRS of $\mathcal{N}_2^{\sigma_2}(v)$, $\mathcal{N}^*$ is a (σ_1, σ_2)-ACRS of $\mathcal{N}_1^{\sigma_1}(u)$ and $\mathcal{N}_2^{\sigma_2}(v)$. $\qquad\square$

Proof of Lemma 12. We prove that every σ_1-CRS of $\mathcal{N}_1^{\sigma_1}(u)$ is a σ_1'-CRS of $\mathcal{N}_1^{\sigma_1'}(u)$. An analogous argument shows that every σ_1'-CRS of $\mathcal{N}_1^{\sigma_1'}(u)$ is a σ_1-CRS of $\mathcal{N}_1^{\sigma_1}(u)$, that every σ_2-CRS of $\mathcal{N}_2^{\sigma_2}(v)$ is a σ_2'-CRS of $\mathcal{N}_2^{\sigma_2'}(v)$, and that every σ_2'-CRS of $\mathcal{N}_2^{\sigma_2'}(v)$ is a σ_2-CRS of $\mathcal{N}_2^{\sigma_2}(v)$. This implies that every (σ_1, σ_2)-ACRS of $\mathcal{N}_1^{\sigma_1}(u)$ and $\mathcal{N}_2^{\sigma_2}(v)$ is a (σ_1', σ_2')-ACRS of $\mathcal{N}_1^{\sigma_1'}(u)$ and $\mathcal{N}_2^{\sigma_2'}(v)$, and vice versa. This in turn implies that every (σ_1, σ_2)-MACRS of $\mathcal{N}_1^{\sigma_1}(u)$ and $\mathcal{N}_2^{\sigma_2}(v)$ is a (σ_1', σ_2')-MACRS of $\mathcal{N}_1^{\sigma_1'}(u)$ and $\mathcal{N}_2^{\sigma_2'}(v)$, and vice versa, that is, that $D^*[u, v, \sigma_1, \sigma_2] = D^*[u, v, \sigma_1', \sigma_2']$.

So let $\mathcal{N}$ be a σ_1-CRS of $\mathcal{N}_1^{\sigma_1}(u)$, let $\mathcal{P}_1$ be a σ_1-consistent acyclic cherry partition of $\mathcal{N}_1$ such that $\mathcal{N}_1^{\sigma_1} = \mathcal{N}_1 - \mathcal{P}_1$, let $\mathcal{P}_2$ be a σ_1'-consistent acyclic cherry partition of $\mathcal{N}_1$ such that $\mathcal{N}_1^{\sigma_1'} = \mathcal{N}_1 - \mathcal{P}_2$, and let $\mathcal{P}_3$ be a σ_1-consistent acyclic cherry partition of $\mathcal{N}_1^{\sigma_1}(u)$ such that $\mathcal{N} = \mathcal{N}_1^{\sigma_1}(u) - \mathcal{P}_3$. Next, let $\mathcal{P}_1' = \mathcal{P}_1 \cap C(\mathcal{N}_1(r_u))$, $\mathcal{P}_2' = \mathcal{P}_2 \setminus C(\mathcal{N}_1(r_u))$, and $\mathcal{P} = \mathcal{P}_1' \cup \mathcal{P}_2' \cup \mathcal{P}_3$. We prove that

(i) $\mathcal{P}$ is a σ_1'-consistent partial cherry partition of $\mathcal{N}_1$ and
(ii) $\mathcal{N}$ is a pendant subnetwork of $\mathcal{N}' := \mathcal{N}_1 - \mathcal{P}$, that is, u is the bottom endpoint of a bridge in $\mathcal{N}'$, and $\mathcal{N}'(u) = \mathcal{N}$.

By the definition of $\mathcal{N}_1^{\sigma_1'}$, (i) implies that there exists a σ_1'-consistent partial cherry partition $\mathcal{P}'$ of $\mathcal{N}_1^{\sigma_1'}$ such that $\mathcal{N}' = \mathcal{N}_1^{\sigma_1'} - \mathcal{P}'$. Since σ_1' exposes u, u is the bottom endpoint of a bridge in $\mathcal{N}_1^{\sigma_1'}$. Thus, by Lemma 2, $\mathcal{P}'' := \mathcal{P}' \cap C(\mathcal{N}_1^{\sigma_1'}(u))$ is a partial cherry partition of $\mathcal{N}_1^{\sigma_1'}(u)$ and $\mathcal{N}_1^{\sigma_1'}(u) - \mathcal{P}'' = \mathcal{N}'(u)$. Since $\mathcal{P}'' \subseteq \mathcal{P}'$ and $\mathcal{P}'$ is σ_1'-consistent, so is $\mathcal{P}''$. This makes $\mathcal{N}'(u)$ a σ_1'-CRS of $\mathcal{N}_1^{\sigma_1'}(u)$. By (ii), $\mathcal{N}'(u) = \mathcal{N}$, that is, $\mathcal{N}$ is a σ_1'-CRS of $\mathcal{N}_1^{\sigma_1'}(u)$, as claimed.

To prove (i), we need to show that $\mathcal{P}$ is an acyclic cherry partition of $\mathcal{N}_1$ and that it is σ_1'-consistent. To prove that $\mathcal{P}$ is an acyclic cherry partition of $\mathcal{N}_1$, we need to prove that the paths in $\mathcal{P}$ are arc-disjoint, that every endpoint of a path in $\mathcal{P}$ is either a leaf of $\mathcal{N}_1$ or an internal vertex of another path in $\mathcal{P}$, and that $G_\mathcal{P}$ is acyclic. To see that the paths in $\mathcal{P}$ are arc-disjoint, observe that no path $P_1 \in \mathcal{P}_1' \cup \mathcal{P}_3$ can share an arc with a path $P_2 \in \mathcal{P}_2'$ because $E(P_1) \subseteq E(\mathcal{N}_1(r_u))$ and $E(P_2) \cap E(\mathcal{N}_1(r_u)) = \emptyset$. No path $P_1 \in \mathcal{P}_3$ shares an arc with a path $P_2 \in \mathcal{P}_1'$ because $\mathcal{P}_1' \subseteq \mathcal{P}_1$, $\mathcal{N}_1^{\sigma_1} = \mathcal{N}_1 - \mathcal{P}_1$, and $\mathcal{P}_3$ is an acyclic cherry partition of $\mathcal{N}_1^{\sigma_1}(u) \subseteq \mathcal{N}_1^{\sigma_1}$. Thus, all paths in $\mathcal{P}$ are arc-disjoint.

Next consider an endpoint v of a path $P \in \mathcal{P}$. If $P \in \mathcal{P}_1'$, then note that, by Lemma 2, $\mathcal{P}_1'$ is an acyclic cherry partition of $\mathcal{N}_1(r_u)$. Thus, v is either a leaf of $\mathcal{N}_1$ or internal to another path in $\mathcal{P}_1' \subseteq \mathcal{P}$. If $P \in \mathcal{P}_3$, then v is either a leaf of $\mathcal{N}_1^{\sigma_1}(u)$ or an internal vertex of a path in $\mathcal{P}_3 \subseteq \mathcal{P}$. If v is a leaf of $\mathcal{N}_1^{\sigma_1}(u)$, then it is either a leaf of $\mathcal{N}_1$ or internal to a path $P' \in \mathcal{P}_1$ because $\mathcal{N}_1^{\sigma_1} = \mathcal{N}_1 - \mathcal{P}_1$. Since $v \in \mathcal{N}_1^{\sigma_1}(u)$ and $u \in B_u$, this path P' belongs to $C(\mathcal{N}_1(r_u))$ and, thus, to $\mathcal{P}_1 \cap C(\mathcal{N}_1(r_u)) = \mathcal{P}_1' \subseteq \mathcal{P}$. Finally, assume that $P \in \mathcal{P}_2'$. Since $\mathcal{P}_2' \subseteq \mathcal{P}_2$ and $\mathcal{P}_2$ is a partial cherry partition of $\mathcal{N}_1$, v is either a leaf of $\mathcal{N}_1$ or an internal vertex of a path $P' \in \mathcal{P}_2$. If $P' \in \mathcal{P}_2'$, then $P' \in \mathcal{P}$. Otherwise, we must have $v = r_u$ because r_u being the bottom endpoint of a bridge in $\mathcal{N}_1$ implies that it is the only vertex that can belong to cherry paths in $\mathcal{P}_2' = \mathcal{P}_2 \setminus C(\mathcal{N}_1(r_u))$ and in $\mathcal{P}_2 \cap C(\mathcal{N}_1(r_u))$. However, since $P \in \mathcal{P}_2$ and $\mathcal{P}_2$ is a cherry partition of $\mathcal{N}_1$, this implies that $\mathcal{N}_1^{\sigma_1'} = \mathcal{N}_1 - \mathcal{P}_2$ cannot contain any vertex in $\mathcal{N}_1(r_u)$, a contradiction because $u \in \mathcal{N}_1(r_u)$ and σ_1' exposes u. Thus, the case when $P' \in \mathcal{P}_2 \setminus \mathcal{P}_2'$ cannot arise. This finishes the proof that every endpoint of a path in $\mathcal{P}$ is a leaf of $\mathcal{N}_1$ or an internal vertex of another path in $\mathcal{P}$.

To see that $G_\mathcal{P}$ is acyclic, note that there are no cycles involving paths from only one of $\mathcal{P}_1'$, $\mathcal{P}_2'$, and $\mathcal{P}_3$ because $\mathcal{P}_1 \supseteq \mathcal{P}_1'$, $\mathcal{P}_2 \supseteq \mathcal{P}_2'$, and $\mathcal{P}_3$ are acyclic cherry partitions. There is no cycle involving paths in $\mathcal{P}_1' \cup \mathcal{P}_3$ and in $\mathcal{P}_2'$ because no endpoint of a path in $C(\mathcal{N}_1(r_u)) \supseteq \mathcal{P}_1' \cup \mathcal{P}_3$ is an internal vertex of any path in $C(\mathcal{N}_1) \setminus C(\mathcal{N}_1(r_u)) \supseteq \mathcal{P}_2'$. Finally, there is no cycle involving cherry in $\mathcal{P}_1'$ and $\mathcal{P}_3$ but not in $\mathcal{P}_2'$ because, for every endpoint v of a path in $\mathcal{P}_1'$, all descendant arcs of v are covered by paths in $\mathcal{P}_1$ and, thus, cannot belong to any path in $\mathcal{P}_3$ because $\mathcal{P}_1'$ and $\mathcal{P}_3$ are arc-disjoint. Thus, $G_\mathcal{P}$ is acyclic and $\mathcal{P}$ is an acyclic cherry partition of $\mathcal{N}_1$.

To finish the proof of (i), we need to prove that $\mathcal{P}$ is σ_1'-consistent. Since $\mathcal{P}_1$ and $\mathcal{P}_3$ are σ_1-consistent and $\mathcal{P}_1' \subseteq \mathcal{P}_1$, $\mathcal{P}_1'$ and $\mathcal{P}_3$ do not contain any path $P \in R(\mathcal{N}_1)$ with $\sigma_1(P) = 0$. Since $\mathcal{P}_1' \cup \mathcal{P}_3 \subseteq C(\mathcal{N}_1(r_u))$ and σ_1 and σ_1' agree on $R(\mathcal{N}_1(r_u))$, this shows that $\mathcal{P}_1' \cup \mathcal{P}_3$ does not contain any path $P \in R(\mathcal{N}_1)$ with $\sigma_1'(P) = 0$. The same is true for $\mathcal{P}_2'$ because $\mathcal{P}_2' \subseteq \mathcal{P}_2$ and $\mathcal{P}_2$ is σ_1'-consistent.

Thus, $\mathcal{P}$ does not contain any path $P \in R(\mathcal{N}_1)$ with $\sigma_1'(P) = 0$. Since $\mathcal{P}_2$ is σ_1'-consistent, it contains every path $P \in R(\mathcal{N}_1)$ with $\sigma_1'(P) = 1$. If this path belongs to $R(\mathcal{N}_1) \setminus R(\mathcal{N}_1(r_u))$, then it belongs to $\mathcal{P}_2' \subseteq \mathcal{P}$. If it belongs to $R(\mathcal{N}_1(R_u))$, then $\sigma_1(P) = \sigma_1'(P) = 1$. Thus, as $\mathcal{P}_1$ is σ_1-consistent, it contains P, as does $\mathcal{P}_1' = \mathcal{P}_1 \cap C(\mathcal{N}_1(r_u))$. Thus, $\mathcal{P}$ contains all paths in $R(\mathcal{N}_1)$ with $\sigma_1'(P) = 0$. Since it does not contain any paths $P \in R(\mathcal{N}_1)$ with $\sigma_1'(P) = 0$, this proves that $\mathcal{P}$ is σ_1'-consistent.

To prove (ii), observe that our proof of (i) shows in fact that $\mathcal{P}_1' \cup \mathcal{P}_2'$ is an acyclic cherry partition of $\mathcal{N}_1$. Since $\mathcal{P}_3$ is an acyclic cherry partition of $\mathcal{N}_1^{\sigma_1}(u)$ and $\mathcal{N}_1^{\sigma_1}(u) \setminus \mathcal{P}_3 = \mathcal{N}$, it suffices to prove that $\mathcal{N}_1^{\sigma_1}(u)$ is a pendant subnetwork of $\mathcal{N}_1 - (\mathcal{P}_1' \cup \mathcal{P}_2')$. Since σ_1 exposes u, $\mathcal{N}_1^{\sigma_1}(u)$ is a pendant subnetwork of $\mathcal{N}_1^{\sigma_1} = \mathcal{N}_1 - \mathcal{P}_1$. By Lemma 2, we have $\mathcal{N}_1(r_u) - \mathcal{P}_1' = \mathcal{N}_1^{\sigma_1}(r_u)$, that is, $\mathcal{N}_1^{\sigma_1}(u)$ is a pendant subnetwork of $\mathcal{N}_1(r_u) - \mathcal{P}_1'$. Since $\mathcal{P}_2' \cap C(\mathcal{N}_1(r_u)) = \emptyset$ and $\mathcal{N}_1(r_u)$ is a pendant subnetwork of $\mathcal{N}_1$, this implies that $\mathcal{N}_1^{\sigma_1}(u)$ is a pendant subnetwork of $\mathcal{N}_1 - (\mathcal{P}_1' \cup \mathcal{P}_2')$, as required. This proves (ii) and, thus, finishes the proof. $\square$

Proof of Lemma 13. We use induction on the length of any path from $D^*[\rho_1, \rho_2, \bot, \bot]$ to $D^*[u, v, \sigma_1, \sigma_2]$ in the dependency graph to prove the lemma. For $D^*[\rho_1, \rho_2, \bot, \bot]$, the lemma clearly holds. For the inductive step, we consider a required entry $D^*[u, v, \sigma_1, \sigma_2]$ that satisfies the lemma, and prove that the lemma holds also for all its out-neighbours in the dependency graph.

So let $D^*[x, y, \sigma_1', \sigma_2']$ be such an out-neighbour. If $D^*[u, v, \sigma_1, \sigma_2]$ depends on it via Eq. (1), then $x = u$ and σ_1' is a B_u-extension of σ_1. Thus, σ_1' fixes all cherry paths in $R(B_u)$ and, since σ_1 does not fix any cherry paths in proper descendant blobs of B_u, neither does σ_1'. By an analogous argument, $y = v$ and σ_2' fixes all cherry paths in $R(B_v)$ and none of the cherry paths in proper descendant blobs of B_v. If $D^*[u, v, \sigma_1, \sigma_2]$ depends on $D^*[x, y, \sigma_1', \sigma_2']$ via Eqs. (2) and (5), then $\sigma_1' = \sigma_1$ and B_x is a descendant blob of B_u. Since σ_1 fixes all or no cherry paths in B_u and no cherry paths in any proper descendant blob of B_u, this implies that σ_1' fixes all or no cherry paths in B_x and no cherry paths in any proper descendant blob of B_x. By an analogous argument, σ_2' fixes all or no cherry paths in B_y and no cherry paths in any proper descendant blob of B_y. This finishes the inductive step and, thus, the proof of the lemma. $\square$

References

1. Bansal, M.S., Kellis, M., Kordi, M., Kundu, S.: Ranger-dtl 2.0: rigorous reconstruction of gene-family evolution by duplication, transfer and loss. Bioinformatics **34**(18), 3214–3216 (2018)
2. Bean, W.J., Jr., Cox, N.J., Kendal, A.P.: Recombination of human influenza a viruses in nature. Nature **284**(5757), 638–640 (1980)
3. Berry, V., Scornavacca, C., Weller, M.: Scanning phylogenetic networks is NP-hard. In: International Conference on Current Trends in Theory and Practice of Informatics, pp. 519–530. Springer (2020)
4. Bruchhold, S., Weller, M.: Exploiting low scanwidth to resolve soft polytomies. In: Kozik, J., Wolff, A. (eds.) SOFSEM 2026: Theory and Practice of Computer Science, pp. 332–346. Springer Nature Switzerland, Cham (2026)

5. Cardona, G., Pons, J.C., Ribas, G., Coronado, T.M.: Comparison of orchard networks using their extended μ-representation. IEEE/ACM Trans. Comput. Biol. Bioinf. **21**(3), 501–507 (2024)
6. Erdős, P.L., Semple, C., Steel, M.: A class of phylogenetic networks reconstructable from ancestral profiles. Math. Biosci. **313**, 33–40 (2019)
7. Gambette, P., Berry, V., Paul, C.: The structure of level-k phylogenetic networks. In: Annual Symposium on Combinatorial Pattern Matching, pp. 289–300. Springer (2009)
8. Hellmuth, M., Schaller, D., Stadler, P.F.: Clustering systems of phylogenetic networks. Theory Biosci. **142**(4), 301–358 (2023)
9. Hills, M.J., Hall, L.M., Messenger, D.F., Graf, R.J., Beres, B.L., Eudes, F.: Evaluation of crossability between triticale (x triticosecale wittmack) and common wheat, durum wheat and rye. Environ. Biosaf. Res. **6**(4), 249–257 (2007)
10. van Iersel, L., Janssen, R., Jones, M., Murakami, Y.: Orchard networks are trees with additional horizontal arcs. Bull. Math. Biol. **84**(8), 76 (2022)
11. van Iersel, L., Janssen, R., Jones, M., Murakami, Y., Zeh, N.: A unifying characterization of tree-based networks and orchard networks using cherry covers. Adv. Appl. Math. **129**, 102222 (2021)
12. van Iersel, L., Jones, M., Weller, M.: Embedding phylogenetic trees in networks of low treewidth. Discrete Math. Theor. Comput. Sci. **25**(Discrete Algorithms) (2023)
13. Jacox, E., Chauve, C., Szöllősi, G.J., Ponty, Y., Scornavacca, C.: Eccetera: comprehensive gene tree-species tree reconciliation using parsimony. Bioinformatics **32**(13), 2056–2058 (2016)
14. Janssen, R., Murakami, Y.: On cherry-picking and network containment. Theoret. Comput. Sci. **856**, 121–150 (2021)
15. Jones, M., Schestag, J.: Parameterized algorithms for diversity of networks with ecological dependencies. In: 20th International Symposium on Parameterized and Exact Computation (IPEC 2025). Leibniz International Proceedings in Informatics (LIPIcs), vol. 358, pp. 11:1–11:21. Schloss Dagstuhl – Leibniz-Zentrum für Informatik (2025)
16. Kelk, S., Scornavacca, C.: Constructing minimal phylogenetic networks from softwired clusters is fixed parameter tractable. Algorithmica **68**(4), 886–915 (2014)
17. Landry, K., Teodocio, A., Lafond, M., Tremblay-Savard, O.: Defining phylogenetic network distances using cherry operations. IEEE/ACM Trans. Comput. Biol. Bioinf. **20**(3), 1654–1666 (2022)
18. Landry, K., Tremblay-Savard, O.: Fast calculation of cherry distance on level-1 orchard networks: optimization, heuristic and implementation. In: RECOMB International Workshop on Comparative Genomics, pp. 157–177. Springer (2025)
19. Landry, K., Tremblay-Savard, O., Lafond, M.: A fixed-parameter tractable algorithm for finding agreement cherry-reduced subnetworks in level-1 orchard networks. J. Comput. Biol. **31**(4), 360–379 (2024)
20. Lin, Y., Rajan, V., Moret, B.M.: A metric for phylogenetic trees based on matching. IEEE/ACM Trans. Comput. Biol. Bioinf. **9**(4), 1014–1022 (2011)
21. López Sánchez, A., Lafond, M.: Predicting horizontal gene transfers with perfect transfer networks. Algorithms Molecular Biol. **19**(1), 6 (2024)
22. Nakhleh, L., Ringe, D.A., Warnow, T.: Perfect phylogenetic networks: a new methodology for reconstructing the evolutionary history of natural languages. Language **81**(2), 382–420 (2005)
23. Robinson, D.F., Foulds, L.R.: Comparison of phylogenetic trees. Math. Biosci. **53**(1–2), 131–147 (1981)

24. Scornavacca, C., Weller, M.: Treewidth-based algorithms for the small parsimony problem on networks. Algorithms Molecular Biol. **17**(1), 15 (2022)
25. Van Iersel, L., Kelk, S., Lekic, N., Whidden, C., Zeh, N.: Hybridization number on three rooted binary trees is EPT. SIAM J. Discret. Math. **30**(3), 1607–1631 (2016)

Machine Learning in Genomics

DipGNNome: Diploid *de Novo* Genome Assembly with Geometric Deep Learning and Beam-Search

Martin Schmitz[1,2(✉)], Lovro Vrček[2], Kenji Kawaguchi[1], and Mile Šikić[2,3(✉)]

[1] School of Computing, National University of Singapore, Singapore, Singapore
[2] Genome Institute of Singapore, A*STAR, Singapore, Singapore
schmitz.kessenich@gmail.com
[3] Faculty of Electrical Engineering and Computing, University of Zagreb, Zagreb, Croatia

Abstract. *De novo* genome assembly remains a central challenge in computational biology, particularly for diploid genomes where maternal and paternal haplotypes must be accurately resolved. Existing assemblers achieve impressive results through carefully designed heuristics, yet modern deep learning methods remain largely unexplored in the diploid setting.

We present `DipGNNome`, the first deep learning–based framework for diploid *de novo* genome assembly. Our approach formulates genome assembly as an edge classification and graph traversal problem, given haplotype-aware assembly graphs. We train a graph neural network (GNN) to guide contig construction as the layout phase in an Overlap-Layout-Consensus genome assembly pipeline. To enable this, we establish a novel pipeline for generating diploid graphs with ground-truth edge labels, providing the first systematic way to produce training data for machine learning models in this domain. This framework creates a foundation for applying and extending graph-based deep learning to diploid assembly. `DipGNNome` creates assemblies comparable to SotA and demonstrates the feasibility of deep learning for diploid assembly and introduces a paradigm that bridges algorithmic genomics with graph representation learning.

Our code, dataset and trained model are openly available at https://github.com/lbcb-sci/DipGNNome.

Keywords: *De novo* genome assembly · Graph Neural Networks · Diploid assembly

1 Introduction

The genome is encoded in sequences of four nucleotides—adenine (A), guanine (G), cytosine (C), and thymine (T), that together form deoxyribonucleic acid (DNA). DNA is organized into chromosomes, each spanning millions to hundreds

M. Lafond (Ed.): RECOMB-CG 2026, LNBI 16569, pp. 317–338, 2026.
https://doi.org/10.1007/978-3-032-26891-4_15

of millions of nucleotides. A genome typically consists of multiple chromosomes, often present in pairs or higher ploidy levels, where homologous chromosomes are similar but not identical. Sequencing technologies make it possible to extract genomic information from these chromosomes, producing large collections of short, unordered fragments known as reads. Reconstructing the full genome from such reads is the central problem of genome assembly. Reference-based assembly methods align reads to a reference genome. While effective when a high-quality reference is available, this approach can introduce bias and obscure genuine structural variation between the target and reference genomes. In contrast, *de novo* assembly reconstructs genomes directly from reads, without relying on a reference, thereby avoiding these biases.

De novo assembly is a cornerstone of computational genomics, with applications spanning foundational research in biology, evolutionary biology, and personalized medicine. Nurk et al. (2022) [12] produced the first complete telomere-to-telomere (T2T) human genome, followed by numerous high-quality assemblies for humans and other species. These efforts combine complementary sequencing technologies, sophisticated assembly algorithms, and extensive manual curation by large consortia. The progress of recent assemblers has been driven primarily by algorithmic innovations. SotA assemblers like `hifiasm` [3] and `Verkko` [14] exploit the high accuracy of PacBio HiFi reads and the ultra-long range of Oxford Nanopore reads to generate phased, near-complete T2T assemblies. Yet, despite these achievements, current approaches remain purely algorithmic and do not incorporate modern machine learning techniques that have transformed other domains.

In this work, we present `DipGNNome`, the first deep learning–based assembler capable of reconstructing diploid genomes directly from raw assembly graphs, augmented with trio information. `DipGNNome` takes overlap graphs as input and performs the Layout step of an Overlap-Layout-Consensus pipeline. Unlike existing methods, `DipGNNome` integrates GNNs with haplotype-aware graph processing to jointly reconstruct maternal and paternal haplotypes, bridging the gap between classical algorithmic assemblers and emerging machine learning approaches.

2   Related Work

GNNome [18] marks the first deep learning–based genome assembler. It trains a GNN to classify assembly graph edges as correct or incorrect and then applies a greedy pathfinding algorithm to construct contigs from the scored edges. Similarly, Simunovic et al. [15] focus on the layout of de Bruijn graphs. However, these approaches are limited to producing haploid assemblies.

Diploid assembly poses additional challenges. Most organisms, including humans, carry two homologous copies of each chromosome. While homozygous regions are nearly identical, heterozygous regions can harbor substantial divergence. Distinguishing sequencing errors from true variants and correctly phasing both haplotypes remains a central difficulty in diploid assembly. Recent years

have seen the development of deep learning–based phasing methods such as
GAEseq [6], CAECseq [5], NeurHap [19], and ralphi [1]. These approaches highlight the potential of machine learning for haplotype resolution, but they operate
in a reference-based setting and do not address *de novo* assembly.

Strategies for *de novo* diploid assembly usually incorporate external information to resolve haplotypes. A prominent example is trio binning [7], which
leverages parental short reads to identify haplotype-specific k-mers and guide
separation during assembly, as implemented in tools such as Verkko [14] and
hifiasm [3]. Our method builds on this paradigm by combining haplotype-aware
assembly graphs with parental k-mer annotation and graph-based machine learning.

3   Overview

Our contributions are threefold: (1) we introduce the first publicly available
pipeline for constructing machine learning–ready assembly graphs with ground-
truth, enabling supervised training in the diploid setting, and share our dataset;
(2) we develop an assembly algorithm that combines model predictions with a
beam-search strategy to efficiently traverse long, string-like graphs with limited
branching, a design that may generalize to other path-finding tasks; and (3) we
show that a GNN-based assembler can achieve assembly results close to SotA
methods while following a fundamentally different, learning-driven paradigm.

Fig. 1 outlines the three stages of our approach: (A) **Data processing**, where
HiFi reads are assembled into a unitig graph with hifiasm, simplified, and
annotated with haplotype information using trio-derived k-mers; (B) **Synthetic
training**, where simulated diploid reads generate labeled graphs for supervised
GNN training via edge classification; and (C) **Genome assembly**, where real
data are processed as in step (A), edges are scored by the trained GNN from step
(B), and a beam-search reconstructs phased maternal and paternal haplotypes.

The remainder of the paper is organized as follows: The next three sections
describe the three parts of the pipeline, (A), (B), and (C). Section 4 details the
data processing pipeline, Sect. 5 describes GNN training, and Sect. 6 presents the
genome assembly algorithm. Section 7 evaluates DipGNNome on real and simulated
datasets, comparing it to the SotA. Section 8 concludes with a discussion of
implications, limitations, and future directions.

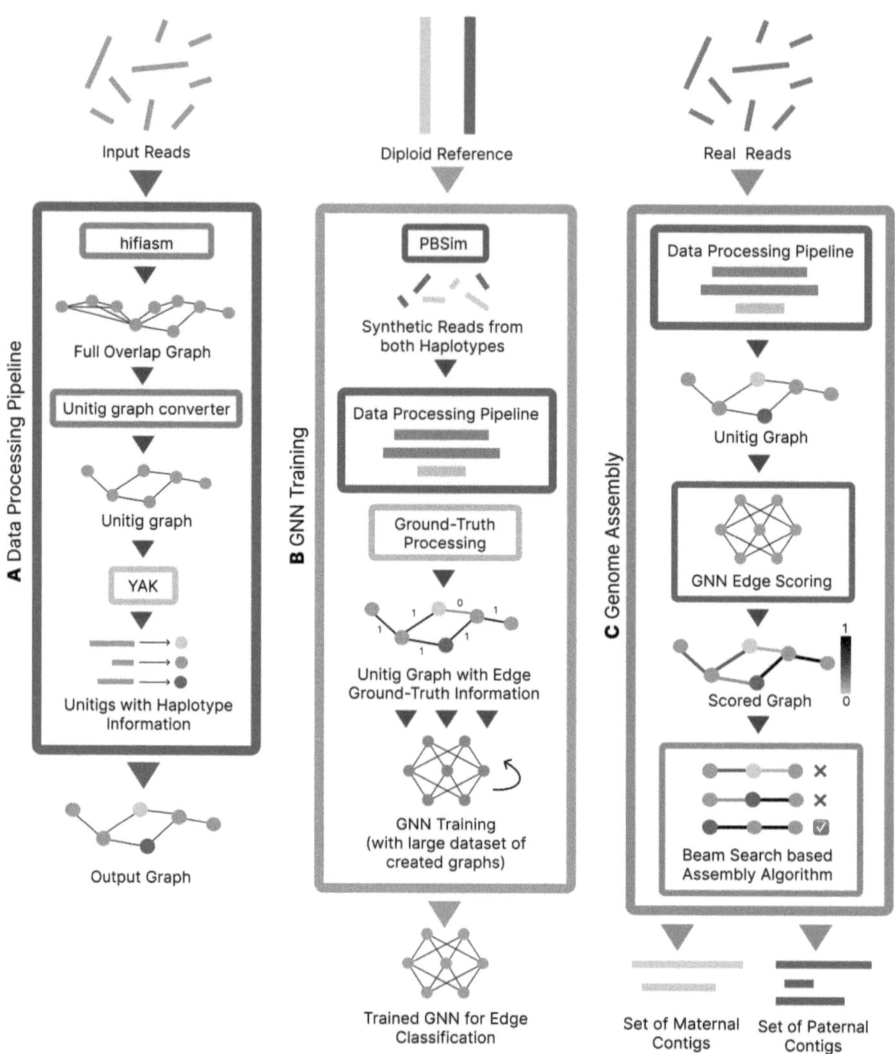

Fig. 1. Overview of the `DipGNNome` pipeline. **A: Data processing.** HiFi reads are assembled into a unitig graph with **hifiasm**, simplified, and annotated with haplotype information using trio-derived k-mers. **B: Synthetic training data.** Simulated diploid reads generate labeled graphs, enabling supervised training of a GNN for edge classification. **C: Genome assembly.** Real data is processed as described in **A**, edges are scored by the trained GNN from **B**, and a beam-search based algorithm reconstructs phased haplotypes.

4   Data Processing Pipeline

The following section describes how we construct graphs as summarized in Fig. 1 **A**. Given a set of reads (real or synthetic) as input, the pipeline consists of the following steps:

4.1   Create Raw Overlap Graph (Step 1)

We start with an overlap graph, where each node represents a read in the dataset and each edge represents an overlap. Any tool that can produce such a graph is possible to be used here. For our experiments, we use `hifiasm` v0.25 [4] with default parameters.

4.2   Convert Raw Graph to Unitig Graph (Step 2)

Given the raw overlap graph, we search for transitive edges using an algorithm based on Myers' transitive edge reduction algorithm [11]. An edge $e_t = a \rightarrow c$ is considered transitive if edges $e_1 = a \rightarrow b$ and $e_2 = b \rightarrow c$ both exist. To relax this rule, we introduce an $\epsilon = 0.12$ parameter (inspired by Raven [17]), which allows transitive edges to be removed only if the indirect path length lies between $d(1 - \epsilon)$ and $d(1 + \epsilon)$ (where d is the direct path). After the removal of transitive edges, unbranching chains of nodes are merged into so-called *unitigs*. This significantly reduces the size and complexity of the graph. The resulting graph is called a unitig graph.

4.3   Add Trio Binning to Unitigs Using yak (Step 3)

We use Illumina short reads from both parents of the genome of interest and apply `yak` [4] trio binning. `yak` identifies unique k-mers in both maternal and paternal short reads. Then it traverses the unitigs in the graph and marks unique k-mers as identified before. We leverage this counts of unique parental k-mer hits, $k_m(v)$ and $k_p(v)$. These values provide an informative representation of each unitig, capturing both the haplotype assignment (maternal vs. paternal) and the strength of the heterozygosity (the absolute count $k_m(v)$ and $k_p(v)$ may reflect how heterozygous a unitig is).

4.4   Add Node and Edge Features (Step 4)

We enrich the graph with both node and edge features. Edge features include the *overlap length* (number of base pairs shared between two reads). Node features include *read support* (the number of raw reads contributing to a given unitig, with read-level coverage computed by `hifiasm` and aggregated in unitigs as a length-weighted average) as well as *in-degree/out-degree* of the nodes.

5   GNN Training

This Section explains the GNN training pipeline as shown in Fig. 1 **B**. Given a set of reference genomes, we create a synthetic training dataset with ground-truth and use this to train a GNN.

5.1   Sample Reads

We simulate $40\times$ coverage reads ($20\times$ per haplotype) using `PBSim3` [13], sampling independently from maternal and paternal references of each chromosome. The resulting FASTQ files are merged into one, with reads annotated by position, strand, chromosome, and haplotype.

5.2   Compute Ground-Truth

Training our GNN requires labeled assembly graph edges. We first mark all edges connecting different chromosomes (translocations) or opposite strands (inversions) as false. Translocation edges are additionally saved as a separate list to train a second translocation classifier. Positional information determines whether an edge represents a valid suffix–prefix overlap. This requires assigning coordinates to unitigs.

For diploid genomes, differences in haplotype coordinates between homologous loci complicate verification. To resolve this, we translate coordinates between haplotypes using `Liftover` [16] and refine missing values by leveraging information from multiple reads within unitigs. Overlaps are then validated in both coordinate systems, and an edge is labeled correct if it is consistent in at least one haplotype. This procedure yields reliable supervision signals for robust model training. A full description of our ground-truth assigning algorithm can be found in Appendix A.

5.3   GNN Model

We train a SymGatedGCN model with SymBCELoss, both introduced by GNNome [18], using the ground-truth edge labels described above. SymGatedGCN extends GatedGCN [2] and produces d-dimensional node and edge embeddings for the assembly graph. A jointly trained MLP classifier uses these embeddings to predict (i) whether an edge is incorrect (i.e., not part of the optimal assembly) and (ii) whether it represents a translocation. The model is optimized using binary cross-entropy loss and stochastic gradient descent.

To mitigate oversmoothing, we apply PairNorm [23] after each GNN layer. For node representations $V = [v_1, \ldots, v_n] \in \mathbb{R}^{n \times d}$, PairNorm centers, normalizes, and rescales features:

$$V' = s \cdot \frac{V - \mu(V)}{\mathrm{mean}_i \| v_i - \mu(V) \|_2}, \tag{1}$$

where $\mu(V) = \frac{1}{n}\sum_{i=1}^{n} v_i$ and $s = 1$. This preserves relative feature distances and improves training stability.

Node features $x_i \in \mathbb{R}^{d_v}$ and edge features $z_{ij} \in \mathbb{R}^{d_e}$ are projected into d-dimensional embeddings through the GNN. The MLP classifier then combines edge and incident node embeddings to produce the two prediction scores, which are used for graph pruning and downstream diploid assembly.

6 Genome Assembly Algorithm

6.1 Problem Definition

We formulate the *de novo* genome assembly problem as a path-finding task in a directed graph $G = (V, E)$, where each node can be traversed exactly once.
Nodes (V): Each node $v \in V$ represents a unitig and is associated with the following attributes:

- $l_n(v)$: Length of the genomic sequence represented by the unitig.
- $\mathrm{kmer}_m(v), \mathrm{kmer}_p(v)$: Counts of unique k-mers assigned to the maternal and paternal haplotypes, respectively.

Edges (E): Each edge $e = (u, v) \in E$ represents an overlap between two unitigs and has the following attributes:

- $S_G(e)$: The model output, interpreted as the probability that the connection is incorrect. Due to the sigmoid activation in the output layer, $S_G(e) \in [0, 1]$.
- $l_e(e)$: Length of the sequence contributed by v when extending from u to v.

Note, the attributes listed here differ from those used during GNN training. Certain features used for training are no longer directly relevant, as their information is captured in the model predictions. Conversely, the assembly algorithm requires additional attributes that were not important for predicting edge quality but are essential for constructing contigs. Additionally, the graph has the following additional properties:

Complement Relationships. For each node v, its complement $v \oplus 1$ represents the reverse complement sequence, where $\oplus$ represents a bit-wise XOR (for example $0 \oplus 1 = 1$ and $1 \oplus 1 = 0$). This relationship is important throughout the algorithm. For each node v, node $v \oplus 1$ is guaranteed to exist in the graph, and for each edge (u, v), an edge $(v \oplus 1, u \oplus 1)$ exists, which imposes a symmetry on the entire graph.

Component Decomposition. The graph is decomposed into weakly connected components, with small components filtered out. Each remaining component is processed independently. This means we can assume for the algorithm that the input graph is weakly connected. Before running the assembly algorithm, we prune the graph using the model's *translocation edge score*. Translocation edges are defined as edges that connect nodes from different chromosomes (i.e., the source node lies on one chromosome and the target node on another). Specifically, we remove all edges with a translocation prediction $p_t > c_t$, where c_t denotes the translocation cut threshold (set to $c_t = 0.5$ in our experiments). We also prune edges with edge score $S_G(e) > 0.9$, to remove edges with very poor model predictions entirely.

6.2   Assembly Algorithm

The full genome assembly consists of two independent passes: one with haplotype $h = m$ (maternal), and one with $h = p$ (paternal). In each pass, the algorithm traverses weakly connected components of the graph and penalizes paths based on unique k-mer counts (Sect. 4.3) from the opposite haplotype. Specifically, nodes with $kmer_p(v) > 0$ are penalized in the maternal pass, and nodes with $kmer_m(v) > 0$ are penalized in the paternal pass.

Each weakly connected component is traversed independently and over several iterations. A path is computed, the component is reduced by the set of visited nodes, and the process is repeated until no valid path longer than a threshold can be found or the component becomes too small. One path search is initiated by sampling n random edges (u, v) in the graph. For each sampled edge, we perform two beam search runs: one forward search from v and one reverse search from $u \oplus 1$ (complement of the source). The reverse path is then flipped and modified to use complement nodes. Finally, we concatenate both paths and select the longest among the n samples. The complete process is shown as pseudocode in Appendix B.

The core of this assembly algorithm is the `AssemblyBeamSearch` function, a path-finding approach based on beam search—an algorithm that maintains a set of k candidate paths (the "beams") and iteratively extends them by considering the k best resulting paths. Our assembly algorithm includes several diploid-genome-assembly-specific adaptations to traditional beam search. The goal is to maximize an assembly heuristic, based on unitig lengths, haplotype information, and model scores. The approach is shown in detail as Pseudocode in Appendix C.

We incorporate several changes to the default beam-search. First, we **save the best intermediate beam** during the search. This is important because it may happen that only negative-scoring edges are available at some point, leading to decreasing path scores. While we do not want to prevent the agent from taking negative edges (to ensure contiguity and because the predictions of the model are not always reliable), we also want to avoid letting the search degrade indefinitely. To address this, we track the best beam seen so far and retain it, even if the current beams degrade. This way, we can output high-scoring sub-paths without requiring the beam to reach the end of a component.

We also explicitly **prevent complement pairs** from being included in a single beam. Each beam maintains a visited set that includes all its nodes and their complements; any node present in this set is excluded from further beam extensions.

Additionally, we perform **beam merging** to reduce redundancy. The directed assembly graph tends to have string-like regions where many beams overlap. When two beams reach the same node, we retain only the higher-scoring one and discard the other. This conserves memory and runtime while ensuring that the best-scoring paths are preserved. Appendix D shows the complete Pseudocode of the beam merging strategy.

To mark visited nodes and their neighbors, we call `MarkVisited`, which includes each node, its complement, and the 2-hop neighborhood of both. This prevents re-traversal of nodes and the traversal of nodes that represent the same sequence in the genome as already visited ones.

6.3   Beam Score Heuristic

The beam search algorithm evaluates each candidate edge $e = (u, v)$ using a score $S(e)$ that balances the amount of novel sequence added to the assembly, haplotype consistency, and model-predicted edge confidence. The overall score is defined as:

$$S(e) = L(v) - K(v) - M(e) \tag{2}$$

The three components are defined as follows:

Length Reward $(L(v))$: This term gives a reward for additional base pairs added to the sequence from passing this edge, such that longer contigs get higher scores than shorter ones. Let $l(v)$ be the total length of the sequence associated with node v, and $l(e)$ the overlap length between nodes u and v. Then the net new sequence added by edge e is $l(v) - l(e)$, and the reward is computed as:

$$L(e) = \alpha \cdot (l(v) - l(e)) \tag{3}$$

where α is a scaling hyperparameter.

K-mer Penalty $(K(v))$: To promote haplotype consistency during diploid assembly, we penalize edges that introduce unique k-mers associated with the *non-target* haplotype. Let $k_{\neg h}(v)$ denote the number of such unique k-mers in node v, where $\neg h$ represents the haplotype opposite to the current assembly target h. We scale this penalty by the proportion of node v that is newly added by edge e, assuming a roughly uniform distribution of k-mers along the sequence (a simplification that holds well enough in practice):

$$K(e) = \beta \cdot k_{\neg h}(v) \cdot \frac{l(v) - l(e)}{l(v)} \tag{4}$$

The hyperparameter β controls the weight of this penalty in the overall beam score.

Model Score Penalty ($M(e)$): Edges are penalized based on their model score $m(e) \in [0, 1]$. To discourage high-confidence anomalous connections strongly, we use a quadratic penalty:

$$M(e) = \gamma \cdot m(e)^2 \tag{5}$$

where γ controls the sensitivity of the penalty to prediction uncertainty.

Putting all components together, the final beam score for edge $e = (u, v)$ is:

$$S(e) = \alpha \cdot (l(v) - l(e)) - \beta \cdot k_{\neg h}(v) \cdot \frac{l(v) - l(e)}{l(v)} - \gamma \cdot m(e)^2 \tag{6}$$

We conducted an ablation study demonstrating that each component of the beam search scoring function is critical for optimal performance. Omitting the model score penalty reduces contiguity, removing the k-mer penalty eliminates the ability to properly phase the assembly, and excluding the length reward results in highly fragmented assemblies. The full results can be found in Appendix E.

7   Evaluation

We evaluate our methods' capacity on various real genomes. Our model is trained on synthetic human data based on a single reference genome: I002C [8]. For evaluation, we focus exclusively on real-world data and assess assembly quality across a range of datasets consisting of a different human genome and 6 primate species genomes. All experiments are conducted at the full diploid genome level, using datasets for which high-quality reference genomes are available.

7.1   Training Dataset Overview

We construct a training dataset, consisting of graphs generated from all chromosomes of the human genome I002C, with about 7.7M nodes and 10.5M edges across training and validation splits, and an average degree of 2.7. Roughly 85% of edges are labeled as correct, while the remaining 15% correspond to false edges. A detailed breakdown of the training dataset statistics is provided in Appendix G.

7.2   Evaluation Dataset Overview

The evaluation set comprises seven genomes: The human (*H. sapiens*) reference HG002 by Nurk et al. 2022 [12], and the six T2T ape references created by Yoo et al. 2024 [21] : *P. paniscus* (bonobo), *G. gorilla* (gorilla), *P. troglodytes* (chimpanzee), *S. syndactylus* (siamang), *P. abelii* (Sumatran orangutan), and *P. pygmaeus* (Bornean orangutan).

For *P. troglodytes*, *S. syndactylus*, *P. abelii*, and *P. pygmaeus*, no parental short-read data were available. Instead, we generated unique k-mers directly from the reference genomes. To approximate read input, we duplicated each reference sequence file 30 times and treated them as read files. The resulting "synthetic" YAK files were then used in place of real-world unique k-mer files typically derived from Illumina parental data. All tested methods use these YAK files for fair comparison. The exact files and coverages used for the evaluation dataset are detailed in Appendix H.

7.3 Training Setup

To allow for mini-batch training on large graphs, we partition each graph using METIS into subgraphs containing at most 40,000 nodes (which is approximately the graph size to fit the full GPU memory). Each resulting subgraph is treated as a training batch. More training details, such as the final model configuration and the hardware used, can be found in Appendix I.

7.4 Genome Assembly Results

We benchmark three approaches: a greedy variant of DipGNNome, the full DipGNNome algorithm, and hifiasm. The greedy variant is identical to DipGNNome except that it replaces beam-search with a purely greedy expansion strategy, always selecting the next best node (according to the same scoring metric as the beam-search-based algorithm).

Table 1 shows the results including assembly length, duplication rate (Rdup), contiguity (NG50 and NGA50), and haplotype error rates (YAK Switch Error and Hamming Error) for paternal (P) and maternal (M) haplotypes. Values are shown as P/M (paternal/maternal). Hamming and switch error rates were computed using yak [4], while all other metrics (length, duplication rate, NG50, NGA50) were computed using minigraph [9]. Lower duplication and error rates indicate better assembly quality, while higher NGA50 reflects better (correct) contiguity. NG50 and total length alone do not necessarily indicate better or worse assembly quality. However, differences between NG50 and NGA50 display misassemblies.

Across most genomes, hifiasm achieves the lowest Hamming error rates and is generally on par with or superior to DipGNNome across the majority of haplotypes, particularly in terms of phasing accuracy. Nonetheless, DipGNNome remains competitive, often reaching comparable NGA50 values and in some cases surpassing hifiasm.

Most notably, on the orangutan genomes, DipGNNome shows clear advantages: for *P. pygmaeus*, it achieves substantially higher NGA50 for both haplotypes while maintaining comparable or slightly improved Hamming error, and for *P. abelii*, it also improves contiguity (NGA50) for one haplotype despite a moderate increase in error rates. Overall, while hifiasm remains the stronger assembler, these results highlight the potential of machine learning–guided diploid assembly. The greedy variant consistently performs worse, with lower NGA50 and higher error rates, confirming the importance of our beam-search-based algorithm for robust path exploration.

Table 1. Comparison of `DipGNNome` and `hifiasm` on different genomes. The '*' marks genomes, where the parental k-mers are created synthetically.

Genome	Metric	DipGNNome Greedy	DipGNNome Beam-Search	hifiasm
H. sapiens	Length (Mbp)	2840.0 / 2946.1	2856.4 / 2970.3	2938.5 / 3032.3
	Rdup (%)	**0.1** / **0.1**	0.2 / 0.3	0.8 / 0.2
	NG50 (Mbp)	35.7 / 25.7	63.0 / 65.2	56.0 / 59.4
	NGA50 (Mbp)	33.6 / 25.7	48.4 / 49.3	**49.7** / **58.6**
	Switch Err (%)	1.1 / 1.2	1.1 / 1.2	**0.8** / **1.0**
	Hamming Err (%)	2.2 / 2.5	1.9 / 3.0	**0.8** / **0.8**
P. paniscus	Length (Mbp)	3114.5 / 2934.2	3125.5 / 2947.4	3204.4 / 3066.0
	Rdup (%)	2.0 / **0.8**	1.9 / 1.0	**1.2** / 1.1
	NG50 (Mbp)	52.6 / 48.5	94.0 / 70.3	100.1 / 63.0
	NGA50 (Mbp)	35.2 / 39.6	50.1 / 42.3	**55.2** / **47.1**
	Switch Err (%)	0.3 / 1.8	0.3 / 1.7	**0.2** / **1.5**
	Hamming Err (%)	1.3 / 2.3	1.7 / 3.1	**0.1** / **1.3**
G. gorilla	Length (Mbp)	3366.2 / 3267.0	3389.8 / 3281.6	3528.4 / 3351.6
	Rdup (%)	**1.1** / 1.4	1.2 / 1.4	1.2 / **1.1**
	NG50 (Mbp)	69.0 / 44.3	108.1 / 100.6	94.7 / 82.6
	NGA50 (Mbp)	45.7 / 33.6	**56.1** / **48.9**	55.5 / 48.6
	Switch Err (%)	0.3 / **0.2**	0.3 / **0.2**	**0.2** / 0.2
	Hamming Err (%)	1.1 / 1.0	1.4 / 1.0	**0.2** / **0.2**
*P. troglodytes**	Length (Mbp)	3056.0 / 2938.1	3080.7 / 2956.3	3149.0 / 3032.9
	Rdup (%)	**0.1** / **0.1**	0.5 / 0.2	**0.1** / **0.1**
	NG50 (Mbp)	72.1 / 74.8	136.8 / 136.6	126.0 / 121.9
	NGA50 (Mbp)	62.2 / 68.0	**102.8** / 76.1	101.9 / **121.9**
	Switch Err (%)	**0.1** / **0.2**	**0.1** / **0.2**	**0.1** / **0.2**
	Hamming Err (%)	0.3 / 0.4	0.7 / 1.3	**0.1** / **0.1**
*S. syndactylus**	Length (Mbp)	3131.6 / 3029.6	3133.5 / 3038.1	3230.0 / 3127.6
	Rdup (%)	0.2 / **0.1**	0.3 / 0.3	**0.1** / 0.7
	NG50 (Mbp)	67.6 / 50.7	90.8 / 69.6	84.9 / 75.4
	NGA50 (Mbp)	55.4 / 45.8	56.1 / **58.6**	**76.1** / 38.7
	Switch Err (%)	**0.1** / **0.2**	**0.1** / **0.2**	**0.1** / **0.2**
	Hamming Err (%)	0.3 / 0.4	0.5 / 0.6	**0.1** / **0.1**
*P. abelii**	Length (Mbp)	3297.8 / 3487.0	3332.7 / 3522.1	2967.5 / 3361.0
	Rdup (%)	10.5 / 12.7	11.1 / 13.6	**2.3** / **7.5**
	NG50 (Mbp)	55.9 / 53.7	79.8 / 80.8	40.1 / 94.2
	NGA50 (Mbp)	43.0 / 40.6	**45.7** / 57.5	34.1 / **65.3**
	Switch Err (%)	6.5 / 4.6	6.5 / 4.6	**4.8** / **3.8**
	Hamming Err (%)	6.3 / 4.4	6.7 / 4.4	**5.0** / **3.8**
*P. pygmaeus**	Length (Mbp)	3166.4 / 3272.6	3176.3 / 3304.0	3092.5 / 3212.2
	Rdup (%)	5.8 / **5.8**	6.2 / 6.4	**2.7** / 6.5
	NG50 (Mbp)	51.8 / 52.5	108.2 / 100.9	65.8 / 85.0
	NGA50 (Mbp)	45.6 / 51.6	**78.9** / **92.2**	55.7 / 47.4
	Switch Err (%)	7.3 / 8.9	7.2 / 8.8	**6.9** / **8.7**
	Hamming Err (%)	7.7 / 9.6	**7.5** / **9.0**	7.8 / 10.7

8   Conclusion

This work presents the first successful application of deep learning to diploid *de novo* genome assembly, demonstrating that GNNs can effectively resolve complex assembly graphs. Building on ideas from `GNNome` [18], we introduce improvements to both the data processing pipeline and the assembly algorithm.

We establish a framework for generating diploid graphs with ground-truth edge correctness when read coordinates originate from distinct reference systems for the same loci. This enables the construction of training-ready diploid graphs in DGL or PyG format. In addition, we propose a beam-search-based path-finding algorithm with strategies such as beam merging, which is particularly effective for navigating string-like graph structures.

Across most genomes, `DipGNNome` achieves results comparable to `hifiasm`, and in one case (*P. pygmaeus*) surpasses the current state of the art. Orangutan genomes are known to be especially challenging due to a higher number of acrocentric chromosomes and extensive tandem segmental duplications [22], suggesting that our approach may be particularly useful in difficult assembly settings where existing methods begin to struggle.

While contiguity and switch error rates are competitive, `DipGNNome` shows higher Hamming error, likely due to the absence of a polishing stage. Unlike `hifiasm`, which integrates read-level refinement, our method operates directly on the raw graph and identifies paths in a Hamiltonian-like manner without reusing nodes or introducing additional edges. This highlights a key trade-off between structural simplicity and sequence-level accuracy.

Performance is also influenced by training data diversity. The current model is trained only on human genomes, which may limit generalization to more distant species. Nevertheless, we observe successful transfer to other primates, and the framework can be readily extended to incorporate more diverse training data.

Finally, `DipGNNome` depends on the quality of the input graph. Since it selects only among existing edges, missing connections can lead to fragmentation or incomplete assemblies. Addressing this limitation, along with improving generalization across species and sequencing technologies, represents a natural direction for future work.

Overall, these results indicate that learning-based approaches are a viable and flexible complement to existing assembly methods, with clear potential for further improvements as models, training data, and integration strategies continue to evolve.

A   Coordinate Translation and Ground-Truth Verification

Diploid Coordinates. Some reads lie within homozygous regions of the diploid genome, meaning that both the maternal and paternal references contain the same sequence at these positions. However, since the maternal and paternal genomes are not identical (due to structural variations occurring elsewhere) the coordinates of a homozygous position differ between the two haplotypes.

Consider an edge between two reads, where one read originates from one haplotype and the other from the opposite haplotype. As with any edge, we must verify whether it represents a valid overlap based on the reads' positional information. However, because each haplotype has its own coordinate system, it is not sufficient to compare positions directly. Instead, we must convert the coordinates from one haplotype to the other's system.

Liftover. We transfer coordinates using `liftover`, specifically the Nextflow-Liftover (`NF-LO`) pipeline [16]. Given two reference genomes, `NF-LO` generates chain files that encode coordinate mappings between them. These chain files can be used with `UCSC Liftover` or, in our case, the Python-based `py-liftover` tool [10] to translate coordinates between haplotypes. For each read, we store both its original coordinates and its translated coordinates in the alternate haplotype's system. If `py-liftover` fails (due to the absence of a corresponding mapping in the other haplotype) the translated coordinate is set to `None`.

During unitig construction, when multiple reads are merged, some unitigs may include reads from both haplotypes. In such cases, we refine or infer missing coordinates using the structure of the unitig. Specifically, we use the postfix lengths of reads within the unitig to compute their relative offset from the end of the unitig, which allows us to estimate a coordinate in the alternate haplotype even when `liftover` coordinates are unavailable. Figure 2 shows an example of this advanced coordinate computation.

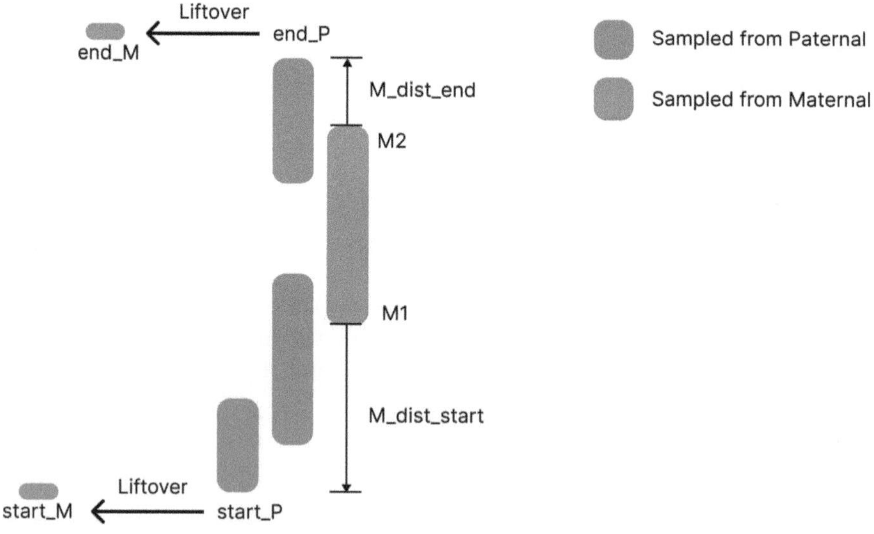

Fig. 2. Coordinate transfer. A unitig consists of multiple reads, and the paternal start and end coordinates are well defined here. Maternal coordinates can be determined either by lifting over the paternal endpoints or, if `liftover` fails, by computing them as $start\_M = M1 - M\_dist\_start$ and $end\_M = M2 + M\_dist\_end$.

Ground-Truth Verification. When checking for edge correctness, we determine whether either the paternal or maternal coordinate system shows a valid overlap. If a correct overlap exists in at least one haplotype, we mark the edge as non-relocation (i.e., not false). For edges where we cannot determine correctness—due to missing coordinate mappings or ambiguity—we assign a separate label, unknown, and exclude these edges from training. Missing coordinates only occur if a unitig consists of only reads of a single haplotype and additionally converting via liftover to the other haplotype fails. An overlap can only not be evaluated if this is the case for both unitigs, connected over an edge, and happens very rarely.

B   Full Assembly Pseudocode

Input: Graph $G = (N, E)$, number of samples per round n, beam width k,
minimum component length $C_{\min}$, minimum path length $P_{\min}$
Output: Set of full diploid assemblies
contigs $\leftarrow \{m : [], p : []\}$;
foreach *haplotype $h \in \{m, p\}$* **do**
    components $\leftarrow$ weakly connected components of G;
    foreach *component $G_c \in$ components* **do**
        while $|G_c| \geq C_{\min}$ **do**
            $P \leftarrow []$;   // Set of candidate paths
            for $i = 1$ **to** n **do**
                $(u, u') \leftarrow$ random edge from E;
                $(p_1, v_1, s_1) \leftarrow$ AssemblyBeamSearch$(G_c, u', k, h, \{u, u \oplus 1\})$;
                $(p_2, v_2, s_2) \leftarrow$ AssemblyBeamSearch$(G_c, u \oplus 1, k, h, v_1)$;
                $p_2 \leftarrow$ ReverseComplementPath(p_2);
                $p_{\text{full}} \leftarrow p_2 + p_1$;
                $v_{\text{full}} \leftarrow v_1 \cup v_2$;
                if *LengthInBasePairs(p_{full})* $\geq P_{\min}$ **then**
                    Append $(p_{\text{full}}, v_{\text{full}})$ to P;
                end
            end
            $(p_{\text{best}}, v_{\text{best}}) \leftarrow$ longest path in P;
            Append p_{best} to contigs(h);
            MarkVisited(G_c, v_{best});
            Remove all nodes in v_{best} and their incident edges from G_c;
        end
    end
end
return *contigs*

Algorithm 1: Full Assembly: Complete diploid genome assembly algorithm that performs independent maternal and paternal passes through weakly connected components, using beam search with bidirectional path extension

`ReverseComplementPath`(path) replaces nodes n with their complements $n \oplus 1$ and then reverses the path. For example, the path with node IDs $(5, 10, 2)$ becomes $(2 \oplus 1, 10 \oplus 1, 5 \oplus 1) = (3, 11, 4)$.

`LengthInBasePairs`(path) takes a path of unitigs and computes the total length in base pairs by summing all unitig lengths and subtracting overlap lengths:

$$\texttt{LengthInBasePairs}(\text{path}) = \sum_{i=1}^{n} \text{len}(u_i) - \sum_{i=1}^{n-1} \text{ovl}(u_i, u_{i+1})$$

where path $= [u_1, u_2, \ldots, u_n]$.

C   AssemblyBeamSearch Pseudocode

Input: Graph component $G_c = (N, E)$, starting node s, beam width k,
   haplotype $h \in \{m, p\}$, visited set V
Output: Best assembly path and visited nodes
$B_{\text{new}} \leftarrow \{([s], V, 0)\}$; // `Each beam is a tuple: (path, visited set, total`
  `score)`
best_beam $\leftarrow ([s], V, 0)$ **while** $B_{new} \neq \emptyset$ **do**

> $B_{\text{old}} \leftarrow B_{\text{new}}$;
> $B_{\text{new}} \leftarrow \emptyset$;
> **foreach** $(p, V, s) \in B_{old}$ **do**
>
>> $u \leftarrow$ last node in p;
>> **foreach** $edge\ (u, u') \in E$ **do**
>>
>>> **if** $u' \notin V$ **and** $u' \oplus 1 \notin V$ **then**
>>>
>>>> $p_{\text{new}} \leftarrow p + u'$;
>>>> $V_{\text{new}} \leftarrow V \cup \{u', u' \oplus 1\}$;
>>>> $s_{\text{new}} \leftarrow s + \texttt{EdgeScore}(u, u', h)$;
>>>> Add $(p_{\text{new}}, V_{\text{new}}, s_{\text{new}})$ to B_{new};
>>>
>>> **end**
>>
>> **end**
>
> **end**
> $B_{\text{new}} \leftarrow \texttt{MergeBeams}(B_{\text{new}})$;   // `Filter dominated beams`
> $\text{Sort}(B_{\text{new}})$;   // `Sort beams by total score in descending order`
> Prune B_{new} to top k beams;
> **if** *any beam in* B_{new} *is better than* best_*beam* **then**
>
>> best_beam $\leftarrow$ best beam in B_{new};
>
> **end**

end
  return best_beam;

Algorithm 2: AssemblyBeamSearch: Beam search algorithm for genome assembly that maintains k candidate paths, excludes visited nodes and complements, and returns the highest-scoring path

D   MergeBeams Pseudocode

Input: Set of beams B
Output: Merged set of beams B
$B_{\text{new}} \leftarrow \emptyset$;
foreach $b \in B$ **do**
    $(p, v, s) \leftarrow b$;
    $u \leftarrow$ last element of p;
    keep $\leftarrow$ True;
    foreach $(p', v', s') \in B \setminus \{b\}$ **do**
        if $u \in p'$ **then**
            $i_u \leftarrow$ index of u in p';
            $p'' \leftarrow p'[0 : i_u + 1]$;
            $s'' \leftarrow$ score of p'';
            if $s > s''$ **then**
                replace p'' in (p', v', s') with (p, v, s);
            end
        else
            keep $\leftarrow$ False;
        end
        break;
    end
    end
    if *keep* **then**
        add (p, v, s) to B_{new};
    end
end
return new;

Algorithm 3: MergeBeams: Merges overlapping beams by retaining only the highest-scoring path when multiple beams reach the same node

E   DipGNNome Ablation Study

In this section, we compare the main results of DipGNNome with results after removing components from the beam search score heuristic.

The heuristic is:

$$S(e) = L(v) - K(v) - M(e) \tag{7}$$

Length reward L, minus K-mer penalty K, minus Model Score penalty M. We compute assemblies without either of these three components and compare with the main results (Table 2).

Table 2. Ablation study of `DipGNNome`. The final column shows the full Beam Search model (beam width = 10).

	Metric	No Len. Pen.	No k-mer Pen.	No Model Scores	Beam Search
H. sapiens	Length (mB)	1546.3 / 1631.1	2959.7 / 3028.3	**2999.3 / 3095.5**	2856.4 / 2970.3
	Rdup (%)	**0.1 / 0.2**	1.2 / 1.5	1.4 / 1.2	0.2 / 0.3
	NG50 (mB)	0.2 / 0.2	52.9 / 45.2	78.4 / 76.2	63.0 / 65.2
	NGA50 (mB)	0.2 / 0.2	38.9 / 37.0	47.0 / 46.4	**48.4 / 49.3**
	Switch (%)	1.0 / **0.8**	0.9 / 1.5	1.0 / 1.2	1.1 / 1.2
	Hamm (%)	**0.9 / 0.7**	21.1 / 34.3	1.7 / 2.6	1.9 / 3.0
P. paniscus	Length (mB)	2057.6 / 1813.7	3404.1 / **3253.8**	3175.4 / 2993.1	**3162.5** / 2980.8
	Rdup (%)	5.6 / **0.3**	8.7 / 7.6	2.4 / 1.5	2.3 / 1.3
	NG50 (mB)	0.9 / 0.4	127.2 / 94.9	77.8 / 64.0	109.0 / 64.9
	NGA50 (mB)	0.8 / 0.4	**56.9 / 58.8**	45.6 / 39.6	57.4 / 50.0
	Switch Err (%)	**0.5** / 1.8	0.9 / 0.9	0.3 / 1.6	0.3 / 1.6
	Hamming Err (%)	**1.6 / 1.6**	27.0 / 22.2	1.9 / 2.9	2.3 / 3.6
G. gorilla	Length (mB)	1856.6 / 1854.3	3727.8 / 3763.6	3649.4 / 3533.7	**3485.9 / 3411.4**
	Rdup (%)	**0.5** / 1.1	7.2 / 8.4	2.1 / 2.4	1.7 / 1.8
	NG50 (mB)	0.3 / 0.2	116.6 / 105.0	116.6 / 116.1	108.7 / 78.6
	NGA50 (mB)	0.1 / 0.0	**60.0 / 63.8**	60.0 / 56.0	60.0 / 56.1
	Switch Err (%)	0.2 / **0.1**	0.2 / 0.2	0.3 / 0.2	0.3 / 0.2
	Hamming Err (%)	**0.3 / 0.2**	30.3 / 27.2	1.8 / 1.3	1.6 / 1.2

F   Parameter Configuration

Table shows the parameters used in all our experiments for the Assembly Algorithm (Table 3).

In all configurations, c_b denotes the *cutting threshold* for removing edges classified as "bad edges," while c_t denotes the *cutting threshold* for removing

Table 3. Beam search algorithm configuration and heuristic parameters.

Assembly Algorithm Parameters	
Beam width (k)	5
Samples (n)	100
Min. contig length ($P_{\min}$)	100k
Min. component length ($C_{\min}$)	25
α	0.0001
β	3
γ	25
c_b	> 0.9
c_t	> 0.5

edges classified as "translocation edges.'n' Higher values of c_b or c_t make the pruning step more selective, keeping more edges in the graph.

G   Training Dataset Overview

This dataset represents complete genome graphs, created out of reads of all chromosomes of the respective organisms. Table 4 summarizes its key statistics. We train on graphs from the human genome I002C only.

The training portion contains 6.86 million nodes connected by 9.38 million edges, while the validation set comprises 0.85 million nodes and 1.17 million edges. The combined dataset has 7.71 million nodes and 10.54 million edges, with an average degree of about 2.74, indicating slightly sparser connectivity than the chromosome-level dataset (3.02).

Across both splits, 84.56% of the edges are labeled as correct, while the remaining 15.46% are false edges. Among the false edges, relocation edges are the most frequent (9.51% of all edges), followed by inversion edges (4.03%), translocation edges (1.91%), and a negligible fraction of unknown edges (0.01%).

H   Real Data Details

We use the diploid T2T assembly of HG002 [20]. We include Gorilla, Bonobo, Siamang, Chimpanzee, and Orangutan as described in [21]. Coverage is computed from genome size and read depth (Table 5).

Table 4. Comprehensive Dataset Statistics (Edge Type Breakdown)—Full Genome Dataset

Metric	Training	Validation	Combined
Graph Structure			
Total Nodes	6,855,076	851,970	7,707,046
Total Edges	9,375,111	1,168,891	10,544,002
Average Degree	2.735	2.744	2.735
Edge Classes			
Correct Edges	8,102,855	1,098,974	9,201,829 (84.56%)
Total False Edges	1,272,256	69,917	1,628,805 (15.46%)
False Edge Distribution			
Relocation Edges	890,937	111,089	1,002,026 (9.51%)
Unknown Edges	424	53	477 (0.01%)
Inversion Edges	377,113	47,557	424,670 (4.03%)
Translocation Edges	178,994	22,638	201,632 (1.91%)

Table 5. Read datasets for human and ape genomes

Species	File	Depth
HG002	m64004_210224_230828	4.77×
	m64014_210227_165255	5.05×
	m64015e_210223_010616	5.32×
	m64015e_210224_100310	5.09×
	Total	**20.23×**
Gorilla	m54329U_210319_174352	4.32×
	m54329U_211102_230231	6.03×
	m54329U_211107_082940	5.15×
	m64076_230112_193924	4.26×
	Total	**19.76×**
Bonobo	m54329U_210509_045858	1.70×
	m54329U_210510_223954	2.12×
	m54329U_210512_043859	1.92×
	m54329U_211104_082916	6.88×
	m64076_210509_004533	2.17×
	m64076_210511_192234	1.91×
	m64076_210802_234415	4.36×
	Total	**21.06×**
Siamang	m54329U_210828_003431	4.30×
	m54329U_210829_112929	5.33×
	m54329U_210901_104742	4.92×
	m54329U_210902_233521	4.11×
	m54329U_210904_103226	1.58×
	Total	**20.24×**
Chimpanzee	m54329U_220226_122930	5.95×
	m54329U_220304_132403	6.54×
	m64076_210810_005444	4.27×
	m64076_210813_021703	4.02×
	Total	**20.78×**
Bornean Orangutan	m54329U_210506_061718	3.96×
	m54329U_220303_022849	6.30×
	m64076_210430_224715	3.91×
	m64076_210505_002126	3.81×
	m64076_210506_062052	3.71×
	Total	**21.69×**
Sumatran Orangutan	m64076_210219_011735	4.17×
	m54329U_210228_013345	3.16×
	m54329U_210404_020346	2.72×
	m54329U_220227_232630	5.76×
	m54329U_220313_173323	6.77×
	Total	**22.58×**

I   Training Configuration

The final model configuration uses 8 GNN layers and an embedding size of 512 for both node and edge features. A dropout rate of 0.3 is applied to reduce overfitting, and Pair Normalization is used throughout the network to enhance training stability. The model is optimized using the Adam optimizer with a learning rate of 0.0001.

Training is conducted on a single GPU (NVIDIA A100-PCIE-40GB), using mini-batch gradient descent and early stopping based on the validation loss. All models are implemented in PyTorch.

References

1. Battistella, E., Maheshwari, A., Ekim, B., Berger, B., Popic, V.: ralphi: a deep reinforcement learning framework for haplotype assembly. In: Sankararaman, S. (eds.) International Conference on Research in Computational Molecular Biology, pp. 349–353. Springer, Cham (2025). https://doi.org/10.1007/978-3-031-90252-9_37
2. Bresson, X., Laurent, T.: Residual gated graph convnets. arXiv preprint arXiv:1711.07553 (2017)
3. Cheng, H., Asri, M., Lucas, J., Koren, S., Li, H.: Scalable telomere-to-telomere assembly for diploid and polyploid genomes with double graph. Nat. Methods 1–4 (2024)
4. Cheng, H., Concepcion, G.T., Feng, X., Zhang, H., Li, H.: Haplotype-resolved de novo assembly using phased assembly graphs with hifiasm. Nat. Methods **18**(2), 170–175 (2021)
5. Ke, Z., Vikalo, H.: A convolutional auto-encoder for haplotype assembly and viral quasispecies reconstruction. Adv. Neural. Inf. Process. Syst. **33**, 13493–13503 (2020)
6. Ke, Z., Vikalo, H.: A graph auto-encoder for haplotype assembly and viral quasispecies reconstruction. In: Proceedings of the AAAI Conference on Artificial Intelligence, vol. 34, pp. 719–726 (2020)
7. Koren, S., et al.: De novo assembly of haplotype-resolved genomes with trio binning. Nat. Biotechnol. **36**(12), 1174–1182 (2018)
8. LBCB-SCI: Telomere-to-telomere diploid Indian Genome (2024). https://github.com/lbcb-sci/I002C. Accessed 28 Dec 2024
9. Li, H., Feng, X., Chu, C.: The design and construction of reference pangenome graphs with minigraph. Genome Biol. **21**, 1–19 (2020)
10. Luu, P.L., Ong, P.T., Dinh, T.P., Clark, S.J.: Benchmark study comparing liftover tools for genome conversion of epigenome sequencing data. NAR Genomics Bioinf. **2**(3), lqaa054 (2020)
11. Myers, E.W.: The fragment assembly string graph. Bioinformatics **21**(suppl_2), ii79–ii85 (2005)
12. Nurk, S., et al.: The complete sequence of a human genome. Science **376**(6588), 44–53 (2022)
13. Ono, Y., Asai, K., Hamada, M.: Pbsim: Pacbio reads simulator–toward accurate genome assembly. Bioinformatics **29**(1), 119–121 (2013)
14. Rautiainen, M., et al.: Telomere-to-telomere assembly of diploid chromosomes with verkko. Nat. Biotechnol. **41**(10), 1474–1482 (2023)
15. Šimunović, M., Šikić, M., Bankevich, A.: Gnndebugger: GNN based error correction in de bruijn graphs. bioRxiv, pp. 2025–05 (2025)
16. Talenti, A., Prendergast, J.: nf-lo: a scalable, containerized workflow for genome-to-genome lift over. Genome Biol. Evol. **13**(9), evab183 (2021)

17. Vaser, R., Šikić, M.: Time-and memory-efficient genome assembly with raven. Nat. Comput. Sci. **1**(5), 332–336 (2021)
18. Vrček, L., Bresson, X., Laurent, T., Schmitz, M., Kawaguchi, K., Šikić, M.: Geometric deep learning framework for de novo genome assembly. Genome Res. **35**(4), 839–849 (2025)
19. Xue, H., Rajan, V., Lin, Y.: Graph coloring via neural networks for haplotype assembly and viral quasispecies reconstruction. Adv. Neural. Inf. Process. Syst. **35**, 30898–30910 (2022)
20. Yang, C., et al.: The complete and fully-phased diploid genome of a male han chinese. Cell Res. **33**(10), 745–761 (2023)
21. Yoo, D., et al.: Complete sequencing of ape genomes. bioRxiv pp. 2024–07 (2024)
22. Yoo, D., et al.: Complete sequencing of ape genomes. Nature 1–18 (2025)
23. Zhao, L., Akoglu, L.: Pairnorm: Tackling oversmoothing in GNNS. In: International Conference on Learning Representations (2020). https://openreview.net/forum?id=rkecl1rtwB

GNNome-Decision: Enhancing GNN Training for *de Novo* Genome Assembly by Targeting Decision Nodes

Martin Schmitz[1,2(✉)], Lovro Vrček[2], Kenji Kawaguchi[1], and Mile Šikić[2,3(✉)]

[1] School of Computing, National University of Singapore, Singapore, Singapore
[2] Genome Institute of Singapore, A*STAR, Singapore, Singapore
schmitz.kessenich@gmail.com
[3] Faculty of Electrical Engineering and Computing, University of Zagreb, Zagreb, Croatia

Abstract. *De novo* genome assembly, the reconstruction of complete DNA sequences from reads without the use of a reference genome, remains one of the most challenging and fundamental problems in computational biology. A common method for *de novo* genome assembly involves creating an assembly graph of reads and defining a graph traversal that represents the genomic sequence. Recently, the first deep learning-based method, GNNome, was proposed to tackle this problem. Starting from an assembly graph, GNNome performs *de novo* assembly in two steps: binary edge classification and a greedy walk. However, we observe that the decisions of the greedy agent only matter in 0.86% of nodes. In this paper, we develop an objective function based on margin-ranking loss for GNN training that focuses on these decision nodes, effectively aligning the training objective with the performance of the downstream task of greedy pathfinding. Furthermore, we introduce a modification to the dataset creation pipeline, which increases the fraction of decision nodes by more than tenfold to 9.35%, strongly enhancing the information density in the training dataset. Trained on only human data, our model improves the NGA50 score compared to GNNome on the CHM13 human genome from 111.0 Mb to 115.8 Mb, while achieving similar assembly quality on three non-human real genomes, consistently increasing assembly completeness and decreasing duplicated genes.

1 Introduction

To retrieve genomic data from a cell, one must use one of the various sequencing technologies. The result of sequencing is a large dataset of small, unordered parts of the genome, known as reads. Reads are extracted from random areas of the genome. After sequencing, information about the position and chromosome from which a particular read is sampled is lost. *De novo* genome assembly involves reconstructing the sequenced genome from the set of reads without incorporating references from similar genomes. The absence of references can help with revealing unique genetic information and structural variations that might be absent

M. Lafond (Ed.): RECOMB-CG 2026, LNBI 16569, pp. 339–353, 2026.
https://doi.org/10.1007/978-3-032-26891-4_16

or misrepresented in reference genomes. Additionally, in many cases, reference genomes are not available or not perfectly reliable.

The release of the first telomere-to-telomere human genome assembly by [10] marked a significant milestone for the field, providing an accurate reference for the human genome for the first time, along with the original sequencing data. This complete haploid reference is 3.1 billion bases long, and the full assembly graph created by Hifiasm [3] consists of 2.4 million nodes and 18.5 million edges.

In addition to the large size of graphs, the complexity of repetitive regions and sequencing errors makes genome assembly a very challenging problem. Recently, several other high-quality references from different genomes have been constructed, such as HG002 [15], while more are being developed within the Human Pangenome Project [21] and several other projects, such as the Vertebrate Genomes Project [14], the Bovine Pangenome Consortium [16] and the Darwin Tree of Life Project [9]. Historically, the development of *de novo* assemblers has been guided by numerous heuristics tailored to specific sequencing technologies. However, the availability of these high-quality datasets has paved the way for the application of deep learning techniques to address the problem of *de novo* genome assembly.

GNNome [18] is the first method to demonstrate how to use this approach, introducing a pipeline where a graph neural network is trained on synthetic data sampled from these high-quality references. Modern *de novo* genome assemblers construct graphs where nodes and edges represent DNA sequences. The assembly can be computed through a walk over these assembly graphs. GNNome divides the problem of finding an assembly from the graph into two parts: a binary edge-classification problem and a graph traversal. For edge classification, GNNome labels edges as positive if they are part of an optimal assembly walk, and negative if they are not. The model is trained using binary cross-entropy for edge classification. The assembly is then constructed through a greedy graph traversal.

We observe a discrepancy between edge classification and greedy graph decoding tasks. Higher accuracy in edge classification does not always translate to better performance in greedy traversal. The model with the best performance on the downstream task may not have the lowest validation loss.

In this paper, we investigate unifying both objectives into one. Changing the pipeline to an end-to-end training pipeline that trains directly on the result of the greedy walk would require more complex frameworks, such as reinforcement learning, which typically come with disadvantages like low sample efficiency and difficult convergence behavior. Additionally, evaluating genome assemblies takes several minutes for a single traversal, making these options less appealing. Instead, we develop an objective function that closely resembles the objective of the greedy walk agent while remaining a differentiable function that can be used for supervised Graph Neural Network (GNN) training. This objective function does not require any specific genome assembly graph assumptions.

Furthermore, we show that the information density of the graphs used in the GNNome dataset is low. We present a method to create graphs with higher information density for training, compared to the original full-chromosome assembly graphs. The reads, constructing these graphs, are sampled from only the centromere and subcentromeric regions of the chromosomes; thus, we call these graphs centromere graphs.

We show that both the new training objective and the new dataset lead to improvements in training a GNN for *de novo* genome assembly. The resulting model generalizes to real data graphs that are two orders of magnitude larger than the proposed synthetic-data-based graphs in the training dataset.

2   Related Work

De Novo Genome Assemblers Modern *de novo* genome assembly efforts are anchored in two predominant frameworks: Overlap-Layout-Consensus (OLC) and de Bruijn graph-based assemblers. Leading de Bruijn graph assemblers, such as Verkko [13], Flye [7], and LJA [1], and leading OLC assemblers, such as GNNome [18], Hifiasm [3], and Raven [17], utilize one or multiple long-read technologies like PacBio HiFi and ONT Nanopore ultra-long reads, which produce reads of a few tens of kilobasepairs to megabasepairs in length.

In this paper, we focus on OLC-based assemblers. OLC assemblers involve three key phases: graph construction from overlapping reads (overlap), graph simplification and path-finding (layout), and the correction of single base pairs and details in the final walk (consensus). During the overlap phase, assembly graphs are constructed where reads are represented as nodes and overlaps between reads as edges. In the layout phase, a path and thereby a DNA sequence are constructed, and it is corrected in the consensus phase.

While most state-of-the-art assemblers develop individual approaches to each of these phases, GNNome [18] leverages existing assembly graphs from assemblers like Raven [17] and Hifiasm [3], focusing solely on the layout phase through machine learning to replace traditional heuristic-based layout phases. GNNome's layout phase is divided into binary edge classification and graph traversal. For binary classification, it generates synthetic training sets representing complete chromosomes from the HG002 Maternal reference using PBSim3 [11], which mimics the length and error profiles of target technologies (PacBio HiFi or ONT Nanopore). This data allows for the accurate computation of ground truth labels for edges, with labels indicating whether an edge is part of an optimal assembly path or not. This ground-truth data is then used to train a GNN on binary edge classification. The GNN is trained to classify whether edges in unseen graphs are likely part of an optimal decoding or not. The model is trained using a binary cross-entropy loss function, augmented by a symmetry auxiliary loss. GNNome employs SymGatedGCN, an extension of GatedGCN [2] for directed graphs, utilizing a bidirectional message-passing procedure. During inference, the trained model scores all edges in a graph. Graph traversal is executed greedily, starting from a random node and sequentially choosing the next node based on the GNN's scoring to form contiguous sequences, called contigs. The final assembly consists of one or multiple walks, forming one or multiple contigs.

3　Decision Nodes and Edge Types

Our first key observation is that the greedy agent does not make meaningful decisions at every node. In fact, the agent often faces choices between multiple equally good or equally detrimental edges. Important decisions occur only when some options are better than others. We refer to nodes where the agent makes these important decisions as "decision nodes."

We formally define a decision node as a node with at least two outgoing edges, where at least one edge is preferable to the other.

There are two cases, where one outgoing edge is preferable to another. The first case is when one edge is part of an optimal assembly walk and the other edge is not. The second case is when both edges are not part of an optimal decoding, but one edge would cause more harm to the assembly than the other. Given a decision node, we group the edges into two categories: correct-decision-edges, which are edges from the best edge types among the outgoing edges, and wrong-decision-edges, which are edges from the other types. Figure 2 shows graphically the difference between decision nodes and non-decision nodes as well as correct-decision and wrong-decision edges. These differences between various non-optimal edges are not captured by binary classification. We extend the categorization of edges by adding different types of non-optimal edges, named after the misassembly type that results from the traversal of the respective edges as follows (Fig. 1):

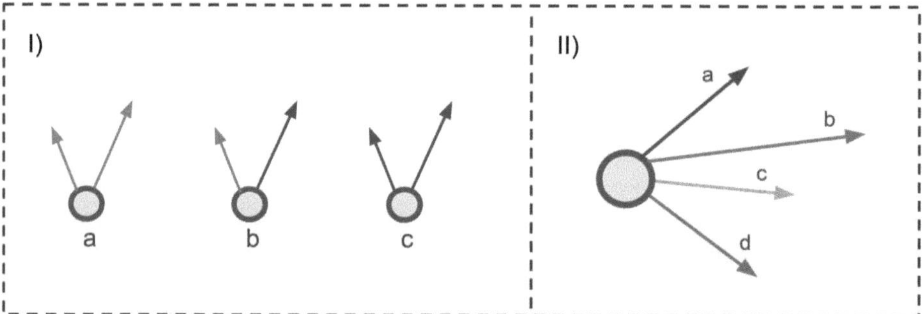

Fig. 1. I) green denotes correct, red denotes incorrect edge. b and c are decision nodes. c is one, because different incorrect edges are not equally detrimental for the final assembly. II) b and d are correct-decision edges. The greedy agent will take a correct decision if and only if one of the correct-decision edges is the highest-scoring out edge. (Color figure online)

1. **Dead-end/Tip**: The connection between two nodes is correct, but taking the edge leads to a traversal that does not maximize the length.
2. **Relocation Edge**: The edge connects two nodes whose corresponding DNA sequences do not overlap on the reference, causing some part of the sequence to be skipped (forward relocation) or repeated (backward relocation).

3. **Inversion Edge**: The edge connects two nodes from complementary DNA strands.
4. **Translocation Edge**: The edge connects two different chromosomes.

These different types of non-optimal edges are not equally disruptive for the assembly. A dead-end edge leads to a non-optimal walk length but does not assemble two incorrect sequences. A relocation edge connects two sequences that should not connect, potentially leading to marginal errors, while an inversion or translocation can disrupt the entire assembly. Thus, we rank edge categories from good to bad in the following order: correct-edge, dead-end, forward relocation, backward relocation, inversion, translocation. Figure 2 shows the different types of edges graphically.

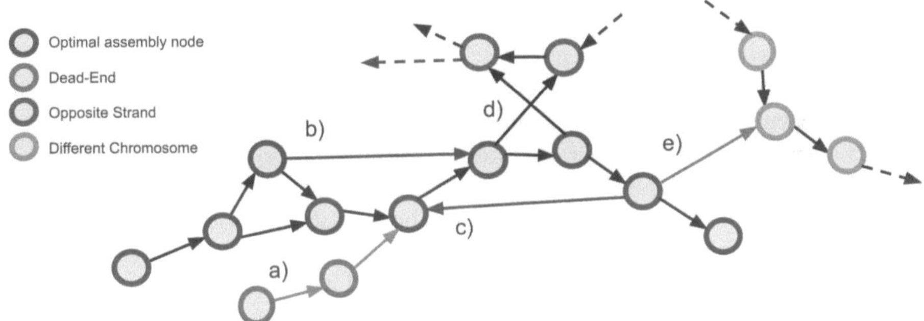

Fig. 2. Different types of malicious edges in an assembly graph. a) shows tips (correct but not part of an optimal length assembly walk). b) and c) are both relocation edges, while b) is a forward relocation and c) a backward relocation. d) shows inversion edges and e) shows a translocation edge.

4   Centromere Graphs

Our second key observation is that the fraction of decision nodes in the dataset is very low. To estimate the number of decision nodes accurately, we created a dataset consisting of a graph for each chromosome from chr1 to chr22 of the Maternal HG002 human genome. We sampled synthetic reads using PBSim [11] and generated an unsimplified overlap graph with Hifiasm [3]. We then counted the total number of nodes and the fraction of decision nodes. We found that out of roughly 3 million nodes, the fraction of decision nodes is 0.86%. Upon further analysis of the graphs, we observed that most decision nodes are clustered around specific areas of the graph. These areas roughly correspond to the centromere and subcentromeric regions of the chromosomes. These regions are known to be the most complex parts of a chromosome, consisting of various types of satellite and tandem repeat structures. The diploid human reference HG002 includes annotations that indicate the locations of these specific complex regions. We are interested in increasing the information density and complexity of the dataset and thus increasing the amount of decision nodes. In order to do so, we aimed to extract continuous intervals from the reference that are as complex as possible. We selected the largest annotated interval on the reference and merged it with neighboring annotated intervals that were no more than 20 kilobases (kb) apart. This resulted in a continuous interval representing the most complex area of the graph, which includes the centromere and subcentromeric regions, mainly composed of different complex repeat structures. We then added 100 kb to both sides of the interval.

We extracted this centromere interval from the reference and used it as a centromere reference for sampling reads. The length of the extracted interval is chromosome-specific and ranges from 2.3% to 55.6%, with a mean of 11.09% of the whole sequence of the respective chromosomes and a standard deviation of 10.31%. The complete list of extracted centromere-interval lengths, positions, and fractions is available in Appendix C. Using the centromere references, we created another set of graphs for chr1 to chr22, to compare it with the original one. As shown in Table 1, the resulting graphs are, on average, less than one-tenth the size of the original graphs and contain roughly only ten percent fewer decision nodes. Additionally, due to their smaller size, these complete graphs fit on a single GPU and do not need to be partitioned by METIS or other graph partitioning algorithms, as was done by GNNome, which leads to another speedup in training. Figure 3 provides a simplified sketch comparing centromere graphs to full-chromosome graphs.

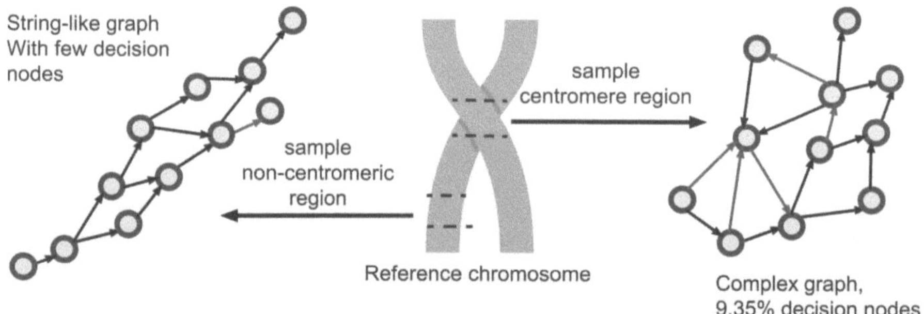

Fig. 3. Full-chromosome assembly graphs have large string-like components full of nodes with only correct out-edges. Our centromere graphs, however, do not include many of these components, resulting in more complex structures and a higher number of decision nodes.

Table 1. Graph Statistics

1xchr1-22	Nodes	Decision Nodes	% Decision Nodes
Full Graphs	3,016,944	25,870	0.86
Centromere Graphs	244,552	22,869	9.35

5   Correct-Decision Objective

Let N be the set of nodes in the graph. Then, let N_{dec} be the set of decision nodes. The greedy agent takes a correct decision at a decision node if and only if the highest-scoring edge is a correct-decision edge. Let c_{dec} be the number of decision nodes where this holds. To measure the performance of a model, we introduce the correct-decision quotient q_{dec}, which is the proportion of correct decisions c_{dec} relative to the total number of decisions made by the greedy agent in the graph.

$$q_{\text{dec}} = \frac{c_{\text{dec}}}{|N_{\text{dec}}|} \tag{1}$$

Next, we design an objective function aimed at training a model to maximize the correct-decision quotient in the dataset. Given a single decision node n, its outgoing edges E_n can be split into a set of correct-decision edges E_{n+}, and a set of wrong-decision edges E_n-. We compute the decision-ranking-loss (DRLoss) as follows:

$$L_{LR} = \frac{1}{|N_{\text{dec}}|} \sum_{n \in N_{\text{dec}}} \max \left(0, \max_{e \in E_{n-}} p_e - \max_{e \in E_{n+}} p_e + \text{m} \right) \tag{2}$$

where p_e is the model prediction for an edge e and m is the margin hyperparameter ($m = 1.0$). This loss function describes the margin ranking loss between the highest-scoring correct-decision edge and the highest-scoring wrong-decision edge of the observed decision node. Intuitively, it pushes the best correct-decision edge score to be at least m higher than the best wrong-decision edge score. This encourages the model to assign the highest score to a correct-decision edge, thereby improving the correct-decision quotient.

6   Experiments

We conduct experiments to validate our two primary contributions: (1) the centromere-graph dataset and (2) the new training objective. To evaluate the first contribution, we train our model on three distinct datasets. The first dataset, the GNNome dataset (G), serves as our baseline and is the same dataset used by GNNome. It includes 90 assembly graphs sampled from six different chromosomes of HG002 maternal, with 15 instances each of chromosomes 1, 3, 5, 9, 12, and 18. The second dataset, referred to as the full-genome dataset (F), assesses the performance improvement gained by sampling from the entire

Table 2. Dataset Statistics

Dataset	Number of Graphs	Number of Nodes	Number of Edges
GNNome Dataset	90	$15,849,102$	$167,219,058$
Full-Genome Dataset	138	$18,245,970$	$192,451,548$
Centromere Dataset	1,380	$16,303,010$	$165,733,672$

diploid genome of HG002. Instead of repeatedly sampling a few chromosomes, this dataset includes three instances of each chromosome from HG002, covering chromosomes 1 to 22, and chromosomes X and Y. The third dataset, the centromere dataset (C), comprises the same set of chromosomes as the full-genome dataset but focuses exclusively on the centromere regions, sampled 30 times each, as described in Sect. 4. All three datasets are comparable in size concerning total nodes and edges, as detailed in Table 2. The validation dataset remains constant across all runs and datasets. It includes specific chromosomes from the paternal HG002, with one instance each of chromosomes 4, 7, 10, 13, 16, 19, and 22. Additionally, there are five instances each of the centromeric regions of chromosomes 1, 5, 9, 15, 17, and 20.

To evaluate the second contribution, our decision-node-based training objective, we train models using both the GNNome training objective and the decision-node training objective for each of the three datasets. All models are trained with identical hyperparameters. Implementation details are given in Appendix A, training details are available in Appendix B. During training, we evaluate the correct-decision-quotient on the validation dataset after each epoch, saving the model with the highest score. We then assess the edge scores of several real data full-genome graphs, constructed by Hifiasm, using each of the trained models. These scores are used to create the assemblies via a greedy decoding. The real genomes tested are the same ones used by GNNome: CHM13 [10], the C57BL/6J strain of *M. musculus* [5], the Col-0 strain of *A. thaliana* [19], and the *G. gallus* maternal genome (bGalGal1 isolate from the Vertebrate Genome Project). Our experiments are limited to HiFi assemblies based on Hifiasm graphs. The assembly graphs range from 146,000 to 3.7 million nodes and from 1.1 million to 44 million edges. Detailed information about reference sizes, read coverage, and assembly graph sizes can be found in the appendix (Table 5).

To access the quality of the assemblies, we use several standard metrics to measure contiguity, quality, and completeness. LG90, NGA50, and NG50 compute the contiguity of the assemblies, while Complete, Duplicated, and QV give additional information about the assembly quality. Details about the metrics used and how we compute them can be found in Appendix D.

6.1   Experiment Results

Table 3 shows the results of our experiments. The following paragraphs describe these results for each genome.

H. Sapiens. Using DRLoss and the centromere dataset leads to an increase in NGA50 from 111.0 in basic GNNome to 115.8. Additionally, the overall completeness of the genome assembly increases from 99.53% to 99.8%, while LG90 improves slightly.

M. musculus. There is a slight decrease in NGA50 using DRLoss trained on the centromere dataset, compared to basic GNNome. Despite this, the assembly shows enhanced completeness and a significant reduction in duplications (from 3.3% to 1.94%), which underscores an improvement in accurately representing genomic uniqueness without redundant sequences.

A. thaliana. The genome assemblies remain mostly constant, regardless of the used model. This hints that the graph has limitations in terms of contiguity. Notably, there is a slight improvement in genome completeness from 99.89% to 100% using the centromere dataset or DRLoss-trained models.

G. gallus. There is a noticeable improvement in the L90 and NGA50 metrics, using the centromere dataset or DRLoss. The best contiguity scores here are produced by GNNome trained with the centromere dataset with NGA50 of 10.8, compared to NGA50 of 10.3 by the DRLoss model trained on the centromere dataset, and 10.1 by basic GNNome. Additionally, duplications in models trained with the DRLoss and the centromere dataset are consistently much lower than those in GNNome, while completeness is slightly lower. Overall, we see an increase in contiguity, characterized by L90, NG50, and NGA50 metrics across the datasets while using centromere training data and the DRLoss objective function. Additionally, we observe a decrease in duplicated genes and an increase in the completeness of assemblies.

7   Discussion

7.1   Limitations

Decision Quotient. In *de novo* genome assembly, errors at the end of a contig are less disruptive than errors in the middle. Additionally, certain nodes are more likely to be visited by a greedy agent than others. The current decision quotient metric does not account for these differences in node importance. To enhance its predictive value for assembly quality, this metric could be extended with importance weighting. These weights could also be incorporated into the loss function to better guide the model during training.

Assembly Graphs. Our approach is constrained by the quality of the input assembly graphs, which in this case are generated by hifiasm. These graphs include edges that represent overlaps, but only the best overlaps are included. Consequently, the graph may be incomplete, sometimes lacking correct links, resulting in fragmented graphs or graphs without a valid end-to-end path. Although our model performs well in predicting edge quality, it cannot accurately assemble the genome if critical edges are missing. The current graphs, optimized for heuristic-based algorithms, could be improved for deep learning methods by including more comprehensive edge information.

Computational Cost. Our proposed method is computationally simpler than GNNome [18], as it operates on smaller graphs, does not require METIS partitioning, and avoids masking steps. However, in the current implementation, the assembly graph must first be generated by running hifiasm [3] end-to-end, after which our pipeline is applied. Consequently, our method introduces additional computational overhead on top of the hifiasm runtime. An end-to-end integration that eliminates this redundancy represents an important direction for future work.

7.2   Conclusion

Training on small, complex graphs with high information density has proven beneficial, even though these graphs represent only about 10% of the genetic information. We demonstrate that a margin-ranking loss objective performs better than binary-edge classification for the complex task of *de novo* genome assembly. This performance improvement likely stems from the alignment of the objective with the greedy agent's downstream task and the model's ability to rank different suboptimal edges.

As described by GNNome [18], the contiguity achieved by current state-of-the-art assemblers on human data is far closer to the highest possible contiguity than on other eukaryotic species. This indicates that the overlap algorithms for graph construction are fine-tuned to human data, leaving more space for improvement compared to other genomes. Our performance boost is also most pronounced for human data, which might be caused by the better-constructed input graph. However, the model also generalizes well to data from other species. Additionally, incorporating more diverse training data from various species could further enhance generalization across different genomes.

Note that all evaluated genomes were also considered in GNNome [18], where it was shown to perform on par with or better than state-of-the-art assemblers. As our evaluation follows the same setup, this implies that our approach achieves comparable or better performance as well.

The resulting model, trained on graphs with fewer than 50,000 nodes, generalizes well to much larger graphs with over 3 million nodes. Our approach simplifies the pipeline compared to the GNNome method by eliminating the need for node masking, auxiliary losses, and METIS graph partitioning. The results show that at the same time, the assembly quality rises. While there are some more possibilities to improve the pipeline for haploid assembly, the new insights and modifications of the pipeline can also pave the way for applying deep learning techniques to more complex tasks like diploid and polyploid assembly.

A   Implementation

All implementations are written in Python. For the machine learning algorithms, we use Pytorch [12] and Deep Graph Library (DGL [20] as well as Torch-Scatter, one of the core libraries of Pytorch-Geometric [4]. Models are trained on a single Nvidia A100 GPU with 40 gigabytes of memory. Training runs need around 24 h to complete (Table 4).

B   Training Details

C   Centromere Intervals

This section lists all intervals of centromeres from chm13. The centromere interval can include subcentromeric area and other non-centromeric regions, as long as they are connected to the centromere as described in Sect. 4 (Table 6).

D   Evaluation Metrics

The LG90 metric represents the smallest number of contigs required to cover 90% of the genome. NG50 and NGA50, which measure the contiguity of the assembly, are computed with minigraph [8]. Contiguity refers to how well the genome assembly is pieced together, with higher values indicating longer contiguous sequences without gaps. For NGA50, the assembly is compared to the reference, and contigs are split if assembled incorrectly before the lengths are measured. The "Complete" percentage reflects the proportion of reference single-copy genes found in the assembly, whereas the "Duplicated" percentage indicates how many of these genes are aligned to multiple positions within the assembly. Both metrics are computed using Compleasm [6], which relies on lineage-specific single copy genes to determine their presence and duplication. QV, or per-base consensus accuracy, is calculated using yak (v0.1-r56, https://github.com/lh3/yak) by comparing k-mers in contigs to those found in highly accurate short reads (for *G. gallus*, due to the unavailability of short reads, QV was computed using PacBio HiFi reads instead.)

Table 3. Assembly results for GNNome (binary classification) loss models, and DRLoss model. The results for GNNome with dataset G are taken from the original paper [18]. Datasets: GNNome dataset (G), Full-Genome dataset (F), and Centromere Dataset (C). Size, NG50, and NGA50 are given in million base pairs. Complete, duplicated, and are given in percentage. QV is a quality value.

Species	Model	Dataset	Size	LG90	NG50	NGA50	Complete	Duplicated	QV
CHM13	GNNome	G	3,051	31	111.3	111	99.53	**0.71**	**54.24**
		F	3,050	32	90.2	88.8	**99.8**	**0.71**	51.49
		C	3,052	**29**	**118.6**	111.1	**99.8**	0.76	52.6
	DRLoss	G	3,051	36	86.1	86.1	**99.8**	**0.71**	49.3
		F	3,051	30	111.3	111	**99.8**	**0.71**	51.39
		C	3,050	30	115.8	**115.8**	**99.8**	**0.71**	49.78
M. musculus	GNNome	G	2,643	**140**	**23**	**19.3**	99.62	3.3	**45.4**
		F	2,614	156	21	15.8	**99.79**	**1.94**	45.1
		C	2,511	150	22	17.8	99.74	**1.94**	44.2
	DRLoss	G	2,609	170	20.3	14.8	**99.79**	**1.94**	42.7
		F	2,611	154	21.6	18.3	**99.79**	**1.94**	43.1
		C	2,610	145	22.9	18.5	**99.79**	**1.94**	42.9
A. thaliana	GNNome	G	139	**13**	**12.4**	**12.4**	99.89	1.09	52.08
		F	140	14	9.5	9.4	99.8	**0.71**	51.5
		C	139	**13**	**12.4**	**12.4**	100	1.07	51.4
	DRLoss	G	139	**13**	**12.4**	**12.4**	100	1.09	49.7
		F	140	**13**	**12.4**	**12.4**	100	1.07	**52.4**
		C	140	**13**	**12.4**	**12.4**	100	1.09	51.46
G. gallus	GNNome	G	1,114	135	10.8	10.1	**95.79**	2.99	**49.35**
		F	1,073	132	**11.4**	10.3	95.63	1.64	45.18
		C	1,060	**125**	11.3	**10.8**	95.61	1.28	45.7
	DRLoss	G	1,074	151	9.8	9.3	95.64	1.67	44.25
		F	1,072	132	11.3	10.2	95.61	**1.06**	45.39
		C	1,072	129	11.3	10.3	95.62	1.58	44.43

Table 4. Hyperparameters of the model

Parameter	Value
Model	
Dimensionality of latent space	64
Number of GNN layers	8
Batch normalization	True
Dropout rate	0.2
Training	
Optimizer	Adam
Number of epochs	100
Learning rate	1×10^{-4}

Table 5. Real data Genome statistics for evaluation assemblies

Genome	Length	Read coverage	No. Nodes	No. Edges	avg. Degree
Mouse	2,728	25x	3,747,706	44,601,962	11.90113686
Chicken	133	30x	454,874	2,518,004	5.535607663
CHM13	3,054	32x	2,356,878	18,523,036	7.859140779
Arabidopsis	1,053	35x	146,242	1,143,378	7.818396904

Table 6. Reference chromosome length and extracted centromere interval (C.I.) position and length.

Chr	Length	C.I. Length	C.I. Start	C.I. End	% C.I.
chr1_M	244,022,132	18,201,522	121,678,665	139,880,187	7.459
chr1_P	252,060,642	26,091,591	121,771,791	147,863,382	10.351
chr2_M	242,114,530	6,632,018	88,956,753	95,588,771	2.739
chr2_P	241,873,600	8,058,261	87,280,393	95,338,654	3.332
chr3_M	201,098,049	6,247,239	90,141,403	96,388,642	3.107
chr3_P	201,513,670	6,670,796	90,223,869	96,894,665	3.310
chr4_M	191,670,063	4,485,939	49,121,928	53,607,867	2.340
chr4_P	192,384,391	5,111,350	49,075,132	54,186,482	2.657
chr5_M	183,262,196	6,000,151	46,045,111	52,045,262	3.274
chr5_P	188,875,663	11,889,167	46,069,851	57,959,018	6.295
chr6_M	174,742,646	9,223,723	57,662,817	66,886,540	5.278
chr6_P	174,411,946	8,333,071	58,177,045	66,510,116	4.778
chr7_M	160,955,131	10,023,708	56,609,479	66,633,187	6.228
chr7_P	160,104,498	9,346,094	56,729,421	66,075,515	5.837
chr8_M	146,740,790	5,877,993	43,073,678	48,951,671	4.006
chr8_P	146,786,846	5,732,023	43,340,708	49,072,731	3.905
chr9_M	143,803,312	36,935,057	38,468,639	75,403,696	25.684
chr9_P	131,519,477	24,650,818	38,462,839	63,113,657	18.743
chr10_M	135,912,142	5,533,379	38,477,166	44,010,545	4.071
chr10_P	135,711,693	6,115,180	38,478,676	44,593,856	4.506
chr11_M	135,235,829	7,537,659	48,653,151	56,190,810	5.574
chr11_P	133,990,234	6,280,507	48,665,321	54,945,828	4.687
chr12_M	133,580,598	4,467,302	33,970,404	38,437,706	3.344
chr12_P	133,573,629	4,434,732	34,005,069	38,439,801	3.320
chr13_M	113,335,067	8,353,372	9,625,296	17,978,668	7.371
chr13_P	109,226,285	13,875,440	0	13,875,440	12.703
chr14_M	108,656,944	12,960,264	8,567,881	21,528,145	11.928
chr14_P	105,800,994	14,306,696	4,966,830	19,273,526	13.522
chr15_M	100,881,410	13,583,877	5,710,241	19,294,118	13.465
chr15_P	96,257,017	9,754,349	5,724,253	15,478,602	10.134
chr16_M	90,398,456	13,764,651	32,888,347	46,652,998	15.227
chr16_P	93,601,116	16,724,719	33,150,574	49,875,293	17.868
chr17_M	83,770,189	12,586,213	15,534,497	28,120,710	15.025
chr17_P	84,843,465	13,387,248	15,536,295	28,923,543	15.779
chr18_M	78,908,821	5,742,588	14,037,517	19,780,105	7.277
chr18_P	80,778,415	7,414,779	14,037,683	21,452,462	9.179
chr19_M	61,317,360	10,281,826	19,837,847	30,119,673	16.768
chr19_P	61,355,730	10,416,199	19,853,484	30,269,683	16.977
chr20_M	66,071,324	7,329,809	25,731,149	33,060,958	11.094
chr20_P	67,035,426	8,168,502	25,738,918	33,907,420	12.185
chr21_M	47,311,730	6,808,759	7,910,254	14,719,013	14.391
chr21_P	44,312,601	6,755,846	4,976,001	11,731,847	15.246
chr22_M	53,395,392	18,019,583	8,063,143	26,082,726	33.747
chr22_P	49,430,498	18,419,483	3,713,463	22,132,946	37.263
chrX	154,341,406	4,089,736	57,309,780	61,399,516	2.650
chrY	62,425,491	34,749,083	27,345,647	62,094,730	55.665

References

1. Bankevich, A., Bzikadze, A.V., Kolmogorov, M., Antipov, D., Pevzner, P.A.: Multiplex de bruijn graphs enable genome assembly from long, high-fidelity reads. Nat. Biotechnol. **40**(7), 1075–1081 (2022)
2. Bresson, X., Laurent, T.: Residual gated graph convnets (2018)
3. Cheng, H., Asri, M., Lucas, J., Koren, S., Li, H.: Scalable telomere-to-telomere assembly for diploid and polyploid genomes with double graph. Nat. Methods 1–4 (2024)
4. Fey, M., Lenssen, J.E.: Fast graph representation learning with pytorch geometric. arXiv preprint arXiv:1903.02428 (2019)
5. Hon, T., et al.: Highly accurate long-read hifi sequencing data for five complex genomes. Sci. Data **7**(1), 399 (2020)
6. Huang, N., Li, H.: compleasm: a faster and more accurate reimplementation of busco. Bioinformatics **39**(10), btad595 (2023)
7. Kolmogorov, M., Yuan, J., Lin, Y., Pevzner, P.A.: Assembly of long, error-prone reads using repeat graphs. Nat. Biotechnol. **37**(5), 540–546 (2019)
8. Li, H., Feng, X., Chu, C.: The design and construction of reference pangenome graphs with minigraph. Genome Biol. **21**, 1–19 (2020)
9. Of Life Project Consortium, D.T.: Sequence locally, think globally: the darwin tree of life project. Proc. Nat. Acad. Sci. **119**(4), e2115642118 (2022)
10. Nurk, S., et al.: The complete sequence of a human genome. Science **376**(6588), 44–53 (2022)
11. Ono, Y., Hamada, M., Asai, K.: Pbsim3: a simulator for all types of pacbio and ont long reads. NAR Genomics Bioi. **4**(4), lqac092 (2022)
12. Paszke, A., et al.: Pytorch: An imperative style, high-performance deep learning library. Adv. Neural Inf. Process. Syst. **32** (2019)
13. Rautiainen, M., et al.: Telomere-to-telomere assembly of diploid chromosomes with verkko. Nat. Biotechnol. 1–9 (2023)
14. Rhie, A., et al.: Towards complete and error-free genome assemblies of all vertebrate species. Nature **592**(7856), 737–746 (2021)
15. Rhie, A., et al.: The complete sequence of a human y chromosome. Nature **621**(7978), 344–354 (2023)
16. Smith, T.P., et al.: The bovine pangenome consortium: democratizing production and accessibility of genome assemblies for global cattle breeds and other bovine species. Genome Biol. **24**(1), 139 (2023)
17. Vaser, R., Šikić, M.: Time-and memory-efficient genome assembly with raven. Nat. Comput. Sci. **1**(5), 332–336 (2021)
18. Vrček, L., Bresson, X., Laurent, T., Schmitz, M., Kawaguchi, K., Šikić, M.: Geometric deep learning framework for de novo genome assembly. Genome Res. **35**(4), 839–849 (2025)
19. Wang, B., et al.: High-quality arabidopsis thaliana genome assembly with nanopore and hifi long reads. Genomics Proteomics Bioinf. **20**(1), 4–13 (2022)
20. Wang, M., et al.: Deep graph library: A graph-centric, highly-performant package for graph neural networks. arXiv preprint arXiv:1909.01315 (2019)
21. Wang, T., et al.: The human pangenome project: a global resource to map genomic diversity. Nature **604**(7906), 437–446 (2022)

FIREFLY: PHYlogeny-Informed REpresentation Learning to Estimate PHyLogenetic dIstances

Meijun Gao[1], Byungho Lee[1,2], and Kevin J. Liu[1,2,3](✉)

[1] Department of Computer Science and Engineering, Michigan State University,
East Lansing, MI, USA
`kjl@msu.edu`
[2] Ecology, Evolution, and Behavior Program, Michigan State University,
East Lansing, MI, USA
[3] Genetics and Genome Sciences Program, Michigan State University,
East Lansing, MI, USA

Abstract. Phylogenetic distance estimation and distance-based phylogeny reconstruction are well-studied cornerstone topics in phylogenetics. Classical approaches for both utilize mathematical or probabilistic graphical models of biomolecular sequence evolution. But model violations can occur and model-based analysis can be impacted as a result. Recent advances in statistical machine learning using deep neural networks provide an alternative in the form of representation learning. Newer applications of deep learning to phylogenetic distance estimation have followed. A number of challenges in this area remain, since state-of-the-art methods are often restricted to pairwise or subset-based analyses and retain other simplifying assumptions.

In fact, classical model-based methods and representation learning are orthogonal and, as we show, their combination can be greater than the sum of its parts. We bridge these different approaches by synthesizing mathematical and logical constraints with statistical machine learning – an approach from physics-informed machine learning (PIML). Our algorithmic solution takes the form of a Transformer-based framework for learning phylogeny-informed representations directly from MSAs, which we apply to the task of phylogenetic distance estimation. The result is FIREFLY, a computational framework for "PHYlogeny-informed REpresentation learning to estimate PHyLogenetic dIstances").

We benchmarked FIREFLY's performance against other state-of-the-art methods using simulated and empirical datasets. We found that FIREFLY improves both pairwise distance estimation accuracy and downstream phylogenetic inference compared with state-of-the-art methods. The gains are particularly pronounced under high indel rates and on estimated MSAs, where alignment errors and gap-induced uncertainty

---

Supplementary Information The online version contains supplementary material available at https://doi.org/10.1007/978-3-032-26891-4_17.

are most severe. Our results highlight the value of integrating phylogeny-based inductive bias into deep representation learning and suggest that MSA-level modeling offers a robust foundation for evolutionary inference under challenging conditions.

Keywords: Phylogenetic distance estimation · Multiple sequence alignment · Phylogeny-informed learning · PIML · Deep representation learning

1   Introduction

Phylogenetic inference aims to reconstruct evolutionary relationships among biological sequences and plays a central role in comparative genomics and molecular evolution. In a standard phylogenetic pipeline, homologous sequences are first assembled into a multiple sequence alignment (MSA), after which a phylogenetic tree is inferred under an explicit event-based model of sequence evolution using statistical or mathematical optimization [3,23,25]. These optimization-based methods are computationally intensive due to repeated optimization score calculation and the need to explore a tree space that grows super-exponentially with the number of taxa [3]. As a computationally efficient alternative, distance-based methods reconstruct phylogenies from estimated pairwise evolutionary distances [8,11]. Despite their differences in formulation and computational cost, both optimization-based and distance-based approaches rely on simplifying assumptions about the evolutionary process, such as site independence and error-free input alignments. In practice, insertions and deletions are typically ignored or treated as missing data rather than explicitly modeled [24]. Consequently, phylogenetic inference can be sensitive to alignment uncertainty and upstream alignment errors, particularly when MSAs are noisy or approximately estimated [7,22].

An alternative to these classical approaches has emerged thanks to recent advances in machine learning: an increasing number of studies have investigated the use of deep learning methods in phylogenetics, as reviewed in [14]. Several likelihood-free methods formulate topology inference as a classification problem over tree topologies [19,27]. However, because the number of unrooted tree topologies grows rapidly with the number of taxa, these methods are restricted to quartets (trees with four leaves) and must be combined heuristically to infer larger trees. Benchmark studies have shown that, under challenging scenarios such as long branches or short alignments, these methods can underperform classical likelihood or distance-based approaches [26]. Other likelihood–free frameworks, including GAN-based methods, require retraining for each dataset and do not scale beyond a modest number of taxa [18]. Phyloformer [15] instead predicts pairwise evolutionary distances from MSAs using Transformer encoders and reconstructs trees via distance-based methods, enabling scalability to larger datasets. However, its reliance on deterministic pairwise signals makes it sensitive to noise in estimated MSAs, and its pairwise representations may discard shared evolutionary context captured at the MSA level [2,4], particularly in gap-rich alignments.

To help address these limitations and explore new directions for applications of deep learning in phylogenetics, our work draws inspiration from recent advances in MSA-based representation learning for protein modeling, where deep networks are pretrained on large MSAs and transferred to downstream tasks [9,16]. While such approaches primarily focus on relationships between sites, phylogenetic inference concerns relationships between sequences. In this study, we introduce a phylogeny-informed representation learning framework built on a DNA-MSA Transformer encoder augmented with a learnable phylogenetic attention bias. By operating on full MSAs, the model learns sequence-level representations that capture global relational structure across taxa, which cannot be recovered from pairwise-only features. We combine a pretrained transformer encoder with a supervised distance estimation module to infer pairwise evolutionary distances, which are subsequently used for downstream tree reconstruction. A key component of our framework is a learnable phylogenetic bias module that is integrated directly into the column-wise attention mechanism. Unlike fixed or hand-crafted phylogenetic priors, this bias is learned end-to-end from data and adapts naturally to alignment noise. It is activated during joint training with the distance estimation module, allowing the model to stabilize cross-sequence attention while remaining flexible. Through extensive simulation studies and empirical analyses, our framework consistently outperforms Phyloformer and achieves performance comparable to, or exceeding, state-of-the-art likelihood-based methods on both true and estimated MSAs. Importantly, it remains robust in challenging regimes with high gap content, where prior deep learning approaches and likelihood-based methods often degrade.

Our contributions can be summarized as follows:

- **MSA-level representation learning.** We introduce an MSA-level representation learning framework that encodes global evolutionary relationships among taxa, rather than relying on pairwise statistics derived from raw alignments.
- **Learnable phylogenetic inductive bias.** We propose a data-adaptive phylogenetic bias that guides attention toward sequences based on their evolutionary relationships, providing a flexible alternative to fixed evolutionary priors.
- **Robustness under realistic alignment noise.** By modeling full MSAs and learning adaptive phylogenetic structure, our approach substantially improved robustness in downstream phylogenetic inference under alignment errors and gap-induced uncertainty commonly encountered in estimated MSAs.

2   Method

Problem Formulation and Framework Overview. Given a DNA multiple sequence alignment (MSA) with N taxa and alignment length L, our overall goal is to learn phylogeny-aware representations of MSAs that can support downstream phylogenetic inference tasks. In this work, we focus on pairwise evolutionary distance inference, followed by distance-based phylogenetic tree reconstruction.

Accurate phylogenetic representation learning requires modeling both site-wise dependencies along sequences and taxon-wise relationships across aligned sequences, while remaining robust to alignment noise. We therefore propose a Transformer-based framework that operates directly on full MSAs to learn sequence-level representations capturing global relational structure across taxa, and serves for downstream phylogenetic inferences. The framework consists of two stages. In the first stage, a shared Transformer encoder is pretrained using a masked language modeling (MLM) objective to learn general evolutionary patterns from MSAs. In the second stage, the pretrained encoder is jointly optimized with a supervised distance prediction module to infer a pairwise evolutionary distance $D \in \mathbb{R}^{N \times N}$ from an input MSA, where each entry D_{ij} represents the evolutionary distance between taxa i and j. During joint supervised training of the encoder and the distance estimation module, a learnable phylogenetic bias module is activated to stabilize cross-sequence attention and guide representation learning under alignment noise. The inferred distance matrix is subsequently used for phylogenetic tree reconstruction via a distance-based method.

We first describe the Transformer encoder architecture and the masked language model pretraining used in the first stage. We then introduce the learnable phylogenetic bias module and the evolutionary distance prediction module, followed by a description of the overall training procedure. An overview of the proposed framework is shown in Fig. 1.

Transformer Encoder Architecture. Our encoder follows the MSA Transformer architecture proposed in [16], as illustrated in Fig. 1. It consists of six stacked attention blocks, each equipped with eight attention heads. Each block adopts an axial attention design comprising tied row attention, column attention, and a feedforward layer with residual connections and layer normalization. Row attention models dependencies across alignment sites within each sequence, with parameters shared across all sequences via tied row attention for efficiency [16]. Column attention captures cross-sequence interactions at each alignment site, operating independently at each column of the MSA.

Since the ordering of taxa in an MSA is arbitrary, we do not include row-wise positional embeddings. Positional biases along the site dimension are also omitted in the current implementation. Additional architectural details are provided in the Supplementary Material.

Transformer Encoder Pretraining with Masked Language Modeling. We represent a DNA multiple sequence alignment with N taxa and L aligned sites as a matrix,

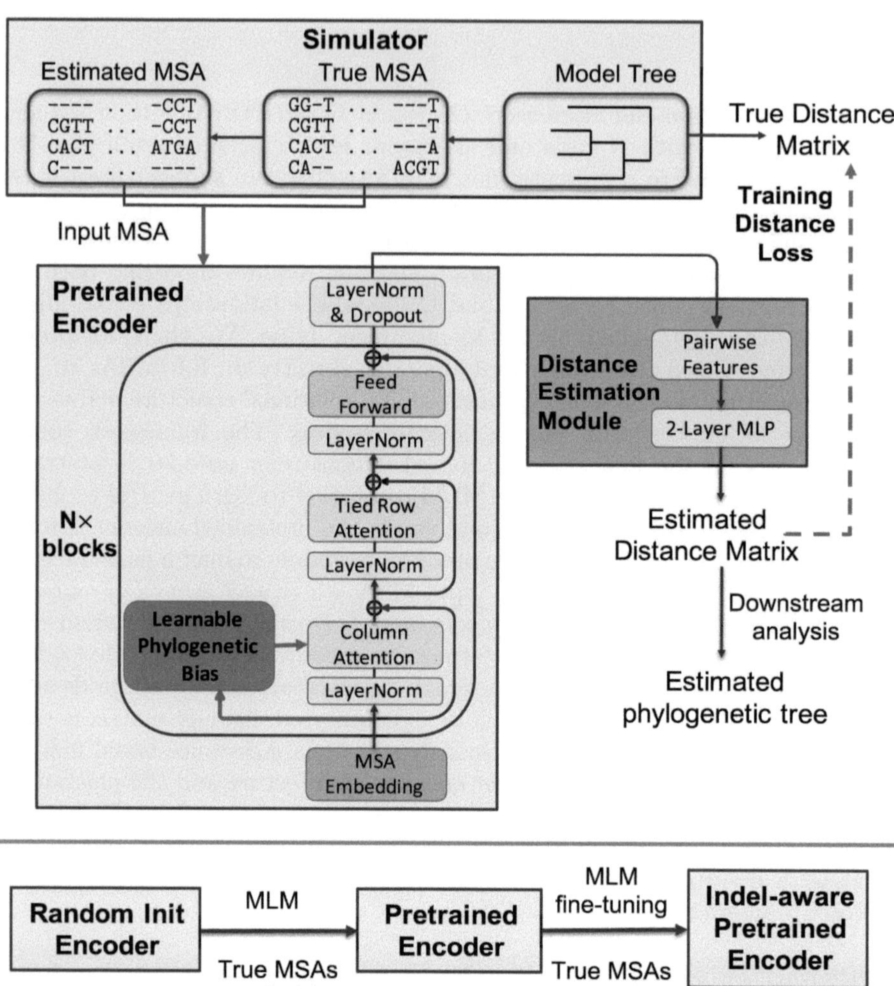

Fig. 1. Overview of the FIREFLY framework for downstream evolutionary pairwise distance inference and phylogenetic tree inference. A simulator is used to generate phylogenetic trees together with true and estimated MSAs under specified evolutionary models. The framework consists of two stages. In the first stage, a Transformer encoder is pretrained using masked language modeling on true MSAs, first without indels and then fine-tuned on MSAs with indels to obtain an indel-aware encoder. The encoder architecture is based on the MSA Transformer [16], with a learnable phylogenetic bias module that is activated only during supervised distance learning. In the second stage, a distance estimation module is trained in a supervised manner using both true and estimated MSAs. Pairwise features are extracted from encoder hidden states and fed into a lightweight MLP to predict pairwise evolutionary distances, which are supervised using true distances derived from the model tree. The resulting pairwise evolutionary distances can serve as input for downstream phylogenetic tree reconstruction using distance-based inference methods.

$X \in \mathbb{R}^{N \times L}$, where each entry corresponds to one of four nucleotides (A, C, G, T) or gap characters ("–").

To mitigate the sparsity and high dimensionality of one-hot encoding and to preserve the column-wise homology structure of MSAs, we avoid subword tokenization (e.g., BPE) and k-mer encodings. Instead, each nucleotide or gap symbol is assigned a unique identifier from a fixed vocabulary and mapped to a learnable embedding via a shared embedding layer. This approach preserves site-level alignment while providing a compact and computationally efficient embedding.

We pretrain the Transformer encoder using a self-supervised masked language modeling (MLM) objective [16]: a subset of alignment entries is masked, and the model is trained to reconstruct the masked nucleotides from both row-wise (within-sequence) and column-wise (across-taxa) context. The training loss follows the standard cross-entropy formulation, with details provided in the Supplementary Material.

In this work, we adopt a similar column–row masking strategy in [16]. Columns are selected with probability P_{col}; within each selected column, a fraction P_{row} of rows are masked while ensuring that at least K_{min} rows remain unmasked. This constraint prevents the model from receiving an insufficient observable signal within a column and stabilizes training. More details are provided in the Supplementary Materials.

Learnable Phylogenetic Bias Module. The attention mechanism is largely data-driven and, by default, does not favor biologically structured interactions without additional inductive bias. In evolutionary MSAs, sequences are related through shared ancestry, which induces hierarchical and non-exchangeable dependencies across taxa. More closely related taxa are expected to exhibit stronger and more consistent correlations across sites than more distantly related ones. These relationships are structured by an underlying phylogenetic tree rather than being uniform across all sequence pairs. In contrast, column-wise attention infers cross-sequence similarity at each site directly from observed alignment patterns. Without additional constraints, this makes the attention scores sensitive to alignment noise, particularly in gap-rich or estimated MSAs.

More broadly, this idea of combining statistical deep learning with logical, mathematical and/or physical constraints has gained traction in the topic known as physics-informed machine learning (PIML) [13]. Rather than learning constraints and laws from scratch, incorporating basic structural information about a problem or domain under study can yield significant efficiencies. An intriguing corollary hypothesis suggests benefits in the other direction as well: it is possible that statistical representation learning can also augment mathematical and logical models in helpful ways and possibly also offset model misspecification, although rigorous testing of this hypothesis would be needed for a given PIML application. Our goal is to implement this idea for the estimation task under study. To achieve this goal, we introduce a learnable phylogenetic bias module. This module encodes pairwise evolutionary relationships directly from learned taxa-level representations, without relying on external distance estimates, and integrates them into the column-wise attention mechanism to guide cross-sequence interactions.

For each taxon, we summarize its representation across all alignment sites by averaging the encoder hidden states along the site dimension. Given the encoder output $\mathbf{H} \in \mathbb{R}^{N \times L \times d}$, where d is the hidden dimension, this yields a taxa-level representation $\mathbf{R} \in \mathbb{R}^{N \times d}$. More expressive aggregation strategies are left for future work. We then apply M independent learned weight matrices to $\mathbf{R}$, each projecting it into a pair of query and key matrices, yielding $\mathbf{Q}, \mathbf{K} \in \mathbb{R}^{M \times N \times d_h}$, where $d_h = d/M$ and M is the number of heads (multi-head attention). Using multiple independent transformations allows the model to capture different aspects of inter-sequence relationships simultaneously, analogous to how multiple sequence alignment tools consider different types of sequence similarity at once. For each head, we compute pairwise similarity scores between sequences as the phylogenetic bias $\mathbf{S}^{(m)} \in \mathbb{R}^{N \times N}$,

$$ S_{ij}^{(m)} = \frac{1}{\sqrt{d_h}} \left\langle \mathbf{Q}_i^{(m)}, \mathbf{K}_j^{(m)} \right\rangle, \quad i, j \in 1, \dots, N. \tag{1} $$

where m indexes the attention heads, and $\langle \cdot, \cdot \rangle$ denotes the dot product between two vectors. We enforce symmetry to reflect evolutionary relationships and remove self-interactions by zeroing the diagonal. The bias is then scaled by a set of learned head-specific positive coefficients $\boldsymbol{\alpha} \in \mathbb{R}^M$, yielding $\mathbf{B}^{(m)} = \alpha_m \mathbf{S}^{(m)}$ for each attention head. The resulting bias $\mathbf{B} \in \mathbb{R}^{M \times N \times N}$ is added to the column-attention logits prior to softmax normalization. This encourages sequences with closer evolutionary relationships to receive higher attention weights during column-wise aggregation.

This design introduces phylogeny-informed structure into attention in a flexible and computationally efficient manner. We further include an oracle experiment using ground-truth evolutionary distances as a column-attention bias to validate the bias design; results are reported in the Supplementary Materials.

Pairwise Evolutionary Distance Prediction and Phylogenetic Tree Inference. To estimate pairwise evolutionary distances, we design a distance estimation module that operates on the encoder output $\mathbf{H}$. For each pair of taxa (i, j), we construct symmetric pairwise features that capture complementary aspects of their relationship. Specifically, we use the element-wise absolute difference $|\mathbf{H}_i - \mathbf{H}_j|$, element-wise product $\mathbf{H}_i \odot \mathbf{H}_j$, and element-wise sum $\mathbf{H}_i + \mathbf{H}_j$. These features are concatenated and passed to a lightweight two-layer feedforward network to predict the pairwise evolutionary distance matrix. The prediction is supervised by a distance loss $\mathcal{L}_{\text{dist}}$ that measures the discrepancy between predicted and true pairwise distances; the specific formulation is provided in Sect. 3 ("Performance evaluation criteria"). Additional details of the distance estimation module are provided in the Supplementary Materials.

Given the predicted pairwise distances from MSAs, phylogenetic trees can be reconstructed using distance-based methods such as FastME [11]. For this downstream analysis, we compare FIREFLY followed by FastME with a maximum-likelihood–based approach (FastTree) and with Phyloformer followed by FastME.

Model Training. We adopt a staged training strategy for the encoder and the evolutionary distance estimation module. In the first stage, the encoder is pretrained using a masked language modeling (MLM) objective on 10,000 simulated true MSAs without indels, in order to learn fundamental sequence- and column-level representations. It is then further pretrained on 10,000 true MSAs containing indels under the same MLM objective to adapt the representations to alignment uncertainty.

In the second stage, the evolutionary distance estimation module is trained jointly with the pretrained encoder on 20,000 MSA–tree pairs, consisting of 10,000 true simulated MSAs and 10,000 corresponding estimated MSAs. During this stage, the parameters of the distance estimation module, the pretrained encoder, and the phylogenetic bias module are jointly optimized. The phylogenetic bias module is activated only during supervised distance training. All training experiments use MSAs with a fixed number of taxa ($N = 50$) and alignment length ($L = 1024$).

The supervised training objective for evolutionary distance estimation module is defined as

$$\mathcal{L} = \mathcal{L}_{\text{dist}}^{\text{true}} + w_{\text{est}}\, \mathcal{L}_{\text{dist}}^{\text{est}},$$

where $\mathcal{L}_{\text{dist}}^{\text{true}}$ and $\mathcal{L}_{\text{dist}}^{\text{est}}$ denote the distance loss $\mathcal{L}_{\text{dist}}$ computed on true and estimated MSAs, respectively; see Sect. 3 ("Performance evaluation criteria") for more details. The weighting factor w_{est} is scheduled across training epochs: it is set to 0.5 for the first five epochs and then gradually increased to 1.0. This schedule allows the model to first emphasize cleaner evolutionary signals and progressively adapt to noise introduced by estimated alignments. Additional training details are provided in the Supplementary Materials.

3   Performance Benchmarking Study

Simulation Datasets. Training and evaluation data are generated using a simulation-based pipeline designed to produce paired true and estimated MSAs, along with their corresponding phylogenetic trees, under controlled evolutionary scenarios. We first generate ultrametric phylogenies under a birth–death process, which are subsequently perturbed to obtain non-ultrametric trees following the procedure described in [20]. Each tree is then rescaled to match the empirical tree length distributions observed in public phylogenetic databases, consistent with the protocol used by Phyloformer [15].

Given each rescaled phylogeny, a true MSA is simulated using INDELible [5] under a finite-sites nucleotide substitution model with gamma-distributed among-site rate heterogeneity, incorporating insertion and deletion (indel) events. Substitution processes follow the general time-reversible (GTR) model [17], with base frequencies and substitution rate parameters derived from empirical rice datasets [6]. We adopt a medium gap-length distribution as described in

[12]. An estimated MSA is then generated from the corresponding true MSA using a multiple sequence alignment algorithm.

All training datasets consist of MSAs with 50 taxa and an indel rate of 0.02, while datasets with 10 to 100 taxa and other indel rates (0.01 and 0.03) are used for evaluation. For all experiments, alignment lengths are truncated or padded to 1024 sites. Unless otherwise specified, all estimated MSAs used for both training and evaluation are generated using MAFFT [10], a widely used multiple sequence alignment method. Additional details on data simulation, parameter settings, and summary statistics of the datasets are provided in the Supplementary Materials.

Empirical Datasets. We evaluate two downstream phylogenetic analysis tasks on three intronic RNA datasets – IGIC2, IGID, and IGIE – from the Comparative RNA Website (CRW) database [21]. The remaining intronic RNA datasets are excluded because their alignments exceed 6,000 sites, beyond the practical input size supported by FIREFLY under current hardware constraints. Each dataset provides curated reference alignments and unaligned sequence data; maximum-likelihood reference trees from [21] are used in place of unknown ground-truth phylogenies. Estimated MSAs are constructed from the unaligned sequence files using MAFFT. For IGIE, which contains 249 taxa and 2,751 sites, the full alignment exceeds the processing limit and is randomly partitioned into five sub-alignments with their corresponding induced subtrees. For both downstream evaluation tasks, analyses are conducted on estimated alignments. Dataset summary statistics are provided in the Supplementary Materials.

Performance Evaluation Criteria. We evaluate pairwise evolutionary distance estimation using the mean absolute error (MAE) and mean relative error (MRE). Let $d_{i,j}$ denote the true evolutionary distance between taxa i and j, and $\hat{d}_{i,j}$ the corresponding predicted distance. The MAE and MRE are defined as

$$\text{MAE} = \frac{1}{|\mathcal{N}|} \sum_{(i,j)\in\mathcal{N}} \left| \hat{d}_{i,j} - d_{i,j} \right|, \quad \text{MRE} = \frac{1}{|\mathcal{N}|} \sum_{(i,j)\in\mathcal{N}} \frac{\left| \hat{d}_{i,j} - d_{i,j} \right|}{d_{i,j}},$$

where $\mathcal{N} = (i,j) \mid 1 \leq i < j \leq N$ denotes the set of all unordered taxon pairs.

The distance loss $\mathcal{L}_{\text{dist}}$ used for supervised training is defined as a weighted combination of these two evaluation criteria,

$$\mathcal{L}_{\text{dist}} = \text{MAE} + \lambda \text{MRE},$$

with $\lambda = 0.3$ in all experiments.

For a reconstructed phylogenetic tree, the normalized Robinson–Foulds (RF) distance is used to evaluate the topological accuracy. In a phylogenetic tree, each branch induces a bipartition of the leaf set. Let A and B denote the sets of bipartitions in the true tree T and the estimated tree T^*, respectively. The normalized RF distance is defined as

$$NRF_{\text{norm}}(T, T^*) = (|A| + |B|)^{-1}(|A \cup B| - |A \cap B|),$$

which measures the fraction of discordant bipartitions between the two trees.

For a distance matrix induced by a phylogenetic tree, satisfaction of the four-point condition is expected, as it reflects the fundamental additivity property of tree-based evolutionary distances. To directly evaluate this structural property, we introduce the four-point residual as an additional assessment to quantify deviations from tree additivity. For any quartet of distinct taxa $\{i, j, k, l\}$, let $d_T(\cdot, \cdot)$ denote the pairwise path length in tree T, and define $S_1 = d_T(i, j) + d_T(k, l)$, $S_2 = d_T(i, k) + d_T(j, l)$, and $S_3 = d_T(i, l) + d_T(j, k)$. The four-point residual for this quartet is

$$\Delta_T(i, j, k, l) = \max\{S_1, S_2, S_3\} - \mathrm{mid}\{S_1, S_2, S_3\},$$

where $\mathrm{mid}(\cdot)$ denotes the median. For an additive tree metric, the two largest sums are equal for every quartet, yielding $\Delta_T = 0$. Larger values indicate stronger violations of tree additivity. To reduce computational cost, we randomly sample 2000 quartets and compute the four-point residual score by averaging $\Delta_T(i, j, k, l)$ over the sampled quartets.

4   Results

Performance on True MSAs with Indels. We evaluate FIREFLY on two downstream phylogenetic tasks using true MSAs with indels: pairwise evolutionary distance inference and phylogenetic tree reconstruction.

We first assess distance estimation accuracy using mean absolute error (MAE). As shown in Fig. 2 (top right), FIREFLY outperforms Phyloformer across most taxa sizes, with the exception of small MSAs ($N \leq 30$). The performance gap increases as the number of taxa grows. Compared with Fast-Tree, FIREFLY achieves comparable or lower MAE and consistently shows reduced variance. We next evaluate phylogenetic tree reconstruction accuracy using the normalized Robinson–Foulds (NRF) distance. As shown in Fig. 2 (top left), FIREFLY combined with FastME (FIREFLY+FastME) achieves comparable or lower NRF distances than both Phyloformer combined with FastME (Phyloformer+FastME) and FastTree across all taxa sizes. While NRF distance increases for all methods as the number of taxa grows, the relative trends differ: the advantage of FIREFLY+FastME over FastTree decreases with the number of taxa, whereas its advantage over Phyloformer+FastME increases.

Beyond topology and distance accuracy, we assess the additivity of the inferred distance matrices using the four-point residual. This analysis is restricted to Phyloformer and FIREFLY, since FastTree is not a distance-based method and distances derived from inferred trees are naturally additive. As shown in Fig. 2 (bottom), FIREFLY consistently yields lower four-point residuals with smaller variance than Phyloformer across all taxa sizes, indicating closer adherence to tree additivity.

Overall, these results indicate that on true MSAs with indels, FIREFLY achieves competitive accuracy in pairwise distance estimation and phylogenetic tree topology inference, while exhibiting improved conformity to tree additivity in the predicted pairwise distances.

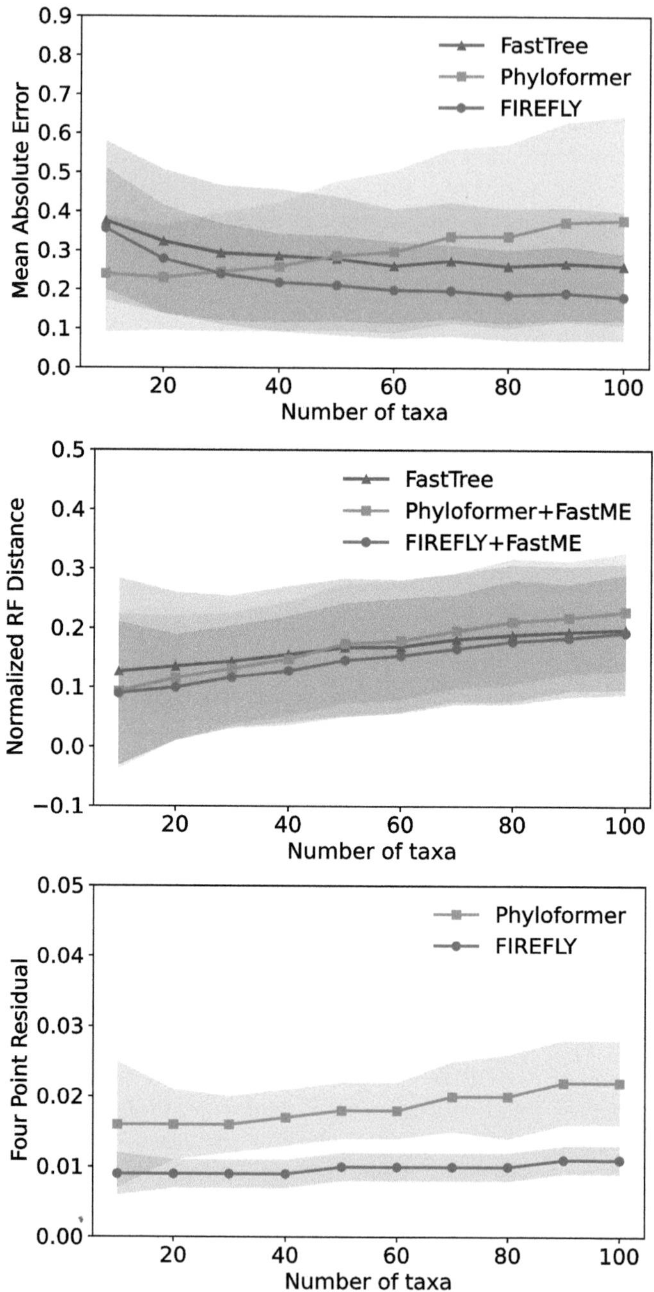

Fig. 2. Performance on simulated true MSAs across different numbers of taxa. Results are shown for three evaluation assessments: normalized Robinson-Foulds (NRF) distance, mean absolute error (MAE) of pairwise distances, and four-point residual. Curves (and shaded regions) report average (and standard deviation) across replicates.

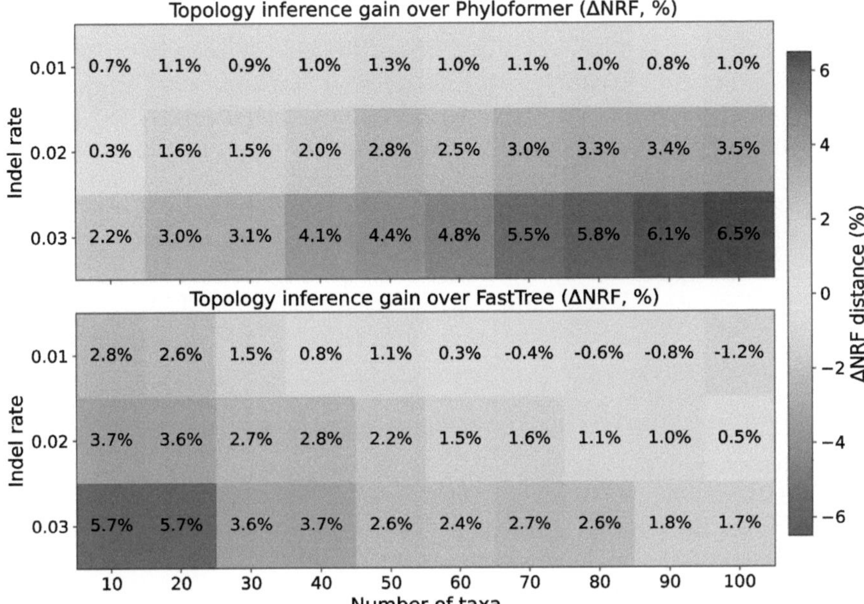

Fig. 3. **Phylogenetic tree topology inference gain of FIREFLY+FastME over two baselines on simulated true MSAs with different indel rates.** Heatmaps report the mean performance gain of FIREFLY in phylogenetic topology inference, measured by the difference in normalized Robinson–Foulds (NRF) distance across varying indel rates and numbers of taxa. Positive values (red) indicate improved topology inference by FIREFLY relative to the corresponding baseline. The upper panel compares FIREFLY+FastME with Phyloformer+FastME, while the lower panel compares FIREFLY+FastME with FastTree. (Color figure online)

Robustness under High Indel Rates. To evaluate the robustness to insertion/deletion (indel) events, we compared the downstream phylogenetic tree topology inference performance of FIREFLY+FastME with Phyloformer+FastME and FastTree across increasing indel rates and varying numbers of taxa. Figure 3 reports the mean topology inference performance gain, measured as the difference in normalized Robinson–Foulds (NRF) distance, where positive values indicate improved accuracy of FIREFLY.

Across all experimental settings, FIREFLY+FastME consistently outperforms Phyloformer+FastME. At low indel rates ($\rho = 0.01$), the gains are modest and relatively stable across taxa sizes, typically below 1.5%. As the indel rate increases, the advantage of FIREFLY and becomes more pronounced and grows with the number of taxa. At a high indel rate ($\rho = 0.03$) and $N = 100$ taxa, the performance gain exceeds 6%. Compared with FastTree, a similar trend is observed with respect to indel rates: the topology inference gain increases as the indel rate rises. However, a different pattern emerges regarding the number of taxa. In this case, the gain decreases as the number of taxa increases.

For example, at $\rho = 0.03$ and small taxa sizes ($N = 10, 20$), the gain exceeds 5%, while it is smaller for larger trees. This difference is expected. Both Phyloformer+FastME and FIREFLY+FastME rely on two-step, distance-based tree construction. In contrast, FastTree performs direct likelihood-based optimization over the tree space and is specifically designed for large numbers of taxa. As tree size increases, the advantage of improved distance estimation is partially offset by the limitations of distance-based tree reconstruction.

Overall, the increasing performance gain with higher indel rates indicates that FIREFLY is particularly effective at mitigating the detrimental effects of frequent indel events. We attribute this robustness to its full-MSA representation learning strategy, which better leverages gap patterns and alignment structure under high indel conditions.

Robustness to Alignment Errors on Estimated MSAs with Indels. Most existing deep learning–based phylogenetic analysis methods are evaluated on simulated true MSAs, where sequences are perfectly aligned. While this setting is useful for theoretical experimentation, it does not reflect typical real-world analyses, where MSAs are estimated and inevitably contain alignment errors. Such alignment errors can distort evolutionary signals and propagate into downstream phylogenetic analysis, as shown in prior studies [1,7]. Therefore, assessing robustness to alignment errors is critical for evaluating practical applicability. For this purpose, we evaluated FIREFLY on estimated MSAs containing indels.

As shown in Fig. 4, the impact of alignment errors varies across methods and evaluation metrics. For NRF distance and four-point residual, all methods show performance degradation on estimated MSAs compared to true MSAs. For MAE, Phyloformer exhibits pronounced degradation on estimated MSAs, reflecting strong sensitivity to alignment noise, while FIREFLY and FastTree show slight improvement, suggesting greater robustness in distance estimation. For phylogenetic tree reconstruction, the degraded distance estimates of Phyloformer lead to reduced topology accuracy. As a result, the performance gap between Phyloformer+FastME and FIREFLY+FastME increases on estimated MSAs compared with true MSAs. For example, at $N = 100$ taxa, the NRF difference increases from 3% to 5%. In contrast, the performance gap between FIREFLY+FastME and FastTree diminishes, and the two methods achieve comparable accuracy across taxa sizes. This trend likely reflects the increased impact of alignment errors on two-step distance-based pipelines. Most notably, alignment errors lead to substantially larger four-point residual values and increased variance for both methods. However, the relative advantage of FIREFLY over Phyloformer also becomes more pronounced under estimated MSAs, indicating improved robustness of FIREFLY to alignment-induced noise.

Overall, these results indicate that the advantages of FIREFLY extend beyond idealized settings and are most evident under realistic conditions with estimated MSAs. In the supplementary material, we further observe consistent trends across estimated MSAs generated by alternative alignment methods.

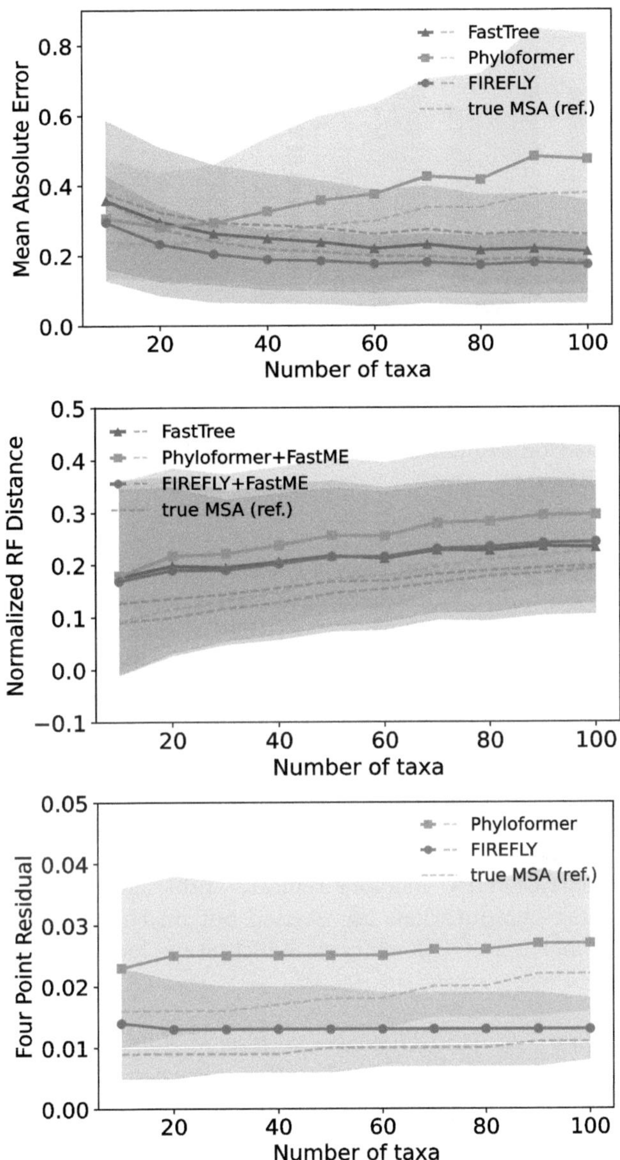

Fig. 4. Performance comparison on estimated MSAs across different numbers of taxa. Results are shown for the same assessments and layout as in Fig. 2, but evaluated using estimated MASs from the simulated datasets.

Computational Runtime and Memory Usage. Similar to other deep learning–based approaches, FIREFLY enables substantially faster inference after training than full maximum-likelihood or Bayesian methods. We assessed computational efficiency by measuring runtime as a function of the number of taxa in

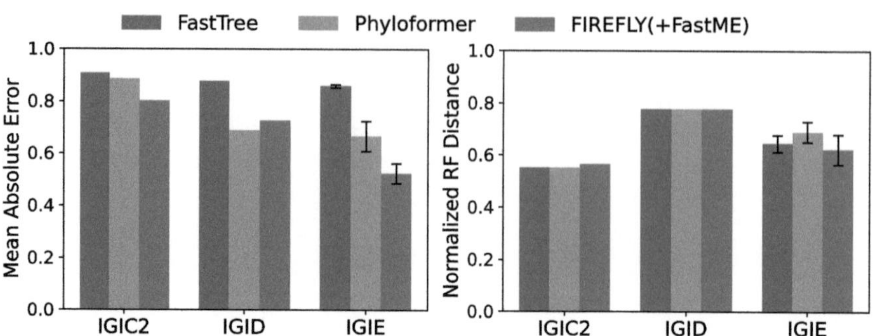

Fig. 5. Performance comparison on CRW intronic RNA datasets. The left panel shows mean absolute error (MAE) of pairwise distance prediction, and the right panel shows the normalized Robinson–Foulds distance (NRFD) of downstream tree inference, evaluated on estimated alignments. Error bars denote standard deviation across IGIE partitions.

Fig. 6. Phyloformer achieves the fastest runtime across all settings, while Fast-Tree incurs substantially higher computational cost due to explicit likelihood evaluation and tree search procedures. FIREFLY introduces a moderate runtime overhead relative to Phyloformer, reflecting the additional cost of learning representations from the full MSA rather than from pairwise-averaged inputs, but remains one to two orders of magnitude faster than FastTree. Importantly, FIREFLY maintains sub-second runtimes even for trees with 100 taxa, demonstrating favorable scaling and practical applicability.

As shown in Fig. 7, GPU memory increases as the number of taxa grows for both deep learning models. The increase is steeper for Phyloformer than for FIREFLY. In contrast, CPU memory remains stable across taxa sizes for both models, since most computations are carried out on the GPU rather than the CPU. FastTree only uses CPU memory, which stays low and increases slightly with taxa.

Ablation Study on the Learned Phylogenetic Bias Module. We conducted ablation experiments to evaluate the contribution of the learned phylogenetic bias module. Results show that the bias term provides increasing benefit as the number of taxa and indel rate grow, yielding up to 37% reduction in distance estimation error. Furthermore, different attention heads learn complementary correlation patterns with the true evolutionary distances, suggesting that the multi-head formulation captures diverse aspects of inter-sequence relationships. Details are presented in Supplementary Appendix Subsects. 3.2 and 3.3.

Performance on Empirical Datasets. We compare the performance of FastTree, Phyloformer, and FIREFLY on three CRW intronic RNA datasets using estimated alignments. Figure 5 reports the mean absolute error (MAE) for pairwise distance prediction and the normalized Robinson–Foulds (NRFD) distance for downstream phylogenetic tree inference. Error bars are shown only for IGIE,

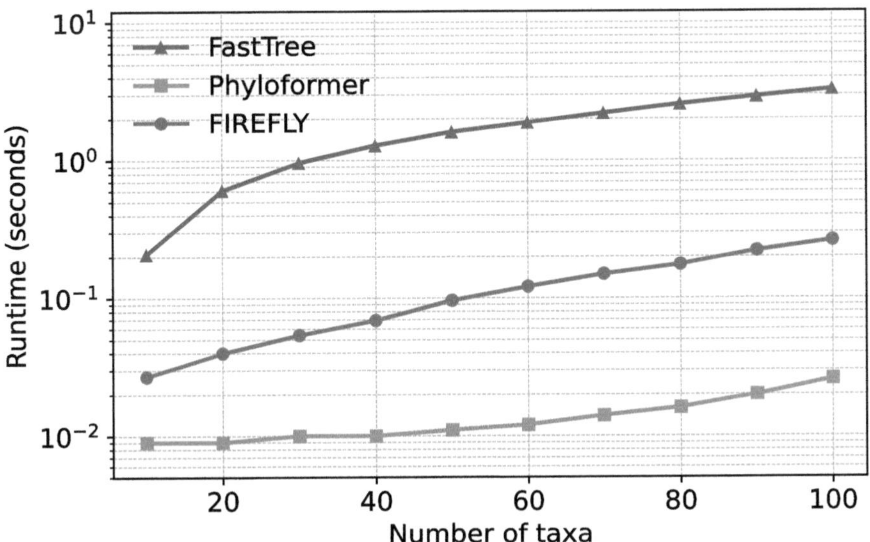

Fig. 6. Comparison of downstream phylogenetic tree inference runtime across different methods. Execution time for tree reconstruction pipelines evaluated on simulated MSAs with 1024 sites. For deep learning–based approaches, the runtime includes distance prediction followed by FastME tree inference, excluding model initialization and weight loading.

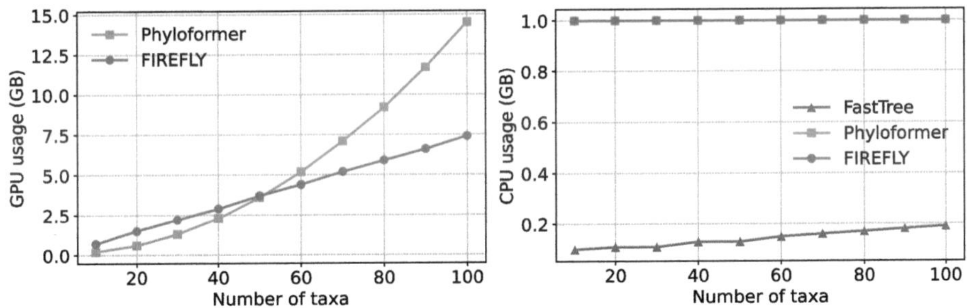

Fig. 7. Computational memory usage across different numbers of taxa. Left: GPU memory usage for deep learning models (Phyloformer and FIREFLY). Right: CPU memory usage for all methods.

where multiple partitions are available; the other datasets consist of a single instance.

Across datasets, FIREFLY achieves the lowest MAE in most cases, indicating more accurate pairwise distance prediction, except for IGID, where Phyloformer performs slightly better. The performance gap is particularly pronounced on IGIE, which remains more challenging despite partitioning due to its larger taxon sampling and higher gappiness compared to the other datasets. For downstream tree inference, all methods yield comparable NRFD values on

IGIC2 and IGID. In contrast, on the more challenging IGIE dataset, FIREFLY produces trees with lower NRFD, particularly compared to Phyloformer, indicating improved robustness under increased alignment noise and larger taxon sampling.

5   Discussion

Across experiments on both true and estimated MSAs, as well as under varying indel rates, FIREFLY consistently improves pairwise distance estimation and phylogenetic tree reconstruction. These gains are most pronounced for MSAs with larger numbers of taxa and higher levels of noise, arising from alignment errors and higher indel events.

Comparisons between true and estimated MSAs highlight the practical value of this design. On estimated MSAs, where alignment errors introduce noise and weaken column-wise evolutionary signals, FIREFLY shows a clearer advantage over Phyloformer and achieves performance comparable to maximum likelihood estimation. This robustness is particularly important in realistic phylogenetic settings, where true alignments are unavailable, and estimation error is unavoidable. The improvement over Phyloformer underscores the importance of full-MSA representation learning for phylogenetic analysis. By modeling evolutionary structure at the alignment level, FIREFLY avoids the systematic biases associated with purely pairwise learning strategies, which have been shown to affect phylogenetic comparison and construction [2,4]. These systematic biases become more severe in the presence of alignment errors.

Across simulations with increasing indel rates, FIREFLY shows progressively larger performance gains. Higher indel rates reflect more complex evolutionary processes and lead to alignments with a larger fraction of gap-related uncertainty. In such settings, traditional likelihood-based phylogenetic methods typically treat gaps as missing data, limiting their ability to exploit evolutionary signals carried by indel patterns. In contrast, FIREFLY learns representations directly from full MSAs and integrates both sequence content and gap structure through row- and column-wise context. By comparison, Phyloformer relies on pairwise sequence representations, which compress the alignment and lose indel-related structure. Together, these factors explain why FIREFLY substantially outperforms both FastTree and Phyloformer in high-indel regimes, highlighting the benefit of MSA-level representation learning.

More broadly, this work highlights the value of integrating modern representation learning with rules and constraints based on first principles. By incorporating phylogenetically motivated inductive bias directly into the model architecture, FIREFLY moves beyond purely data-driven approaches and better aligns learned representations with evolutionary structure.

6   Conclusion

In this study, we introduced FIREFLY, a Transformer-based framework that learns phylogeny-informed representations directly from multiple sequence alignments and supports downstream phylogenetic analyses. By integrating a learnable phylogenetic bias into MSA representation learning, FIREFLY improves both evolutionary distance estimation and downstream phylogenetic tree inference, particularly in settings characterized by high indel rates and alignment uncertainty. Using simulated and empirical datasets, we conduct extensive performance benchmarking and ablation experiments, and we demonstrate that explicitly modeling phylogenetic structure within deep representations leads to more robust and accurate inference. These findings underscore the potential of combining modern representation learning with classical evolutionary knowledge to advance phylogenetic analysis in complex evolutionary regimes.

We conclude with thoughts on future research directions. We hypothesize that FIREFLY's phylogeny-informed representation learning can be applied not only to phylogenetic distance estimation and tree reconstruction, but also to other tasks in computational phylogenetics. More generally, FIREFLY provides a case study of an important insight from physics-informed machine learning (PIML): logic and theory synergizes with purely statistical representation learning. This insight can be leveraged throughout computational biology and bioinformatics and beyond.

Acknowledgments. This research has been supported in part by the National Science Foundation (DBI-2144121, DBI-2214038, and CCF-1714417 to KJL). All computational experiments and analyses were performed on the MSU High Performance Computing Center, which is part of the MSU Institute for Cyber-Enabled Research.

Disclosure of Interests. The authors have no competing interests to declare.

Data Availability. Data and scripts used in this study are publicly available under an open copyleft license at https://gitlab.msu.edu/liulab/firefly-study-data-and-scripts.

References

1. Ashkenazy, H., Sela, I., Levy Karin, E., Landan, G., Pupko, T.: Multiple sequence alignment averaging improves phylogeny reconstruction. Syst. Biol. **68**(1), 117–130 (2019)
2. Dunn, C.W., Zapata, F., Munro, C., Siebert, S., Hejnol, A.: Pairwise comparisons across species are problematic when analyzing functional genomic data. Proc. Natl. Acad. Sci. **115**(3), E409–E417 (2018)
3. Felsenstein, J.: Evolutionary trees from DNA sequences: a maximum likelihood approach. J. Mol. Evol. **17**(6), 368–376 (1981)
4. Felsenstein, J.: Phylogenies and the comparative method. Am. Nat. **125**(1), 1–15 (1985)
5. Fletcher, W., Yang, Z.: INDELible: a flexible simulator of biological sequence evolution. Mol. Biol. Evol. **26**(8), 1879–1888 (2009)
6. Gao, M., Liu, K.J.: Statistical analysis of GC-biased gene conversion and recombination hotspots in eukaryotic genomes: a phylogenetic hidden Markov model-based approach. In: Proceedings of the 12th ACM Conference on Bioinformatics, Computational Biology, and Health Informatics, pp. 1–24 (2021)

7. Gao, M., Wang, W., Liu, K.J.: The impact of gene sequence alignment and gene tree estimation error on summary-based species network estimation. In: Proceedings of the 13th ACM International Conference on Bioinformatics, Computational Biology and Health Informatics, pp. 1–17 (2022)
8. Gascuel, O.: BIONJ: an improved version of the NJ algorithm based on a simple model of sequence data. Mol. Biol. Evol. **14**(7), 685–695 (1997)
9. Jumper, J., et al.: Highly accurate protein structure prediction with AlphaFold. Nature **596**(7873), 583–589 (2021)
10. Katoh, K., Standley, D.M.: MAFFT multiple sequence alignment software version 7: improvements in performance and usability. Mol. Biol. Evol. **30**(4), 772–780 (2013)
11. Lefort, V., Desper, R., Gascuel, O.: FastME 2.0: a comprehensive, accurate, and fast distance-based phylogeny inference program. Mol. Biol. Evol. **32**(10), 2798–2800 (2015)
12. Liu, K., et al.: SATe-II: very fast and accurate simultaneous estimation of multiple sequence alignments and phylogenetic trees. Syst. Biol. **61**(1), 90 (2012)
13. Meng, C., Griesemer, S., Cao, D., Seo, S., Liu, Y.: When physics meets machine learning: a survey of physics-informed machine learning. Mach. Learn. Comput. Sci. Eng. **1**(1), 20 (2025)
14. Mo, Y.K., Hahn, M.W., Smith, M.L.: Applications of machine learning in phylogenetics. Mol. Phylogenet. Evol. **196**, 108066 (2024)
15. Nesterenko, L., Blassel, L., Veber, P., Boussau, B., Jacob, L.: Phyloformer: fast, accurate, and versatile phylogenetic reconstruction with deep neural networks. Mol. Biol. Evol. **42**(4), msaf051 (2025)
16. Rao, R.M., et al.: MSA transformer. In: International Conference on Machine Learning, pp. 8844–8856. PMLR (2021)
17. Rodriguez, F., Oliver, J., Marin, A., Medina, J.: The general stochastic model of nucleotide substitution. J. Theor. Biol. **142**, 485–501 (1990)
18. Smith, M.L., Hahn, M.W.: Phylogenetic inference using generative adversarial networks. Bioinformatics **39**(9), btad543 (2023)
19. Suvorov, A., Hochuli, J., Schrider, D.R.: Accurate inference of tree topologies from multiple sequence alignments using deep learning. Syst. Biol. **69**(2), 221–233 (2020)
20. Szöllõsi, G.J., Höhna, S., Williams, T.A., Schrempf, D., Daubin, V., Boussau, B.: Relative time constraints improve molecular dating. Syst. Biol. **71**(4), 797–809 (2022)
21. Wang, W., Hejasebazzi, A., Zheng, J., Liu, K.J.: Build a better bootstrap and the RAWR shall beat a random path to your door: phylogenetic support estimation revisited. Bioinformatics **37**(Supplement_1), i111–i119 (2021)
22. Warnow, T.: Standard maximum likelihood analyses of alignments with gaps can be statistically inconsistent. PLoS Currents **4**, RRN1308 (2012)
23. Warnow, T.: Computational Phylogenetics: An Introduction to Designing Methods for Phylogeny Estimation. Cambridge University Press (2018)
24. Wong, K.M., Suchard, M.A., Huelsenbeck, J.P.: Alignment uncertainty and genomic analysis. Science **319**(5862), 473–476 (2008)
25. Yang, Z., Rannala, B.: Bayesian phylogenetic inference using DNA sequences: a Markov Chain Monte Carlo method. Mol. Biol. Evol. **14**(7), 717–724 (1997)
26. Zaharias, P., Grosshauser, M., Warnow, T.: Re-evaluating deep neural networks for phylogeny estimation: the issue of taxon sampling. J. Comput. Biol. **29**(1), 74–89 (2022)
27. Zou, Z., Zhang, H., Guan, Y., Zhang, J.: Deep residual neural networks resolve quartet molecular phylogenies. Mol. Biol. Evol. **37**(5), 1495–1507 (2020)

Detecting Outlier Subtrees of Gene Trees Using SPR Moves and Machine Learning

Alan K. Mayer[1], Shayesteh Arasti[2], and Siavash Mirarab[3]($\boxtimes$)

[1] Department of Mathematics, UC San Diego, San Diego, CA 92093, USA
[2] Department of Computer Science and Engineering, UC San Diego, San Diego, CA 92093, USA
[3] Department of Electrical and Computer Engineering, UC San Diego, San Diego, CA 92093, USA
smirarab@ucsd.edu

Abstract. Abundant discordance among gene trees is widely documented, but the causes of this heterogeneity are varied. Discordance among estimated gene trees can stem from real sources such as coalescent processes, hybridization, and horizontal gene transfer (HGT). It can also stem from errors in data, such as hidden paralogy, mistaken homology, bad alignment, and contamination. While some of these processes create stochastic and subtle changes in gene tree topologies (e.g., human closer to gorilla than to chimp), others can produce unexpected patterns (e.g., guinea pig sister to gorilla). Given a large number of gene trees and a median species tree, one could attempt to automatically find these outliers among gene trees. In this paper, we develop a method that uses quartet-based subtree-prune-and-regraft (SPR) moves, paired with gradient-boosted decision trees, to predict whether parts of a gene tree disagree with species trees in unusual ways. We show that our method, GBOD, is quite accurate in finding HGT events, but less so in other scenarios. Nevertheless, this combination of machine learning and phylogenetic features provides a promising framework for outlier detection.

Keywords: Phylogenomics · machine learning · quartet-based methods · SPR moves · outlier detection · horizontal gene transfer (HGT)

1 Introduction

Phylogenomic analyses that infer species trees and gene trees from large collections of loci sampled across the genome have become standard [37]. While much attention has been paid to methods for inferring species trees and gene trees, a key challenge for practitioners is that input data often do not fully conform

---

Supplementary Information The online version contains supplementary material available at https://doi.org/10.1007/978-3-032-26891-4_18.

M. Lafond (Ed.): RECOMB-CG 2026, LNBI 16569, pp. 373–391, 2026.
https://doi.org/10.1007/978-3-032-26891-4_18

to the assumptions of these methods. On the one hand, large-scale datasets are often riddled with errors [23], stemming from incorrect homology and orthology [28,34], sequence contamination [9,14], and alignment error [15,43]. Moreover, gene trees have their own added set of errors [33] due to the impact of fragmentary data [12,29], sequence model misspecification [13,26], and long branch attraction. Besides these errors, individual parts of the genomes can undergo unusual and unaccounted modes of evolution, such as recombination suppression [20], horizontal gene transfer in unexpected places [40], further creating modeling deficiencies. Regardless of cause, abnormalities in data impact gene trees and the species trees by extension [18,22].

In response to these difficulties, many phylogenomic analyses adopt a wide range of filtering strategies, many operating on alignments [30,36]. These methods tend to over-filter [24,38] and lack the full evolutionary context to detect which patterns of discordance are expected or unusual [43]. Others have noted that errors and unmodeled processes often lead to unusual placements of taxa on gene trees, beyond what is expected based on typical causes of gene tree discordance such as incomplete lineage sorting (ILS). For example, a large number of unexpected topologies among mammalian gene trees (e.g., a guinea pig going with gorilla) was used to cast doubt on the validity of some earlier datasets, or even the whole notion of a gene tree [33].

In response to these observations, several methods have been developed to look for *outliers* in the gene trees [e.g., 4,7,16]. The guiding principle of these methods is that while some forms of discordance are expected for each dataset, some gene trees are so widely different that they can only be explained by errors or unmodeled processes (e.g., HGT in eukaryotes). We refer to these as outliers, remaining agnostic as to the exact cause. In this terminology, calling part of a gene tree an outlier simply means that the discordance it creates is above and beyond what is implied by the species tree or the distribution of gene trees. While automatic outlier detection methods exist and have been adopted increasingly, benchmarking shows that they all have room for improvement [4,7]. Note that these methods do not look for outlier gene trees (an easier task), but rather, individual species or clades in individual gene trees that appear misplaced.

One way to formalize outliers in this context is to examine subtree prune and regraft (SPR) moves that remove a clade and regraft it onto another branch. Since SPRs are powerful, applying a few of them can create many forms of outliers. For outlier detection, we may expect that a highly erroneous clade would have to move very far on the gene tree to agree with a reference tree, such as an estimated species tree. Luckily, using an algorithm we recently developed, for each clade in a gene tree, one can determine the optimal SPR move that maximizes the quartet agreement with a reference tree in $O(n \log^2(n))$ time, amortized over all clades [3]. Thus, we can compute a quartet-based SPR (QSPR) for each clade of each gene tree in addition to several properties of the QSPR move (e.g., how far, what improvement in quartet score, etc.) in a scalable fashion.

As we will see, while QSPR does have some signal regarding outliers, interpreting that signal is far from trivial. How far is far enough? What improvement in quartet score is significant? The answers depend on the context in non-trivial ways. To address this challenge, we resort to machine learning, aiming to train a model that can determine whether a particular QSPR is outside the normal range. However, machine learning poses its own challenges, including a lack of access to labeled training data or knowledge of the underlying error mechanisms. Moreover, QSPR statistics and their interpretation are likely dataset dependent, necessitating adjustments to the criteria per dataset. To address these challenges, we design a method called Gradient Boosted Outlier Detection (GBOD) based on cross-validation and data augmentation with positively labeled samples, trained for a given dataset. We show that QSPR statistics, when interpreted through this machine learning approach, have enough signal to detect outliers with levels of accuracy that rival existing methods. Our machine learning approach can, in the future, be combined with features beyond QSPR (e.g., those extracted using existing methods) to further improve accuracy.

2   Material and Methods

2.1   Notations

Let $T = (V_T, E_T)$ be a rooted binary tree, where V_T and E_T denote the sets of vertices and edges of T, respectively. For an edge $e \in E_T$, we use $l(e)$ to denote the length of e. Let $L_T \subset V_T$ be the leaf set of T. A *quartet* of T on four taxa $\{a, b, c, d\} \subseteq L_T$ is defined as the unrooted topology of T restricted to these four taxa. For a tree T with n leaves, there are $\binom{n}{4}$ such quartets. We say that two trees T_1 and T_2 *share a quartet* on $\{a, b, c, d\}$ if the quartets of T_1 and T_2 on these taxa have the same topology (among the three possible topologies). The *quartet score* of T_1 and T_2 is defined as the number of quartets shared between them. For each vertex $v \in V_T$, we define $C_T(v)$ as the subtree below v in T, and use the simpler notation $C(v)$ when T is clear from context. With a slight abuse of notation, we use $L_{C_T(v)}$ to denote the set of leaves below v in T. A subtree prune and regraft (SPR) move of a subtree $C_T(v)$ to an edge $(u, u') \in E_T$ is defined as follows: we prune $C_T(v)$ from T by removing the edge above v, and then regraft $C_T(v)$ onto the edge (u, u') by introducing a new vertex w, removing (u, u'), and adding the edges (u, w), (w, u'), and (w, v), resulting in a new tree T'. We call an SPR move of $C_T(v)$ to an edge $e \in E_T$ a *quartet-based* SPR (QSPR) with respect to a tree T_2 if, among all possible SPR moves of $C_T(v)$, this move results in the highest quartet score between T' and T_2. When trees are unrooted, we define two SPRs per edge, placing each side of its bipartition on the other side.

2.2   GBOD Overview

Motivation. We observed that when outliers are introduced to a subtree in a simulated dataset, the distance of the optimal QSPR moves on these subtrees is generally higher than that of typical subtrees (Fig. 1a). However, these distributions also have much overlap. Thus, this signal alone cannot definitively discriminate outliers from non-outliers. Moreover, the interpretation of the signal depends on the context. For example, even absent outliers, QSPR distances

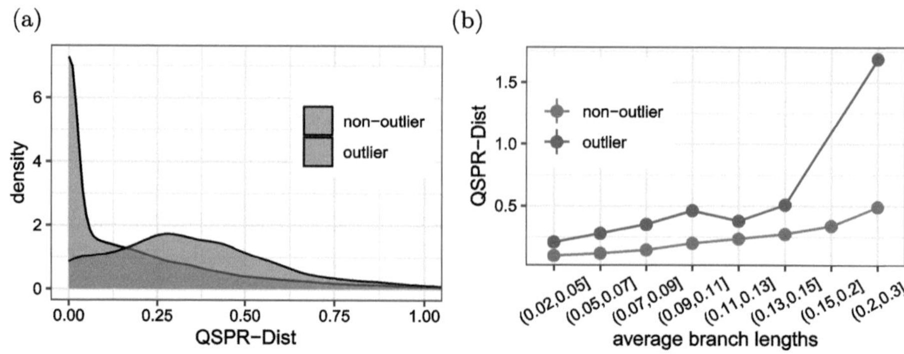

Fig. 1. (a) The distribution of how far a subtree moves with an optimal SPR move (QSPR-Dist in Table 1), for outlier and non-outlier subtrees in the simulated S200-perturbed dataset (Sect. 3.1). A subtree $C(v)$ is considered an outlier if all its leaves are outliers, and a non-outlier if all its leaves are non-outliers. (b) The correlation between the average branch lengths for a gene tree and the QSPR-Dist for outlier and non-outlier subtrees.

correlate with average tree branch length (Fig. 1b). Thus, to make sense of QSPR results, context needs to be considered. However, what information needs to be added and how it should affect our classification is not obvious. Instead of mathematical modeling of QSPR, we turn to the tools that machine learning offers to automatically learn meaningful relationships between the QSPR move and other features of the tree, in order to determine which parts are anomalous.

GBOD Framework. GBOD takes as input a set of gene trees and aims to train models that can learn to identify outlier subtrees in the gene trees based on QSPR statistics. The outliers are subtrees that have placements that are highly unusual given the other gene trees. Since the errors are unknown, we cannot train a supervised model on this data directly. To enable training, we use augmentation (Fig. 2): We inject the gene trees with known anomalies via SPR moves, train a model to identify these introduced outliers, and use the trained model to infer the anomalous parts of the original gene trees.

Leave-out mechanism: The unknown anomalies in the original gene trees we aim to detect persist in the augmented gene trees used as training data. Therefore, a procedure that naively trains a model on an augmented gene tree and uses the same model for inference on the original tree will be biased toward classifying the unknown anomalies as typical, since it was labeled as such in the training data. To remedy this, we adopt a procedure similar to k-fold cross-validation: We partition our original gene trees into k folds. Errors are added in $k - 1$ folds, and inference is done on the left-out fold (Fig. 2). Thus, inference is done on gene trees not seen during training. For computational efficiency, we reuse the augmented gene trees for training across different folds when possible.

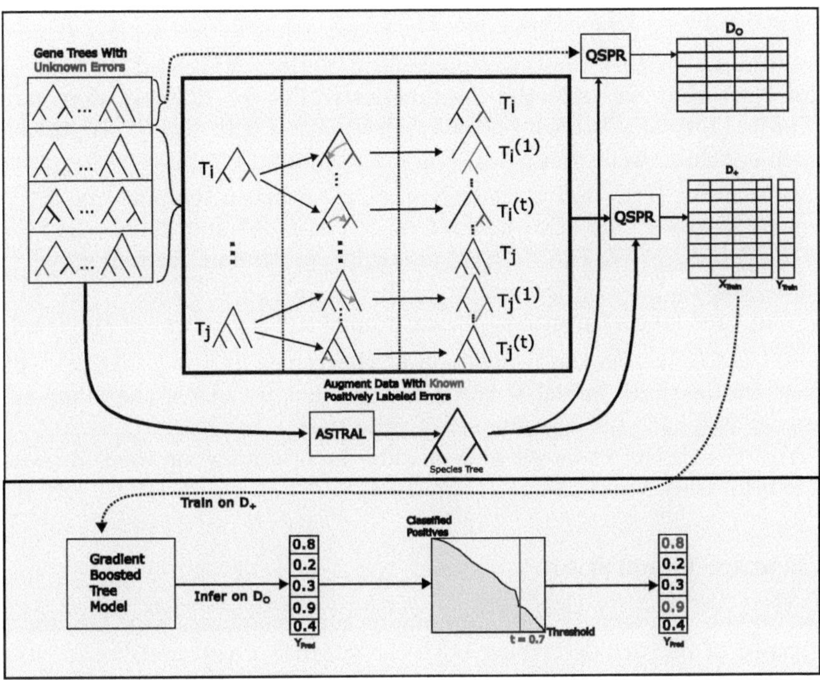

Fig. 2. An overview of the Gradient Boosted Outlier Detection pipeline. GBOD begins by partitioning gene trees into k folds. Then, $k-1$ folds have augmented errors introduced via random SPR moves, creating the augmented dataset D_+. We then use QSPR [3] with respect to the ASTRAL species tree to measure how far each clade is from its optimal position. Results are used to assign a feature vector to each clade of each gene tree (Fig. 3). We train our model on D_+ and infer outliers on the original, unperturbed gene trees in the left-out fold, D_o. Finally, we automatically threshold these predictions to classify outliers.

The GBOD pipeline has four steps, which will be detailed in what follows: 1) creating positively labeled samples for training the model, 2) measuring a set of features for each tree or clade in a tree, 3) training a regression model for each clade to compute the probability that a randomly selected leaf under that clade is anomalous, and 4) using this model to infer outliers in the left out fold, with automatic adjustments to decide what threshold should indicate outliers.

2.3   Creating Labeled Positive Samples for Training

On a set of N rooted gene trees, $G = \{T_1, T_2, \ldots, T_N\}$, we begin by perturbing the data to inject known, labeled errors, which will be used to train a model applied downstream to detect errors in the original unperturbed data (from left-out fold). This augmentation transforms the problem into a supervised learning task well-suited for classical machine learning methods. In particular, for each gene tree, T_i, we augment it t times to get a set of gene trees with known errors

$G'_i = \{T_i^{(1)}, T_i^{(2)}, \ldots, T_i^{(t)}\}$. Each $T_i^{(j)}$ is constructed by choosing a source $e = (u, v)$ and a destination $e' = (u', v')$ on T_i at random, weighted by their branch lengths. Note that this selection is different from how HGT is often modeled where probability of a branch becoming destination is inversely proportional to the path length between that branch and the source. We then perform an SPR move, by pruning $C(v)$ and regrafting on e', resulting in the new tree $T_i^{(j)}$. This placement divides e' into two new edges $e'_1 = (u', w)$ and $e'_2 = (w, v')$, where $l(e'_1) + l(e'_2) = l(e')$. We choose $l(e'_1)$ uniformly at random from $[0, l(e')]$.

Let $L_+ = L_{C_{T_i}(v)}$ be the set of all leaves in the subtree $C_{T_i}(v)$ that are moved and thus labeled positive. For each node $n \in V_{T_i^{(j)}}$, we set the label $y_n = |L_{C(n)} \cap L_+|/|L_{C(n)}|$ as the proportion of the leaves in $C(n)$ that are outliers. The machine learning model is trained to predict y_n. After the augmentation process, we will have the training set of gene trees $\mathcal{G} = G \cup G'_1 \cup \ldots \cup G'_t$, where $|\mathcal{G}| = N \times (t+1)$. Note that we also include the original gene trees in $\mathcal{G}$, and we set $y_v = 0$ for all vertices of T_i.

2.4   Feature Generation

For each node, v in tree T, we define the feature vector x_v. This feature vector is composed of metrics describing both the subtree, $C(v)$, and the full tree, T. A comprehensive list of features with brief descriptions is presented in Table 1.

The main signal is provided by features generated from a QSPR move using a dynamic programming algorithm [3], which identifies the optimal position for each subtree relative to a reference tree. A natural choice for this reference tree is an estimated species tree, S, which can be thought of as a summary of the gene trees. For a subtree $C(v)$ of a gene tree T, and the species tree S, QSPR identifies the optimal SPR move on $C(v)$ such that the number of quartets shared between the updated T and S is maximized. The primary subtree-specific signals that our model may learn from is the distance, both topological and in terms of path lengths (i.e., considering branch lengths), that $C(v)$ moved from its original position on the gene tree to its new position, as well as many derived features. An example of a few possible SPR moves, along with their associated features, can be found in Fig. 3. For each vertex v, we perform QSPR both for $C(v)$ and its complement (i.e., the clade below v if the tree is rerooted somewhere inside $C(v)$). Therefore, we obtain two different feature vectors for each vertex v, corresponding to the bipartition defined by the branch above v. All other features, with the exception of clade size, are defined on the gene tree as a whole. These provide context for the model about the tree in which this clade was found, giving it more information to interpret the QSPR values.

2.5   Modeling and Predicting

Let $x_i^{(j)}$ represent a feature vector for subtree j in the unperturbed gene trees $T_i \in G$. Recall that we have no associated ground truth $y_i^{(j)}$ on these gene trees, but hope to train a model that gives meaningful outputs when evaluated

Table 1. Features used by GBOD. The first set is outputs of QSPR [3], which runs in $O(kn^2 \log^2(n))$ time for all features across all k gene trees, each with n leaves. Others are calculated in $O(kn)$ time.

name	W†	C‡	Description
QSPR-Nodes	F	C	The number of nodes a clade moves by QSPR to S
QSPR-Nodes Norm.	F	C	QSPR-nodes / Avg. Node Height
QSPR-Dist	T	C	The branch distance a clade moves by QSPR to S
QSPR-Dist Norm.	T	C	QSPR-Dist / Avg. Branch Height
Root Dist Before	T	C	Distance of clade to root before QSPR.
Root Dist After	T	C	Distance of clade to root after QSPR.
Root Dist Change	T	C	Root Dist After − Root Dist Before
Quartet Score Diff.	F	T	Change in gene tree quartet score vs. S after SPR.
Clade Size	F	C	The number of leaves in the clade.
Avg. Node Height	F	T	Average topological height of each vertex in the gene tree.
Std. Node Height	F	T	Standard Deviation of topological height of each vertex.
Avg. Branch Height	T	T	Average of height of each vertex.
Std. Branch Height	T	T	Standard Deviation of height of each vertex.
Avg. Branch Length	T	T	Average length of all branches in tree.
Tree diameter	T	T	Longest path between any two nodes
Branch Length Sum	T	T	Sum of all branch lengths
Treeness	T	T	Sum of internal branch lengths / Sum of all lengths
Cherry Proportion	F	T	Proportion of leaves sister to another leaf.
Colless Index	F	T	Sum of difference in child clade sizes.
Sackin Index	F	T	Sum of the depths (edges from the root) of all leaves
Tree Height	T	T	Maximum of the depths of all leaves

$\dagger$: F: Branch lengths of gene trees are ignored; T: otherwise.
$\ddagger$: C: Defined per clade; T: defined per tree.

on these feature vectors. To train these models from our perturbed data, we adopt a procedure similar to k-fold cross-validation. Our unperturbed and randomly ordered gene trees, $G = \{T_1, T_2, \ldots, T_N\}$ are partitioned into k folds, $G^{(1)}, \ldots, G^{(k)}$. This induces a partition on the perturbed dataset, G' where if $G^{(i)} = \{T_{k \cdot i}, T_{k \cdot i + 1}, \ldots, T_{k \cdot i + N/k}\}$, then $G'^{(i)} = \bigcup_{k \cdot i \leq j \leq k \cdot i + N/k} G'_j$. Then for a fold, i, we have labeled training data: $D_+^{(i)} = \bigcup_{j \neq i} G'^{(j)}$, and original unperturbed data: $D_\circ^{(i)} = G^{(i)}$. We train our model, $\phi^{(i)}$, on labeled pairs of data, $(x_j, y_j) \sim D_+^{(i)}$, where x_j is the feature vector for some subtree in an augmented gene tree of $D_+^{(i)}$, and y_j is its associated label. We then apply the trained model $\phi^{(i)}$ on the unlabeled data, $x_j \sim D_\circ^{(i)}$ to get $\hat{y}_j = \phi^{(i)}(x_j)$. Thus, test gene trees are absent from the training dataset, even as perturbed gene trees.

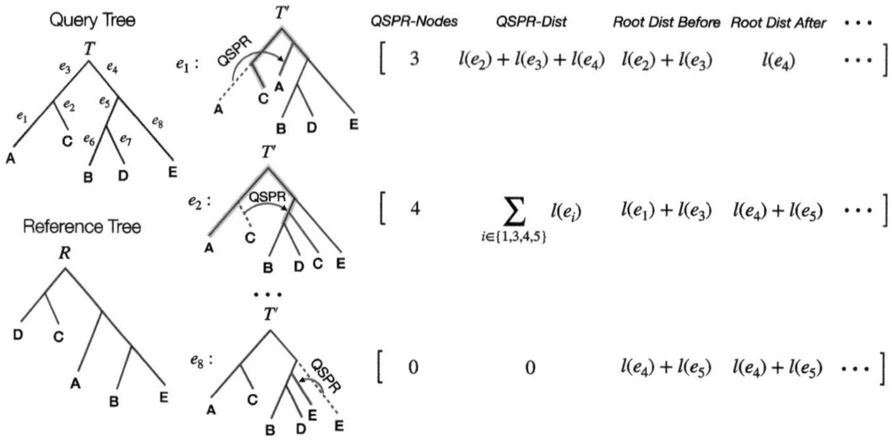

Fig. 3. Using QuartetSPR [3] to extract QSPR features. For each subtree C in the query (gene) tree, we find its optimal placement with respect to the reference (species) tree. We then compute QSPR features of C, e.g., QSPR-Nodes, QSPR-Dist, etc., based on the original and the optimal placement of C.

For $\phi^{(i)}$, we opt to use gradient boosted decision trees. These are ensemble machine learning models that train by iteratively building weak decision trees that correct the errors of the existing ensemble. Gradient boosted decision trees are computationally efficient to train compared to other methods. Additionally, they have been shown to often outperform deep models on tabular data [31]. While gene trees are not inherently tabular, the featurization and regression on each subtree independently of each other tabularize our graph-structured data, lending itself well to models optimized for working on this kind of data. The model is trained via logistic regression with the standard cross-entropy loss.

2.6   Thresholding

For each original gene tree, we now have a prediction for each subtree on the proportion of its leaves that are outliers. We need a threshold to use these predictions to determine which taxa are believed to be erroneous with sufficient confidence. We designed an automatic thresholding strategy that analyzes the proportion of leaves we remove at each threshold. Once a threshold is determined, a leaf is removed from the gene tree if it is contained in any subtree whose confidence is above that threshold. The thresholding strategy analyzes the *predicted* positive rate as a function of the decision threshold. Intuitively, our scheme aims to numerically find a point with high derivative for this function, corresponding to a value at which we get a relatively large increase in the number of filtered leaves. Formally, let the proportion of leaves removed at some threshold t be $F(t)$ where $F(0) = 1$ and $F(1) = 0$. For some small value, δ, and configurable parameter controlling how large the spike is, $f_{\max}$, we gradually decrease t from 1 to 0 and look for when:

$$F(t - \delta) - F(t) > f_{\max} \ .$$

We then opt to cut at the first threshold t that satisfies this condition. By default, we will decrease t in intervals of 0.01 at a time with $\delta = 0.01$ and $f_{\max} = 0.005$.

3 Experiments

We evaluate GBOD on two simulated datasets and one biological dataset. The simulated datasets, which will be discussed in detail below, each have known outliers defined on the leaves of some of the gene trees. We use ASTRAL-III [42] for species tree inference from the original unperturbed gene trees to use as a reference for computing QSPR features. GBOD is then trained and tested separately for each dataset. When training the model in each fold, we use the left-out augmented data for the testing fold to determine early-stopping (without knowledge of true outliers). We evaluate our method on the simulated datasets both by analyzing the F_1 score on each of the datasets, which takes into account our automatic thresholding strategy, as well as the Receiver Operating Characteristic (ROC) curves, which are agnostic to our thresholding strategy. In our setting, a true positive is an outlier detected as such, and a false positive is a normal sequence marked as an outlier. On the biological dataset, we evaluate the effect that removal of the labeled outliers has on species tree inference.

We use $k = 10$ folds in the experiments. When computationally feasible, we augment each gene tree, T_i, 99 times, to get an augmented dataset size 100 times larger than the original. When this is too computationally demanding, we augment each gene tree 9 times. When training our gradient boosted tree models, we opt to use the XGBoost Python package [6]. In each fold during model training, we use the augmented gene trees associated with the left-out fold to perform early stopping to prevent overfitting.

3.1 Datasets

HGT Carnivora Simulated Dataset. We simulated a dataset, following the same exact procedure as the simulated Carnivora dataset used to test PhylteR [7]. This dataset is simulated using SimPhy [17] and uses a 53-taxa Carnivora tree [2] as a reference species tree. In each replicate, 500 gene trees were simulated with the rate of HGT being 1e-8 and varying levels of ILS. Each gene tree in this simulation contains at most one HGT event, and all taxa affected by such HGT events are considered outliers, and it is of interest if our procedure is able to detect these HGT events. From the 500 simulated trees, the first 100 gene trees were retained due to runtime restrictions. We only analyzed the most general case in which we do not limit the number of outliers per gene, so up to 53 can undergo an HGT event (on average: 1.25 outliers per gene tree).

S200-Perturbed Dataset. We use a second dataset described in earlier work [4]. This dataset starts with a 201-taxon SimPhy-simulated dataset [17] with gene trees that undergo ILS. Errors are then algorithmically introduced into the sequence alignment by performing between one and four rounds of perturbation to the sequence data (not gene trees). In a single round of perturbation, $k \sim \text{Poisson}(40)$ genes are modified uniformly at random, and in each of the genes, a small proportion ($t \sim \text{Beta}(10, 90)$) of the taxa are chosen to be modified. The taxa chosen to be modified are sampled according to a strategy that leads them to be closer to each other on the trees than sampling taxa uniformly at random would, but they are not monophyletic. Once the taxa for the round are selected to be modified, the species tree is randomly rerooted, and the root of the tree becomes the "source" taxon. Some portion of the alignment of the source taxon replaces the portions at the same sites in each of the taxa chosen to be modified, leading to chimeric sequences between the chosen taxa and the source taxon. This procedure introduced between 0 and 81 perturbations to each gene (3.2 perturbations in expectation). All perturbed sequences in each gene tree are marked as (unknown) outliers for that gene tree.

Once alignments are perturbed, gene trees are re-inferred with FastTree II [25] and these gene trees become the input to our method. We omit two of the replicates due to extremely large polytomies in some of the original gene trees. This dataset was too large to run with 99 augmentations on each gene tree, so we opted to run it with 9 augmentations per gene tree instead.

Biological Data. We study six real datasets. We use a mammalian dataset [32] with 37 species and 424 genes that is known to have errors identified in previous studies [11,19]. We use the gene trees inferred by RAxML [35] from a reanalysis of this dataset [18]. There are two datasets studying Xenacoelomorpha, one studied by Cannon *et al.*, [5], which we call XenCannon, and one studied by *Rouse et al.*, [27], which we call XenRouse. XenCannon has 78 species and 213 genes, while XenRouse has 26 species and 393 genes. Our next dataset is on frogs [8] and contains 164 species and 95 genes. We study a dataset focused on plants and algae [41] containing 104 species and 852 genes. Lastly, we use the trancriptomic insect dataset [21] containing 144 species and 1478 genes. Only the insect and plant datasets were too large to run GBOD with 99 augmentations of each gene tree, so we opted to run it with 9 instead.

3.2   Method Comparison

We compare GBOD to PhylteR [7] and TCMM [4]. PhylteR detects local outliers in gene trees by representing each gene tree as a distance matrix on the leaves. Then, using DISTATIS [1], a generalization of multidimensional scaling built to compare distance matrices, PhylteR represents each taxa by a point in a low-dimensional Euclidean space, called the compromise space. By doing this, PhylteR can identify which leaves in which gene trees exhibit abnormally large deviations from expectations in this compromise space and classify them as outliers. TCMM provides a way to transform the branch lengths of the gene trees according to some reference tree. It is parameterized by a parameter λ, which determines how much it changes branch lengths. TCMM's output can be used as input to PhylteR to control its ability to detect either topological (with $\lambda = 0$) or branch length outliers (with very large λ), or some compromise between the two. We run TCMM with its default $\lambda = 1$ value, looking for both topological and branch length outliers.

PhylteR and TCMM identify outliers using a modified Tukey's method that defines outliers as values above the third quartile, adjusted for distributional skew. The aggressiveness of this method is controlled by the parameter k: lower values of k result in more taxa being classified as outliers. We evaluate how true and false positive rates change as we vary this k parameter, yielding ROC curves. GBOD differs in that it directly predicts a confidence score for each subtree. We then make a determination of outliers by considering some threshold, t. We classify a taxon as being an outlier if it is contained in any subtree whose confidence is above this threshold. By default, this threshold is determined automatically.

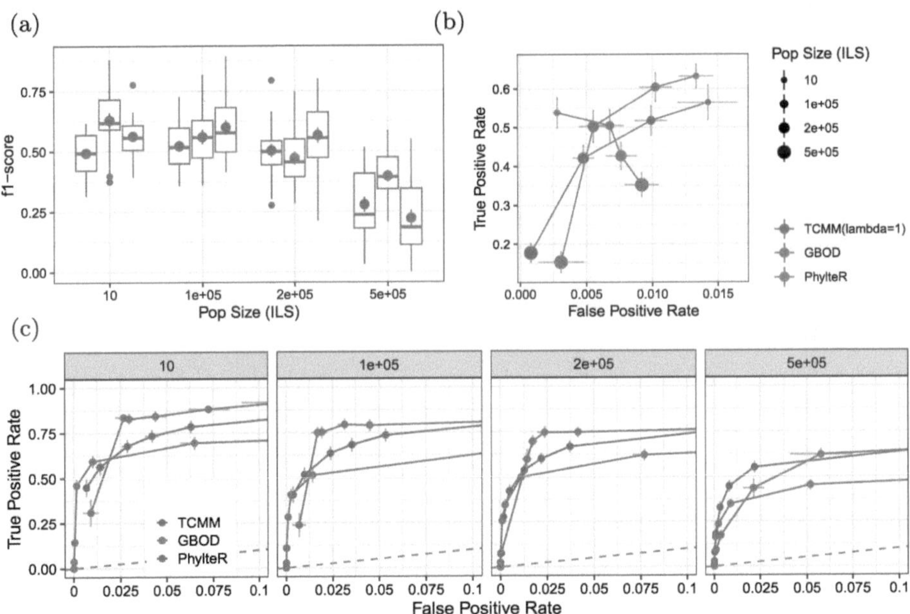

Fig. 4. (Carnivora Simulated Dataset) (a) Comparing GBOD to PhylteR and TCMM on various levels of ILS using F_1-score. Mean and standard error are shown on top of the box plots. (b) False Positive Rate (FPR) versus True Positive Rate (TPR) for each method across various levels of ILS. (c) ROC curve for each method across all levels of ILS, showing FPR < 0.1. See Fig. S1 for full figure without restrictions and Table S1 for area under curve (AUC).

4   Results

4.1   Simulated Datasets

HGT Detection on the Carnivora Dataset. With default settings (automatic thresholding for GBOD, $k = 3$ for PhylteR and TCMM, and $\lambda = 1$ for TCMM), the relative F_1 accuracy heavily depends on the ILS level (Fig. 4a). In a setting with almost no ILS (population size set to 10), our method outperforms both TCMM and PhylteR. However, the F_1 score of GBOD consistently degrades with the increase in ILS, exhibiting both a lower TPR and higher FPR as ILS increases (Fig. 4b). These together lead to a consistent decrease in F_1 across ILS levels for GBOD (Fig. 4a). This contrasts the behavior of both TCMM and PhylteR, which suffer from lower TPR but enjoy lower FPR as well with higher ILS(Fig. 4b). Thus, in their default mode, these methods simply find fewer outliers with high ILS, and as a result, they can potentially maintain, or even slightly improve, F_1 score as ILS increases to the next two levels (Fig. 4a). Nevertheless, at the highest ILS level, these methods find very few outliers, leading to GBOD having a substantially higher F_1 than the alternatives, achieved through much better recall. We note that our choice of threshold is not necessarily optimal in these cases, as other thresholds with similar recall but better precision exist (Fig. S2).

When exploring various thresholds, several patterns emerge. Across ILS levels, GBOD displays a lower Area Under the Curve (AUC) than either PhylteR or TCMM (Table S1). Importantly, this drop in AUC comes from GBOD's poorer performance only at significantly high levels of FPR (Fig. S1). In the context of outlier detection, outliers are assumed to be quite rare, so in practice, each of these methods will choose thresholds corresponding to low false positive rates. Thus, this AUC cannot be taken as the best comparison method. Focusing on cases where FPR is below 10%, we see that GBOD is able to achieve lower FPR rates, and for very low FPR (e.g., below 2%), is better than or competitive with other methods (Fig. 4c). In all conditions except for the low ILS case, our automatic choice of threshold flags too many outliers compared to the threshold that maximizes F_1 (Fig. S2), leaving room for better threshold tuning.

Error Detection in the S200 Dataset. We next turn to outliers caused by data errors (chimera), as simulated in the S200-perturbed dataset. On this dataset, none of the methods were particularly effective at finding outliers in the default settings, detecting 7–10% of errors on average and resulting in low F_1 scores (Fig. 5). The FPR was also low, showing that the errors introduced were simply missed in most cases in the background of real heterogeneity introduced by ILS.

Among the methods, GBOD has the lowest F_1 score, followed closely by PhylteR and then TCMM, with the slightly lower F_1 compared to PhylteR being due to higher FPR, which offsets its slightly higher recall. When examining the ROC curve, we observe that TCMM with better choices of threshold could find close to 70% of errors with an acceptable FPR (7%). In contrast, GBOD and

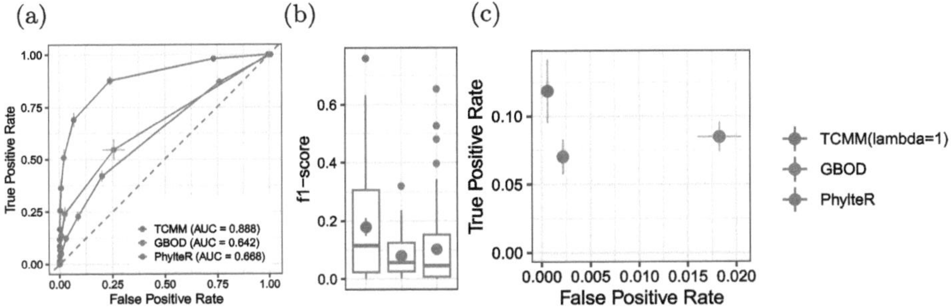

Fig. 5. (S200-Perturbed Dataset) (a) ROC curve comparing all methods. Area under curve (AUC) values are shown for each method. (b) Comparing GBOD to PhylteR and TCMM using F_1-score. Mean and standard error are shown on top of the box plots. (c) False Positive Rate (FPR) versus True Positive Rate (TPR) for each method. Mean and standard error are shown.

PhylteR achieve only about 25% error detection if FPR is controlled at 10% or lower. And between the two, GBOD performs slightly worse across all thresholds. Therefore, the GBOD model does not appear to have been able to learn and generalize from our augmented outliers to the signal in the underlying errors introduced to the perturbed sequences on this dataset. The lower accuracy is only mildly affected by the reduced number of perturbations we were forced to perform; using 99 augmented trees on a subset of 10 replicate simulated datasets improved accuracy only slightly (Fig. S3).

4.2   Biological Data

Across the six biological datasets we studied, removal of taxa classified by GBOD as outliers improves the gene trees' similarity to the species tree in all cases (Fig. 6b and Fig. S4). In the Mammal and XenRouse datasets, while TCMM and PhylteR do not identify many outlier taxa in each gene tree (Fig. 6a), and subsequently do not see a large change in RF distance, GBOD's clade-based approach identifies an order of magnitude more taxa as outliers. We also observe that it is in these two datasets that we see the largest normalized RF improvement across all methods and datasets, followed by the plants dataset, where TCMM and PhylteR lead to higher RF improvements than GBOD (Fig. 6b). For frogs, all methods achieve similar levels of improvement, though PhylteR removes far more. On insects, PhylteR and GBOD achieve the best improvement, both removing far more than TCMM.

In general, GBOD tends to find different outliers compared to both TCMM and PhylteR (Fig. 6a). In each dataset except Frogs, only 0.5%-10% of the outliers GBOD identified were shared by either TCMM or PhylteR. For example, in the insect dataset, PhylteR and GBOD remove almost disjoint sets of outliers, yet they achieve an almost identical increase in normalized quartet score. On the

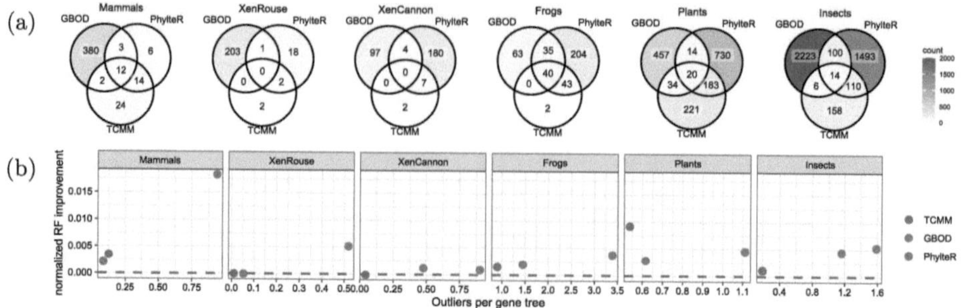

Fig. 6. (Biological Dataset) (a) Venn diagrams for the outliers marked by the three methods for each dataset, and their intersection. The number of outliers is shown on each partition. (b) The average number of outliers per gene tree versus the improvement in normalized RF distance (nrf) between the gene trees and the estimated species tree, defined as the difference between nrf after removing detected outliers from the gene trees and the nrf of original unfiltered gene trees with respect to the species tree. Above the dashed line shows improvement over the original gene trees.

other hand, in the Frogs dataset, the majority (54%) of the outliers GBOD identified were also identified by PhylteR. While PhylteR removed many more taxa than GBOD or TCMM, the increase in normalized quartet score was extremely similar across all three methods.

5 Discussion and Future Work

We introduced a general machine learning framework with a novel scheme for augmenting gene trees with positively labeled errors, which can be used to train models to recognize subtrees with unusual placements. This task is more computationally demanding than detecting outlier gene trees as a whole. The goal of our GBOD model is to detect which exact subtrees of a gene tree are positioned in anomalous positions with respect to a species tree. These questions would be easy to answer using QSPR moves if gene trees were always extremely similar to the species tree; any QSPR to a sufficiently distant place would be anomalous. However, true gene tree discordance due to normal processes such as ILS makes it harder to distinguish outliers from normal heterogeneity (Fig. 1). The goal of the machine learning component is to distinguish the two.

Our simulations indicated that GBOD is successful in making such distinctions when outliers are due to HGT, but less so for chimeric sequences. On biological data, GBOD detected outliers that were quite different from the existing methods. In particular, it appeared that many of the issues with the known problematic mammals dataset were detected by GBOD. These data are extremely unlikely to have HGT. High effectiveness of GBOD on these data likely indicates that *some* forms of data errors can be detected by GBOD. Thus, the low recall of GBOD on the S200-perturbed dataset may be specific to its particular form of introducing errors by creating chimeric sequences. Since that mode of error introduction is not necessarily the most realistic scenario, future work should further explore other methods of simulating errors in the data.

The underlying augmentation strategy we use here is implemented using random SPR moves to change gene trees. SPR were chosen because they allow general modifications to the tree topology and are easy to implement. Our results showed both the promise and limitations of this approach. On the dataset with HGT outliers, which are similar though not identical to SPR moves [39], the results were strong. On the S200-perturbed dataset, where errors were not similar to SPR, the results were far less strong. On this dataset, TCMM, which accounts for branch lengths more explicitly than GBOD performed far better. Thus, our results demonstrate that training a model to recognize SPRs alone has inductive biases that carry over to inference. Future work can expand on the data augmentation mechanisms used. Our choice of SPR-based data augmentation methods was motivated by their ease of application. A simple extension is to apply multiple rounds of SPR (one on top of another) to change gene trees more dramatically. Other approaches to explore in the future could include p-Edge-Contract-and-Refine moves [10] and tree-bisection and reconnection. Finally, we performed between 10 and 100 rounds of augmentation per gene tree; preliminary results show that the number of rounds does impact accuracy (Fig. S5). However, augmenting further will increase the running time of feature extraction in our current implementation. With some preprocessing, we may be able to perform QSPR moves on SPR-adjacent gene trees without recomputing all the counters, saving much of the running time.

Beyond data augmentation, both the choice of model and features are easily expanded upon. While our results showed that QSPR moves do have the signal to detect at least some types of outliers, there is no reason that our machine learning framework could not incorporate other features, such as results of TCMM and PhylteR, as extra features. Such combinations seem promising and should be explored in the future. In particular, our current method is mostly focused on topological outliers, but branch length outliers can be equally informative, motivating adding such features.

Gradient boosted trees, while computationally efficient to train, work best on tabular data. As such, in this framework, we tabularize the inherently tree-based data by assigning each internal node a feature vector and performing regression on each vector independently. A natural next step, then, is to directly incorporate trees into the model. To this point, the augmented, positively labeled clades are created in a dependent manner (e.g., all smaller subtrees of a moved clade are labeled as outliers); however, due to our tabularization of the data, our current framework ignores these dependencies and treats each subtree as an independent training point. A modification to this framework that more naturally lends itself to modeling the tree structure may significantly improve some results.

Acknowledgement. This study was funded by NIH MIRA grant 1R35GM142725.

Disclosure of Interests. The authors have no competing interests to declare that are relevant to the content of this article.

References

1. Abdi, H., O'Toole, A., Valentin, D., Edelman, B.: DISTATIS: the analysis of multiple distance matrices. In: Proceedings of the IEEE Computer Society: International Conference on Computer Vision and Pattern Recognition, pp. 42–42 (2005). https://doi.org/10.1109/CVPR.2005.445

2. Allio, R., et al.: High-quality carnivoran genomes from roadkill samples enable comparative species delineation in aardwolf and bat-eared fox. eLife **10**, e63167 (2021). https://doi.org/10.7554/eLife.63167

3. Arasti, S., Mirarab, S.: Median quartet tree search algorithms using optimal subtree prune and regraft. Algorithms Mol. Biol. **19**(1), 12 (2024). https://doi.org/10.1186/s13015-024-00257-3

4. Arasti, S., Tabaghi, P., Tabatabaee, Y., Mayer, Alan K., Mirarab, S.: Optimal tree metric matching enables phylogenomic branch length reconciliation. Syst. Biol. bioRxiv: 2023.11.13.566962 (2026). https://doi.org/10.1101/2023.11.13.566962 (in press)

5. Cannon, J.T., Vellutini, B.C., Smith, J.r., Ronquist, F., Jondelius, U., Hejnol, A.: Xenacoelomorpha is the sister group to nephrozoa. Nature **530**(7588), 89–93 (2016)

6. Chen, T., Guestrin, C.: XGBoost: A scalable tree boosting system. In: Proceedings of the 22nd ACM SIGKDD International Conference on Knowledge Discovery and Data Mining, pp. 785–794. KDD '16, Association for Computing Machinery, New York, NY, USA (2016). https://doi.org/10.1145/2939672.2939785

7. Comte, A., et al.: PhylteR: efficient identification of outlier sequences in phylogenomic datasets. Mol. Biol. Evol. **40**(11), msad234 (2023). https://doi.org/10.1093/molbev/msad234, https://academic.oup.com/mbe/article/doi/10.1093/molbev/msad234/7330000

8. Feng, Y.J., et al.: Phylogenomics reveals rapid, simultaneous diversification of three major clades of gondwanan frogs at the cretaceous-paleogene boundary. Proc. Natl. Acad. Sci. **114**(29), E5864–E5870 (2017)

9. Francois, C.M., Durand, F., Figuet, E., Galtier, N.: Prevalence and implications of contamination in public genomic resources: a case study of 43 reference arthropod assemblies. G3 **10**(2), 721–730 (2020). https://doi.org/10.1534/g3.119.400758, http://g3journal.org/lookup/doi/10.1534/g3.119.400758

10. Ganapathy, G., Ramachandran, V., Warnow, T.: Better hill-climbing searches for parsimony. In: Algorithms in Bioinformatics, pp. 245–258 (2003). https://doi.org/10.1007/978-3-540-39763-2_19

11. Gatesy, J., Springer, M.S.: Phylogenetic analysis at deep timescales: unreliable gene trees, bypassed hidden support, and the coalescence/concatalescence conundrum. Mol. Phylogenet. Evol. **80**, 231–266 (2014). https://doi.org/10.1016/j.ympev.2014.08.013, http://www.ncbi.nlm.nih.gov/pubmed/25152276

12. Hosner, P.A., Faircloth, B.C., Glenn, T.C., Braun, E.L., Kimball, R.T.: Avoiding missing data biases in phylogenomic inference: an empirical study in the landfowl (Aves: Galliformes). Mol. Biol. Evol. **33**(4), 1110–1125 (2016). https://doi.org/10.1093/molbev/msv347, https://academic.oup.com/mbe/article-lookup/doi/10.1093/molbev/msv347

13. Jeffroy, O., Brinkmann, H., Delsuc, F., Philippe, H.: Phylogenomics: the beginning of incongruence? Trends Genet. **22**(4), 225–231 (2006). https://doi.org/10.1016/j.tig.2006.02.003, iSBN: 0168-9525

14. Laurin-Lemay, S., Brinkmann, H., Philippe, H.: Origin of land plants revisited in the light of sequence contamination and missing data. Curr. Biol. (2012). https://

doi.org/10.1016/j.cub.2012.06.013, iSBN: 1879-0445 (Electronic) r0960-9822 (Linking)

15. Li-San Wang, Leebens-Mack, J., Wall, P.K., Beckmann, K., de Pamphilis, C.W., Warnow, T.: The impact of multiple protein sequence alignment on phylogenetic estimation. IEEE/ACM Trans. Comput. Biol. Bioinf. **8**(4), 1108–1119 (2011). https://doi.org/10.1109/TCBB.2009.68, http://www.ncbi.nlm.nih.gov/pubmed/21566256

16. Mai, U., Mirarab, S.: TreeShrink: fast and accurate detection of outlier long branches in collections of phylogenetic trees. BMC Genomics **19**(S5), 272 (2018). https://doi.org/10.1186/s12864-018-4620-2, https://bmcgenomics.biomedcentral.com/articles/10.1186/s12864-018-4620-2, iSBN: 9783319679785

17. Mallo, D., De Oliveira Martins, L., Posada, D.: SimPhy: phylogenomic simulation of gene, locus, and species trees. Syst. Biol. **65**(2), 334–344 (2016). https://doi.org/10.1093/sysbio/syv082, http://sysbio.oxfordjournals.org/content/early/2015/12/04/sysbio.syv082.short?rss=1

18. Mirarab, S., Bayzid, M.S., Boussau, B., Warnow, T.: Statistical binning enables an accurate coalescent-based estimation of the avian tree. Science **346**(6215), 1250463–1250463 (2014). https://doi.org/10.1126/science.1250463, http://www.sciencemag.org/cgi/doi/10.1126/science.1250463

19. Mirarab, S., Bayzid, M.S., Warnow, T.: Evaluating summary methods for multilocus species tree estimation in the presence of incomplete lineage sorting. Syst. Biol. **65**(3), 366–380 (2016). https://doi.org/10.1093/sysbio/syu063, iSBN: 1063-5157

20. Mirarab, S., et al.: A region of suppressed recombination misleads neoavian phylogenomics. Proc. Natl. Acad. Sci. **121**(15), e2319506121 (2024)

21. Misof, B., et al.: Phylogenomics resolves the timing and pattern of insect evolution. Science **346**(6210), 763–767 (2014)

22. Patel, S.: Error in phylogenetic estimation for bushes in the tree of life. J. Phylogenet. Evol. Biol. **01**(02), 110 (2013). https://doi.org/10.4172/2329-9002.1000110, http://www.esciencecentral.org/journals/error-in-phylogenetic-estimation-for-bushes-in-the-tree-of-life-2329-9002.1000110.php?aid=154

23. Philippe, H., Vienne, D.M.d., Ranwez, V., Roure, B., Baurain, D., Delsuc, F.: Pitfalls in supermatrix phylogenomics. Eur. J. Taxonomy (2017). https://doi.org/10.5852/ejt.2017.283

24. Portik, D.M., Wiens, J.J.: Do alignment and trimming methods matter for phylogenomic (UCE) analyses? Syst. Biol. (2020). https://doi.org/10.1093/sysbio/syaa064, https://doi.org/10.1093/sysbio/syaa064

25. Price, M.N., Dehal, P.S., Arkin, A.P.: FastTree-2 – approximately maximum-likelihood trees for large alignments. PLoS ONE **5**(3), e9490 (2010). https://doi.org/10.1371/journal.pone.0009490, http://www.pubmedcentral.nih.gov/articlerender.fcgi?artid=2835736&tool=pmcentrez&rendertype=abstract

26. Reddy, S., et al.: Why do phylogenomic data sets yield conflicting trees? Data type influences the avian tree of life more than taxon sampling. Syst. Biol. **66**(5), 857–879 (2017). https://doi.org/10.1093/sysbio/syx041

27. Rouse, G.W., Wilson, N.G., Carvajal, J.I., Vrijenhoek, R.C.: New deep-sea species of xenoturbella and the position of xenacoelomorpha. Nature **530**(7588), 94–97 (2016)

28. Salichos, L., Rokas, A.: Evaluating ortholog prediction algorithms in a yeast model clade. PLoS ONE (2011). https://doi.org/10.1371/journal.pone.0018755, iSBN: 1932-6203 (Electronic) r1932-6203 (Linking)

29. Sayyari, E., Whitfield, J.B., Mirarab, S.: Fragmentary gene sequences negatively impact gene tree and species tree reconstruction. Mol. Biol. Evol. **34**(12), 3279–3291 (2017). https://doi.org/10.1093/molbev/msx261, http://dx.doi.org/10.1093/molbev/msx261

30. Sela, I., Ashkenazy, H., Katoh, K., Pupko, T.: GUIDANCE2: accurate detection of unreliable alignment regions accounting for the uncertainty of multiple parameters. Nucleic Acids Res. **43**(W1), W7–W14 (2015). https://doi.org/10.1093/nar/gkv318, https://academic.oup.com/nar/article-lookup/doi/10.1093/nar/gkv318

31. Shwartz-Ziv, R., Armon, A.: Tabular data: deep learning is not all you need. Information Fusion **81**, 84–90 (2022). https://doi.org/10.1016/j.inffus.2021.11.011, https://www.sciencedirect.com/science/article/pii/S1566253521002360

32. Song, S., Liu, L., Edwards, S.V., Wu, S.: Resolving conflict in eutherian mammal phylogeny using phylogenomics and the multispecies coalescent model. Proc. Natl. Acad. Sci. U.S.A. **109**(37), 14942–14947 (2012). https://doi.org/10.1073/pnas.1211733109

33. Springer, M.S., Gatesy, J.: The gene tree delusion. Molecular Phylogenetics and Evolution **94**(Part A), 1–33 (2016).https://doi.org/10.1016/j.ympev.2015.07.018, http://www.sciencedirect.com/science/article/pii/S1055790315002225, iSBN: 1095-9513 (Electronic) r1055-7903 (Linking)

34. Springer, M.S., Gatesy, J.: On the importance of homology in the age of phylogenomics. Syst. Biodivers. **16**(3), 210–228 (2018). https://doi.org/10.1080/14772000.2017.1401016, https://www.tandfonline.com/doi/full/10.1080/14772000.2017.1401016

35. Stamatakis, A.: RAxML version 8: A tool for phylogenetic analysis and post-analysis of large phylogenies. Bioinformatics **30**(9), 1312–1313 (2014). https://doi.org/10.1093/bioinformatics/btu033, arXiv: 10.1093/bioinformatics/btu033 ISBN: 1367-4811

36. Steenwyk, J.L., Buida, T.J., Li, Y., Shen, X.X., Rokas, A.: ClipKIT: A multiple sequence alignment trimming software for accurate phylogenomic inference. PLoS Biol. **18**(12), e3001007 (2020). https://doi.org/10.1371/journal.pbio.3001007, https://dx.plos.org/10.1371/journal.pbio.3001007

37. Steenwyk, J.L., Li, Y., Zhou, X., Shen, X.X., Rokas, A.: Incongruence in the phylogenomics era. Nat. Rev. Genet. **24**(12), 834–850 (2023). https://doi.org/10.1038/s41576-023-00620-x, https://www.nature.com/articles/s41576-023-00620-x

38. Tan, G., et al.: Current methods for automated filtering of multiple sequence alignments frequently worsen single-gene phylogenetic inference. Syst. Biol. **64**(5), 778–791 (2015)

39. Tannier, E., Tricou, T., Benali, S., De Vienne, D.M.: HGTs are not SPRs: in the presence of ghost lineages, series of horizontal gene transfers do not result in series of subtree pruning and regrafting. Mol. Biol. Evol. **42**(6), msaf128 (2025). https://doi.org/10.1093/molbev/msaf128, https://academic.oup.com/mbe/article/doi/10.1093/molbev/msaf128/8154856

40. Van Etten, J., Bhattacharya, D.: Horizontal gene transfer in Eukaryotes: not if, but how much? Trends Genet. **36**(12), 915–925 (2020). https://doi.org/10.1016/j.tig.2020.08.006, https://linkinghub.elsevier.com/retrieve/pii/S0168952520302067

41. Wickett, N.J., et al.: Phylotranscriptomic analysis of the origin and early diversification of land plants. Proc. Natl. Acad. Sci. **111**(45), E4859–E4868 (2014)

42. Zhang, C., Rabiee, M., Sayyari, E., Mirarab, S.: ASTRAL-III: polynomial time species tree reconstruction from partially resolved gene trees. BMC Bioinformatics **19**(S6), 153 (2018). https://doi.org/10.1186/s12859-018-2129-y, https://bmcbioinformatics.biomedcentral.com/articles/10.1186/s12859-018-2129-y

43. Zhang, C., Zhao, Y., Braun, E.L., Mirarab, S.: TAPER: pinpointing errors in multiple sequence alignments despite varying rates of evolution. Methods Ecol. Evol. **12**(11), 2145–2158 (2021). https://doi.org/10.1111/2041-210X.13696, https://onlinelibrary.wiley.com/doi/10.1111/2041-210X.13696

"Frustratingly Easy" Domain Adaptation for Cross-Species Transcription Factor Binding Prediction

Mark Maher Ebeid[1,2,3], Ali Tuğrul Balcı[1,2,3], Maria Chikina[1,2,3], Panayiotis V. Benos[1,2,3,4], and Dennis Kostka[1,2,3(✉)]

[1] Department of Computational and Systems Biology, University of Pittsburgh, Pittsburgh, PA, USA
[2] School of Medicine and Joint Carnegie Mellon, University of Pittsburgh, Pittsburgh, PA, USA
[3] Computational Biology, University of Pittsburgh, Pittsburgh, PA, USA
kostka@pitt.edu
[4] Department of Epidemiology, University of Florida, Gainesville, FL, USA

Abstract. How DNA sequence encodes gene regulation remains a central challenge in regulatory genomics. Transcription factors (TFs) are key mediators of this process, binding to specific sequence motifs to control gene expression. Yet, predicting where they bind from sequence alone remains a challenging problem. A cross-species angle offers two complementary benefits: it tests whether trained models have learned conserved, biochemically grounded rules that generalize across species, and it enables binding prediction in species where experimental data is scarce. Key challenges in this context are that TF binding sites undergo rapid evolutionary turnover, and that there are systematic distributional differences between species' genomes. We present MORALE, a domain adaptation framework for cross-species TF binding prediction. By aligning the first and second moments of sequence embeddings across species during training, MORALE learns species-invariant representations without adversarial training, additional parameters, or architectural changes. Applied to liver ChIP-seq data from two species (human, mouse) and five species (adding rhesus macaque, rat, and dog), MORALE consistently matches or outperforms gradient reversal (the adversarial baseline) across all TFs, and avoids the performance degradation below the no-adaptation baseline that gradient reversal can exhibit. In the five-species setting, MORALE surpasses a human-only model, demonstrating that moment alignment can unlock cross-species generalization that neither multi-species training nor adversarial adaptation achieves alone. MORALE also recovers TF binding motifs more faithfully than the adversarial approach, suggesting its representations capture biologically meaningful sequence features. As a closed-form, parameter-free regularizer, MORALE integrates into any embedding-based sequence model.

Keywords: Evolutionary Genomics · Gene Regulation · Transcription Factor Binding · Sequence To Function Modeling · Cross-Species Comparison · Domain Adaptation

© The Author(s), under exclusive license to Springer Nature Switzerland AG 2026
M. Lafond (Ed.): RECOMB-CG 2026, LNBI 16569, pp. 392–410, 2026.
https://doi.org/10.1007/978-3-032-26891-4_19

1   Introduction

Genomic regulatory activity is largely governed by transcription factors (TFs) that bind to DNA and influence gene expression. Sequence-to-function models, typically deep neural networks, have become a cornerstone for predicting TF binding and regulatory activity from sequence, enabling downstream interpretation, functional element discovery, and in silico sequence design [4,6–8,14,24,26]. A key open question is the extent to which these models learn universal biochemical principles that generalize across species.

Training on multi-species data offers a promising path toward such generalization. However, TF binding sites undergo rapid evolutionary turnover (even between closely related species), making cross-species transfer non-trivial [29]. At the same time, TF DNA-binding domains are highly conserved across species, suggesting a shared regulatory vocabulary that multi-species models could exploit [3,9,11,16]. Realizing this potential requires methods that can bridge systematic differences between species' genomes' nucleotide distributions.

Cochran et al. [10] addressed this with a gradient reversal layer (GRL) approach: an adversarial discriminator penalizes the model for learning species-specific features, encouraging a species-invariant representation. While effective in some settings, GRL requires a separate discriminator branch, substantially increasing model complexity, and, as we show, can degrade performance below a no-adaptation baseline.

We propose MORALE, which instead aligns the first and second moments of sequence embeddings across species [31] using a closed-form operation that requires no additional parameters and integrates into any embedding-based model. We compare MORALE against GRL on human–mouse TF binding prediction across four TFs, and extend the evaluation to a five-species setting. MORALE consistently matches or outperforms GRL and learns a robust species-invariant feature set.

2   Materials and Methods

2.1   Data

Two-Species. Following Cochran et al. [10], we processed ChIP-seq data for CTCF, HNF4α, RXRA, and CEBPA in human and mouse liver (ENCODE: ENCSR000CBU, ENCSR911GFJ, ENCSR098XMN; ArrayExpress: E-TABM-722; GEO: GSM1299600). Sequences were tiled into 500-bp windows (50-bp overlap), ENCODE blacklist regions removed [2], and aligned to GRCh38/GRCm38 with BowTie2 [19]. Peaks were called with multiGPS v0.75 [22]. Windows covering a peak center were labeled "bound." Chr1/Chr2 were held out for validation/testing; sex chromosomes excluded.

Multi-species. We used published liver ChIP-seq for CEBPA, FOXA1, ONECUT1, and HNF4α across five mammals (human, rhesus macaque, mouse, rat, dog) [5] (ArrayExpress E-MTAB-1509), tiled into 1000-bp windows. Validation/test chromosomes were chosen via a linear program to approximate a target

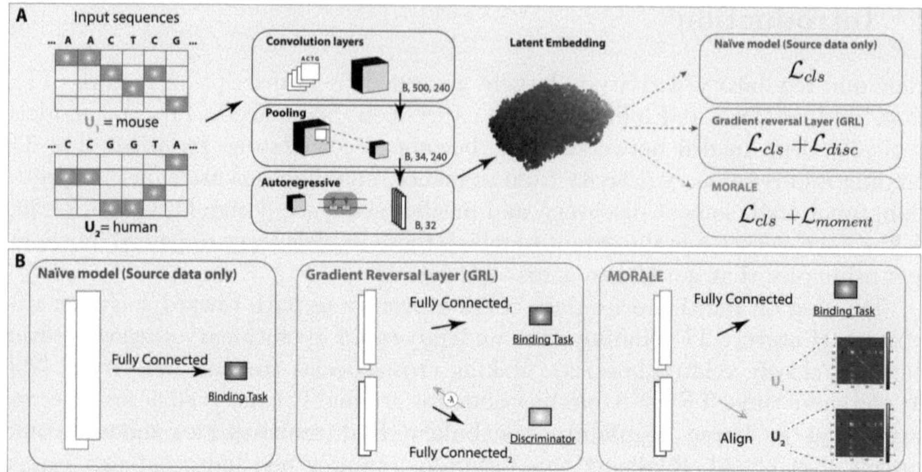

Fig. 1. Overview of MORALE and comparison with alternative domain adaptation approaches. (A) Input sequences from source and target species are one-hot encoded and passed through convolution, pooling, and autoregressive layers to produce a latent embedding, which feeds all downstream tasks. Three training procedures are compared: (1) source-only (naive baseline), (2) gradient reversal (GRL), and (3) moment alignment (MORALE). **(B)** Architectural detail of each approach. In the source-only model, only source embeddings are passed to the classification head. GRL adds an adversarial discriminator branch that predicts species of origin; gradients from this branch are reversed to discourage species-specific representations. MORALE forgoes the discriminator entirely, instead adding a moment alignment loss between source and target embeddings directly to the training objective.

fraction of positive windows while maintaining class balance (R package lpSolve; see Tables 7, 8).

2.2 Model Architecture

Two-Species. We use the architecture of Cochran et al. [10]: Conv(240 × 20bp) → MaxPool(15) → LSTM(32) → dense head (613K params). GRL adds an adversarial branch (9.5M params total). MORALE adds no parameters; moments are computed on the 32-d LSTM embedding via ADAPT [23].

Multi-species. We adapted a bidirectional GRU from gReLU [18]: Conv(240 × 20bp, ReLU) → MaxPool(16) → BiGRU(240) → FC×2 → Conv1×1(63ch) → GlobalAvgPool (63-d embedding, 960K params).

2.3 Hyperparameter Tuning

For GRL, to determine λ, we search a grid of values from 0.0 to 10.0 with a step size of 0.50 and select the maximal auPRC value for each transcription factor.

For mouse-to-human we select 1.5, 6.0, 0.5, 7.5 for CTCF, CEBPA, HNF4α, and RXRA, respectively. For human-to-mouse direction, we select 6.5, 8.5, 10.0, 1.0.

For MORALE we use a range from 0 to 10 with a step size of 1. In the mouse-to-human direction we select 4, 7, 8, and 8 for CTCF, CEBPA, HNF4α, and RXRA, respectively. In the human-to-mouse direction, we select 4, 8, 6, and 7. For the multi-species analysis we use a grid between 1 and 8 with a step-size of 1.

2.4   Training and Evaluation

Two-species models used TensorFlow 2.18 [1], Adam (lr=1e-3), 15 epochs, batch size 400 (200 bound/unbound from source) plus 400 background per species for domain adaptation. Multi-species used PyTorch 2.5.1 [25], same optimizer, mini-batches of 200 bound/unbound (balanced across sources) plus 250 background (50/species). Performance was assessed by auPRC (scikit-learn [27]).

2.5   Attribution and Downstream Analyses

Per-nucleotide attribution scores were computed via expected integrated gradients (CREsted [17]). Motifs were discovered with TF-MoDISco lite [30] on up to 2,000 high-confidence bound sites (sigmoid ≥ 0.98) and annotated with Tomtom [13]. Repeat analyses used RepeatMasker tracks from UCSC [28]; phylogenetic trees were built with phyloT v2 [20].

3   Results

3.1   Approach

Our approach, MORALE, is a straightforward generalization of the Deep CORAL method [31], applied to TF binding data. We assume labeled (sequence) data from n different source domains and unlabeled data from a target domain; in our application, the domains are species. We also assume an encoder model (see Section 2.2), mapping input sequences $x^{i|j}$ to vector embeddings $z^{i|j} \in \mathbb{R}^d$, where $i = 1, \ldots, m_j$ and $j = 1, \ldots, n+1$ (including the target domain), and m_j is the number of sequences in the j-th domain. Vector representations are input to a classification model predicting the sequence label (e.g., whether the input sequence binds a specific TF), with an associated loss (e.g., cross entropy) denoted $\mathcal{L}_{\text{label-classification}}$.

Given a mini-batch of $b < m_j$ sequence representations $\{z^{i|j}\}_{i=1}^b$, we denote the sample mean by $\hat{\mu}^j$ and the sample covariance matrix by $\hat{C}^j$. For a mini-batch containing sequences from each domain, the MORALE loss is calculated as

$$\mathcal{L}_{\text{MORALE}} = \frac{2}{n(n+1)} \sum_{l \neq k} \left(||\hat{C}^l - \hat{C}^k||_F^2 + ||\hat{\mu}^l - \hat{\mu}^k||_2^2 \right),$$

quantifying the difference in first and second moments between all domain pairs. Here $|| \cdot ||_F$ denotes the Frobenius norm. We estimate $\hat{\mu}^j$ and $\hat{C}^j$ for $\mathcal{L}_{\text{MORALE}}$

396   M. M. Ebeid et al.

using the background portion of the mini-batch (described in Section 2.1), which
contains an equal number of unlabeled sequences sampled from each domain
(source(s) and target). For source domains, the vector representations $\{z^{i|j}\}_{i=1}^{b}$
are additionally used for label classification.

The MORALE loss is added to the label classification loss, encouraging
sequence representations that are moment-aligned across domains:

$$\mathcal{L} = \mathcal{L}_{\text{label-classification}} + \lambda \mathcal{L}_{\text{MORALE}}. \tag{1}$$

We note that gradient reversal layers (GRL) are a successful adversarial
approach to encouraging domain-invariant representations [12,15,21]. Briefly, a
domain classifier and associated loss for predicting embeddings' domain of origin
(e.g., cross entropy) is added to the model for an overall loss of

$$\mathcal{L} = \mathcal{L}_{\text{label-classification}} + \mathcal{L}_{\text{domain-classification}}. \tag{2}$$

Backpropagated gradients from $\mathcal{L}_{\text{domain-classification}}$ are penalized by a factor of
$-\lambda$ (for $\lambda > 0$, hence "gradient reversal"), encouraging representations that are
uninformative about the domain of origin.

In our analyses we compare MORALE (Eq. 1) with the GRL approach (Eq. 2)
and find it generally outperforms GRL. We also note that MORALE does not
require additional parameters beyond the encoder and label classification model,
whereas GRL requires additional design and parameterization of the domain
classifier. Figure 1 summarizes all three approaches.

3.2 Cross-Species TF-Binding Prediction Between Human and Mouse

MORALE Improves Cross-Species TF Binding Prediction Performance. First, we applied our framework to re-analyze a dataset introduced by
Cochran et al. [10], in which the binding of four TFs (CTCF, HNF4α, RXRA,
and CEBPA) was assayed in liver tissue from humans and mice, giving two
domains.

We assess TF binding site prediction for each species as target, with the
other as source. Test set performances are summarized in Fig. 2. We compare four models: source-only (no domain adaptation), target-trained (upper
bound), MORALE, and GRL. Domain adaptation methods have access to target embeddings but not target labels. As expected, source-only performs worst
and target-trained performs best, with domain adaptation methods in between.
Interestingly, for CTCF with human as target, GRL performs worse than the
no-adaptation baseline, while MORALE does not suffer this drop. Likewise,
with mouse as target, GRL falls below the source baseline for three of four TFs
(CEBPA, HNF4α, RXRA); MORALE does not. For TFs where GRL does outperform the source baseline, MORALE either matches or exceeds GRL. Numerical values are in Table 1.

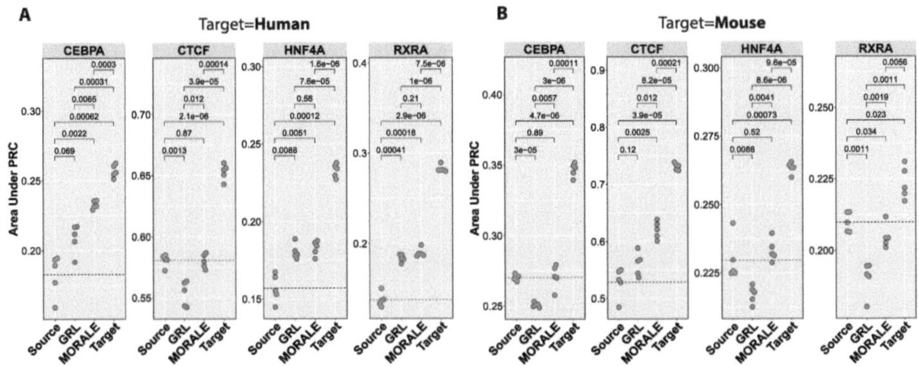

Fig. 2. Moment alignment improves cross-species TF binding site predictions. Prediction performance is shown for four models: (1) source-only (red), (2) gradient reversal (green), (3) MORALE (blue), and (4) target-trained (purple), across four TFs. **(A)** Results when adapting a mouse-trained model to human data. MORALE outperforms or matches GRL in each case without suffering degradation. **(B)** The same analysis in the reverse direction, human-to-mouse. Performance degradation is persistent under GRL, while MORALE meets or exceeds the source baseline and outperforms GRL.

Upon examining results (Table 1), we observe that in the mouse-to-human direction the GRL model makes more positive predictions overall, but with disproportionately more false positives compared with MORALE. This leads to worse performance, especially for CTCF. MORALE has lower false positive rates (e.g., 1.38% versus GRL's 1.85% for CTCF) while suffering only slightly increased false negative rates (0.079% vs. 0.068%, again for CTCF). In the human-to-mouse direction this pattern holds for CTCF. For CEBPA and HNF4α, MORALE has more true positives and fewer false positives than GRL, and consequently shows better performance. For RXRA, MORALE's performance is close to the source model, while GRL performs somewhat worse.

From these analyses, we conclude that MORALE improves over GRL for cross-species TF binding prediction on this dataset. Next, we compare the quality of predictions in more detail.

MORALE Improves Cross-Species TF Binding Prediction Quality. To quantify and assess the quality of TF binding predictions, we compare how well MORALE and GRL adapt the source-trained model toward the target-trained model, using the latter as a reference. We use two complementary metrics, both based on per-nucleotide importance scores from post-hoc attribution analysis (see Sect. 2.5).

First, we compare the correlation of GRL/MORALE importance scores with those of the target model, focusing on sequences where the source model disagrees with the target: differential false positives (dFPs), where the source model wrongly predicts TF binding but the target model does not; and differential false

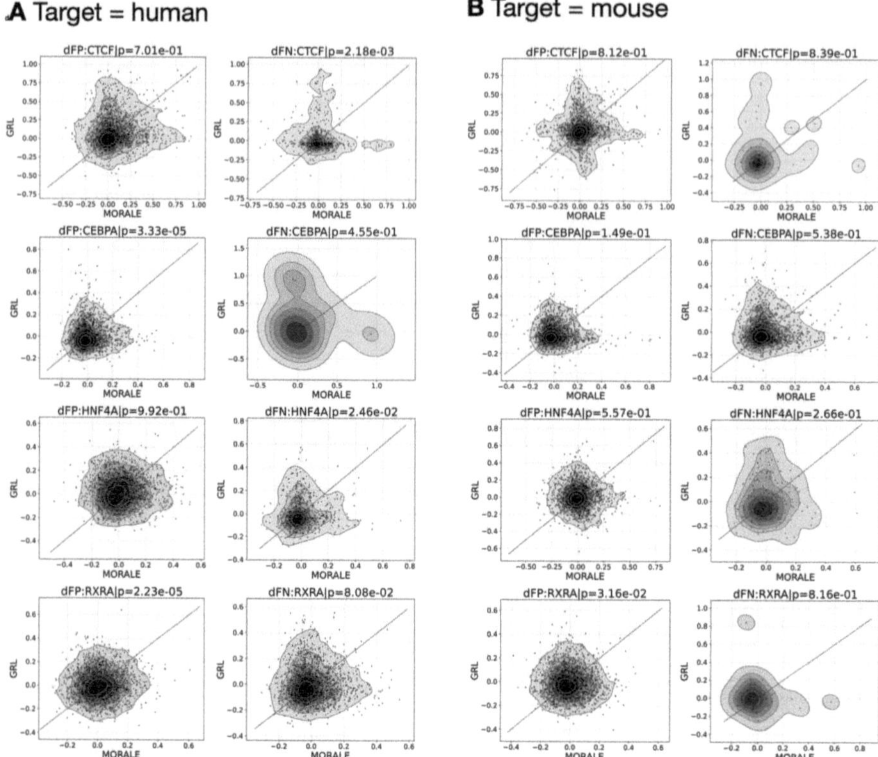

Fig. 3. **MORALE attribution scores align more closely with the target model than GRL.** Pearson correlation coefficients between MORALE's importance scores and the target model (x-axes) and GRL's importance scores and the target model (y-axes), for differentially false positive sites (dFPs, where the source model incorrectly predicts binding but the target model does not) and differentially false negative sites (dFNs, where the source model misses a binding event correctly predicted by the target model). Panel A shows mouse-to-human and panel B human-to-mouse. P-values for a one-sided Wilcoxon Rank Sum Test are indicated for each plot.

negatives (dFNs), where the source model misses a binding event that the target model correctly predicts. Figure 3 shows contour plots of these correlation coefficients, stratified by TF and sequence type (dFP/dFN). Higher correlation indicates that the domain adaptation method better reproduces the importance scores of the target-trained model.

For mouse-to-human and dFP sequences, MORALE importance scores are significantly more correlated with the target model than GRL scores for CEBPA and RXRA, while results are comparable for CTCF and HNF4α. For dFN sequences, MORALE's correlation is significantly closer to the target for RXRA and HNF4α. For human-to-mouse and dFPs, MORALE scores are more correlated with the target for RXRA, while other comparisons are not significant.

Table 1. We display the true positives (TPs), false positives (FPs), false negatives (FNs) with the auPRC across four TFs and three methods in this study. In order to generate this confusion matrix across five-folds of data, we average the sigmoid value over the folds and compute the relevant prediction types w.r.t. the ground truth label, percentages are with respect to the complete dataset. We observe that the GRL more often predicts positives than either the source model, or MORALE, however this comes at much higher occurrence of FPs, leading to performance degradation as in CTCF when the target is human. MORALE, on the other hand, takes a much more conservative step towards the target model, finding a middle ground that strictly results in improvement or meeting baselines.

Target=**Human**

TF	TPs (%)			FPs (%)			FNs (%)			auPRC		
	GRL	MORALE	Source	GRL	MORALE	Source	GRL	MORALE	Source	GRL	MORALE	Source
CTCF	**0.413**	0.402	0.401	1.847	**1.384**	1.397	**0.068**	0.079	0.08	0.553	0.58	**0.581**
CEBPA	0.542	0.545	**0.547**	10.436	**9.77**	11.938	0.048	0.045	**0.044**	0.209	**0.233**	0.183
HNF4A	**0.605**	0.578	0.574	9.137	**8.496**	10.806	**0.136**	0.164	0.168	0.181	**0.183**	0.157
RXRA	**1.133**	1.035	1.076	9.981	**8.023**	13.181	**0.468**	0.566	0.526	0.184	**0.19**	0.137
Average	**0.673**	0.64	0.649	7.85	**6.918**	9.33	**0.18**	0.213	0.205	0.282	**0.296**	0.264

Target=**Mouse**

TF	TPs (%)			FPs (%)			FNs (%)			auPRC		
	GRL	MORALE	Source	GRL	MORALE	Source	GRL	MORALE	Source	GRL	MORALE	Source
CTCF	**0.7**	0.699	0.697	3.937	**3.237**	3.591	**0.042**	0.044	0.046	0.56	**0.62**	0.528
CEBPA	0.988	**0.996**	0.952	7.158	6.66	**5.732**	0.335	**0.328**	0.372	0.251	**0.271**	0.27
HNF4A	0.824	**0.828**	0.813	11.026	10.388	**9.802**	0.122	**0.118**	0.133	0.217	**0.233**	0.23
RXRA	0.831	0.827	**0.843**	17.891	**16.039**	19.208	0.087	0.091	**0.075**	0.19	0.205	**0.21**
Average	0.836	**0.837**	0.826	10.003	**9.081**	9.583	0.146	**0.145**	0.156	0.304	**0.332**	0.31

Table 2. We display the performance (auPRC) for windows that overlap both SINE and LINE repeats, and their average. We observe that MORALE performs competitively in the performance across windows that contain species-specific repeats in the mouse-to-human direction, and outperforms the GRL in the human-to-mouse direction. Notably, for HNF-4α and RXRα, the GRL does a convincingly better job windows that overlap SINE elements in human, which would be Alus in the majority.

Target=**Human**

TF	LINE			SINE			Average		
	GRL	MORALE	Source	GRL	MORALE	Source	GRL	MORALE	Source
CTCF	0.149	**0.173**	0.17	0.181	**0.191**	0.177	0.165	**0.182**	0.173
CEBPA	0.039	**0.042**	0.034	0.04	**0.04**	0.027	0.039	**0.041**	0.031
HNF4A	**0.043**	0.04	0.03	**0.048**	0.034	0.023	**0.046**	0.037	0.026
RXRA	**0.058**	0.056	0.036	**0.062**	0.052	0.029	**0.06**	0.054	0.033
Average	0.072	**0.078**	0.068	**0.083**	0.079	0.064	0.077	**0.079**	0.066

Target=**Mouse**

TF	LINE			SINE			Average		
	GRL	MORALE	Source	GRL	MORALE	Source	GRL	MORALE	Source
CTCF	0.103	**0.124**	0.111	0.102	**0.129**	0.104	0.102	**0.126**	0.107
CEBPA	0.077	0.08	**0.082**	0.097	0.103	**0.108**	0.087	0.092	**0.095**
HNF4A	0.04	0.043	**0.045**	0.061	**0.063**	0.062	0.051	**0.053**	0.053
RXRA	0.027	**0.029**	0.026	0.041	**0.042**	0.035	0.034	**0.036**	0.03
Average	0.062	**0.069**	0.066	0.075	**0.084**	0.077	0.068	**0.077**	0.071

Overall, MORALE improves upon GRL in highlighting meaningful sequence positions after domain adaptation, and we have not observed the reverse.

Second, we identified sequence motifs from attribution scores using TF-MoDISco [30] (see Sect. 2.5). Figure 4 summarizes our findings. For CTCF binding sites, the target model's most frequently found motif corresponds to CTCF, which is recapitulated by MORALE. GRL, like the source-only model, finds a larger fraction of motifs corresponding to other TFs. In panels B to E, the target-trained, MORALE, and source-only models all find CTCF as their most frequent motif match. Surprisingly, GRL shifts the motif distribution such that poorly matching motifs become more frequent.

Finally, we explored model performance on test windows overlapping LINE and SINE repeat elements, following Cochran et al. [10], who reported that GRL reduces over-prediction at such sites in the mouse-to-human direction. Results are in Table 2. Indeed, GRL outperforms both MORALE and the source model for HNF4α and especially RXRA in this direction. Nevertheless, MORALE outperforms GRL for CTCF and CEBPA. Furthermore, in the human-to-mouse direction, MORALE outperforms GRL for all four TFs and achieves the best average performance for three of the four.

Overall, MORALE outperforms GRL in most comparisons and rarely performs worse, demonstrating robust domain adaptation across multiple evaluation criteria.

3.3 Learning TF-Binding in Human by Leveraging Data Across Five Mammals

Next, we explored whether multi-species training combined with domain adaptation can improve TF binding prediction in human. Using liver ChIP-seq data from rhesus macaque, mouse, rat, and dog, we evaluate generalization to human test data across four TFs (FOXA1, HNF4α, HNF6, and CEBPA), comparing three approaches: human-only training, multi-species training without adaptation, and multi-species training with MORALE. Results are in Fig. 5. We make two observations. First, as in the two-species case, a source-only model trained on a single non-human species cannot approach human-only performance. However, a plain multi-species model trained on all five species surpasses the human-only model for all four TFs—a result not seen in the two-species setting. Second, applying MORALE to the multi-species model further increases performance, making MORALE the top performer for every TF. This demonstrates MORALE's broad applicability to settings with more than two domains.

We then examined the contribution of individual species. The phylogenetic tree in Fig. 6 shows the evolutionary relationships between species. Based on evolutionary distance, we would expect rhesus macaque holdout to have the largest effect on human prediction, followed by mouse, rat, and dog. This expectation holds for three of four TFs (CEBPA, FOXA1, HNF4α), where rhesus macaque holdout indeed has the largest impact. However, for HNF6 the expected trend does not hold: rat (rn7) holdout has an impact comparable to rhesus macaque, likely because rn7 is enriched for bound sites relative to the other species for this TF (seeTable 8).

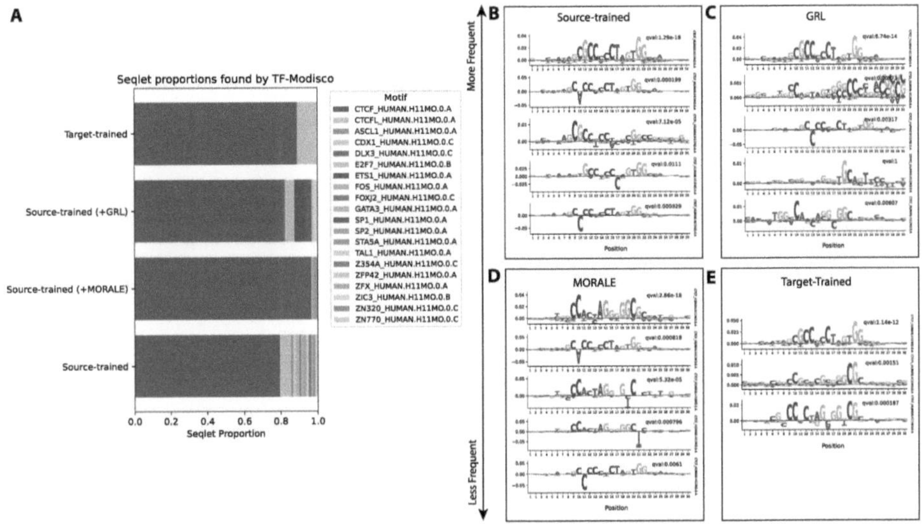

Fig. 4. MORALE discovers de-novo motifs more similar to CTCF. De-novo motifs were identified from attribution scores across source, source-adapted, and target models, using 2,000 randomly sampled bound sites for CTCF. **(A)** Proportions of all motifs found across the four models. The target model finds only CTCF (and its paralog CTCFL). Source-trained and GRL-adapted models find CTCF as their primary motif but at a lower proportion than MORALE, which reports nearly exclusively CTCF. **(B–E)** Top 5 de-novo motifs for the source-trained, GRL-adapted, MORALE-adapted, and target-trained models respectively, with TomTom match annotations and p-values. The top match of MORALE strongly resembles the established CTCF motif with a significant q-value.

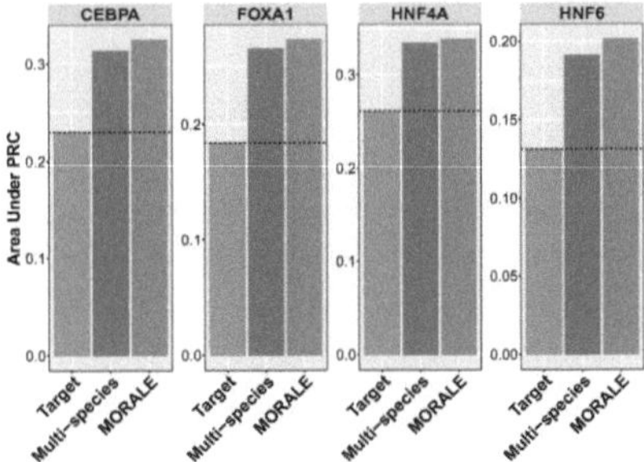

Fig. 5. MORALE attains higher performance when training on multiple source species and predicting in human as the target. Across all four TFs, MORALE consistently increases performance compared to multi-species training without domain adaptation. Unlike the two-species case, training on multiple source species allows the multi-species model to surpass the human-only model—previously unattained.

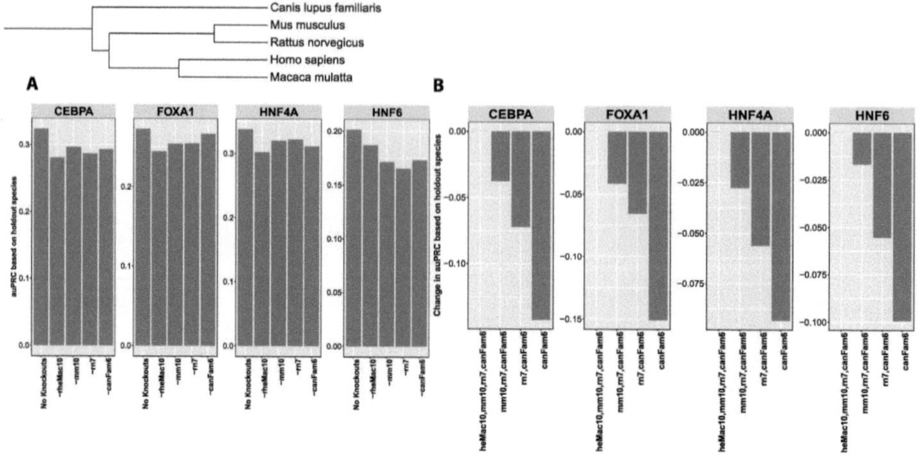

Fig. 6. Species contributions to multi-species performance gains are species-dependent but consistently beneficial. We quantify the effect of holding out species individually **(A)** and in groups **(B)** on model performance under MORALE, with human as the target. In **(A)**, 'No knockout' is the model trained on all source species; each subsequent bar removes one species. In **(B)**, species are removed successively from left to right, from all included to a single source species, illustrating that more source species consistently aids performance.

The group holdout analysis in panel B confirms this picture. Successively holding out rhesus macaque, then mouse, then rat leads to a monotonic drop in performance, emphasizing that each additional source species contributes to the gain over the human-only model.

4   Discussion

We develop and apply MORALE, a domain adaptation framework for cross-species TF binding prediction based on moment alignment of sequence embeddings. The core idea is to encourage a species-invariant latent representation by aligning the first and second moments of embeddings across all pairs of domains (species) during training—without requiring an adversarial component, additional parameters, or extra architectural decisions. This contrasts with gradient reversal (GRL), the prevailing approach, which requires a separate discriminator branch that substantially increases model complexity.

We evaluate MORALE in two settings. In the two-species case, we train on liver ChIP-seq data from one species (human or mouse) and predict TF binding in the other, for four TFs (CTCF, HNF4α, RXRA, CEBPA). Here, MORALE consistently matches or outperforms GRL, and critically avoids the performance degradation below the no-adaptation baseline that GRL exhibits for several TF/direction combinations. In the five-species case, using liver data from rhesus macaque, mouse, rat, and dog to predict human TF binding, MORALE improves over both the no-adaptation multi-species model and—notably—over a human-only model, a result not observed in the two-species setting. This suggests that the combination of diverse training signal and moment alignment unlocks cross-species generalization that neither approach achieves alone.

Beyond predictive performance, MORALE recovers sequence motifs that more closely match known TF binding motifs compared to GRL, indicating that its invariant representations capture biologically meaningful features rather than superficial domain differences. MORALE's simplicity—it integrates into any embedding-based model without modifying gradient computation—makes it broadly applicable to cross-species regulatory genomics and, more generally, to multi-domain sequence modeling tasks where domain-invariant representations are desirable.

Several limitations deserve mention. First, MORALE introduces a regularization weight λ that requires tuning; though the same is true of GRL, and optimal values varied across TFs and adaptation directions for both methods. This means performance gains may depend on access to held-out target data for validation, which may not always be available. Second, despite consistent improvements over GRL, a substantial gap between domain-adapted and target-trained models remains in the two-species setting, indicating that moment alignment alone cannot fully bridge species-specific differences when only a single source species is available—and that further methodological advances are needed. Third, all experiments use liver tissue and TFs with relatively strong, conserved binding motifs. Whether MORALE generalizes to tissues with more divergent regulatory landscapes, or to TFs with weaker sequence preferences, remains an open question.

In summary, MORALE is a simple, parameter-efficient domain adaptation method that improves cross-species TF binding prediction by aligning sequence embedding distributions across species. It is robust to the performance degradation seen with adversarial approaches, scales naturally to multiple source species, and recovers more accurate sequence motifs—making it a practical and interpretable tool for cross-species regulatory genomics.

Acknowledgments. We thank Shaun Mahony for the helpful discussions and the processed data used in this work. This work was partly supported by the University of Florida and the University of Pittsburgh School of Medicine.

Disclosure of Interests. The authors have no competing interests to declare that are relevant to the content of this article.

A   Appendix

See Figs. 7, 8, 9, 10, 11, Tables 3, 4, 5 and 6.

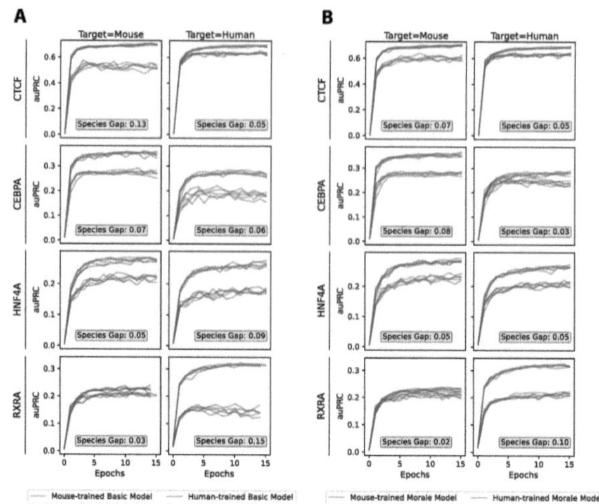

Fig. 7. The species gap is closed during training when using MORALE. We display the training, in the two-species case, of the five-fold cross validation performance over 15 epochs between the basic model, and MORALE. The gap between the peak performance between the source-on-target and the target-on-target models are annotated at the bottom of each plot. In **(A)** we have the basic model across each TF-target pair, and in **(B)** we display the same information, but with MORALE.

Table 3. The confusion matrix for mouse-adapted models when evaluating on the test set (Chr2) from human. The table displays the percentage of true positives (TPs), false positives (FPs), true negatives (TNs), and false negatives (FNs) for each TF. They were calculated as a ratio over all the windows in the test set and attach the auPRC value based on model type.

TF	TPs (%)				FPs (%)				TNs (%)				FNs (%)				auPRC			
	Source	GRL	MORALE	Target	Source	GRL	MORALE	Target	Source	GRL	MORALE	Target	Source	GRL	MORALE	Target	Source	GRL	MORALE	Target
CTCF	0.401	0.413	0.402	0.435	1.397	1.847	1.384	1.894	98.122	97.672	98.135	97.625	0.08	0.068	0.079	0.046	0.581	0.553	0.58	0.653
CEBPA	0.547	0.542	0.545	0.501	11.938	10.436	9.77	5.676	87.472	88.974	89.64	93.733	0.044	0.048	0.045	0.089	0.183	0.209	0.233	0.257
HNF4A	0.574	0.605	0.578	0.616	10.806	9.137	8.496	6.716	88.452	90.121	90.762	92.543	0.168	0.136	0.164	0.126	0.157	0.181	0.183	0.233
RXRA	1.076	1.133	1.035	1.364	13.181	9.981	8.023	11.665	85.218	88.418	90.376	86.734	0.526	0.468	0.566	0.237	0.137	0.184	0.19	0.283

Table 4. The confusion matrix for human-adapted models when evaluated on the test set (Chr2) from mouse. The table displays the percentage of true positives (TPs), false positives (FPs), true negatives (TNs), and false negatives (FNs) for each TF. We include them as a ratio over all the windows in the test set and attach the auPRC value based on model type.

TF	TPs (%)				FPs (%)				TNs (%)				FNs (%)				auPRC			
	Source	GRL	MORALE	Target	Source	GRL	MORALE	Target	Source	GRL	MORALE	Target	Source	GRL	MORALE	Target	Source	GRL	MORALE	Target
CTCF	0.697	0.7	0.699	0.683	3.591	3.937	3.237	1.514	95.666	95.32	96.02	97.743	0.046	0.042	0.044	0.06	0.528	0.56	0.62	0.731
CEBPA	0.952	0.988	0.996	1.126	5.732	7.158	6.66	8.387	92.944	91.518	92.016	90.289	0.372	0.335	0.328	0.198	0.27	0.251	0.271	0.346
HNF4A	0.813	0.824	0.828	0.824	9.802	11.026	10.388	8.419	89.252	88.027	88.666	90.635	0.133	0.122	0.118	0.122	0.23	0.217	0.233	0.263
RXRA	0.843	0.831	0.827	0.763	19.208	17.891	16.039	8.612	79.873	81.191	83.043	90.469	0.075	0.087	0.091	0.155	0.21	0.19	0.205	0.223

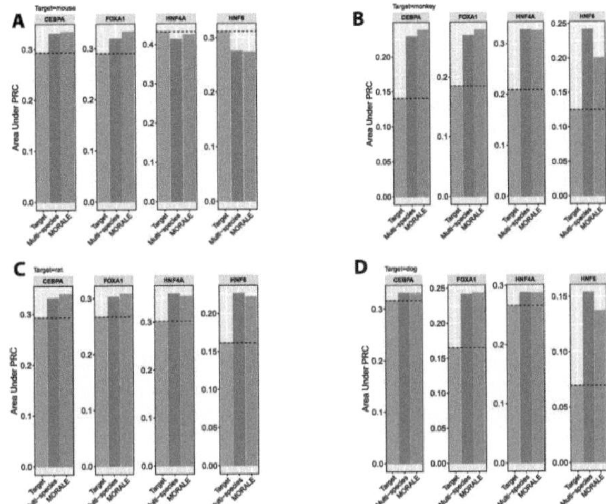

Fig. 8. We display performance across all targets in the multi-species case. In **(A)** we show the performance across all 4 transcription factors when the target is mouse, **(B)** displays monkey, **(C)** displays rat, and **(D)** display dog.

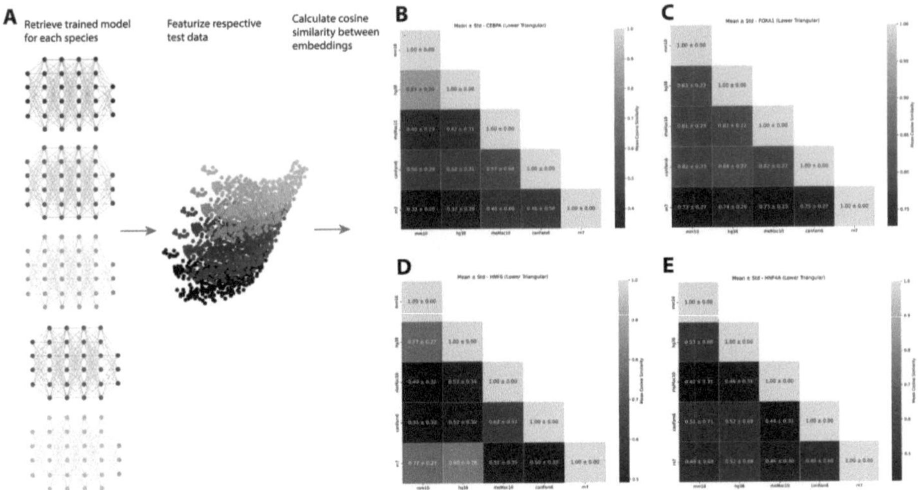

Fig. 9. We construct heatmaps based to understand relatedness of learned embeddings through the different species models. We do so for each TF under study in the multi-species case. For each TF we use the models trained to predict in each species and run the test data through the feature extractor in our models to capture the embeddings. In **(B)** we show the lower triangular for CEBPA, **(C)** FOXA1, **(D)** HNF6, and **(E)** HNF4α.

Table 5. We compare across all repeat types (with at least 500 instances in our test (human) chromosome, Chr2) in the source (mouse) genome between the two mouse-adapted models. The last row is the average auPRC across all repeat types

TF	DNA			LINE			Low complexity			LTR			Simple repeat			SINE			Unknown			Average		
	GRL	MORALE	Source	GRL	MORALE	Source	GRL	MORALE	Source	GRL	MORALE	Source	GRL	MORALE	Source	GRL	MORALE	Source	GRL	MORALE	Source	GRL	MORALE	Source
CTCF	0.174	0.209	**0.216**	0.149	**0.173**	0.17	0.119	0.142	**0.146**	0.082	0.097	**0.1**	0.154	**0.193**	0.191	0.181	**0.191**	0.177	0.321	0.405	**0.44**	0.168	0.202	**0.206**
CEBPA	0.047	**0.051**	0.044	0.039	**0.042**	0.034	0.044	**0.049**	0.041	0.031	**0.033**	0.03	0.045	**0.047**	0.038	0.04	**0.04**	0.027	0.052	**0.055**	0.043	0.042	**0.045**	0.037
HNF4A	**0.056**	0.056	0.044	**0.043**	0.04	0.03	0.058	**0.061**	0.049	0.033	**0.036**	0.03	**0.062**	0.059	0.046	**0.048**	0.034	0.023	0.087	**0.109**	0.105	0.055	**0.056**	0.047
RXRA	0.073	**0.078**	0.055	**0.058**	0.056	0.036	0.087	**0.093**	0.07	0.047	**0.053**	0.041	**0.086**	0.085	0.058	**0.062**	0.052	0.029	0.11	**0.139**	0.109	0.074	**0.079**	0.057
Average	0.087	**0.098**	0.09	0.072	**0.078**	0.068	0.077	**0.086**	0.076	0.048	**0.055**	0.05	0.087	**0.096**	0.083	**0.083**	0.079	0.064	0.142	**0.177**	0.174	0.085	**0.096**	0.086

Table 6. We compare across all repeat types (with at least 500 instances in our test (mouse) chromosome, Chr2) in the source (human) genome between the two human-adapted models. The last row is the average auPRC across all repeat types

TF	DNA			LINE			Low complexity			LTR			Simple repeat			SINE			Unknown			Average		
	GRL	MORALE	Source	GRL	MORALE	Source	GRL	MORALE	Source	GRL	MORALE	Source	GRL	MORALE	Source	GRL	MORALE	Source	GRL	MORALE	Source	GRL	MORALE	Source
CTCF	0.151	**0.18**	0.168	0.103	**0.124**	0.111	0.128	**0.148**	0.133	0.086	**0.108**	0.097	0.121	**0.147**	0.127	0.102	**0.129**	0.104	0.122	**0.156**	0.138	0.116	**0.142**	0.125
HNF4A	0.061	0.065	**0.068**	0.04	0.043	**0.045**	0.059	**0.062**	0.061	0.042	0.046	**0.047**	0.058	**0.059**	0.058	0.061	**0.063**	0.062	0.085	0.075	0.081	0.058	0.059	**0.06**
RXRA	0.044	**0.049**	0.043	0.027	**0.029**	0.026	0.047	**0.048**	0.043	0.028	**0.031**	0.027	0.038	**0.038**	0.033	0.041	**0.042**	0.035	0.058	**0.067**	0.058	0.04	**0.044**	0.038
Average	0.09	**0.102**	0.1	0.062	**0.069**	0.066	0.08	**0.087**	0.082	0.057	**0.066**	0.063	0.076	**0.084**	0.078	0.075	**0.084**	0.077	0.091	**0.111**	0.105	0.076	**0.086**	0.082

Table 7. The binding site information for the four transcription factors used in the two-species case. The following quantities are listed: the number of peaks called across the entire genome; the number of called peaks within the filtered window set, merged if within 500 bp of each other; the number of windows in the filtered window set labeled bound due to peak overlap; the fraction of the filtered window set labeled bound; and the database accession ID (ENCODE, GEO, or ArrayExpress). The size of the filtered window sets for the mouse and human genomes were 41883806 and 48742577, respectively.

TF	Species	Raw Peaks	Filtered Peaks	Bound Windows	Frac. Bound	Accession ID
CTCF	Mouse	32006	28943	296117	0.71%	ENCSR000CBU
	Human	29067	26477	270100	0.55%	ENCSR911GFJ
CEBPA	Mouse	62636	48812	566945	1.35%	E-TABM-722
	Human	32243	28545	298066	0.61%	E-TABM-722
HNF4A	Mouse	44800	36540	415846	0.99%	E-TABM-722
	Human	42766	34714	387077	0.79%	E-TABM-722
RXRA	Mouse	46443	33751	404284	0.97%	GSM1299600
	Human	95085	71032	854289	1.75%	ENCSR098XMN

Table 8. The binding site information for the four transcription factors used in the multi-species case. The following quantities are listed: the number of peaks called across the entire genome; the number of called peaks within the filtered window set, merged if within 1000 bp of each other; the number of windows in the filtered window set labeled bound due to peak overlap; the fraction of the filtered window set labeled bound; and the database accession ID (ArrayExpress).

TF	Species	Raw Peaks	Filtered Peaks	Bound Windows	Frac. Bound	Accession ID
CEBPA	Mouse	50263	32751	830115	1.80%	E-MTAB-1509
	Human	34253	26749	615953	1.16%	E-MTAB-1509
	Rhesus Macaque	11600	9985	214440	0.40%	E-MTAB-1509
	Dog	44749	32816	780102	1.77%	E-MTAB-1509
	Rat	50851	37010	900363	1.84%	E-MTAB-1509
FOXA1	Mouse	66728	38683	1071971	2.32%	E-MTAB-1509
	Human	36454	27406	651070	1.22%	E-MTAB-1509
	Rhesus Macaque	30546	22421	532725	1.00%	E-MTAB-1509
	Dog	24316	18151	436461	0.99%	E-MTAB-1509
	Rat	59983	37940	993292	2.02%	E-MTAB-1509
HNF4A	Mouse	135057	54343	1762041	3.82%	E-MTAB-1509
	Human	50611	34022	856878	1.61%	E-MTAB-1509
	Rhesus Macaque	32331	21628	535077	1.01%	E-MTAB-1509
	Dog	69264	37839	1049132	2.38%	E-MTAB-1509
	Rat	52694	33640	891098	1.82%	E-MTAB-1509
HNF6	Mouse	57255	38899	966248	2.09%	E-MTAB-1509
	Human	17021	14378	311320	0.59%	E-MTAB-1509
	Rhesus Macaque	9425	8238	174525	0.33%	E-MTAB-1509
	Dog	9283	7687	168142	0.38%	E-MTAB-1509
	Rat	22686	18058	407416	0.83%	E-MTAB-1509

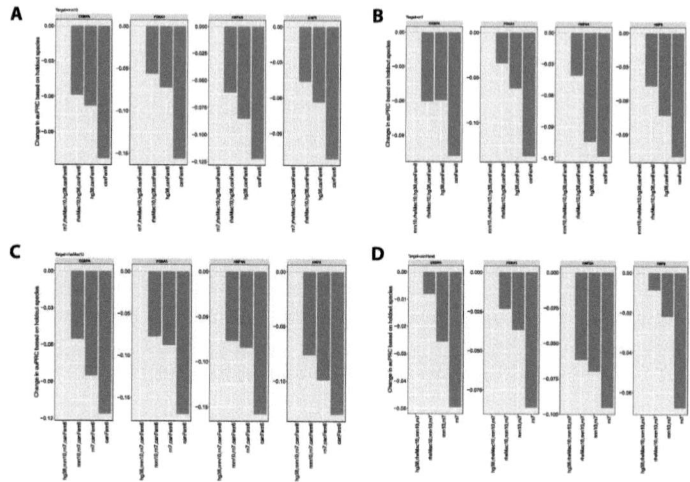

Fig. 10. We include group holdouts under each species (other than human) as the target species.

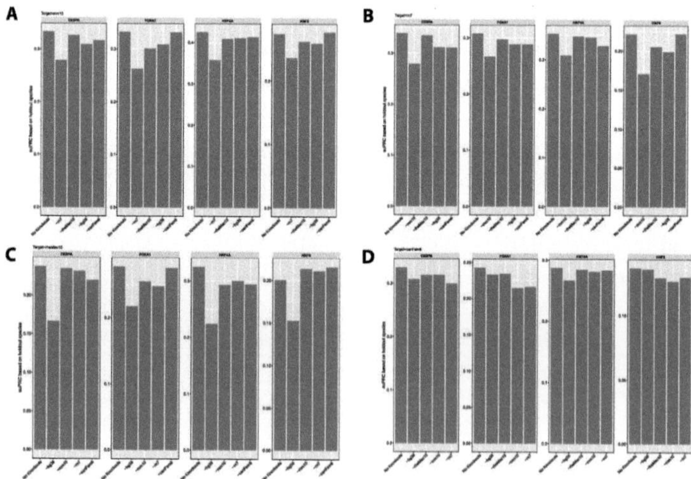

Fig. 11. We include per species holdout under each species (other than human) as the target species.

References

1. Abadi, M., et alX.: TensorFlow: Large-scale machine learning on heterogeneous systems (2015). https://www.tensorflow.org/, software available from tensorflow.org
2. Amemiya, H.M., Kundaje, A., Boyle, A.P.: The ENCODE blacklist: identification of problematic regions of the genome. Sci. Rep. **9**(9354), 1–5 (2019). https://doi.org/10.1038/s41598-019-45839-z
3. Avsec, Ž, et al.: Effective gene expression prediction from sequence by integrating long-range interactions. Nat. Methods **18**, 1196–1203 (2021). https://doi.org/10.1038/s41592-021-01252-x
4. Avsec, Ž, et al.: Base-resolution models of transcription-factor binding reveal soft motif syntax. Nat. Genet. **53**, 354–366 (2021). https://doi.org/10.1038/s41588-021-00782-6
5. Ballester, B., et al.: Multi-species, multi-transcription factor binding highlights conserved control of tissue-specific biological pathways. eLife (2014). https://doi.org/10.7554/eLife.02626
6. Brennan, K.J., et al.: Chromatin accessibility in the Drosophila embryo is determined by transcription factor pioneering and enhancer activation. Dev. Cell **58**(19), 1898–1916.e9 (2023). https://doi.org/10.1016/j.devcel.2023.07.007
7. Brixi, G., et al.: Genome modeling and design across all domains of life with evo 2. bioRxiv (2025). https://doi.org/10.1101/2025.02.18.638918, https://www.biorxiv.org/content/early/2025/02/21/2025.02.18.638918
8. Chen, K.M., Wong, A.K., Troyanskaya, O.G., Zhou, J.: A sequence-based global map of regulatory activity for deciphering human genetics. Nat. Genet. **54**, 940–949 (2022). https://doi.org/10.1038/s41588-022-01102-2
9. Chen, L., Fish, A.E., Capra, J.A.: Prediction of gene regulatory enhancers across species reveals evolutionarily conserved sequence properties. PLoS Comput. Biol. **14**(10), e1006484 (2018). https://doi.org/10.1371/journal.pcbi.1006484
10. Cochran, K., Srivastava, D., Shrikumar, A., Balsubramani, A., Hardison, R.C., Kundaje, A., Mahony, S.: Domain-adaptive neural networks improve cross-species prediction of transcription factor binding. Genome Res. **32**(3), 512–523 (2022). https://doi.org/10.1101/gr.275394.121
11. Dalla-Torre, H., et al.: Nucleotide transformer: building and evaluating robust foundation models for human genomics. Nat. Methods pp. 1–11 (2024). https://doi.org/10.1038/s41592-024-02523-z
12. Ganin, Y., et al.: Domain-adversarial training of neural networks. J. Machine Learn. Res. **17**(59), 1–35 (2016). http://jmlr.org/papers/v17/15-239.html
13. Gupta, S., Stamatoyannopoulos, J.A., Bailey, T.L., Noble, W.S.: Quantifying similarity between motifs. Genome Biol. **8**(2), 1–9 (2007). https://doi.org/10.1186/gb-2007-8-2-r24
14. Hu, Y., et al.: Multiscale footprints reveal the organization of cis-regulatory elements. Nature **638**, 779–786 (2025). https://doi.org/10.1038/s41586-024-08443-4
15. Javanmardi, M., Tasdizen, T.: Domain adaptation for biomedical image segmentation using adversarial training. In: 2018 IEEE 15th International Symposium on Biomedical Imaging (ISBI 2018), pp. 04–07. IEEE. https://doi.org/10.1109/ISBI.2018.8363637
16. Kelley, D.R.: Cross-species regulatory sequence activity prediction. PLoS Comput. Biol. **16**(7), e1008050 (2020). https://doi.org/10.1371/journal.pcbi.1008050
17. Kempynck, N., et al.: CREsted: Cis regulatory element sequence training, explanation, and design (2024)

18. Lal, A., Gunsalus, L., Nair, S., Biancalani, T., Eraslan, G.: gReLU: A comprehensive framework for DNA sequence modeling and design. bioRxiv p. 2024.09.18.613778 (2024). https://doi.org/10.1101/2024.09.18.613778
19. Langmead, B., Salzberg, S.L.: Fast gapped-read alignment with Bowtie 2. Nat. Methods **9**, 357–359 (2012). https://doi.org/10.1038/nmeth.1923
20. Letunic, I.: phyloT : a phylogenetic tree generator (2025), https://phylot.biobyte. de. Accessed 19 Mar 2025
21. Mahmood, F., Chen, R., Durr, N.J.: Unsupervised reverse domain adaptation for synthetic medical images via adversarial training. IEEE Trans. Med. Imaging **37**(12), 2572–2581 (2018). https://doi.org/10.1109/TMI.2018.2842767
22. Mahony, S., et al.: An integrated model of multiple-condition ChIP-Seq data reveals predeterminants of Cdx2 binding. PLoS Comput. Biol. **10**(3), e1003501 (2014). https://doi.org/10.1371/journal.pcbi.1003501
23. de Mathelin, A., Deheeger, F., Richard, G., Mougeot, M., Vayatis, N.: Adapt: Awesome domain adaptation python toolbox. arXiv preprint arXiv:2107.03049 (2021)
24. Pampari, A., et al-regulatory sequence syntax, transcription factor footprints and regulatory variants. bioRxiv p. 2024.12.25.630221 (2025). https://doi.org/10.1101/2024.12.25.630221
25. Paszke, A., et al.: PyTorch: An Imperative Style, High-Performance Deep Learning Library. arXiv (2019). https://doi.org/10.48550/arXiv.1912.01703
26. Patel, A., Singhal, A., Wang, A., Pampari, A., Kasowski, M., Kundaje, A.: DART-Eval: A Comprehensive DNA Language Model Evaluation Benchmark on Regulatory DNA. arXiv (2024). https://doi.org/10.48550/arXiv.2412.05430
27. Pedregosa, F., et al.: Scikit-learn: Machine learning in Python. J. Mach. Learn. Res. **12**, 2825–2830 (2011)
28. Perez, G., et al.: The UCSC genome browser database: 2025 update. Nucleic Acids Res. **53**(D1), 1243–1249 (2025). https://doi.org/10.1093/nar/gkae974
29. Schmidt, D., et al.: Five-Vertebrate ChIP-seq reveals the evolutionary dynamics of transcription factor binding. Science **328**(5981), 1036–1040 (2010). https://doi.org/10.1126/science.1186176
30. Shrikumar, A., et al.: Technical Note on Transcription Factor Motif Discovery from Importance Scores (TF-MoDISco) version 0.5.6.5. arXiv (2018). https://doi.org/10.48550/arXiv.1811.00416
31. Sun, B., Saenko, K.: Deep CORAL: Correlation Alignment for Deep Domain Adaptation. arXiv (2016). https://doi.org/10.48550/arXiv.1607.01719

Author Index